How Mitchell Energy & Development Corp. Got Its Start And How It Grew

An Oral History and Narrative Overview

UPDATED

Joseph W. Kutchin

Universal Publishers • USA • 2001

How Mitchell Energy & Development Corp Got Its Start and How It Grew

Universal Publishers/uPUBLISH.com
USA • 2001

ISBN: 1-58112-663-8

www.upublish.com/books/kutchin.htm

Library of Congress Catalog Card Number: 200109197

If you would see the man's
monument, look around.

A tribute to
Christopher Wren
by his son

Table of Contents

Foreword 2001

This amounts to a Foreword to the Foreword written when the first version of *How Mitchell Energy & Development Corp. Got Its Start and How It Grew* was published early in 1998.

Real estate development, and especially that of The Woodlands, no longer plays a role at Mitchell Energy & Development. The company's sole business now is energy. But George Mitchell remains intensely interested in the successful development of the new community he created. Both the energy and the real estate components of what once was MEDC are prospering, separately, not exactly the case a few years ago.

Now seemed to be an opportune time to talk about what's been happening. Epilogue 1 brings things up to date and has been added to the narrative part of the original book. It reports on developments that seemed out of reach just three years earlier. In Epilogue 2, George Mitchell peers about 20 years into the future. Otherwise, except for some tightening here and there, the substance of the original is unchanged.

What happens next? No one can say. But I do know that it has been a good race, and being one of the runners alongside George Mitchell (well, actually out of breath and several paces behind) has been my good fortune.

Joseph W. Kutchin
The Woodlands, Texas
April 25, 2001

Foreword 1998

First, something about the company. For a number of years, it was kind of a hybrid, but once again it is a pure play in energy. It started small, in 1946, as a gas and oil wildcatter. By the end of its first 50 years, it had grown into one of the nation's big independent energy producers: natural gas and natural gas liquids particularly, and, to a lesser extent, oil.

Along the way, it became a real estate company, too—more exactly, the developer of The Woodlands, a new community north of Houston that comprises 25,000 acres. But after a strategic decision in the late 1990s, The Woodlands was sold in 1997, and Mitchell Energy & Development Corp. (MEDC) left the real estate business. (In fact, while The Woodlands earned a high public profile, real estate never accounted for as much as 20 percent of MEDC revenues.)

My association with the company started in 1972. I had been a senior officer in a Chicago-based public relations agency for many years when MEDC became a client, having just sold shares to the public. George Mitchell and Morris B. Rotman, chairman of the agency, knew and admired each other. They met through the Young Presidents' Organization.

I headed the agency account team. Joyce Gay was MEDC's PR vice president in the early '70s, a period of intense growing pains for the company, with the energy and real estate operations vying for always-scarce financial and human capital. Joyce chose to leave the company. At the suggestion of B. F. "Budd" Clark, then executive vice president, George Mitchell invited me to take over Joyce's job, and I moved from Chicago to Houston early in 1974.

It was a good marriage. I found a fast-moving, fast-growing company which thrived on excitement and innovation. We even seemed to flourish on the lack of discipline and structure. After serving a diversity of clients over 20 years in the agency, I felt then and still do that George and Budd were the smartest businessmen I'd ever worked with. George gave the impression of being everyone's sweet-natured, self-effacing uncle. Budd appeared to be gruff, acerbic and impatient. But George is not necessarily all that kindly and avuncular, and Budd has turned out to be considerate and gentle, a better friend to company employees than many would guess. With both men, I've developed a relationship that is something more than employer-employee but less than close personal friendship. My experience is that they complement each other remarkably well. George has always believed that anything is possible, Budd is less sanguine. Things usually come to earth somewhere in between, but closer to where George is standing. Both are truly good men. And both will be uncomfortable at reading these words of respect and affection.

I was public relations vice president from 1974 until 1989. I am sure I had one of the best public relations jobs in America. First, George Mitchell turned out to have a remarkable feel for both PR and publicity. Our department was an important part of the company. George had ideas, he was accessible and knew the value of controversy. He tried to talk as fast as he thinks (still does), and parsing his rhetoric was and is a bear. The press loved him.

But diversity was the real kick. Next to the major oil companies, we were pipsqueaks, but we were widely heard on matters ranging from the need for an oil import fee to energy conservation. Even though a relatively small energy producer, Mitchell Energy & Development created or refined a number of advanced technologies, and the public relations department was successful in obtaining recognition for the company's accomplishments. Because of George Mitchell's wide-ranging interests, our department became involved in programs designed to develop public appreciation of issues concerned with resource depletion, population growth and environmental degradation. We introduced the company to and developed relationships in the financial community. And although the real estate division had its own PR staff, corporate energy and real estate worked closely together.

Because of health concerns, I retired early, in December 1989, at age 60. But barely had all the paperwork been completed before I found myself as founding president of the Cynthia Woods Mitchell Pavilion, an extremely popular outdoor entertainment center in The Woodlands. That was an especially restless retirement. I really intended to quit in 1993 but, at George Mitchell's request, soon found myself working on corporate archives and then this history.

Anyone close to the company would quickly agree that Mitchell Energy & Development Corp. has been the length and shadow of George Mitchell. He has set a pace and a tone that gives MEDC its personality and character. He's been a big factor wherever he's focused his diverse interests: energy production, national energy policy, creation of a new community designed to help deal with the nation's urban ills, reinvigorating the economy of his hometown of Galveston, pure and applied research, public education, the arts. It's an endless list.

In recent years, W. D. "Bill" Stevens, a veteran of Exxon, has assumed responsibilities as president and chief operating officer of MEDC. George maintains a heavy schedule of day-to-day involvement, but inevitably, things are changing. It is a different company now from the freewheeling outfit that I knew in the '70s. Undoubtedly it's time for such a change.

In 1995, George asked me to prepare an oral history. No historian, and not necessarily a completely objective party, I agreed. So that's what the following pages are. I interviewed many people who've had a role in Mitchell Energy & Development Corp.'s evolution. These conversations comprise a chronicle of what have been the creative forces in the company over the past half-century. One interview may contradict parts of another. There are occasional fragmentary thoughts that don't spin out to completion. Memories sometimes fuzz over. But if I've been successful, these recollections will tell how an important American enterprise was created, how it prospered in the second half of 20th century America, and how a hard-driving, creative, concerned son of immigrants served as the catalyst for making good things happen for many, many people.

* * * * * * *

I've prepared a consolidated story of the company, which precedes the interview transcripts that occupy most of the following pages. My sources have been the interviews themselves, corporate annual reports, newspaper stories, correspondence and a number of reference works and data bases. I must acknowledge that *The Woodlands, New Community Development, 1964-1983* by George T. Morgan, Jr., and John O. King, published in 1987 by Texas A&M University Press, was especially helpful in providing information about the early years of The Woodlands. The late Dr. Morgan was professor of history at the University of Houston and published widely on the conservation movement and U.S. labor history. Dr. King, now retired, was a professor at the U of H who specialized in the U.S. petroleum industry.

Finally, I wish to acknowledge the pivotal roles played by Adora Kutchin, my wife, who served as editor; and Jackie Nolan and, especially, Bea Forse, who handled the daunting job of transcribing the free-ranging interviews. John Weaver, with whom I've worked for more than 20 years, was responsible for design, and Ed Dyess and Ron Batista of MEDC's Graphics Department set the type and finetuned some of the graphics. All, with inexhaustible patience, contributed beyond the call of duty, and I thank them.

Joseph W. Kutchin
The Woodlands, Texas
February 18, 1998

A Risktaker's Advice:

'When You're Convinced You're Right, Go For It'

In the scheme of things, the fact that the Cynthia Woods Mitchell Pavilion has overhead fans to comfort the audience on sultry summer nights is hardly more than a minor footnote. But the fan story helps explain why Mitchell Energy & Development Corp. (MEDC) is a successful New York Stock Exchange company rather than a small, shoot-from-the-hip wildcatter.

The $10 million Pavilion, an outdoor entertainment center in The Woodlands, Texas, opened on April 27, 1990. It was a huge success from the very beginning, with performances by popular contemporary entertainers and eminent classical arts groups. It was paid for mainly by contributions from the George P. (for Phydias) Mitchell family and The Woodlands Corporation, at that time the real estate subsidiary of MEDC.

But there is this thing about summer evenings on the Texas Gulf Coast: They are warm and humid and, to many people, quite uncomfortable.

Air conditioning such a large outdoor space isn't feasible, although it's been tried elsewhere. So George Mitchell asked Coulson Tough, a senior officer of The Woodlands Corporation, to investigate whether overhead fans could be installed in the soaring, tentlike structure that shelters the Pavilion's permanent seats.

Tough, a professional architect, has spent most of his life getting buildings built on time and within budget. The construction he was responsible for at the Mitchell corporation is vast. He knows how to get things done.

Tough tried every source he could think of. A number of sample fans from one manufacturer would be installed on the structure supporting the Pavilion roof. One model would have a motor strong enough to move air the 40 feet necessary to reach the audience. But the noise it generated was disturbing. A week later, other fans would be tested: Quiet enough, possibly, but there would be no discernible air movement.

And so it went for months. Eventually Tough told Mitchell he had run out of possible suppliers.

And Mitchell said, "Coulson, keep looking."

Tough did keep looking and eventually found a small local manufacturer who, in effect, custom built each of the double-bladed fans that they produced. Installed in time for the 1993 season, the fans are remarkably quiet. And even on summer evenings when the mercury remains in the 80s, they provide a gentle breeze to keep the audience inside the tent comfortable.

If there is anything about George Mitchell, it is that he is determined; his insistence on finding the right fans is just a lesser example. He starts with the qualities that are found in most successful men and women: He's smart and intuitive, energetic, imaginative. But beyond that, he is extraordinarily persistent and foresighted. As a result, in the early 1950s, he began to acquire the lease acreage in North Texas which to this day remains the heart of his company's natural gas production; in the 1960s, through 300 or more transactions, he assembled most of the 25,000-acre block of land that now constitutes The Woodlands, the new community north of Houston (sold in 1997, The Woodlands nonetheless is Mitchell's signal achievement); and in the 1970s, he obtained permission to drill offshore and onshore in his home town of Galveston, an environmentally sensitive resort area for millions on the Texas Gulf Coast.

Since the end of World War II, the story of Mitchell Energy & Development Corp. and of George Mitchell are largely one and the same. For almost all of the company's history, he has been the active, involved chief executive and chief operating officer, although in January 1994, he ceded to W. D. "Bill" Stevens, a veteran of the Exxon Corporation, the positions of president and chief operating officer. Mitchell remains chairman and chief executive.

SAVVAS'S STORY

Savvas Paraskevopoulos, Mitchell's father, was a goatherd in Nestani in Greece's Peloponnesus. Impoverished and illiterate, he saw early that if he stayed home, he could look forward only to a hardscrabble, rural life, and so in 1901, at the age of 20, he emigrated to the United States.

With no skills and virtually no English, he took a job doing manual labor on a railroad gang. When young Savvas went to pick up his pay, the Irish paymaster found "Savvas Paraskevopoulos" more of a mouthful than he could handle. From now on, he said, your name is the same as mine: Mike Mitchell.

Savvas—Mike—later moved to Houston and, with his cousin, ran a pressing shop and shoeshine parlor near the Rice Hotel. Then he settled in Galveston, again with a pressing shop and shoe shine parlor. He saw a picture of his wife-to-be, Katina Eleftheriou, in a Greek language weekly newspaper and immediately fell in love. She was visiting her sister in Florida, and the prospect was that she would soon be married. No matter to Mike: He traveled to Florida and wooed and won Katina (persistence in the Mitchell family has not been unique to George). She, too, was a Peloponnesian, from Argos. Unlike Mike, she had some formal education—six grades—but she never did learn to speak English. "She was a very pretty woman, very warm," George recalls.

George was born on May 21, 1919, the third of four children. Christie was the oldest, then Johnny, then George and then the only daughter, Maria. Money was scarce, but George flourished in the warmth of a closeknit family. Then tragedy struck when he was only 13: His mother died of a stroke at age 44. Shortly after his wife's death, Mike Mitchell suffered a badly shattered leg in an automobile accident. Christie and Johnny were old enough to be on their own, but George and Maria went off to live with relatives.

As with many aspiring immigrant families, so it was with the Mitchells, even after Katina Mitchell's untimely death. For them, the keys to the future lay in education, and no sacrifice was too great to assure a college degree for each of the four children. George was graduated in 1935 from Galveston's Ball High School just as he was turning 16, too young to go off by himself to college. So, during the next year, he took additional studies at Ball, heavy on math and science and Latin. His mother had wanted him to attend Houston's Rice University and then become a doctor. George, in fact, was accepted at Rice.

By this time, George's brother Johnny had earned his chemical engineering degree at Texas A&M University (Christie received his degree from the University of Texas, Maria got hers from Mary Hardin-Baylor) and was working in the oil patch for a predecessor to Exxon Corporation. Johnny arranged a summertime job for George as an oil field roustabout (but wouldn't let him work on the oil rig platforms, which he considered too dangerous for his kid brother). And with that, George's thoughts of becoming a physician ended as he became hooked on the excitement and opportunity of the oil business. The opportunity part required an act of faith: Oil at the time was selling at about $1.20 per barrel, less than one-twenty-fifth of its price in the year 2000. Natural gas was priced so low it was all but being given away.

$29 MONTHLY FOR ROOM, BOARD, LAUNDRY

George enrolled at Texas A&M University in College Station, the breeding ground for much of Texas's, and particularly Houston's, business and energy leadership. Tuition was cheap and living was cheap—$29 a month for room, board and laundry—but that hardly mattered because George had virtually no money. He waited tables in A&M's Sbisa residence hall for 26 cents an hour. He built book cases and sold them to other A&M cadets. He sold candy in the dormitory. He obtained the concession for tailoring. Later he sold stationery to lovesick freshmen. He says, "Very often I didn't have any money if I didn't make it on the candy. Somebody might steal the candy or not pay me for the tailoring. They were about to kick me out two or three times. I would have to go borrow the money from Buddy Bornefeld (a longtime friend and fellow Aggie from Galveston)."

Mitchell says that Bornefeld was the only one in his circle who had a car. "We'd go to Galveston once in a while, and he'd charge a dollar for the round trip. But if the wind was blowing from the north, he'd charge $1.50."

George's father was an occasional source of money. George says, "My father . . . knew everyone in Galveston. I'd send him a wire that I was about to be kicked out of school because I owed Buddy and I couldn't pay him and I owed the school their $29 and they were pretty severe about it—if you didn't pay them in 45 days, you were out.

"So he'd take my wire and he'd go over to Sam Maceo and he'd say, 'OK, my son is at the top of his class and he needs a hundred dollars.' And so Maceo would give him $100, and my father would send me $50 and keep $50."

At A&M Mitchell honed what would become a lifelong passion. He first started playing tennis at age 13 in Galveston, watching good tennis players and learning from them. He became A&M's Number 1 player and captain of the intercollegiate tennis team. He is still playing in his early 80s, although he concedes that wear and tear on his knees has made it impossible for him to play singles.

In the A&M Corps of Cadets, Mitchell was in B Battery of the Field Artillery, just like brother Johnny before him. By the time he graduated, he had been promoted to major and was battalion commander, leading 800 other students.

Mitchell spent long hours at his studies and graduated with high grades, especially in his major. He averaged 23 credit hours a semester, about half again the number of hours required for graduation. At the time, A&M offered no degree in geology. Still, George took every geology course offered, even though technically his degree is in petroleum engineering.

'YOU'D BETTER GO OUT ON YOUR OWN'

He credits his instructors with providing more than just book learning and cites, for instance, Harold Vance, a professor of petroleum engineering. It was Vance, Mitchell says, who provided this thinking: "If you want to go to work for Exxon (Humble at that time), fine, then you can drive around in a pretty good Chevrolet, but if you want to drive around in a Cadillac, you'd better go out on your own some day."

Mitchell received his bachelor's degree in 1939, just before the start of World War II in Europe. The fact that the United States would one day be at war was obvious, and a number of Mitchell's friends accepted offers from the Army. George, instead, took a job with Amoco as an exploitation engineer in Jennings, Louisiana, deep in Cajun country.

He recalls, "I had to roughneck for three months, out of Hackberry. What a miserable place. I can remember going at night, it was raining and sleeting and I had to work on that rig. It was dangerous as hell. I thought, my God, if I have to do this all my life, I'll go jump off this damned rig right now."

His career with Amoco was brief, however. A few months before America's entry into the war in 1941, the Army beckoned. George went into the Corps of Engineers as a second lieutenant and was assigned to the San Jacinto Ordnance Depot in Houston.

With Amoco, he'd learned much about the petroleum business. Now in the Army he learned to manage projects and people. First, as a 21-year-old second lieutenant, he played a role in building the 5,000-acre San Jacinto Ordnance Depot on Houston's Ship Channel. Then he became involved in working on the Dixon gun plant and after that, he was assigned a role in building, first, Bergstrom Air Force Base near Austin, Texas, and then pipelines to Alaska to move oil from the Imperial Field there to the continental United States.

A START AS A CONSULTANT

The Army held on to Mitchell even after the end of the war in 1945. He was among those responsible for winding down the many ordnance contracts the government had committed to. But even though still a soldier, George found time to provide engineering and geology consulting services in Houston. His brother Johnny, in fact, after serving under General George Patton, was released from the Army five months before George. Johnny and H. Merlyn Christie, a well-connected oil business broker and family friend, put together deals, and George provided expertise in geology and engineering. One of their earliest successes was the Caplan Field near Galveston.

Finally, in 1946, when George was 26, he was released from the Army as a captain. He had the option of returning to Amoco but chose not to. For one thing, the advice of his A&M Professor Vance rung in his head: The reward potential is far greater for the entrepreneur. For another, he recognized that he had probably fallen behind the other engineers and geologists at Amoco who had not been in the service.

So, after being mustered out, and with virtually no experience, Mitchell set up shop in Houston as an oil and gas consultant. One of his first clients was Roxoil Drilling, Inc., a small wildcatting company named for one of its partners, Roxie Wright. Roxoil's chief asset was an old drilling rig. Its founding partner was Merlyn Christie.

Also, six of George's friends provided him with a total of $300 a month plus a small override to do geology and engineering. His job was to find good petroleum prospects for them, and he did.

Finally, in early 1947, George bought out Jimmy Gray's one-sixth interest in Roxoil for $9,000, and the company name was changed to Oil Drilling, Inc. Roxoil and Oil Drilling were the predecessor companies to today's Mitchell Energy & Development.

Mitchell says, "When we began we didn't have any money. We'd put a deal together, and if we made a well, we'd go to the bank and borrow money on it. If we made a dry hole, we didn't have much money in it anyway."

Separate from his Oil Drilling activities, Mitchell was consulting geologist for such famous names as Glenn McCarthy, Floyd Karston, Louis and Harry Pulaski, Morris Rauch, Eddie Scurlock. For Oil Drilling, Johnny Mitchell, who by now had a small interest in the company, put together the "Big Nine," a group of businessmen from Houston's and Galveston's Jewish community, including Abe and Bernard Weingarten, who owned supermarkets; Abe Lack, who was in the auto parts business; Jake Oshman of the sporting goods retailers; Irving Alexander, who had interests in real estate and supermarkets; Will Zinn, a Galveston-based lawyer. Later on, as members of the Big Nine chose to slow down the pace of their investments, the Barbara Hutton estate became a participant, as did the Singer Sewing Machine estate, R. E. "Bob" Smith and John Riddell. All played roles as early investors not only in developing the Mitchell company's Boonsville Field in North Texas, which grew to be the company's largest production area, but in Mitchell properties in South Texas, as well.

Mitchell recalls, "Old man Zinn was tough. I'd come in and say, 'Mr.

Zinn, sorry, but that well we drilled got salt water,' and Mr. Zinn would say, 'Hell, I've got all the salt water I need in Galveston and I don't need to spend my money to get salt water.'"

Bob Smith was famed for both his real estate and oil investments. At the beginning of his association with Mitchell, Smith's wealth came primarily from oil production in East Texas. Later, in one of Smith's canniest moves, he purchased large tracts of real estate along Westheimer Road, which has been the spine of Houston's expansion to the west. Eventually, he became a 25 percent participant in all of Oil Drilling's drilling ventures.

George Mitchell, Johnny Mitchell and Merlyn Christie, operating as Oil Drilling, Inc., were headquartered in a single office, with one shared secretary, in Houston's downtown Esperson Building. The Roxie Wright estate still owned a small share of the company. George enjoys telling how he would do the geology of oil and gas prospects, then Johnny and Merlyn would go downstairs to the Esperson Building drug store to sell the deals.

Oil Drilling did well for its owners and for its outside participants. The targets usually were natural gas; even though it sold at only 3 cents per thousand cubic feet, with careful cost control, it was possible to make money.

Mitchell says now, "There were dry holes, too . . . always a scattering of dry holes, of course. We would always try to drill four nearby wells for every wildcat. . . . If you drilled 12 straight dry holes, you were out of business."

A BOOKIE PAYS OFF

Louis Pulaski—often called General Pulaski after the Polish general in America's Revolutionary War—over the years had become both a friend of George Mitchell and an investor in his prospects. One day in 1952, Pulaski called Mitchell and asked that he meet a bookie from Chicago who had a line on some interesting lease acreage in North Texas. Mitchell, with never a minute to spare, was reluctant to take the time, but in deference to his long friendship with Pulaski, he said yes.

George talked to the bookie, then checked out the geology on the acreage, which looked very promising indeed. As it turned out, the men mainly responsible for promoting the North Texas idea were Ellison Miles, at the time a drilling contractor whom George had known at Texas A&M, and John Jackson, a consulting geologist with a degree from the University of Texas. Together they, along with the Big Nine, Bob Smith and several others, purchased leases on the 3,000-acre Hughes Ranch in Wise County. The first well, the D. J. Hughes Number 1, was successful. So were the next

MITCHELL ENERGY CORPORATION
25 LARGEST DISCOVERIES AND FIELD EXTENSIONS
(Cumulative Production as of June 30, 2000***)

Field	State	County	8/8 Cumulative Production Oil (Bbls)	Gas (Mcf)	Discovery Year	MEC Discovery
Boonsville	TX	Jack, Parker, Wise	9,346,790	1,521,932,415	1945	
Personville North	TX	Limestone	670,836	291,314,640	1968	X
Newark East*	TX	Wise	254,231	207,365,603	1981	X
Morris	TX	Jack, Wise, Young	5,706,761	92,763,816	1950	
Palacios	TX	Matagorda	2,171,375	113,673,718	1937	
Madisonville	TX	Madison	1,573,174	107,520,880	1946	
Lafitte's Gold	TX	Galveston	1,619,020	89,291,535	1971 (offshore) & 1973 (onshore)	X
Alvord South	TX	Wise	13,090,894	19,683,776	1956	
Alvord	TX	Wise	6,282,982	46,572,723	1954	
Vienna	TX	Lavaca	2,527,287	54,470,247	1945	
Galveston Island	TX	Galveston	178,711	67,465,002	1949	
Hortense	TX	Polk	1,632,535	54,287,583	1972	X

Leggett**[**]**	TX	Polk	2,146,559	47,972,585	1973	
Caughlin	TX	Wise	8,250,327	10,543,768	1955	
Pinehurst	TX	Montgomery	1,514,836	46,869,533	1942	
Lake Creek**[*]**	TX	Montgomery	1,835,919	33,376,999	1941	
Hell's Hole	CO, UT	Rio Blanco, Uintah	149,138	39,806,011	1951	X
Cap Yates	TX	Jack, Wise	1,626,297	28,450,100	1952	
Alba	TX	Wood	6,232,660	35,864	1948	
Calhoun	LA	Jackson	3,036,197	19,073,042	1948	
Sawyer	TX	Sutton	73,434	30,391,354	1960	
Mesquite Bay	TX	Aransas	1,531,166	19,441,844	1974	X
Chico W	TX	Wise	1,241,048	17,359,651	1948	
Kramberger	TX	Jack	444,543	20,170,684	1971	X
Oaks	TX	Limestone	1,747	19,979,655	1975	X
TOTAL			73,138,467	2,999,813,028		

Total cumulative production from these fields exceeds the equivalent of 3 trillion cubic feet of natural gas.

* *Fields are relatively new to Mitchell Energy with cumulative production expected to grow.*

** *Ike T. Smith #1 well produced 1,947,303 Bbls and 41,828,305 Mcf through June, 2000.*

*** *Excludes production in which Mitchell owned no interest.*

10 consecutive wells. Mitchell and Jackson perceived a huge stratigraphic trap underlaying the entire area, a source of liquids-rich natural gas that even in 2000 accounted for about half of Mitchell Energy & Development's sales of natural gas.

Within 90 days of the Hughes Ranch discovery, Oil Drilling had purchased leases on 300,000 acres of North Texas land at an average of $3 per acre. Mitchell says the dry holes previously drilled in the area by major companies were not really dry holes at all: "We went in and made wells in every one of them." It was a combined engineering and geological answer that worked, thanks to the technology of hydraulic fracturing, which was new then.

In 1953, George and Johnny Mitchell bought out the Roxie Wright estate's interests in Oil Drilling. With Merlyn Christie they created a new company, Christie, Mitchell & Mitchell. Christie owned 50 percent, George Mitchell 37.5 percent and Johnny Mitchell 12.5 percent.

B. F. "Budd" Clark, long-time executive vice president of the company and later vice chairman of the Board, joined Christie, Mitchell & Mitchell in 1956. A native of New York City and graduate of Fordham, he had been associated with petrochemical companies on the Gulf Coast, finding time to earn a Harvard MBA along the way. If George Mitchell has been the company's upbeat, full-speed-ahead yang, Budd Clark has been the yin who's urged a slower, more cautious approach. Together they have been a strong, and volatile, leadership team.

Clark feels that Mitchell's philosophy about asset acquisition—both the original North Texas acreage and subsequently—has been based on two facts, "or to him they were facts: First, there would be inflation as far as the eye could see, and second, the price of gas had to approach equivalency with oil . . ."

Christie, Mitchell & Mitchell sewed up vast reserves in North Texas . . . but not without experiencing some severe indigestion. There was virtually no market for the gas; where it could be sold, prices were extremely low. The demand for capital at cash-poor CM&M was overwhelming, and the company's outside investors began to advocate a more cautious approach. Further, Lone Star Gas, which dominated the area, was playing a tough competitive game in order to have abundant supplies for its market in Dallas.

The solution for CM&M came through its relationship with Natural Gas Pipeline Company of America. Natural wanted to build a pipeline from production in the Texas Panhandle and run it east through Oklahoma to North Texas. From there, gas would be transported to the Chicago area.

Mitchell and two major companies were interested in working with Natural, but the Supreme Court in 1954 held that producers selling gas interstate would be subject to regulation. The two majors immediately lost interest in the deal—as much as possible they wanted to avoid government regulation—but CM&M remained steadfast. With government regulation, tough competitive conditions and a none-too-sturdy market, it was an uphill climb for Natural, but it built its new pipeline and also was able to help CM&M, advancing $7 million to drill North Texas wells and build reserves. The result was a mutually beneficial relationship that lasted for decades.

Natural got its pipeline built and became CM&M's largest customer. Together, in 1957, they entered into a 20-year contract, which was renewed for another 20 years in 1977 after some tough bargaining that shook, but did not destroy, the long-time good relationship. Natural received a dependable source of natural gas for its Chicago market (the Mitchell company for many years supplied 10 percent of Chicago's needs) and CM&M and its successor companies received premium prices and guaranteed "takes" for much of its North Texas production. Finally, the contract was bought out in 1995 by Natural, two-and-a-half years before it expired.

NATURAL GAS LIQUIDS: A SPECIAL NICHE

The North Texas gas was and remains extremely rich in natural gas liquids—ethane, propane, butanes and natural gasoline. In its contract, Natural insisted that the liquids be processed out of the gas stream because once the gas stream entered the pipeline, the liquids would drop out and interrupt normal transmission. To the Mitchell people, this was a welcome requirement. Yes, a big investment would be needed to process the gas and remove the liquids, and the gas itself, with lower energy content, would require "makeup" gas to be restored to pipeline quality. But even after all that, the processed gas and the liquids, which serve as feedstocks for petrochemical manufacture, heating and other applications, had far greater value than unprocessed gas. And thus was born a special niche for Mitchell Energy & Development: Over the years it has become one of the nation's largest producers of gas liquids.

In 1958, the GM&A (for George Mitchell & Associates) Gas Products Plant, which extracts the liquids from natural gas, came on stream in Bridgeport, in the heart of Mitchell's North Texas operations. Warren Petroleum, a company with a long record of successful natural gas processing, would have been willing to build and own the plant, but the Mitchells and Christie preferred otherwise. Warren obtained $4.3 million in financing from a Chicago bank. It was responsible for construction and, in the end, owned 20 percent of the plant. Christie, Mitchell & Mitchell, Bob Smith and other investors owned the remaining 80 percent. In December 1969, CM&M purchased the last of all minority interests in the plant and its related subsidiaries. For many years, Warren marketed the plant's output.

The strong entry into the liquids business was largely completed with the 1963 acquisition of Southwestern Gas Pipeline, Inc. Southwestern served mainly as a gathering system in North Texas, transporting gas from wells throughout the area to customers and to company-owned processing plants.

It was the happy confluence of George Mitchell's thinking with that of Bruce Withers, in the early 1970s a rising executive with Tenneco, that led to the Mitchell company's further plunge into processing.

Withers, a 1950 petroleum and natural gas engineering graduate of Texas A&I University, was happy and yet frustrated at Tenneco. The big, interstate pipeline company was active in processing large gas streams, but Withers felt there was additional opportunity available by stripping the liquids from lesser streams through the use of smaller, portable, less costly plants that could be moved to new locations when a gas supply became depleted. George Mitchell was thinking along identical lines and recruited Withers, who left Tenneco for Mitchell Energy in 1974. Withers was soon joined by Allen J. Tarbutton, a long-time associate at Tenneco. When Withers left Mitchell in 1992, Tarbutton took charge of the company's processing and other service operations.

Within a year of their joining the company, Withers and Tarbutton had received delivery of 10 portable plants, all of them utilizing advanced cryogenic technology. The earlier plants were located along the Southwestern Gas Pipeline system in North Texas, in the Seven Oaks Field, Galveston, Sutton County and Barton Chapel.

As part of the strategy to increase liquids production, Withers steered the company into enlarging its pipeline system in order to make additional gas available for processing. The Winnie Pipeline system and plant near Beaumont, Texas, and the Ferguson Crossing Pipeline Co. in the state's Austin Chalk area were acquired, as was an interest in the Tejas Gas Corporation network. The Southwestern system was expanded. Smaller acquisitions were made in Texas and other states.

At one time, the company had 50-plus of the portable, turbo-expander plants. But as available streams became depleted and the economics and technology of gas liquids recovery changed, the company reduced its emphasis on the smaller plants, and in the late 1990s, while total production remained within range of earlier records, the number of plants in operation decreased and the size of each increased.

MOVING AWAY FROM PARTNERS AND OUTSIDE INVESTORS

Budd Clark's idea is that "George Mitchell did not like to work with partners." And the number of partners and outside investors gradually diminished.

- The Mitchells and Christie in 1959 bought out the interests of John Riddell; the Stephen C. Clark estate, which was related to the Singer Sewing Machine estate; and the Waterford Oil Company, part of the Barbara Hutton estate.
- In 1962, at a cost of $4.2 million cash plus assumption of Merlyn Christie's share of debt on company properties, George and Johnny acquired Christie's interests in the company, which now became Mitchell & Mitchell Gas & Oil Corporation. After riding the Mitchell tiger for more than 15 years, Christie felt he was ready to play a slower game and distance himself from the company's large debt burden. After that transaction, George controlled 70 percent of the new entity, Johnny 30 percent. George later bought one-third of Johnny's interests and, as a result, owned 80 percent to Johnny's 20 percent. Christie, Mitchell & Mitchell became George Mitchell & Associates, Inc., a wholly owned operating subsidiary of the new entity.
- By 1967, all of the Big Nine were bought out.
- In 1968, for $12 million, the Mitchells acquired Bob Smith's interests. Smith had never owned an equity interest in the series of companies that became Mitchell Energy & Development, but through his investments, he was a major player in the North Texas production and acreage, the Bridgeport plant and other assets.
- In fiscal 1970, Gulf Warren's interests in the Bridgeport processing plant were purchased.
- Johnny Mitchell eventually sold all his interests.
- George Mitchell picked up a lot of "partners" when shares were sold to the public in 1972, but he has always maintained a controlling majority of the voting stock.

REAL ESTATE: 'JUST MIGHT HOLD PROMISE'

In the early '60s, Mitchell & Mitchell Gas & Oil was a modest-sized, successful energy independent with a good record of building reserves. Wise County and the surrounding area were (and remain) the company's bread and butter. Moreover, during the early years, the company was successful in other areas. It discovered or extended the Madisonville, Pinehurst, Palacios, Vienna, Buffalo, Buffalo South, King Ranch, La Sal Vieja, Alba and Keeran Fields, among many others. At the same time, the Mitchell company had dabbled in real estate development, mainly through some small developments done in conjunction with Norman Dobbins and Pacesetter Homes. In addition, it had started some resort development on Galveston Island and in 1962 had purchased Pelican Island just offshore Galveston. George Mitchell recognized that domestic energy shortages lay ahead, and that energy prices would increase, but that provided little nourishment at a time when the Texas Railroad Commission permitted oil to be produced only eight days each month, when oil and gas prices were barely above costs, held down to a large extent by growing imports from the Middle East. Real estate, he began to think, just might hold promise for the company.

The plunge into the North Texas natural gas play was a clear example of Mitchell's seeing a main chance and then committing resources in order to capitalize on the opportunity. Another main chance, this one in real estate, presented itself in 1964.

George Mitchell personally owned (and later sold to the company) a parcel of land where The Woodlands Trade Center is located now, across Interstate 45 from the main body of the community. Through Max Newland, a forester who harvested timber on George's land, Mitchell became aware that the Grogan-Cochran Lumber Company was contemplating the sale of its 50,000 heavily wooded acres north and west of Houston. To Mitchell, the land represented an enormous opportunity, although he wasn't quite sure what he would do with it.

The Grogan and Cochran families had owned the acreage, essentially a source for timber, since 1903. Mitchell says there were three main family groups involved, each group (and individuals within each group) with different ideas about what should be done with the property.

Mitchell offered $125 an acre, or $6.25 million for the package, an amount he arrived at by estimating potential timber and gravel sales that could be made to pay the interest on the money he'd need to borrow. The offer was accepted, but the deal didn't become final until Mitchell helped work out settlements on the various lawsuits family members had filed against each other.

Like just about all of Mitchell's business transactions, this one was somewhat more complicated than buying a loaf of bread at the supermarket. As usual, he was short of cash. He borrowed from Bank of the Southwest for a down payment. The Grogan-Cochran shareholders received notes secured by a deed of trust on the land. Loans from Great Southern Insurance Company paid these notes over time. Then there was an agreement with Louisiana Pacific under which that company would make annual purchases of Mitchell's timber, with the proceeds used to pay Great Southern. (Budd Clark estimates that when the value of the timber is factored in, Mitchell actually paid closer to $40 per acre than $125.)

Why invest in the land? "Because," Mitchell says, "it was a good buy," its value enhanced by the on-site timber and gravel. "I hadn't even thought about The Woodlands at that time," he says. "(But it) was close to Houston and it was beautiful land."

During the later 1960s, two seemingly outside forces were coming to bear on the future of George Mitchell and Mitchell Energy & Development: first, the programs of the Young Presidents' Organization (YPO), and second, a growing concern in Washington about the deterioration of America's cities.

For George Mitchell, YPO was the means to an education in the social sciences and liberal arts for which his heavy schedule at Texas A&M had never allowed time. A YPO seminar in the mid-'60s took him to the Watts section of Los Angeles and Bedford-Stuyvesant in Brooklyn, scenes of economic depression, joblessness, poor housing and high crime rates. Mitchell was devastated by what he saw. As neighborhoods had decayed, middle-class whites fled to the suburbs, taking with them much of the stability, economic strength and other qualities that make cities work.

Says Mitchell, "This was not necessarily, or solely, a black and white phenomenon. America's worn-out cities were left to the poor, and the poor lacked the resources to manage.

"Then these wealthy suburbs in effect put up the gates and locked out people who hadn't the wherewithal to enter within and enjoy better schools, cleaner streets, lower crime rates.

"But that didn't stop the suburbanites from taking advantage of the city's great institutions–the museums, the fine universities, the airports and ports, the wonderful research hospitals, the economic base."

In Washington, opportunities were being created. First, it seemed that financial help for beginning to deal with urban blight might be available under Title X of the 1965 housing act. Other legislation followed, including the Urban Growth and New Community Development Act of 1968 and then the 1970 act under a similar name. It was under Title VII of the 1970 act that the Mitchell company was able eventually to obtain a large loan guarantee, a pivotal development in creation of The Woodlands.

The idea for The Woodlands grew gradually. First came the 1964 Grogan-Cochran land acquisition. Of that 50,000 acres, approximately 2,800 ultimately served as the nucleus of The Woodlands land mass (which in 2000 amounted to about 27,000 acres). To Mitchell, it was clear that the path of Houston's growth lay to the north of the city. The potential societal and economic advantages of creating a planned community began to come into focus. But additional land acquisition would be necessary, a means of financing the ambitious idea had to be found, a plan had to be devised.

BUILDING THE DEVELOPMENT TEAM

Members of the original development team were mostly made-over oil and gas men. Morris Thompson, a senior officer and member of the Board of Directors, had been with the company since early on. A petroleum engineer by education and experience, he became a jack of all trades. Jim McAlister joined the company in 1963 with responsibilities primarily in oil and gas evaluation and banking; through the years, his job shifted toward getting the new town under way. Plato Pappas was a young civil engineer with real estate development experience when he came to Mitchell in 1967. Charles Lively started in 1956 and in his early years was a petroleum landman in North Texas, learning that skill from scratch; later he became the man primarily responsible for blocking up the acreage that comprises The Woodlands.

When Pappas joined the company, he assumed engineering responsibilities for Mitchell-Dobbins Pacesetter Homes, which was building 300 to 400 homes a year, and for the Mitchell resort properties in Galveston, Pirates' Beach and Pirates' Cove. His earliest accomplishment was to bring engineering costs in line. Later he played key roles in building The Woodlands' infrastructure.

After the Grogan-Cochran acquisition, the company started to develop small rural subdivisions on fringe acreage in Waller and Grimes Counties, and Charles Lively's first job in real estate was to purchase land parcels necessary to fill out a block and make it contiguous with other land in the subdivision. He soon found himself mainly responsible for obtaining parcels that would become part of The Woodlands—a job that eventually took 300, or as Lively claims, as many as 500, transactions.

Land was owned that provided a southern entrance to the property, but it was clear that access to Interstate 45 on the east would be essential. Acquisition of the Sutton-Mann tract made that possible. Lively and Mitchell sought to buy a 1,000-foot-wide tract, owned by the Sutton and Mann families, that would go west from I-45 for about a mile-and-a-half and connect with the main block that already had been assembled. After three years of tough negotiations, the deal for the 4,000-acre tract was made in 1972.

Karl Kamrath, a Houston architect, had designed George and Cynthia Mitchell's home in Houston and had played a major role in other local building projects. In 1966, on assignment from Mitchell, Kamrath prepared a plan for a 20,000-acre community. While Mitchell has warm feelings for Kamrath's plan, it had little impact on the final concept. Then, again at Mitchell's request, Cerf Ross, another Houston architect, developed a plan for a new community that was submitted to the company in 1969 and was used in presentations to potential lenders. A number of features of the Ross plan survived as The Woodlands developed, but, in fact, both Kamrath's and Ross's ideas yielded to those of subsequent planners.

'FLYING BY THE SEAT OF THEIR PANTS'

McAlister's original duties with Mitchell had to do with energy operations; gradually, he became more involved in real estate and for a period of time, he says, he was the only one assigned to work on the idea of a new community. He describes the atmosphere of the mid-to-late-1960s:

> There just weren't any defined standards up to this time. Everybody was flying by the seat of their pants. You have to remember that George has a lot of unusual qualities. Perception is one of them. But another one is a terrific capacity to recall. . . . He carries files in his head that most people don't remember are in their filing cabinets. The company was operating as a projection of George's personality. Well, we had to come out of that, we were growing, we were getting bigger, we had to have a budget and better-defined investment criteria.

McAlister's recollection is that the idea for a satellite city—to be called Satellite City—grew out of a meeting with Mitchell. The initial thought was that it would be a bedroom satellite of Houston. Ultimately, Cerf Ross, who had been working on some government projects in the West, introduced the idea that under new programs, federal guarantees on loans might be available. George Mitchell and his associates pursued that approach despite many warnings from financial institutions that Mitchell would regret doing business with the government.

Little information was available on how to plan and build a new town. Columbia, Maryland; Reston, Virginia; and Irvine Ranch in California were newly open or under way. Government loan guarantees were in the pipeline for other new communities, including Flower Mound near Dallas. With no body of knowledge about new community development readily available and with computers still in their relative infancy for business use, McAlister and a number of associates set about creating an economic model for the new town.

After meetings with executives in George Romney's Department of Housing and Urban Development, which was responsible for administering the New Communities plan, Mitchell and McAlister started to develop a team with more widely recognized credentials. In 1969, the late Robert Hartsfield, an architect with strong experience in urban planning and environmental protection, was hired away from the Caudill, Rowlett, Scott consulting organization in Houston to be in charge of planning and environmental design. Hartsfield had studied under Ian McHarg at the University of Pennsylvania; he brought the colorful, outspoken Scotsman aboard as one of the chief consultants. McHarg, author of the landmark book *Design With Nature*, is widely regarded as father of the environmental movement. William Pereira, a Los Angeles-based planner with international accomplishments, joined the group. Gladstone Associates of Washington, D.C., became consultants on economics and marketing. Richard P. Browne of Columbia, Maryland, who later joined the Mitchell organization, helped on development, engineering and liaison with HUD. By 1970, the team was complete.

'SATELLITE CITY' GETS A NEW NAME

Where did the name "The Woodlands" come from? More than one person claims parentage, but, clearly, the definitive source was Cynthia Mitchell and the Mitchells' wooded 5,000-acre place in Montgomery County, Texas, which they earlier had named "The Woodlands." More than 100 names for the new community were considered. But in the end, Cynthia recommended "The Woodlands" as perfect for the new project. She volunteered to select another name for their family place: It became "Cook's Branch," after a spring that flows through it.

PACKAGING THE APPLICATION

McAlister and others were doing economic feasibility studies, while at the same time working with HUD. A final decision about whether to seek HUD backing had not been made, although, essentially, the assumption was that the answer to financing needs would be through HUD. Parts of the application package were coming together.

McAlister says, "The hours we worked were humongous. In fact, my family throat doctor was at the building next door and we would work, we were tired, we would get sick. In the winter time, we all would have colds . . . and his cure for your sore throat was gentian violet. . . . They paint your throat with purple stuff and it hurts so bad, tears come down your eyes. We would go over there and keep ourselves going . . ."

Finally, in 1972, the voluminous application—which included the first environmental impact statement ever submitted to HUD—was ready to go to Washington. Up to that point, the largest loan guarantee had been $33 million to Flower Mound, Texas.

Mitchell, McAlister claims, did not feel comfortable in requesting a $50 million guarantee but accepted the consensus of his staff that $50 million was the right amount.

McAlister recalls

> So now comes the hour. We go into HUD. . . . The way I had it set up was, first, to have Jim Veltman get up and go through the ecology, then, second, we went through the social part, which was (David) Hendricks's . . . followed by the presentation of the development plan with all exhibits. At this point, we had never told them (HUD) how much money we were going to request. . . . George was nervous all the way on the airplane going there about the $50 million he wanted. At any rate, my economic presentation was the last part. So I went through the economics and was the last speaker. I had to close and state our request. I said, "And therefore, for the reasons you've just heard, I respectfully request $50 million." Total, total silence for the longest period . . . Mitchell begins to look around and looks at me like, Oh, Oh! . . . I was standing at the podium and the silence just went on. If you know about sales, you know that the first person to speak loses, and I just stood there and thought, "God, I'm going to get fired, my life's over, it's done." Then (Secretary Romney's representative) speaks up and says, "That was a very fine presentation and we will take it under advisement."

THE LOAN GUARANTEE IS APPROVED

On April 4, 1972, Mitchell Energy & Development was notified that the entire $50 million loan guarantee was approved. A syndicate headed by A. G. Becker & Co. purchased the $50 million offering at an interest rate of 7.10 percent, which, because of the government guarantee, was more favorable than what would have been available on the open market. HUD insisted that a substantial part of the $50 million be used to pay off all mortgages on the land and that HUD have the first lien. So, according to Budd Clark, the company did away with 4, 5 and 6 percent loans that it already had and instead had to pay 7.1 percent. On the plus side, the company did gain a 15-year period in which no principal payments were required. Since almost half the money was used to pay off mortgages, much of the loan was not available for development. Because of these restrictions and the ensuing government involvement in Mitchell affairs, the loan guarantee proved to be a mixed blessing. But at the time, the company pursued what appeared to be the best of the options available, and maybe the only one.

Few from the original group remain with Mitchell. Some have gone into business for themselves, some retired, some died. For all it was a rewarding but high pressure period. Was it worth it? McAlister—who left the company in 1972 to become an independent real estate investor—says, "I would fight a bear for Mitchell in the parking lot."

DIFFERENT VIEWS FROM PEREIRA, BROWNE

In the early planning, William Pereira and Richard Browne had conflicting ideas. Browne's recollection is that Pereira advocated creation of a linear commercial urban center like a main street that would go from east to west, with the residential development villages behind and off that main street. Browne, successfully, urged that The Woodlands take advantage of the Columbia experience and build villages off a hierarchy of townwide roads that would provide convenient access while separating the villages from each other. He argued for village centers surrounded by a variety of housing in a series of self-contained communities small enough that people would not feel lost in a sea of suburban sprawl.

He says that the wide use of cul-de-sacs in The Woodlands stems from his belief that "the arithmetic of the group should be as close to that of the original primate hunting troops of man as possible . . . 50 to 60 people living together on a defensible piece of territory. In community building, culs are friendlier and more secure than living (several) blocks off some street where there's no beginning or end. . . . You're not subject to through traffic, the kids are safer . . . you get to know your neighbor. . . ."

WHICH IS BETTER, PINE OR OAK, BIG OR SMALL?

Robert Heineman in 1995 became head of planning for The Woodlands. With architecture and urban design degrees from Rice and Harvard, Heineman started at Mitchell in 1971 as a summertime employee. Right from the beginning, one driving force was that development would be in response to environmental factors, primarily soil permeability and vegetation type. He says, "We dealt with questions such as whether pine trees were better than oak trees, big trees better than small trees . . ."

Saving the forest during development was an innovative part of the original plan, Heineman says. "If you're going to name the project The Woodlands, don't name it for the forest that was there prior to development but which was destroyed in the development process." Now, on a drive-through on the major thoroughfares, "you see trees. In fact, the houses and buildings are often hidden. It gives . . . the distinct impression that the forest is still there . . ."

He says that initial reaction to these concepts was adverse "because people expected manicured landscaping which required clearing of all the understory."

Heineman points out that commercial buildings in already-developed cities often front on major thoroughfares. Not so in The Woodlands. That restriction increases cost to the developer, possibly by as much as 15 percent: Separate "entrance" streets become necessary, and developable land is lost. But, says Heineman, the advantages in marketability and aesthetics more than offset the extra cost. In addition, unlike big-city thoroughfares with their curb cuts for commercial and retail buildings and the frequent traffic lights, The Woodlands' hierarchical road system minimizes traffic congestion. And through the use of cul-de-sacs for residential neighborhoods, through traffic is forced onto major collector streets.

Another innovation was the concept of natural drainage. Ian McHarg's original idea was that, at multi-million-dollar savings, The Woodlands should rely on streams, bayous and rivers to drain the area. He felt that an expensive system to handle storm runoff would be unnecessary and inappropriate, as would be concrete curbs and gutters in residential neighborhoods. Natural drainage would rely on swales, retention ponds and strategic use of land below the flood plain, among other ideas.

That concept has had to be modified, Heineman says. "What has been maintained is the natural drainage system in the major streams as well as open channels which flow from neighborhoods into those streams. Initially, we experimented with open-ditch drainage along residential streets . . . The market acceptance of open-ditch drainage is very poor . . ."

A NEW CULTURE IS INTRODUCED

To a large extent, Mitchell recruited the in-house team to build The Woodlands from the Rouse Company, which was responsible for the new town of Columbia, Maryland, near Washington, D. C. The linchpin was J. Leonard Ivins, who joined Mitchell in 1972 from his position as vice president and director of development at Columbia to become a senior vice president and member of MEDC's Board of Directors. Ivins brought with him a group to take over a wide range of activities, from accounting and finance to general operations. The Ivins group also played a sizable role in shaping the application to HUD and subsequently working with the government agency.

Michael Richmond, who had been with Columbia's outside accounting firm, is president of The Woodlands Operating Company, L.P., formerly The Woodlands Corporation. His recollection of the team from Columbia is:

> It was a flamboyant group. Len Ivins was a visionary, a dreamer, an impact person. The group had a feeling of invincibility. They conceived more from the gut than from the hip, but they weren't great on day-to-day operations. They were partiers, they worked hard and they played hard, and they were spiritual and emotional leaders. They had a dream of making this place special. They weren't good at cost control, they weren't good at day-to-day operations and management, but some people will look back on it and say it was that outlandish commitment early on that gave the project an identity. Others will say that it was a waste of money, excessive expenditures that weren't necessary. And having been there . . . I'd say it was probably in between.

James Rouse, one of the most innovative and widely respected men in real estate, was generous of his own and his organization's time in providing advice and information to get The Woodlands started. But he did become impatient at the talent drain under which experienced, capable men and women left Columbia and headed for Texas. Eventually, he sent an associate to Mitchell to suggest that the recruitment stop. It did.

With Ivins's group in control of real estate activities, a new culture was introduced into the Mitchell corporation. A number of old-time employees were let go; others were uncertain about their futures. Energy people complained in secret and openly about the freewheeling real estate folks while financial and human resources were in short supply in oil and gas exploration and development. The kind of hard-working camaraderie that Jim McAlister described in the early years ceased. G. David Bumgardner, who started with Mitchell in 1956 as mail boy and who later earned a law degree and became general counsel of The Woodlands Corporation, is generally complimentary about Ivins's contributions. But he tells of a conversation with Budd Clark: "It is probably safe to say that Budd Clark was not enamored with the Ivins management style. Nonetheless, Budd remained philosophical and gave me probably the most sage advice I have ever received. After listening to one of my complaints, Budd told me not to worry about it. I asked why, were things going to change? Budd said no, I would get used to it." Paul Wommack, who was general counsel when he retired in 1987 after nearly 20 years with the company, comments favorably about Ivins and the Columbia team, but he does say, "The real estate activities . . . were a foreign thing to the Mitchell company, which had always been an energy company."

A START ON CONSTRUCTION

The real estate construction group moved out of corporate headquarters at One Shell Plaza in downtown Houston to begin development of The Woodlands in 1972; the target date for opening was the fall of 1974. Plato Pappas recalls that the first concrete was put in place on New Year's Eve in 1972 on Woodlands Parkway because Vern Robbins, who was in charge of all construction, wanted to have a 1972 starting date. The builders went through months of bad weather and changes in building plans (not the least by George Mitchell, who, according to one member of his Board of Directors, kept changing specifications for what is now called The Woodlands Executive Conference Center and Resort). Pappas remembers, "For about a month, we were working day and night around the clock" under floodlights. He would sleep for four or five hours, then switch with Robbins, who would return to their shared hotel room for his four or five hours.

On October 19, 1974, The Woodlands opened with a big celebration that attracted 20,000 persons. A key part of the original strategy in getting The Woodlands moving was to establish the community's credibility from the very beginning, to say, in effect, "We're here now and we'll be here tomorrow." On opening day, visitors could tour 19 furnished model homes, including townhomes priced from the upper $20s. Fifty apartments were ready for occupancy. The Woodlands Executive Conference Center and

Resort was open for business, as was The Woodlands Country Club. The Woodlands Athletic Center—with its world class swimming facilities—was near enough completion to serve as an attractive amenity. Two public schools were ready for students. Visitors could stop at a dramatic Information Center. Three office buildings were completed. Drainage and sewage disposal systems, roads and hike-and-bike paths and all utilities, including a state-of-the-art cable television and security system, were in place.

One of the early breakthroughs was the Jack Eckerd Drug Company's 1976 purchase of 26 acres in The Woodlands' new Trade Center, located east of Interstate 45 across from the main part of The Woodlands. By the late 1990s, the Trade Center comprised 900 acres and, with its location on an interstate highway and proximity to rail, air and even sea transportation, was home to distribution centers for 15 companies.

GOLF COMES EARLY TO THE WOODLANDS

In 1974, The Woodlands Corporation signed a 10-year agreement with the Houston Golf Association to host the PGA Houston Open. Earl Higgins joined Mitchell as an accountant in 1967 and later became responsible for real estate property management. An excellent golfer himself and in 1984 president of the Houston Golf Association, Higgins says the Open came to The Woodlands after being in financial trouble at Quail Valley in the Houston area. He says, ". . . Mr. Mitchell, in essence, gave them $50,000 to bail them out of their problems at Quail Valley in exchange for a commitment to move the tournament to The Woodlands in 1975 . . ."

Later, in golf-addicted Texas, the company created a "stadium" Tournament Players Course for the Houston Open, with public attendance for the four-day competition reaching into hundreds of thousands. And as a potent marketing tool in the Houston area, where golf is played year-round, The Woodlands by the late 1990s had 81 holes of public and country club golf courses with more to come, including a new TPC course.

THE PLAN AND HOW IT DEVELOPED

The company's annual report for the year ended January 31, 1972, provides some insight into where things were headed:

> . . . The new town concept emerged from analysis of Greater Houston's expansion requirements. Current projections of growth to almost four million people during the next two decades substantiate both a need for the new town and its potential profitability.
>
> The Woodlands will be environmentally and ecologically con-

> trolled by Mitchell throughout the 20-year development program. At maturity it will have a maximum of 49,000 dwelling units and 150,000 residents. It will be comprised of a number of inner-village communities, each with schools, commercial/business centers, interfaith religious facilities and recreational, cultural and health-care centers. With 2,800 acres dedicated to primary open space, The Woodlands will provide recreational opportunities for all residents: lakes, golf courses, hike and bike trails, swimming centers and other facilities for leisure-time usage. Core of the new town will contain a major commercial/business district and hospital. Subject to ratification by state authorities, the University of Houston will build its North Campus extension . . . on 400 acres which the Company has donated for that purpose. . . .
>
> Priority will be given to ecological preservation of the abundant existing natural life in the area, as well as to architectural design that will be compatible with the rural woodlands environment. . . . As developer of The Woodlands, the Company expects to limit its construction activities to certain income-producing structures such as industrial office and commercial buildings and multifamily housing.

The annual report language in many ways paralleled that in the 20-year development plan submitted to HUD. Instead of 49,000, the HUD submission projected 47,500 residences: 27,500 single family detached and attached, 20,000 multiple family units. The totals included 13,000 units for low- and moderate-income families to be distributed throughout the project. The plan called for utilities to be placed underground, where possible. It contemplated quasi-government and community organizations.

A quarter-century after those words were written, it is remarkable how closely development of the new community has hewed to the original ideas. Not perfectly: Oil embargoes and economic downturns and social changes could not be foreseen. The original 17,000 acres increased to something closer to 27,000 acres. A development period of 20 years proved to be unrealistically optimistic, and it will be another company, not Mitchell, that shepherds the project through to completion. The estimate of area population growth to four million has been just about right. The amount of open space is turning out to be nearly twice the amount originally projected. Environmental and ecological considerations remain paramount. One idea that did not pan out was establishment of a North Campus of the University of Houston, but The Woodlands does have a growing campus of the North Harris Montgomery Community College that, through its unusual University Center, awards degrees from major universities in the area. There are nearly a dozen highly rated public schools.

Immediately after opening, however, there was no assurance that the projections could be met or even that The Woodlands would have a long life. By early 1975, just a few months after the grand opening, Mitchell Energy & Development found itself with severe problems in cash flow and liquidity. The engine that brought the company to its difficult straits was, more than anything, the front-end investment—both budgeted and over budget—required to open The Woodlands. For instance, nearly $30 million was spent on the conference center, information center, swim center and wharf shopping area, well beyond the anticipated amount. Moreover, the Arab oil embargo, which ended in the spring of 1974, had caused a spike in the price of gasoline and a supply shortfall which made living in a suburban location less desirable. The concomitant high rate of inflation added to Mitchell's difficulties. Bank credit was just about unavailable. Skilled building labor in the Houston area was in short supply. Aggravating the problem—some say it was the main source—were the difficulties in dealing with HUD's bureaucracy, the painfully slow pace of the Department's response to urgent problems and its failure to extend help, some of it previously promised, when help was reasonably needed.

Obviously, money was essential. And, the Board of Directors decreed, so was a change in management and operations of the real estate function.

Leland W. Carter, who was president of the company's energy subsidiary, also was named acting president of real estate operations. The real estate staff size was chopped from 365 to 160. Expenses and planned capital investments were cut drastically. Len Ivins and members of his management group left the organization.

Despite the vast and costly differences in style and outlook between company veterans and the new real estate team, George Mitchell regards Ivins as having been extremely innovative, a man whose creative contributions left an indelible mark.

PAYING THE BILLS

Michael Richmond was 24 in 1972 when he joined Mitchell as controller of real estate, just about the time of groundbreaking for The Woodlands. He was in for a bumpy ride.

"When I walked through the door," he says, "we had $26-$27 million left in the bank (since much of the HUD-backed $50 million had been used to pay off outstanding debt on the land mass). . . . But it didn't take us long to spend it."

A month after opening, says Richmond, "we could not pay all the bills that we owed from getting the project opened." He says:

> I'm going to guess it was $9 million (that we owed), and we didn't have a nickel in the bank. Energy (operations) owed some more, too. Every Friday I would go downtown and meet and see who we could pay . . . There were no controls at the very end because everyone wanted to get it open, and there were a lot of people who authorized expenditures, and the bills rolled in later. We talked to as many people as we could, we tried to be honest with them, told them we would try to get them paid . . . We eventually paid everybody out over time.

Compounding the cash flow problems, HUD in January 1975 notified MEDC that the company was in default on its loan agreement. In retrospect, the problems seemed to be those important only to lawyers, accountants and government bureaucrats. HUD had not given prior permission for so-called related party transactions; some subsidiaries were engaged in activities prohibited under the project agreement; some of the proceeds from bonds guaranteed by HUD were being used improperly. The situation was complicated, but both sides were anxious that The Woodlands survive. With unprecedented dispatch and cooperation, the defaults were cured to HUD's satisfaction by mid-March 1975, three months after HUD's formal complaint.

But even in a more positive atmosphere, deep problems remained. There was disagreement on land values, which had the effect of limiting Mitchell's ability to borrow. HUD and Mitchell disagreed on the value of Mitchell equity contributions to The Woodlands. The company insisted that HUD was failing to make promised grants to the community. By early 1977, most of the disagreements had been settled, but to no one's complete satisfaction. Subsequently, however, HUD came through with an increased flow of grants, mainly of the kind that were being provided to other cities and towns throughout the nation.

THE PRESSURE EASES

Even with the HUD problems, the cash flow pressures of preceding months were ameliorated by late spring of 1976: With the oil embargo of 1973 and 1974, gas and oil prices strengthened considerably (the price of oil went from a pre-embargo $4 per barrel to $12 post-embargo, and four years later, following a disruption in supply from Iran, the price moved quickly from $14 per barrel to more than $40). With stronger prices for oil and gas, the energy part of the company flourished. Then, cash and cash commitments of $28 million became available in the early months of 1976 from bank financing, the sale of some energy assets, advance payments on timber sales and partial repayment of utility costs at The Woodlands. Lenders, especially the Chase Manhattan Bank, had imposed tough financial restraints, and for a while, the possibility loomed that George Mitchell would have to cede control of the company to the Chase. But gradually, defaults in loan agreements and in the contract with HUD were cleared up, and the company found banks and insurance companies more forthcoming with the needed financing.

J. Leonard Rogers, a certified public accountant experienced in developing the community of La Jolla, California, took over as vice president and general manager of Mitchell real estate operations late in 1975. He was succeeded two years later by Edward P. Lee, vice president of land development operations at Irvine, California, possibly the nation's most successful new community. Lee was determined to make The Woodlands "a world class development," and was successful in putting the development program on track, setting up long-range operating plans and improving relations with residents. He died in June 1986.

THE WOODLANDS AND HOUSTON

George Mitchell's original plan was that in the early years, community services—such as operation of recreational facilities and contracting for trash collection, fire protection and emergency medical services—would be provided by one or more nonprofit community associations. At the beginning, directors of these associations would be nominated by the developer and would be in the majority; a minority would be elected by the community. Over time, the makeup would change, with the majority to be elected.

But also at the outset, the Mitchell company planned for The Woodlands to be in the extraterritorial jurisdiction (ETJ) of Houston, which would allow Houston to annex The Woodlands and preclude annexation by another city. In Houston's ETJ, future residents of The Woodlands could not incorporate without Houston's permission, which had the effect of giving the Mitchell company several decades to develop its project. George Mitchell gives additional reasons: He has consistently voiced concern about how free-standing suburbs in other parts of the country had largely insulated themselves from, and exacerbated, many of the problems of the inner city. And, he feels, the city provides the main reasons for the suburbs' existence in the first place. The Woodlands and other such communities bear a responsibility to the welfare of the city, he says.

The Houston City Council in November 1971 approved the Mitchell petition to join the city's ETJ. Before long, the newly incorporated City of Shenandoah, immediately adjacent to The Woodlands, opened its jaws wide and moved to annex its new neighbor, but that effort finally went nowhere. As time passed, the concept of being annexed to Houston was not popular with most residents of The Woodlands. The issue became a center of growing controversy as Houston annexed another suburb, Kingwood, in 1996 over protests from that community's residents. Roger Galatas, who became chief executive officer of The Woodlands Operating Company, estimated in 1995 that 95 percent of Woodlands residents would oppose being incorporated into the City of Houston. He says, "I hope that we as a development company and we as a community can preserve the opportunity to continue to develop . . . comprehensively as we've done in the past and not have small city incorporations that fragment the government . . . so that it becomes impossible to implement the comprehensive plan."

BEYOND THE BASICS

From the beginning, the idea of The Woodlands was that it was to be a planned, complete community, rather than another of the helter-skelter housing subdivisions that ring the nation's cities. The original concept contemplated not only the basics of jobs and transportation and streets and sewers, it also looked at residents' spiritual lives, as well as enrichment through the visual and performing arts, educational opportunity and the other requirements of people living together.

Rev. G. Richard Wheatcroft came to Houston from Missouri in 1950 to be the founding minister of St. Francis Episcopal Church, which has grown to be one of Houston's larger Episcopal congregations. George Mitchell and his family were congregants. Rev. Wheatcroft says, "I think I baptized most, if not all, of their children. They would come to church and fill up two full pews with the 10 kids."

The Woodlands Interfaith concept, Rev. Wheatcroft says, grew out of a conversation he had with George Mitchell. Mitchell asked him to come up with a creative response to the issue of land use and money that churches spend for buildings and parking lots, something that would be helpful both to the churches and to the community itself. Says Rev. Wheatcroft:

> . . . (F)inally, one day, (Mitchell) asked if I would be willing to organize and head up a committee of representatives from the major denominations in the Houston area . . . to see what we might come up with in terms of some cooperative planning.
>
> . . . I began to contact heads of judicatories . . . all the majors . . . and got a committee organized, about 15 people. . . . George sent the whole group on a trip to Columbia, Maryland, and Reston, Virginia, to visit those two planned communities to see what they had done in terms of religious institutions. . . .

The committee found that Columbia had come up with the concept of multiple-use buildings for religious groups. They had built one large building that could be used as a sanctuary, and various groups used it on Sunday mornings at different times. Rev. Wheatcroft recalls:

> That had some appeal in terms of land use and the problem that churches spend too much on buildings and parking lots and land and so forth. That was one of George's concerns, too. . . .
>
> We tried to sell that idea to the judicatories in the Houston area and it just wouldn't work. They all wanted their own places. . . . (S)o we ended up in the traditional mode of each church buying its own land and building its own buildings, etc., although several churches like the Baptist church and the Lutheran church on Grogan's Mill Road built next to each other so they could share some parking space. . . . In the long run that didn't work either.

Interfaith's first executive director was Rev. Don Gebert, a Lutheran, who took office in 1975 and resigned in 1984. He was succeeded by Rev. W. D. Broadway, a Baptist minister and long-time member of the Interfaith Board. Interfaith—more formally The Woodlands Religious Community, Inc.—has assisted in the establishment of churches, supported developing social agencies and community organizations, and has created programs to help meet human needs. It runs a visitation program for new residents, an information and referral service and senior citizen and employee assistance programs. The Interfaith facilities serve as a meeting place for newer congregations. For congregations just starting out, Interfaith helps negotiate for church building sites. And among its most important activities is the Child Development Center, which provides day care for some 500 children in The Woodlands area.

A WELCOME FOR THE ARTS

As with religion, The Woodlands has also provided a welcoming climate for the arts. The beginnings go back to the company's 1972 move from the Houston Club Building in downtown Houston to the brand new One Shell Plaza. (Corporate headquarters were moved from the Esperson Building to the then-new Houston Club Building in the mid-1950s, from there to One Shell Plaza and in 1980 to The Woodlands.)

In the cramped quarters of the Houston Club Building, little attention was paid to any decor other than the strictly necessary. One Shell, however, was Houston's newest and most desirable skyscraper, the corporate home of Shell Oil, which had just moved from New York City. The responsibility for coordinating the program that would make the appearance of Mitchell's new offices consistent with corporate aspirations fell on Joyce Gay, who had joined the company in 1957 as a secretary and who was now vice president for public relations.

Mrs. Gay worked with the decorators and exhibited an uncanny ability to anticipate George Mitchell's preferences. "He was always a contemporary man and liked very contemporary things," she says. Some employees groused about the lack of privacy resulting from the use of movable partitions to delineate office space, but the tasteful, modern look of the offices, with their emphasis on works by established regional artists, was hailed by both employees and visitors.

Gay left the company in 1973 and opened her own art gallery; the Mitchell company was an important customer. With the opening of The Woodlands in 1974 and for nearly 15 years after, Mitchell looked to Gay and a small committee to select sculptures for The Woodlands's public spaces. Financed by a small part of land sales to builders, the sculptures add warmth and character to the community. A special lagniappe has been provided by Cynthia Woods Mitchell who has contributed several striking sculptures.

The performing arts, too, have helped set The Woodlands apart, especially as evidenced by The Woodlands Center for the Performing Arts, more popularly known as the Cynthia Woods Mitchell Pavilion. The Pavilion—summer home of the Houston Symphony—has presented works ranging from Shakespeare's "Julius Caesar" to performances by Elton John, from an original opera to programs by Houston Ballet.

The plan submitted by Gladstone Associates, part of the original team of consultants, talks about an outdoor theater (to be built at a cost of $75,000). In 1971, when George Mitchell toured the Columbia new town in Maryland with Richard Browne, they discussed the idea of a pavilion in The Woodlands similar to Columbia's Merriweather Post Pavilion. And,

SCULPTURE IN THE WOODLANDS

Name of Sculpture	Sculptor	Location	Installation Date
The Family	Charles Pebworth	Main Entrance, The Woodlands	1974
Man's Struggle for a Better Environment	Bob Fowler	Grogan's Mill Road Median at Woodlands Parkway	1974
Casten Metal Sculpture	Richard Hunt	Research Forest Drive	1977
Disc II	Richard Rogers	Median, Timberloch Place	1979
Revoluta	Corbin and Dixon Bennett	MND Building	1980
Sculpture (unnamed)	Horace L. Farlowe	Entrance to the Trade Center	1980
Tomorrow	Charles Cropper Parks	New Information Center	1981 (1997 at Info. Center)
Children at Play	Clement Renzi	Panther Creek Village Square	1984
Excalibur	David Hayes	Median, Lake Woodlands Drive, Cochran's Crossing	1984
Rise of the Midgard Serpent*	Marc Rosenthal	Lake Woodlands, Woodlands Parkway	1985

The Family	Patrick Foley	Entrance to Memorial Hospital–The Woodlands	1985
On the Shoulders of Giants	Robert Cook	Research Forest Drive & Grogan's Mill Road	1989
The Dreamer	David Phelps	Median west of East Panther Creek Drive in Woodlands Parkway	1989
Smokedance*	Dale Garman	Cynthia Woods Mitchell Pavilion	1989
The Watch Owl*	Mark Bradford	Pathway to Cynthia Woods Mitchell Pavilion	1993
Mirage II	Ben Woitena	Median, Research Forest Drive west of Gosling Road	1995
Spirit Columns	Jesús Bautista Moroles	Median, Six Pines Drive and Lake Robbins Drive in Town Center	1996
Commemorative Tribute to George P. Mitchell	Jay Hester	Cynthia Woods Mitchell Pavilion	1997
Boy With Hawk	Charles Parks	Grogan's Mill Village Center	1998
Big Barbara	Peter Reginato	Alden Bridge Village Center	1998

* Selected and contributed by Cynthia Woods Mitchell.

several civic leaders in the mid-1980s suggested to George Mitchell that he build an outdoor venue for the fine arts, a thought that Mitchell found attractive but untimely because of the fragile state of the energy market at that time. Building the place would cost money. And no one knew how to run fine arts programs in the black.

Richard Browne recalls, ". . . I took Roger (Galatas) to the 18th green of the TPC course after lunch and walked him up on the hill and said, 'Can you imagine instead of looking down on the golf green that there was a stage down there? Why don't we do this in the Town Center's urban park . . . ?' He said, 'OK, how much will that cost?' I said, 'We could start a little one for a million dollars.'"

The completed pavilion was not "a little one," and its cost, before further expansion, was in the range of $10 million.

Browne got a proposal for a tent structure and a small hill from architectural designer Horst Berger, with whom he had worked before. The cost estimate headed uphill.

Browne set up a meeting with Pace Entertainment, the large Houston-based producer. He recalls, "I presented the concept and Cynthia (Mitchell) said, 'Well, that's lovely'. . . And she turned to Bryan Becker from Pace (later Pace's chief executive officer) and asked, 'What do you think?' He said, 'That's a nice little design, too bad nobody will come . . . it's too small.'" Mrs. Mitchell's response was simple: "We'll make it bigger." The plan became more ambitious, and the costs headed higher as such requirements as an orchestra pit and an orchestra shell and air conditioning on stage were added.

The Houston Symphony, under Maestro Christoph Eschenbach, headlined the opening performance on April 27, 1990, the first of three consecutive nights of super stars (including Frank Sinatra). J. Kirk Metzger was recruited from another venue to be executive director; Joseph Kutchin, recently retired as vice president of communications for MEDC, became president. At the outset, Metzger was the only one directly employed by the Pavilion who had any experience in managing a large entertainment center. In 1993, following Kutchin's resignation, David Gottlieb, who had had associations with the University of Houston and then with The Woodlands Corporation, took over and assumed the duties of chief executive officer, Metzger later leaving.

The Pavilion has been a success from opening night. Originally, capacity was 10,000: 3,000 in permanent seats under the Horst Berger tent, 7,000 on a big, grass berm. Later, largely at the urging of and with contributions from the Pace organization, capacity was increased to 13,000. Further enlargement, to 17,000, started in the autumn of 2000.

A typical season now includes approximately 35 programs by contemporary entertainers and 15 to 20 classical arts programs, 10 of them by the Houston Symphony and others by Houston Ballet and Houston Grand Opera. Education in the performing arts is a strong component of the Pavilion's offerings.

Cash flow from the contemporary programming largely offsets the deficits incurred in presenting the fine arts, whose audiences have grown each year. George and Cynthia Mitchell committed to a $5 million endowment for the Houston Symphony and the Pavilion became the Symphony's summer home in March 1992. No other community in the Southwest, and only a few nationally, provides its residents and neighbors with such an accessible and attractive star quality menu of rock and the classics.

BRINGING HIGHER EDUCATION, HIGH TECH TO THE WOODLANDS

Creating a good mix of jobs, from high tech to unskilled, was one of the earliest goals. George Mitchell had long been aware of the major roles in their communities played by the Research Triangle of North Carolina, Silicon Valley in California and Route 128 around Boston. These concentrations of high tech created wealth and recognition for their entire areas; they made contributions not only locally but to the entire nation; they attracted non-polluting industry and well-educated men and women who could afford, and demanded, the good life for their families. A common denominator among the most successful efforts was the convenient presence of excellent universities which offered not only opportunities for undergraduate, graduate and continuing education, but top rate experts who could be resources for advanced research and consulting.

In 1971, it seemed likely that the University of Houston would establish a campus in The Woodlands that would serve as a focal point for advanced education and development of technological industry. Coulson Tough, the Mitchell employee later in charge of building development at The Woodlands, had been a vice president of the university for several years. His recollection is that in 1968, the Coordinating Board of the Texas College and University System had encouraged the University of Houston to open two new campuses. Tough says he approached George Mitchell, on behalf of the university, with the idea of a U of H campus in the as-yet unopened Woodlands. Mitchell responded enthusiastically: In 1971, he offered a 400-acre site for a U of H north campus, at no charge, and on March 7, 1972, the university's regents accepted.

The University of Houston's purchase of a large downtown building, which became a downtown campus, however, was the main factor in bringing creation of a Woodlands campus to a halt. The Clear Lake campus south of Houston was the first to be established, and the university administration planned that The Woodlands campus would be the second. But the Coordinating Board considered the Downtown campus as the second. The Woodlands site was stymied.

David Gottlieb, a University of Chicago Ph.D. in sociology, came to Houston in 1973 as a U of H professor. Philip Hoffman, president of the university at the time, assigned Gottlieb to work with the Mitchell company to change the Coordinating Board's mind. Among other impediments he found, Gottlieb cites objections from nearby public universities concerned about their own enrollments, as well as opposition from a diversity of elected officials. Many in the state legislature were concerned that costs for education were getting out of hand. In the early '80s, the university and MEDC asked the Coordinating Board to take another look. But it was not long before it became clear that there would be no U of H standalone campus in The Woodlands.

George Mitchell and his company, however, did finally achieve most of what he sought originally, but in the form of a community college which incorporated a number of unusual ideas. With a grant of 10 acres and $20 million from The Woodlands Corporation and $9 million from the North Harris Montgomery Community College District, Montgomery College opened in the summer of 1995. Located just north of The Woodlands, the college offers diverse programs, ranging from health sciences to criminal justice. Then in mid-1996, ground was broken for the college's University Center, Texas's first public/private partnership offering students full bachelors' and masters' degrees from any of six universities—including the University of Houston and Texas A&M.

Even without a University of Houston campus, the essence of the original concept had been achieved: a wide range of educational opportunity in The Woodlands, as well as development of an intellectual community to provide a welcoming atmosphere for area residents and businesses.

A START FOR THE RESEARCH FOREST

On a parallel track, George Mitchell was intrigued by a *Fortune* article in the early 1980s which talked about the start of a high tech center at Stanford University. He asked Tough and Gottlieb to investigate. After further research and a report prepared on assignment by the Arthur D. Little organization, both the Houston Advanced (originally "Area") Research Center (HARC) and The Woodlands Research Forest were established in 1982. Also, George Mitchell and Michael Richmond worked to form The Woodlands Venture Capital Company, whose purpose was to provide seed money to start-up high tech companies, particularly in medicine. Richmond's recollection is that "back then . . . all the technologies that were being created and developed in one of our greatest assets, the Texas Medical Center, the largest medical complex in the world by a factor of maybe two, were being licensed out to other parts of the country. There was an opportunity to try to keep some of that at home, and we worked with Baylor (College of Medicine) and M.D. Anderson (Hospital) and UT Health Science Center and, in effect, through the venture capital company, helped them raise funds to start their first technology transfer companies." The Mitchell company made available a total of $15 million in venture capital, which, according to Richmond's estimate, was leveraged up to some $200 million in investment. By the late 1990s, the Research Forest comprised nearly 30 companies which provided employment for 1,850, including 300 Ph.D's.

The HARC concept is closely intertwined with the idea of the Research Forest. Having observed the role played by universities in developing regional high tech powerhouses, George Mitchell was confident that the same ideas applied at least equally in Texas. That, in turn, led to creation of HARC. In addition to providing the original seed money, Mitchell persuaded the four most eminent universities in the Houston area—Rice, University of Houston, Texas A&M and The University of Texas—to collaborate in a research consortium based in The Woodlands. The idea was that the entire region would benefit through the results of projects that drew from the best minds at the universities.

For the most part, HARC's efforts have been in traditional "hard" science and engineering. But one area of concentration came from George Mitchell's concerns about worldwide environmental, population and resource depletion problems. His interest in these growth issues, which stemmed from an Aspen Institute seminar with the noted Buckminster Fuller, had led to the establishment in 1975 of a conference series at The Woodlands centering first on growth and later on sustainable development. George and Cynthia Mitchell inaugurated the Mitchell Prize for original papers in connection with the 1975 conference; by the late '90s, the Prize, now amounting to $100,000, had been awarded at each major gathering in the conference series.

HARC got off to a shaky start. Its original acting director, Harvey McMains, concentrated on maintaining harmony among the four founding universities and establishing directions. A strong Board of Directors, comprising university representatives and some of the Houston area's most accomplished business people, was recruited. But early in 1984, the now-defunct Houston *Post* published a story questioning McMains's claims about his academic credentials. McMains left HARC, and, at Mitchell's request, Gottlieb took over as interim director. The whole episode was a severe blow to the nascent organization, especially to the way it was perceived in the academic community. Nonetheless, Gottlieb was successful in keeping HARC on an even keel until W. Arthur Porter—a Ph.D. engineer, head of the Texas Engineering Experiment Station at Texas A&M and an early member of the HARC Board—became its president in January 1985.

Early in its history, HARC's largest project was the design of magnets for the proposed Superconducting Super Collider accelerator. Later the institution played a successful coordinating role in the State of Texas bid to be home to the multi-billion-dollar government-financed SSC project. (Congress voted to locate the SSC near Waxahachie, Texas, but ultimately cut off funding, and the entire effort died.)

Through the years, HARC became a pillar of the Research Forest, a center of scientific and engineering research. By 1997, it had developed a campus on 11 of the 100 acres presented to it by Mitchell Energy & Development. Its three buildings comprise 160,000 square feet. Both Porter and Mitchell agree that a record of handling $100 million in research over a 10-year period, from a standing start, is one they're pleased with. But still, they stress, more support from Houston area business and industry is essential to finance operations and retire debt.

* * * * * * *

From the time he became president of real estate operations in November 1977 until his death in June 1986, Edward P. Lee, Jr., played an active and forceful role. The Research Forest and HARC got under way. The Woodlands Hospital was opened. MEDC corporate headquarters were moved to a new 150,000-square-foot building in The Woodlands. New office buildings were completed, the Trade Center was expanded and the John Cooper School, a private college preparatory institution, was opened. In Galveston, the San Luis Hotel and San Luis Condominium, striking embellishments to the city's lure as a resort playground, were completed. (Performance failed to live up to expectations, and the hotel and condominium property was sold in 1996.) The 203-acre Lake Woodlands was completed in 1985, providing both an important recreational amenity and an area for choice residential development. During Lee's tenure, a $1.2 billion road program in The Woodlands area got under way. Its most important element was the $800 million Hardy Toll Road, opened in 1987, which provides express links between The Woodlands and downtown Houston and Intercontinental Airport. But, in a strange twist, despite his creative efforts in bringing The Woodlands to life, Lee is likely to be remembered equally well for his introduction of the annual Christmastime Lighting of the Doves, an electric light spectacular in the Town Center for the entire community.

DEFINING EVENTS, DEFINING IDEAS

Roger Galatas arrived on the scene in 1979. Educated as a geologist, Galatas had been with Exxon's Friendswood Development Company and was recruited to be in charge of residential development of The Woodlands. Following Lee's death, he became president in 1986.

Galatas says his earliest tasks were to help calm down a roiled community. The Woodlands had opened in a soft economy a few years earlier; the developer had encountered severe economic difficulties; there had been a dramatic change in top management and then later, another change at the top; and members of the community then and from that time forward were interested, assertive, outspoken, almost every resident feeling a proprietary interest in The Woodlands. Street lights were not being installed timely. Environmentally concerned residents were disturbed that trees were being displaced to make room for a service station. Builders felt they were being ignored.

Galatas acted to ease the complaints, and he initiated a policy of forthrightly and promptly communicating with interested parties, a policy that has continued in effect. And, with improvement in the Houston area economy, things got better.

To Galatas, a number of defining events and defining ideas have helped set The Woodlands apart from other communities and have established its character. They range from something as seemingly prosaic as establishing the community's first supermarket to something as dramatic as opening The Woodlands regional mall. Taken together, these defining events have transported The Woodlands beyond being just a large real estate development to its position as a uniquely successful and responsible new community.

DEFINING EVENTS IN THE WOODLANDS

1964
- 50,000 acres of forested land north and west of Houston, approximately 2,800 of which will form the nucleus of The Woodlands, are purchased from the Grogan-Cochran Lumber Company. Over the next 10 years, an additional 300-plus transactions are completed to form the original land block of 17,455 acres.

1971
- The first public school, Lamar Elementary, opens.

1972
- HUD agrees to $50 million loan guarantee.

1974
- The Woodlands formally opens on October 19. Grogan's Mill is the first village, followed over time by Panther Creek, Cochran's Crossing, Indian Springs, Alden Bridge.
- The company's real estate division moves to a new office building in The Woodlands.
- Ten-year contract is signed with the Houston Golf Association to host the PGA Houston Open.

1975
- The Interfaith/religious community is started.

1976
- Jack Eckerd Drug Company buys land and develops the first distribution facility in the Trade Center.
- McCullough High School opens.

1978
- Jamail's, the first supermarket, opens in Grogan's Mill Village Center.

1980	• Mitchell Energy & Development moves from downtown Houston to new 150,000-square-foot headquarters.
	• Construction begins on The Woodlands National Bank Building.
1982	• Research Forest and Houston Advanced Research Center are established.
1984	• Agreement is signed to create a Tournament Players Course as site of the Houston Open.
1985	• The Woodlands Hospital opens. A $20 million expansion program is announced in the late '90s.
	• Lake Woodlands is completed and filled with water.
1986	• Harmony Bridge is opened, providing an alternate route into Harris County.
	• South County YMCA opens.
1987	• Hardy Toll Road opens. Completed road offers express access to Intercontinental Airport and downtown Houston.
	• Ventures Technology building opens in the Research Forest.
1988	• John Cooper School opens.
1990	• Cynthia Woods Mitchell Pavilion opens.
	• First 18 holes of the Palmer Golf Course are completed.
1991	• Hughes Christensen moves to its new 245,000-square-foot world headquarters in The Woodlands.

1992

- Magnolia and Conroe Independent School Districts' boundaries are re-aligned. At a cost of $12 million in concessions to Magnolia district, The Woodlands comes entirely under Conroe, allowing for efficiencies in everything from education to operations.
- The $50 million in HUD-guaranteed bonds is retired.

1994

- Allstate Insurance moves to The Woodlands.
- Highway 242, providing east-west access, is opened.
- Cochran's Crossing Shopping Center opens.
- The Woodlands Mall opens in Town Center. The Woodlands Corporation President, Roger Galatas, says Town Center is the downtown for a million people within 20 miles.

1995

- Montgomery College opens.
- Cynthia Woods Mitchell Pavilion capacity is expanded to 13,000 from original 10,000.

1996

- Ground is broken for University Center Building.
- The Woodlands High School opens.
- 30,000-square-foot Shell Learning Center opens.
- Multiscreen Tinseltown Cinema is opened in Town Center.
- 1,053 new homes are sold. For the seventh consecutive year, The Woodlands is Number 1 in new home sales in the region.

1997
- Population is expected to exceed 50,000.
- J.C. Penney announces it will be the fifth anchor store in the mall.
- Construction starts on Alden Bridge shopping center.
- 95-room addition to The Woodlands Executive Conference Center and Resort is opened.
- Plans are announced for a six-story, 150,000-square-foot office building in Town Center.
- The Woodlands is sold for $543 million to a partnership of Crescent Real Estate Equities Company and Morgan Stanley Real Estate Fund II.

1998
- Hewitt Associates purchases 77 acres for million-square-foot campus.
- 232,000-square-foot Mobil building is occupied, 600 jobs are added.
- Town Center One office building is opened, construction starts on Town Center Two.
- Buckalew Elementary School opens in Alden Bridge.
- Alden Landing Apartments are completed.

1999
- The Woodlands' 25th anniversary is celebrated.
- Construction begins on Sterling Ridge Village Center.
- Anadarko Petroleum starts building new 800,000-square-foot office tower, occupies previous Mobil building.

- Construction begins on The Woodlands Waterway.
- Development of the Club at Carlton Woods begins.
- Construction is under way for Jack Nicklaus golf course, Club Windsor fitness facility in Windsor Hills, and Waterway Plaza One, a nine-story 221,000-square-foot office building.

2000

- New home sales reach all-time record of 1,679.
- The Woodlands' Laura Wilkinson wins a gold medal in Olympics 10-meter platform diving.
- The Woodlands High School Highlanders team wins the class 5-A baseball state championship.
- Clearing and design begin for new Gary Player golf course.
- Construction begins on Waterway Plaza Two, a 143,000-square-foot, six-story ofice building.
- Sale of 130 lots in Phase One of Carlton Woods, the first gated community, is complete.
- Branch Crossing Junior High School debuts.
- St. Luke's Episcopal Health System announces it will construct a 263,000-square-foot, 82-bed medical center in The Woodlands.
- Construction begins on expanding capacity of Cynthia Woods Mitchell Pavilion to 17,000.

Galatas feels that The Woodlands has done a good job of conforming to goals set forth in the application for the HUD loan guarantee. For instance, he says, "(O)ne of the goals was to create jobs in the local community, one job per household, and we're pretty much on target. We've got about 16,000 jobs (in late 1995) and we've got about 16,000-17,000 homes. . . .

"The general goal of addressing religious communities is being matched pretty well. We have . . . 20 to 25 churches in The Woodlands today and probably about 30 different congregations."

Affordability of housing? "A great challenge," Galatas says. "The original goal was to have a wide range of housing for all income levels . . . to fully utilize all the subsidized housing programs in HUD. We still do that, but there are just not very many subsidized housing programs left . . . The programs (at the state and national levels) have disappeared by and large, so it makes it very difficult to provide a wide range of housing opportunities that have no subsidy or support."

As for attracting socio-economic diversity, Galatas says, "If you look at suburban planned communities around Houston, The Woodlands probably has a higher percent of minority resident population. We're the only planned community around Houston that has subsidized housing . . . We've made the best effort we can and continue to try to make that effort. We aggressively seek out minority realtors . . . We made the commitment that we would aggressively do things that would attract minorities to move here . . . and I think . . . that opportunity increases as we create jobs at all levels (and through) the community college (and) the outreach programs we have in place."

LET'S GO TO THE MALL

Commercial and industrial development at one time was the responsibility of Michael Richmond. The Woodlands Trade Center across Interstate 45; College Park, at the north end of the development, home to the community college; the Research Forest; and the Town Center, in the east central area, comprise a total of 5,000 acres, an area fertile for future business development. To Richmond, the 800-acre Town Center "is clearly the most exciting commercial development that will occur in The Woodlands," and probably most exciting within the Town Center is the million-square-foot regional shopping mall. Richmond expects that the mall will incubate another million square feet of adjacent retail activity, plus offices, hotels and recreational facilities.

Recognizing early on the fundamental importance of a regional mall, Woodlands people met with the nation's leading mall developers: DeBartolo, Rouse, Hahn, Homart, Federated Development. Homart, at the time a Sears subsidiary, seemed the best prospect, and in 1982, Homart and The Woodlands Corporation signed an agreement under which each would own half of the mall.

Then the Houston economy began to crumble. By mid-decade, some 200,000 or more jobs in the area disappeared. The mall was put on the back burner until the late '80s, when the economy began to revive.

What was needed to make the mall a reality were major anchor tenants—the large department stores that draw shoppers from miles around. Sears was a probable but still not in the bag. Particularly for Foley's, one of Houston's dominant retailers, a major problem was the possibility of "transference," the siphoning of sales by a store from one of its sister stores in a nearby shopping center. Foley's has stores in shopping centers throughout the city.

But transference was less of a concern of Dillard's, which did have stores in the area, but not to the extent of Foley's. Dillard's committed. Sears committed. So did Mervyn's. And, finally, so did Foley's. The Woodlands Mall was a "go."

The mall opened to a celebratory crowd of 80,000 on October 5, 1994, in time for that year's Christmas season. Within two-and-a-half years, J.C. Penney announced plans to join the mall's retail community with a new 147,000-square-foot store. A state-of-the-art multiplex movie theater and string of restaurants and other retail outlets surround the mall. Immediately north, in the Pinecroft Center, is a retail concentration, including a Target store, other national discounters and a Barnes & Noble book store that also serves as a community social center.

"Corporate relocations," Richmond comments, "are the toughest." The main problem has been existing, lower-priced space, most of it in the downtown and Galleria areas of Houston. But, he points out, there are good success stories in The Woodlands, including Shell Oil, Exxon Chemical, Allstate, Tenneco, GeneMedicine and Hughes Christensen. More recent additions include Anadarko Petroleum, Maersk and Hewitt Associates.

ON THE HORIZON: THE WOODLANDS WATERWAY

The next big step will be development of the mile-and-a-half waterway, a Woodlands version of the San Antonio Riverwalk, canals in Amsterdam and similar construction in Perth, Australia, and Singapore. Besides adding to retail and recreational opportunities, the waterway will serve as The Woodlands' paseo, a place to stroll and see and be seen.

According to Richard Browne, the planner, "We had to build a major drainage way through the Town Center because urban development in a natural forest like this really intensifies the runoff. When you build parking lots, rooftops, concrete and hard surfaces, storm water needs a new outfall."

What made the drainage way a virtue instead of just a necessity, according to Browne, was recognition that land in the Town Center and frontage along busy roads, most of them in the eastern part of The Woodlands, all command premium dollars. The completed waterway, running between the mall and Lake Woodlands, with a retention pond at Interstate 45, would provide comparable benefits toward the middle of the community. Browne describes the result as "an urban spine that just sings of real estate value enhancement."

Browne's planning ideas were strongly influenced by his interest in social behavioral science. He says, "I believe that you can't design and build communities just knowing engineering and planning. You must also understand the end users, 'people,' and the things they need to optimize living a good life."

Through a friend, he was introduced to the writings of the eminent American anthropologist Robert Ardrey. Says Browne, citing Ardrey as an authority, "As human beings, as a result of evolutionary forces on our lifestyle over several million years, we still carry a lot of genetic traits that shape our behavior and responses even today." That thinking influenced everything from the size of neighborhoods to the placement of shopping centers.

Browne left The Woodlands in 1996 "before the icing was on the cake"—that is, before he could see it mature—but he maintained interest in its progress, first, as he was working on a new town near Jakarta, Indonesia, and later, from a new business enterprise in Perth, Australia.

Although Browne often had George Mitchell's support, working relationships between him and top management of The Woodlands were rocky from time to time. Sometimes the tensions were constructive, sometimes not. Although Mitchell hasn't articulated it, he's not uncomfortable with dissension, within limits, and he clearly admires Browne's creativity and perseverance. Others, too, acknowledge Browne's contributions, although some say the drummer he hears sometimes seems very distant.

A ROCKY MARRIAGE COMES TO AN END

In April 1983, the often-strained relationship between HUD and the company was ended: Congress had cut off funds for the New Communities program, which had the effect of ending HUD's role in overseeing the Project Agreement. The Mitchell company was still responsible for the loan (which was fully paid off in 1992), and agreed to maintain its affirmative action program and continue to provide housing for low and moderate income families. In all, HUD made loan guarantees to 15 new communities; only The Woodlands was successful, as evidenced by these statistics: In 1996, more than 1,000 residential lots were sold in The Woodlands; for the seventh consecutive year, there were more home sales and starts than in any other planned community in the region; with impetus from the increasingly successful mall, the Town Center was fulfilling its promise as a commercial and entertainment magnet for all of north Houston. It had taken time and many acts of faith, but George Mitchell's determination to build a new and better community was beginning to pay off.

GAS IN LIMESTONE COUNTY, GALVESTON

During the late '60s and early '70s, The Woodlands was the company's high profile operation, but at the corporate level and in energy operations, things also were moving forward at a fast pace.

In 1968, Mitchell Energy discovered gas in the Limestone County area of East Central Texas. The same year, Lafitte's Gold Field onshore and offshore Galveston, Texas, was discovered. In 1969, a discovery was made in Polk County, Texas.

The Limestone County area has grown to be Mitchell's second most prolific area of natural gas production, trailing only North Texas. It is an area of extremely tight rock—that is, the gas-bearing formations are barely permeable and porous enough to allow the gas to flow. The key to success has been continual advance in the art and science of massive hydraulic fracturing.

In 1978, the Muse-Duke Number 1 well in the Limestone County area received national attention when the company, with financial assistance from the Department of Energy, conducted the largest massive hydraulic fracture on record up to that time.

The project involved pumping nearly a million gallons of fluid and 2.8 million pounds of sand down the well bore and out into the tight formation known as the Cotton Valley Lime. Under high pressure, the fluid created the fracture which, when packed with sand to keep it from closing, became a conduit for gas to reach the well bore. As a result of the "frac," production from the Muse-Duke more than doubled, and the well's recoverable reserves were increased by billions of cubic feet.

In the Galveston area, Lafitte's Gold Field has been a good productive area, but its significance is magnified because Galveston is George Mitchell's home town. In Galveston, he's been the prime mover and largest individual investor in reviving the historic Strand district, he owns three hotels, his company has developed the Pirates' Beach and Pirates' Cove resort areas (where he spends almost every weekend) and built the swank San Luis Hotel and Condominium, and he has been active in reviving the annual Mardi Gras celebration. And, in an environmentally sensitive area which serves as a playground for millions, his company has successfully developed oil and gas reserves.

Howard Kiatta, now a successful independent oilman in his own right, played a big role in developing gas reserves in the Galveston area. He joined the Mitchell company as a geologist in 1967 after seven years with Texaco.

His earliest days with Mitchell left him uncomfortable. He recalls, "When I came over here I began to realize that the company had a reputation for slow pay . . . I remember my first experience in trying to order a geological base map to work an area in south Louisiana. I called a company in New Orleans and they said, 'Who do you work for?' When I told them, they said, 'Send us a check and (if the check clears), then we'll send you the map.'"

Another of his early impressions: "I found out that George was very vocal in his discussions with Budd Clark out in the hall . . . I'd hear these guys screaming at each other . . . I thought they were going to kill each other, and it turned out that's just the way they discussed things. They'd stomp off and you would think they were mad and would never speak to each other again, but they were obviously very close."

'A HUGE GEOLOGICAL STRUCTURE' UNDER THE ISLAND

When Kiatta arrived at Mitchell, the company had already acquired some geological data for the area immediately north of Galveston Island. The seismic lines indicated a huge geological structure that appeared to be underneath the island. Soon the company drilled a discovery well onshore in the island's industrial district. Its success was kept very confidential.

More information was necessary. Gathering seismic data usually requires that explosive charges placed at intervals be set off. The echoes are captured and plotted on special equipment, then interpreted by seismologists and geologists. In unpopulated areas, such work usually causes little if any concern. But in Galveston, seismic activity was not going to be well received.

George Mitchell's answer: Instead of dynamite, use Vibroseis, a process using trucks with vibrators that could create a sufficient shock or energy transmission to acquire seismic images from the subsurface. Need to do it in some of the city's business and residential areas? OK, but do it during the early morning hours. Get a police escort to move things along. It worked.

Following the acquisition of as much lease acreage as the company could afford, a number of wells were drilled from a warehouse. Kiatta says, "Those were mostly directional holes. We reached as far south as we could, down toward the beach area with those slant holes, a mile, mile-and-a-half, maybe 7,000 feet. . . . We . . . found a really nice field and developed it very quietly."

Among other things, putting the whole deal together involved obtaining 9,000 leases–some of them as small as one-tenth of an acre–in order to get the needed acreage for developing the field. That job became the responsibility of Jack J. Yovanovich, who left Conoco in 1965 to join Mitchell. He eventually became senior vice president–land for energy operations.

Yovanovich says that in the early '70s, the Galveston play had the full force of the company behind it: Mitchell himself, geologists, the legal department, engineers. All were involved in trying to get permits to drill, both onshore and offshore.

At the same time as the Galveston play, the company had bought some offshore oil and gas leases on the nearby Bolivar peninsula. The City of Crystal Beach then extended its limits into the offshore area and incorporated Mitchell's leases. That meant involvement with city government.

Offshore Crystal Beach became Mitchell's 176 Field. Yovanovich recalls:

> The City Council met on Friday nights at 7 p.m., so any time you had to get a drilling permit, you had to meet with them at that time. I had been meeting with various groups, such as the church groups, men's groups, women's groups, and making talks on the energy crisis and the need to support offshore drilling . . . You might be meeting at church or at a saloon or beer joint–no matter. . . . They ultimately had a referendum on whether they wanted offshore drilling. And George told me that I could never win an election at the height of the environmental movement, and it just made me that much more determined. We won the election.

BUYING THE FORT CROCKETT SITE

In order to get permits for drilling offshore Galveston, Yovanovich, Kiatta and others from Mitchell talked to every business along the city's seawall, with positive results. But then the City Council said: Find some property onshore and drill directionally. The company bought the 23-acre Fort Crockett site, where its San Luis Hotel later was located, and proposed a drillsite there.

Yovanovich says, "We gave a rendition of the property, showing how it would look when it was fully developed, with a screened drillsite and a hotel. However, then a group of people who lived around our new site came before the City Council and protested the proposed location. So the City Council said, 'Go back offshore.'"

George Mitchell says, "(We) asked the Corps of Engineers for a permit for the first well. (W)e had a thousand people protesting. . . . I'll never forget . . . this woman had the same story every time. She'd get on the stand and say, 'It's going to catch fire and burn my children' and tears would run down her eyes. A thousand people. What could I do? I loaded up three busloads of our people to cheer for me and we finally, after two years, got the permit to drill . . ."

That permit allowed the drilling of three wells offshore, provided no oil was produced and the gas was taken ashore in an unpopulated area of the island. Over the years, Lafitte's Gold Field has been a big source of production for the company, has had no negative effect on the Galveston area's environment and has provided important tax revenues for the city.

Kiatta has especially fond memories of the Seven Oaks Field in Polk County, near Livingston, Texas. He took over development of the field shortly after he joined the company. The discovery well blew out. The second well, Number 2 Southland Paper Mills, drilled a few hundred feet to the south, had discouragingly tight sands. The deciding factor in completing the well was that an interstate pipeline ran nearby, so if the well did turn out to be commercially productive, there would be no major expense in connecting it to a pipeline. "Last time I checked," Kiatta says, "the well had made something between 10 and 12 billion cubic feet of gas." Even more important, Number 2 Southland Paper Mills led Mitchell Energy to develop the field, which includes the Ike T. Smith Number 1 well, one of the company's biggest producers ever.

Kiatta is especially pleased with how well the Mitchell company sewed up leases in the area. He says, ". . . a few years ago as an independent, I looked at the area and tried to figure out a way to get back in there and maybe come up with some prospects, but we had done such a good job of tying up the acreage for Mitchell back then that I shut myself out. . . . Mitchell controlled the whole thing."

GOING PUBLIC: A PARTIAL ANSWER TO THE NEED FOR CAPITAL

Early in the decade of the '70s, with the opening of The Woodlands on the horizon and with the fast pace of energy exploration and production, the Mitchell company found itself in a familiar position: Cash was short, debt was high and borrowing capacity was strained.

George Mitchell's recollection is that at that time, "we had fairly large debt, all endorsed by me, and Johnny, too. . . . You couldn't keep going to the bank . . . You can't grow too easily as a private company. Some people have done it but it's very difficult." The options that Mitchell saw were: merge, sell out or go public and still have controlling interest.

Going public seemed to offer at least a partial escape valve. Eastman Dillon, an underwriter, showed an interest in working with MEDC, and the Mitchell organization set about compiling the necessary information. The Eastman Dillon contact submitted the proposal to the appropriate committee in his firm, which voted 24-1 to turn it down. Budd Clark recalls that they said "we should sell assets, reduce debt and then we might qualify." But, he adds drily, "We knew that answer anyway."

Earlier, Shaker Khayatt had called on Mitchell at the suggestion of the company's banking contact at Chase Manhattan Bank. Khayatt at the time was a senior officer with Coggeshall & Hicks, Inc., a New York-based investment banker; now he's a member of the Mitchell Board and head of his own investment firm. After the turndown from Eastman Dillon, Clark called Khayatt. Khayatt, in turn, enlisted Walter Lubanko, then with F. Eberstadt & Co., also a New York investment banker, but larger than Coggeshall. Lubanko was a member of the Mitchell Board and was head of his own firm in New York until his death in 1999.

The first proposal by Khayatt and Lubanko was to make a public offering of units consisting of debt, equity and warrants. That idea was aborted before it came to market: It was, finally, believed to be too complicated, and the market was soft at the time. So what Khayatt calls "a plain vanilla" public offering of 700,000 shares of Mitchell Energy & Development Corp. (Mitchell & Mitchell Gas & Oil Corporation until a half-year earlier) was made on February 24, 1972. The shares were priced at $13.25 ($12.19 net to the company). By day's end, an additional 70,000 shares were sold.

George Mitchell retained ownership of more than 70 percent of the shares (which, over the years, reflecting exercise of stock options by employees and sales of stock by the company and Mitchell himself, had become 60-plus percent in the late 1990s). Johnny Mitchell owned about 11 percent, but later disposed of all of his shares.

Khayatt remembers that on the morning of the offering, "I got a call from my floor partner saying, 'You'd better be careful, the bears are going to maul you.' . . . Now as lead underwriters, we (Coggeshall & Hicks and Eberstadt) had to hold the price. That means when people dumped or shorted stock, we had to buy it."

The offering had been effective in mid-morning and Khayatt and Lubanko went to lunch with Clark and a few other corporate officers. Khayatt and Lubanko got calls from their floor partners, saying the stock was under heavy pressure. According to Khayatt:

> There were no buyers in sight. And that was the time to call in some chits from friends. I called a good friend at Loeb Rhoades. As I was picking up the phone to make other calls, one of our traders said, "Hey, look, look." Across the (Dow Jones) broad tape it came that Loeb Rhoades had entered the sheets for Mitchell Energy & Development one way only—buy. In other words, they were saying, "OK, guys, we will buy anything you can sell." And of course it was Loeb Rhoades's capital that really panicked the shorts.

Developing stock market understanding of the company has not been easy. Khayatt says, "I remember a friend at Travelers saying, 'This is a unique situation, but how do you marry oil and gas into real estate?' It's a very big hurdle . . . we could never get analysts that follow both industries at the same time. You talk to a real estate analyst who looks at Mitchell as a real estate play and then questions why he needs oil and gas. In talking . . . to oil and gas analysts, they ask why do you need real estate, it's only a drag." The financial community's lack of enthusiasm for a company combining energy operations and real estate development was a big factor in the decision 25 years later to leave the real estate business.

But even with the difficulty in winning investor understanding, after stock dividends and splits through the years, one share of the original Mitchell offering in 1972 was equivalent in the year 2000 to more than 10 shares, each with a market price about four times that of the original shares.

Shortly after the initial offering, Mitchell shares were listed on the American Stock Exchange. In June 1992, the company moved to the New York Stock Exchange. A month later, shareholders approved a plan under which each share of common stock was converted into one-half share of voting stock (MND A) and one-half share of nonvoting (MND B). The dual-class stock structure was adopted to facilitate the use of common stock in future financing transactions or in making acquisitions without diluting George Mitchell's voting position. In May 1993, the company sold to the public 5.9 million class B shares and, in addition, Mitchell sold 1 million of his class B shareholdings. Proceeds to the company were used to buy out a drilling partnership owned by the company and a number of lenders and to pay for 1994 drilling costs. In 2000, after no acquisitions had been made with the "B" shares and a consensus that the two classes tended to complicate investment decisions, stockholders voted to return to one class of shares.

WHERE THE MONEY COMES FROM

By its very nature, Mitchell Energy & Development since the beginning has been a capital-intensive company. Drilling for oil and gas and developing real estate—especially a project as ambitious as The Woodlands—require large amounts of money up front, well before there's revenue from energy production or real estate sales. George Mitchell throughout his career has accepted the need for financial leverage pretty much as a condition of the kind of businesses he's in. He doesn't necessarily like it, with the constant need for high payments to service debt (although he and investors in his enterprises have benefited handsomely when the leverage tide runs their way, such as in the late 1970s and early 1980s, when energy prices spiked).

In the annual report covering its first full year as a public company, the year ended January 31, 1973, MEDC's debt was almost four times as great as its stockholders' equity, a high leverage ratio that indicated risk to potential lenders. With the go-for-broke attitude to get The Woodlands open, when financial discipline broke down, most lenders became skittish. Chase Manhattan Bank of New York and First National Bank of Chicago had long provided conventional financing for energy operations, but The Woodlands experience soured the banks and the company on each other, and those relationships were soon ended (although a considerably different Chase than the one the company did business with in the '70s and before is once again Mitchell's lead banker). As for the equity markets, George Mitchell had given up a measure of control when the company sold shares to the public in 1972; however, his aim was to maintain a high degree of ownership, and another offering was not in the cards if it meant his control would be further diluted. And, in fact, there was no further equity offering for more than 20 years. (The sale of the class B shares in 1993 illustrated how the then-new dual-class stock idea was supposed to work: It was successful in raising capital without affecting Mitchell's voting control.)

Money always seemed to become available when it was needed, although not necessarily on the most favorable terms. Early on, the Big Nine and "big name" investors risked money—and benefited because George Mitchell, after all, was an oil and gas finder. A little later, Bob Smith and Warren Petroleum helped persuade the First National Bank of Chicago to provide financing for the large processing plant in Bridgeport, completed in December 1957. Natural Gas Pipeline previously had made $7 million in advance payments on production to get Mitchell started on drilling in North Texas. The purchase of the Grogan-Cochran land package, which included the nucleus of The Woodlands acreage, was creatively financed so that the company had little in out-of-pocket costs. In 1972, the HUD loan guarantee helped make capital available to start development of The Woodlands.

And the 1975 spike in gas and oil prices increased the cash flows of Mitchell's energy operations, alleviating the Woodlands-related bind. Later, money advanced for creation of municipal utility districts in The Woodlands was repaid by the sale of bonds. Soon the investment banking firm of Goldman Sachs, with the strong participation of Mitchell Director Khayatt, arranged a $100 million private placement. Insurance companies became lenders. William Tonery, then Mitchell's chief financial officer, established lines of credit with Manufacturers Hanover Bank of New York and banks in Los Angeles, Philadelphia and other cities.

It was clear that the company was coming into the big leagues financially with its successful early 1987 public offering of $250 million in notes underwritten by First Boston Corporation and Goldman Sachs. Two years later, another $200 million of notes, underwritten by First Boston, were sold. In succeeding years, new public debt was issued at reduced interest rates, to a substantial extent refunding the earlier, higher-cost borrowings. Among other benefits, these offerings consolidated most of the company's debt at the parent company level. And in May 1993, Mitchell Energy & Development received $123.4 million for its public offering of 5.9 million nonvoting class B shares.

Philip S. Smith, long-time president of the MEDC administration and finance division and chief financial officer, had been in charge of Arthur Andersen's oil and gas practice before joining Mitchell in 1980. He describes the company's use of debt:

> George has always had a philosophy of primarily growing the company through internal cash flow and to the extent that cash flow was not sufficient, debt was used.
>
> For years that was primarily accomplished through bank debt, although the company had done some financings with . . . private placements through insurance companies in the late 1970s. That vehicle continued to be used periodically . . . up through the mid-1980s . . . at which time the company—in fact, the whole oil and gas industry . . . —began to undergo financial stress as a result of lower oil and gas prices . . . That caused the private placement market to be more difficult to access and made it desirable to lower the dollar magnitude of the bank debt because it was out of kilter in terms of being too large a percentage of the total debt of the company. As a result, the company first accessed the public debt market in the late 1980s . . .
>
> The company was rated a BB credit for its first offering, which was a couple of notches below investment grade. . . . After that, the public market became the company's primary source of longer-term debt, and private placements were phased out. A number of public debt issues were subsequently done . . . and bank debt became a smaller part of the financing picture, which is customary in large, publicly held companies. Normally a company borrows under its bank revolving lines of credit and then once those borrowings accumulate, they are paid down with proceeds from either public debt financings and/or equity offerings.

Mitchell Energy & Development may not have reached the status of an Exxon or Shell, but the financial community was looking at the company as a more solid citizen than ever before.

NEW DIRECTIONS

The death in November 1993 of F. Don Covey at age 59 hastened changes that had been under consideration. Covey joined Mitchell in 1976 and was president of the exploration and production division. Some felt that he would become president and chief operating officer of the entire corporation. With Covey's death, however, that job as well as the presidency of exploration and production, went to W. D. "Bill" Stevens, a member of the Board and 35-year veteran of Exxon Corporation.

Over time, the company has not been successful when it ventured into businesses where it lacks experience and expertise. It has operated a small refinery and has dabbled in a program to produce nitrogen, for tertiary recovery projects, at its Bridgeport gas liquids plant. It once owned a drilling ship that operated on Lake Erie, and it had retail gasoline stations. It acquired a contract drilling company, but it moved away from that business. It's been interested in sulfur extraction in Canada and Mexico and lignite mining in Texas. For a number of years, it conducted an oilfield supply business. (There's never been any serious interest in drilling overseas; George Mitchell's stock comment has been that he'd be happy to provide one-way transportation to any of his employees who wanted to start foreign operations.) In real estate, too, there were moves that, when all costs and lost opportunities are considered, didn't pan out: land investments in Aspen, Colorado; resort developments in the Houston area and a number of such efforts in Galveston; commercial development on property owned by the company near a regional mall 15 miles south of The Woodlands. For a while, MEDC considered developing a new ski resort in Colorado.

At the time of Covey's death, the company had been in the doldrums for several years. Earnings fluctuated but never came close to the $114 million of 1982. The energy industry was depressed. The end of the favorable contract with Natural Gas Pipeline Company of America loomed. The Woodlands was doing well, but the company's other real estate investments weren't, and there was no near-term prospect of an increase in their value.

Following an asset management study, MEDC in fiscal 1995 embarked on a program in which it identified core holdings both in energy and real estate and concentrated technical and financial resources on their development. Also, the company chose to sell some assets—those having greater value to others and those not likely soon to add to profits. As a result, idle drilling rigs, a natural gas storage facility, a number of gas processing plants and some lesser energy assets were sold (while other interests were expanded, including those in oil and gas properties and natural gas liquids fractionation and gasoline additive plants). The company adopted a strategy of concentrating its exploration efforts in North Texas, East Texas, onshore along the Gulf Coast and in southeastern New Mexico. Other acreage was sold or offered for sale.

In real estate, the decision a year later was to concentrate on developing The Woodlands and withdraw from almost all other real estate activities (an approach that in 1997 was turned around 180 degrees). Large tracts of timberlands in Montgomery County, to the northwest of Houston, were sold, as were the company's interests in the San Luis Resort in Galveston.

At the end of fiscal 1991, MEDC real estate holdings amounted to 60,000 acres; five years later, the comparable number was 22,000 acres; a year-and-a-half after that it was headed for zero.

And in a move to streamline operations, the number of full-time employees was reduced in 1994 and 1995 to 1,950 from 3,100.

CONTRACT WITH 'NATURAL' COMES TO AN END

A milestone in Mitchell history was the agreement with Natural Gas Pipeline Company of America to end the long-term contract under which Natural had bought Mitchell gas at above-market rates since 1957. In exchange for Mitchell's consenting to the early termination, effective July 1, 1995, Natural agreed to make sizable cash payments to MEDC over a period of years. For the first time in four decades, the Mitchell company was selling almost all of its production, instead of about half, at prices dictated by the market. The change was dramatic. Over a two-year period, the average price realized declined nearly 25 percent–but the cash payments from Natural eased the pain.

Further, cash proceeds from asset sales and from the Natural buyout were used to reduce debt by $400 million. That strengthened the balance sheet and brought about a reduction of more than $30 million, or 40-plus percent, in annual interest expense from fiscal 1995 to the amount budgeted for fiscal 1998.

* * * * * * *

The company's movement towards a tighter, more efficient focus in its operations was dealt a severe shock by a Wise County, Texas, court decision on March 1, 1996. In the so-called Bartlett case, a jury found in favor of eight plaintiff groups who claimed that MEDC natural gas operations had affected their water wells. The jury awarded the plaintiffs $4 million in actual damages and a stunning $200 million in exemplary damages. The huge award was depressing to MEDC, in everything from investor perceptions of the company's future through employee morale to future planning.

MEDC marshalled additional technical and legal resources to counter the water well accusations, and the fog began to lift about a year after "Bartlett" with a favorable court decision in another case. Similar lawsuits, each claiming damages of at least $1 million, had been brought by 46 other plaintiff groups. But on May 28, 1997, a Wise county jury after an 11-week trial found that the company was not the cause of water well problems in the "Bailey case," a decision that gave MEDC increased hope that the earlier

Bartlett decision was vulnerable.

And so it was. The Court of Appeals for the Second District of Texas, Fort Worth, handed down a 42-page decision on November 13, 1997. In essence the Appeals Court ruled that most of the claims in the Bartlett case were invalid because they were filed after the statute of limitations had expired, and that, further, the plaintiffs had not proved that Mitchell operations were the cause of their problems. "Consequently," the Court said, "we reverse the trial court's judgment and render judgment that (plaintiffs) take nothing from Mitchell."

MEDC had been victorious, but at a high price. Mitchell people at all levels had been distracted by the suits and their possible impact. Morale, even though it rebounded strongly, had been hurt. And, in tangible financial terms, cost of the litigation amounted to more than $20 million.

THE WOODLANDS IS SOLD

Probably the most surprising event in MEDC history came in the summer of 1997, when the company announced the sale of The Woodlands and some other property to a partnership of Crescent Real Estate Equities Company and Morgan Stanley Real Estate Fund II, L.P. The price was $543 million; after deducting current income taxes and transaction costs, MEDC realized net cash proceeds of $460 million, part of which was used to pay down debt.

Conventional wisdom had it that one day MEDC real estate and energy operations would be split into two companies, with MEDC shareholders retaining proportional interests in each. A main reason for such a split would be what Director Shaker Khayatt had expressed years earlier: The investment community preferred a pure play. Energy securities analysts weren't comfortable with real estate; real estate analysts weren't comfortable with energy.

But splitting into two was easier said than done, principally because of the amount of debt with which an independent real estate company would start its new life, compounded by a continuing voracious demand for capital to complete developing The Woodlands.

The Board's and George Mitchell's difficult decision was that an outright sale would be in the best interests of the shareholders, employees and residents of The Woodlands. The Crescent-Morgan Stanley partnership, high bidder in a competitive auction, had ample capital available to meet The Woodlands's needs, which would continue for some time into the future. Crescent previously had purchased a 75 percent interest in some Mitchell-owned office buildings in The Woodlands, and George Mitchell and management of The Woodlands Corporation had come to know the Crescent people and were comfortable that the same high development standards of the first 23 years would be maintained. Mitchell also expected that employees of The Woodlands Corporation would benefit by becoming associated with a company which has broad and deep involvement in real estate and that, as a result, opened new areas of opportunity for employees.

Beyond that, the real estate market in mid-1997 was stronger than it had been in years, and Crescent's bid was regarded as favorable. The deal provided other benefits, too. In August 1997, MEDC announced how it would use money from the sale: $100 million to repurchase company stock (and thereby increase earnings-per-share); approximately $200 million to retire public debt, which would help MEDC maintain an investment-grade credit rating; and the bulk of the remainder for acquisitions and increasing capital spending for existing programs.

BITTERSWEET FOR GEORGE MITCHELL

For George Mitchell, the sale on July 31, 1997, was the bittersweet culmination of a 31-year labor of love, in some ways comparable to placing a loved child in someone else's care. He has said often that if money had been the only motivator in building The Woodlands, he could simply have subdivided the acreage early on and realized bigger and quicker returns. But to him, The Woodlands was to be a place that showed that the private sector and government could cooperate to build a good community and address some of the nation's urban ills, that nature could be nurtured in making such a community, that a company could profit if it was willing to make the required investment of human and financial resources.

Says Mitchell:

> My personal involvement in seeing The Woodlands done right is no secret. But finally, a lot of factors came together that made it necessary to take a fresh look at things. My wife and I talked it over with our children: Did any of them want to step into the company and undertake a major involvement with The Woodlands? None did; they've followed their own lives and Cynthia and I respect that—in fact, we've always encouraged them to think that way. Everything pointed in the direction we finally took. Am I saddened? Yes, of course. But The Woodlands, I believe, is unique in community-building, and, frankly, I'm proud of what we've accomplished.

Through most of its history, The Woodlands—with an engaging story to tell and effective advertising and public relations—had been the high profile component of MEDC. Hundreds of thousands had visited the conference center and resort; millions had seen the Houston Open golf tournament on network TV; others heard or read about it in other media. Yet, in fact, The Woodlands and other parts of MEDC real estate operations were a comparatively small part of the total company. In the five years ended with fiscal 1997 (January 31, 1997), for instance, real estate operations never accounted for as much as 20 percent of total corporate revenues or operating earnings. MEDC would be different, but not shockingly different; an enterprise that had been 80-plus percent energy would now be 100 percent energy.

After more than a half-century, George Mitchell and employees of Mitchell Energy & Development (it's hard to tell where the man leaves off and the company begins) can look back at the best of times and some not so good. As a geological consultant and oil and gas explorationist, Mitchell has been involved in the drilling of more than 8,000 wells, of which 1,000 were wildcats. He has played a role in discovering nearly 250 new fields, fault blocks and deeper production reservoirs in Texas, Louisiana, Oklahoma and Canada. His company has produced 3 trillion cubic feet of natural gas and has become a major factor in the natural gas liquids industry. Mitchell put together the finest elements of private enterprise and government involvement to create The Woodlands. He founded the Houston Advanced Research Center and the Cynthia Woods Mitchell Pavilion. He's been the prime mover behind the revitalization of Galveston's historic Strand district and chaired a task force of leading Texans to help his beloved Texas A&M plan its future. He's led the Texas Independent Producers & Royalty Owners, is an All American Wildcatter and has an honorary doctorate from the University of Houston. He and Cynthia give the Mitchell Prize to encourage research into environmental and growth issues. They've provided a major endowment to the Houston Symphony. George has received recognition from a broad array of organizations. The list goes on and on.

Most of the unproductive deadwood assets have been pruned, and the company is operating with rewarding efficiency. The Woodlands has been sold, providing cash for the company to intensify and expand its energy operations. The energy markets may forever be cyclical, but the Mitchell company since its beginnings has demonstrated clearly that it can find hydrocarbons and sell them at a profit. Moreover, by excelling in special niches—such as gas liquids production and fractionation, gasoline additive production and natural gas transportation—additional sources of earnings are available.

But there are imponderables. George Mitchell has built a capable management team and recruited a strong Board of Directors. He remains vigorous and fully occupied in the company's business and plans. But what kind of company it will be when Mitchell steps back, and how it will be led, aren't clear.

George Mitchell's attitude, however, always has been that no one has a guarantee on the future. "Get the best information you can," he says. "Be careful, but be optimistic. Set a proper balance between wildcats and development, no matter what business you're in. Then, when you're convinced you're right, go for it."

With that attitude, Mitchell plunged into North Texas, acquiring leases on 300,000 acres of land whose geological potential was unproven. His company was a pioneer in the use of massive hydraulic fracturing, and as a result, opened the prolific Limestone County area of Texas to economic exploration and production. He invested heavily in extracting liquids from natural gas, and MEDC became one of the nation's largest gas liquids producers. He set about to build an environmentally concerned new city on the southern edge of the great piney woods. Planning big, managing risk, operating in an atmosphere of uncertainty about the future? Just part of the game. That's the way George Mitchell sees it.

Epilogue 1

Curtain falls followed by three-year intermission. Curtain up.

George Mitchell makes no bones about where his heart lies: If the stars had been in proper alignment, he'd have sold the energy component of Mitchell Energy & Development Corp. and kept The Woodlands. But the decision to sell real estate and retain energy was forced by the fact that the company faced a $204 million judgment as the result of losing a lawsuit in North Texas, with the possibility of many more suits to come. No prospective buyer of the energy assets would have been willing to assume the burden of a seemingly unlimited series of crushing court judgments. But selling the real estate segment looked like no cinch, either. Highly leveraged with debt as always, it would not be able to stand alone because of the continuing requirement for large amounts of capital, although it could be a fine buy for a cash-rich buyer.

And then as in ancient Greek plays, a god descended and everything turned out fine:

- Mitchell Energy appealed the court judgment and won. It also won all subsequent lawsuits making the same claims, with findings that Mitchell Energy was not the cause of the plaintiffs' problems. The company tightened and focused its operations, the economy strengthened, demand for natural gas and gas liquids exploded, and Mitchell Energy & Development Corp. prospered as it never had before.
- A deep-pockets partnership bought The Woodlands in 1997. Then, with accelerated development, a strong Houston economy and growing acceptance of the notion that The Woodlands was without peer as a place to live and do business in the Houston area, annual revenues of the new community doubled over the three-year period after it was sold.

If anybody's complaining now, late in the year 2000, it's not very loud. George Mitchell acknowledges he has some would-have-been, could-have-been hangovers from the real estate sale, but with Mitchell Energy & Development stock about 20 percent higher than at any time in history, his comments are more musings than complaints. And with the torrid rate of residential sales and burgeoning growth of the Town Center, it's clear that purchasers of The Woodlands made a good buy.

In fact, the end may be in sight. Michael Richmond, president of The Woodlands Operating Company, talks in terms of 10 to 20 years as the time when The Woodlands will be completely built out. And after having put MEDC up for sale solely as an energy company, receiving no realistic offers and retracting the offer to sell, George Mitchell wouldn't be averse to a good, new offer.

A GRIM LANDSCAPE

Immediately after the sale of The Woodlands, however, the landscape was grim for both entities.

For the energy company, earnings took a nosedive for two consecutive years. Prices realized for natural gas, crude oil and condensate and natural gas liquids all headed south. In response, the company cut back on capital spending, and in January 1999, the number of full-time employees was reduced by 21 percent.

At The Woodlands Operating Company, Richmond says that when the sale was closed, "we had no idea what our balance sheet would look like or what kind of liquidity we had, we had no budgets, we were so focused on closing that we didn't realize that we were broke. We had negative working capital, our bank lines were tapped out and none of that was understood until opening day. Yet, there was a lot of optimism that a challenge was ahead of us that would bring new activities and new opportunities."

The confusion brought about by the change in ownership was soon brought under control. First, the sale of some notes receivable made needed cash available quickly. And, most important, the real estate market was growing stronger every day and The Woodlands Operating Company had a sizable number of lots ready for sale. That meant an acceleration of incoming revenues, a cure for almost all ills.

The pop in the energy business took longer to mature. With proceeds from the sale of The Woodlands, Mitchell Energy & Development paid off a considerable amount of debt, bought back some of its own common stock, whose market price failed even to approximate intrinsic value, and beefed up capital spending. After two years of net loss, the company was making a good comeback in fiscal 2000, which essentially coincided with calendar 1999. Then the Board of Directors dropped a bombshell: On October 6, 1999, the Board announced that Goldman, Sachs & Co. and Chase Securities, Inc., had been retained to advise the company "regarding strategic alternatives, including possible transactions that might result in a sale or merger."

THE BOARD SAYS: FORGET IT

The announcement was followed by months of uncertainty on the part of employees and investors. Would jobs be lost? Whose? Would a fair offer come that would reward investors? But on April 6, 2000, the Board said, Forget it. Offers to buy had been received from bottom fishers, but none that fairly valued the company's assets and its prospects. We'll continue as a stand-alone company, the Board of Directors said, but we'll go back to having a single class of stock instead of two classes, which was a confusing factor in the marketplace, and we'll continue to reduce debt and buy back shares on the open market.

In fact, the long-latent worldwide demand for energy had been all-but-visible for years. The end of an economic crisis which had sapped Asia's vitality was in sight, and as the 20th century ended, demand for oil and products in the Far East became robust, as did demand for natural gas in the United States. OPEC nations, disorganized and in disarray for decades, were ready to act in concert again to control the world supply and price of oil. At the same time, America's production of oil continued its steady decline and the nation grew more dependent on imports. Worldwide, and especially in the United States, the harsh impact of most fossil fuels on the environment was increasingly recognized, and demand for natural gas, the least polluting of the fossil fuels, was getting stronger. Further, the upward pressure on prices was compounded because during the soft years, drilling had dropped precipitously, the service industry had lost both skilled and unskilled help, its equipment became degraded. To a large extent, natural gas producers were faced with meeting the surge in demand from a standing start.

IN 2000, NEAR THE TOP OF THE CYCLE

For years, the industry had struggled at the bottom of the cycle. In 2000, it was near the top, feasting after years of famine. For Mitchell, natural gas prices were nearly double what they were when the energy and real estate companies were separated; oil, whose average price the year before had dropped to barely more than $12, was selling at well more than twice that amount; the price realized for natural gas liquids had doubled. Gas production, natural gas liquids production, pipeline throughput—all up substantially.

George Mitchell points out that a number of factors—restrictions on drilling, harder-to-find reserves, uncertainty about the future—"have made it all but impossible for the United States to approach self sufficiency in oil production. In the Mideast, excess capacity is down to two or three million barrels a day, compared with 20 million or 30 million barrels per day 15 years ago, so they're no longer able to make up the shortfall easily." (But," he says, "I think that in two or three years, supply will get better because of the enormous effort to increase production overseas now—West Africa, South America, Caspian Sea, Russia. My hope is that it will catch up with any deficiency we may have.")

In the longer term, Mitchell sees a more balanced market, although one on a higher plane than has been the case in a number of years. "As the industry catches up, more natural gas will be available. Quite possibly, some alternative fuels will become more economical. No one can guess where or how big new discoveries will be, no one can predict the impact of worldwide climate change, the effects of continued technological improvement or ups and downs in the economy."

But, he says, with the Asian crisis coming under control and the economy of the world better, energy demand will remain strong.

"Add to that demand from the new electrical power generation plants being built all over the United States. Their efficiency is about 64 percent, compared with 37 percent for a conventional plant. But the new ones depend on gas.

"They can generate new power at 4 cents a kilowatt, versus 11 cents for conventional plants such as those operated by Houston Lighting and Power.

"The nation needs new supplies of electricity, and gas is by far the preferred fuel. So that's going to help keep demand strong.

"But there will be a steadying of price. Average gas prices will settle somewhere in the $3.50 to $4 area," he says, "and that will be enough to allow profits in the industry and serve as an incentive for more exploration and development.

"Although gas will still be a good buy for the consumer, it will cost more than it has in the past decade."

PRODDED, PROBED, STREAMLINED

George Mitchell is convinced that his company has been prodded, probed, streamlined, trimmed and rationalized so that it will do well in just about any economy. "Our reserves have increased consistently. Our production is up, and that's the heart of higher revenues and higher profits.

We've concentrated on drilling and producing in a few core areas, which has improved our efficiency by orders of magnitude. Our staff is lean and motivated.

"We've long been among the nation's largest producers of natural gas liquids, and we've gotten smarter there, too. Right now, we have six natural gas liquids plants producing as much as more than 30 plants did not too many years ago, and we're constantly increasing our capacity and our technical know-how."

When Mitchell bought in on the North Texas deal introduced to him by the Chicago bookie in the mid-1950s, he was also buying the keys to the kingdom of the 1990s and beyond: the Barnett shale which underlays the productive zones that accounted for half of the company's gas production for decades.

The company's largest acreage position is in North Texas, where it has leases on about 600,000 gross acres. Since early in the '90s, development has been focused on the Barnett shale, which is the source rock for minerals contained in numerous producing horizons in the Fort Worth Basin. The Barnett has been known to be rich in natural gas, but its low porosity and permeability hampered economic development. And then came light sand fracture technology.

Fracturing, to ease the flow of gas through dense rock formations, has long been required in most Mitchell North Texas development programs. But light sand fracturing, which was adopted by Mitchell in the late '90s, both reduces development costs and significantly increases the amount of recoverable reserves. In light sand fracturing, less sand is used and water replaces costly heavy gels to carry the sand to prop open subsurface fractures.

2,000 ADDITIONAL DRILLING LOCATIONS

The company has identified 2,000 additional locations to drill in the Barnett on currently held acreage. Even if there weren't another prospect to drill anywhere else, the locations would be enough to keep the company productively and profitably busy until well into the first decade of the new millennium. And there could be an additional 5,000 locations on top of that.

George Mitchell describes the new success in North Texas as "an engineering feat in itself with the fracture technology we put together." With exploitation of the Barnett, Mitchell predicts increases in gas and liquids production of 20 percent in 2000 and 10 to 15 percent annually for several years after that.

The handmaiden of natural gas production in North Texas is the production of natural gas liquids: propane, ethane, butanes and natural gasoline. Much of North Texas gas is extremely rich in liquids. With Mitchell's increasing production in the region, capacity at the company's largest processing plant, in Bridgeport, Texas, has been under strain. So, in a program completed in the last days of 2000, the company expanded the Bridgeport facility to process an additional 100 million cubic feet of gas daily, a jump of about 50 percent. And then once again, late in the year, it announced another 100-million-foot expansion to be completed by July 2001. The company also initiated a big stepup in its North Texas drilling program, from 6 rigs to 12.

ON THE HORIZON

What's on the horizon for the industry—and for Mitchell?

The United States now consumes about 21 trillion cubic feet of gas each year, but "in 10 years, that demand will be up to 30 trillion cubic feet, an increase of more than 40 percent," Mitchell says. "The resource base is there, and present prices are more than enough to provide good incentives. My guess is that the price of natural gas will level out a couple of dollars or so below its present $6, which will still be enough to encourage drilling. But if it drops to where it has been in the past—say $1.50—well, then, everybody just shuts down and we're back to severe shortage."

As is widely recognized, he says, coal is plentiful, "but no one has successfully and economically found a way to clean it up and, especially, minimize or remove the particulates created in coal's combustion. Maybe one day. If that ever happens, and I'm not confident it will, the nation's energy problems will largely disappear," he says.

"Our biggest weapon is conservation, something that Americans have never become serious about."

He's especially optimistic about breakthroughs in fuel cell technology. "The Houston Advanced Research Center here in The Woodlands has been doing cutting-edge work on that technology, and it has found sponsorship for some pilot projects in the Houston area. Still, at best, that's years away from commercialization."

THE SAME, ONLY MORESO

In the end, Mitchell Energy & Development Corp. will remain the same company it's been, "only moreso," Mitchell says.

"We've learned, we've trimmed, we've focused, we're doing smarter drilling, we're more efficient, we're managing our balance sheet much better, we've reduced debt dramatically and further reductions are in store. It's unlikely that we'll go overseas, and big, good acquisitions are almost impossible to come by.

"So our future growth will come by way of the drillbit. We've got the reserve base, we've got the technical capabilities and proven management, we're in the right business at the right time in the right place. Opportunity may be harder to come by in the future, but it'll still be out there.

"There's an exciting decade ahead."

And yet, always one to keep his options open, he says, "If someone came and offered me a big price for the company, I would look at it again."

SPEED THE PACE OF LOT SALES

At The Woodlands, after the trauma of separating from Mitchell Energy & Development, the initial strategy, in light of the heavy debt load, was to start generating cash, Richmond says. "The way to do that was to accelerate the pace of lot sales and at the same time, speed up the community's job growth, which feeds into potential home sales and retail activity, among other things."

That, in turn, led to a trimming of the time The Woodlands would have until buildout: a new goal of shortening residential from 15 to 12 years and commercial from 18 to 14 or 15, in all knocking three to five years off the time required to develop the community.

That means George Mitchell's 11th child—he and wife Cynthia have 10 of the flesh and blood variety—will be fully grown about 2012 or soon after. Projected population, according to Richmond, will be in the range of 100,000 to 125,000, compared with 65,000 late in the year 2000. The final population is less than once planned because there will be fewer multi-family apartments and rentals than originally forecast. The land mass totals about 27,000 acres, up a little from the time of the sale in 1997, up a lot from the 13,000 when The Woodlands opened in the fall of 1974.

In the land business, he says, "we knew that with so much infrastructure now in place, it would take only 20 cents of capital investment to generate $1 of revenue. The money had already been spent for roads and the acquisition of land and the interest and so on. So we knew that there were big dollars in the sale of land. We just had to get those dollars faster."

It took about two-and-a-half years to ramp up the pace of lot sales from 1,000 annually to 2,000. A main factor was diversification, Richmond says.

"We added higher-end housing, we added lower-end housing. We broadened the product mix to include town homes, condominiums, age-qualified neighborhoods, high-end gated communities, things that we weren't selling before."

But even with the stepup, according to Richmond, "we've always tried to maintain the vision that George Mitchell set for us, to stay with the original design vision and plan."

Were prices cut in order to achieve the faster pace of land sales? Declares Richmond: "No. In fact, we raised them in the normal course of business."

The result is that over a three-year period, revenues went from $125 million a year to $250 million, cash flow from something in the neighborhood of $50 million or $60 million to $120 million. "That extra money enabled us to pay down debt and give the owners a good return on their investment, and they are very happy."

AT THE TOP OF THE LIST

Development of the Town Center is at the top of Richmond's list of big bangs, past, present and future. In the year 2000, the waterway is coming into reality. "It will bring something unique, a higher quality level, that will allow the workers in the Town Center to walk to stores, walk to activities," he says. "The first phase will be done in the autumn of 2001, but the real grand opening will be more like 2002, when a hotel and convention center and Anadarko should be open, along with more of the urban residential development and restaurants and entertainment. So 2002 will be a momentous year.

"We've even now decided to add residential into the Town Center mix, and that will bring a form of urban housing that you see being built in downtown and midtown Houston.

"The retail and entertainment component will continue to grow, with the Cynthia Woods Mitchell Pavilion undergoing its second expansion since opening in 1990. And the mall will undergo further expansion within five years."

In an earlier interview, Richmond said the office market was an especially difficult nut to crack, but now, he says, "For the past two years, we've led the city in new construction, and it's clear that we've arrived at a critical mass of office activity, that we are in a growing market and with that come names like Hewitt, Maersk, Anadarko, Grant Prideco, Allstate, Lyondell, Pantellas, Lexicon, Chevron-Phillips.

"We've built three spec buildings and leased them all ahead of schedule, and a fourth will open by the end of 2001."

Problems? The most immediate one is the broader Houston matter of environment and ozone and the impact on marketing of the region. "If those issues are not resolved," Richmond says, "there's the threat mobility funds will not be made available, although every project we need is committed except one–getting Lake Woodlands Drive across the lake [and that was approved later]."

He acknowledges that to some, The Woodlands appears to be a rich man's suburb, not consistent with some of the commitments originally undertaken to help secure the Federal government loan guarantee of 1972.

A PICTURE THROUGH FAULTY FILTERS

"But that's a picture taken through faulty filters," he insists. "Our average housing price over the past three years has hardly changed. People do see the high end homes, but the amenity that people like most is not the golf courses, it's the parks and pathways, and that's not a rich man's, rich woman's amenity. It's a community amenity.

"We do have subsidized housing out here, but you must recognize that government has completely left the business of subsidizing new housing, and no one can do it without federal support.

"We've tried to deliver affordable housing through condominiums in the $80,000 range, we've converted industrial land in the Trade Center to residential, and we've built duplexes and townhouses to provide more affordable housing.

"I don't think most people appreciate that we have a great deal of affordable housing. They don't see it because it's in the trees, but it exists. But what will probably get most of the attention is the more expensive housing that's built as the community develops. That will get more visibility than the fact that we're still maintaining the average price."

Richmond insists that 26 years after groundbreaking, development remains consistent with the original plan. "There have been modifications, of course, because nobody knew what the changes in technology, retailing, education and so on would be.

"There have been changes in the marketplace, the concept of the village, the concept of diversity in the villages, the concept of what we used to call the Metro Center, now the Town Center, and so on. But I don't consider those as significant.

"The basic fabric is just about the same as it was originally.

"We've provided the institutional services, the churches, the synagogues, the schools, security, shopping, services. The wooded environment is our strength. Trees are the Number 1 reason people move out here. Quality of education, public safety and the amenities of the parks and pathways follow right behind. The Woodlands may not be precisely what we thought at the beginning, but it's close.

"With the culture here now, the symphony, the arts, the regional mall, the neighborhood shopping centers, the community college, it's possible to live a full life in The Woodlands. The infant born in 1974 is rapidly coming of age. It's a wonderful community."

Epilogue 2

An early part of this book tells of the 1901 immigration to the United States of Savvas Paraskevopoulos, George Mitchell's father. In the ensuing years, George was born and grew up in Galveston, attended Texas A&M, served in the U. S. Army during World War II, founded Mitchell Energy & Development Corp., discovered and drilled thousands of gas and oil wells, and created and then sold The Woodlands. Epilogue 1 reviewed developments during the three years following that sale. To really conclude this narrative, after having dealt with the past, it seemed reasonable to look ahead because looking and taking actions based on what he sees are among George Mitchell's defining characteristics.

Mitchell agreed to give his thoughts about what things would be like in the year 2020. That year was chosen somewhat arbitrarily, but no matter: George Mitchell is not one to color inside the lines when his fancy takes him elsewhere. So here are some of his ideas of what The Woodlands and the nation's energy picture will be like in 2020 . . . or before or after.

* * * * * * *

- Houston's annexation of The Woodlands will be remembered as a non-event, if it's thought about at all. "In the year 2020, you'll be able to look back and say the transition was smooth and painless and the quality of life unaffected because men and women of good will worked together to make it that way," says Mitchell. "I have perfect confidence that community leaders in The Woodlands, the developer and Houston officials will take whatever steps are necessary in the next few years to effect change without shock." For instance, his view is that The Woodlands should retain responsibility for schools, traffic and security, and the City should provide for water needs, regional transportation, including transportation to the airport, and sewage disposal. The key, he says, is goodwill and cool heads. "I don't think it's in the cards that the City, once the present 10-year holding period ends, will fail to exercise its right to annex, and that's not bad, considering that without Houston, The Woodlands would have remained undeveloped logging country.

Houston is responsible for so many things that The Woodlands depends on–jobs, a wide range of professional services, the Port, the Medical Center, museums and the arts, big time sports, the airport–that it's only natural that the new community join the City. So what I see now, in 2001, is that cooperation and thoughtful planning on the part of all concerned parties during the next 10 years will solve the problems in advance and that annexation will have no negative impact on the community."

- Mitchell and Michael Richmond are in modest disagreement about how soon The Woodlands will be completed. Richmond thinks 2020, Mitchell would add 10 years beyond that. "There's going to be a lot of office and apartment space to be filled, a lot of housing yet to be built," says Mitchell.
- Growing pains experienced in the early years of the century–such as traffic congestion and large school enrollments–will be "pretty much cured" by 2020, according to Mitchell. "The community will be completely built out, and in the later years of development, new housing will be less dense and there will be more upper end housing built. All of these will ease the pressures of growth."
- Achieving the racial and social mix planned originally simply is not in the cards. "Our goal was to have in The Woodlands a mirror image of Houston's racial and social makeup, but after more than a quarter-century of honest effort, that's been beyond reach and probably will remain so," according to Mitchell. Housing costs, distance from the Houston job market and the absence of government help will make realization of the "mirror image" impossible. Moveins from the North, East and California are making the population "a bit more heterogeneous, but the mix in 2020 will not be much changed from the century's early years: white, well educated, affluent and politically conservative.
- The area's transportation problems will be largely solved. "If I'm wrong on this one, the six million people in the greater Houston area will be in deep trouble," Mitchell says. "Local and national leaders need to bite the bullet now in order to get started on the fast rail service that's needed if the City is to avoid strangulation. Rail service from Houston, and the Woodlands, to Bush Intercontinental is essential; toll roads might help, but it's just impossible to build enough roads to meet future transportation needs." Mitchell expects that inside The Woodlands, adequate roads will be in place by 2020, bus service will be available, and access to and in the Metro Center will be provided by specially designed

vehicles on the waterway/transitway.

- A potential problem in water supply will be averted. "As things now stand, The Woodlands is in good shape, unlike Harris County, and I'm assuming that solid planning in the near future will head off water problems." He believes that as mid-century approaches, The Woodlands will need access to river water or, possibly, to aquifers under the 150,000-acres of the Sam Houston National Forest.
- The University Center of the North Harris Montgomery Community College District, which offers degrees from some of the State's top universities, will blossom further as a model nationally for providing high quality and affordable higher education while reducing the enrollment load at flagship universities.
- By 2020, the Houston area's credentials as a world center of fundamental and applied research will be beyond challenge. "The community's business, civic and scientific leadership has just announced its commitment to establishing a huge biotechnology research center adjacent to the Medical Center, and I think that energy will also be translated into support for the Houston Advanced Research Center in The Woodlands," Mitchell says. "HARC has the potential to provide topnotch applied and basic research while leading the way to sustainable societies. In fact, HARC is at the forefront of developing applications for fuel cell technology, which may play a big role in protecting the environment. But financial support for the organization ultimately must come from Houston's business and philanthropic organizations, which have been slow so far to see HARC's potential."

Mitchell's passion for The Woodlands is high profile, but in fact, energy has always played an equal or, even, a dominant role in his interests. Right from the beginning, he was an "oil finder." He was among the first to warn that the United States was running out of oil. He's been outspoken about dangers inherent in the nation's dependence on energy imports and the need for an effective energy policy. And, not surprisingly, he has ideas about the future. He predicts that by the year 2020:

- The electricity supply shortages of 2001 will be solved, with action by federal and state governments to permit siting of small generating plants fueled by natural gas. The NIMBY–Not in My Backyard–phenomenon as it applies to locating generating plants is the main cause of shortfalls in California in 2001, Mitchell says. "The fever will soon cool, and reasonable compromises will be struck."
- Between 65 and 70 percent of the nation's oil supply will be imported

in 2020, compared with about 55 percent in 2001. "Simply, oil in the United States is very hard and expensive to find, it's necessary in our economy, and a reversal of declining production doesn't seem possible," Mitchell says. Worldwide, he says, reserves are adequate for another 75 years.

- Natural gas, which long sold at a discount to oil on an energy content basis, will attain parity with oil. In 2001, gas often sold at a premium to oil, but that price will settle, although not to its previous low levels. In fact, says Mitchell, natural gas could be priced at $15 per thousand cubic feet (compared with $8 to $10 early in 2001) and still be competitive with other fuels in power generation.
- With energy prices rising everywhere and some fuels in short supply, conservation will carry the day. Motor vehicles and home appliances will be made more efficient. Home insulation practices will improve. Even sports utility vehicles, almost all of them huge gas guzzlers in 2001, will be redesigned for improved mileage.
- Through advances in technology, coal will be cleaned up somewhat for use in power generation, although its use in the United States will be at a much reduced level compared with 2001. Use of indigenous coal in third world and developing countries, however, will continue, adding to the world's environmental problems.
- Global warming appears to be real; strong conservation measures are the best defense. "Worldwide, people need to use less energy, particularly forms that pollute," according to Mitchell, "and that's a big order. Natural gas is best and wherever possible it should be used in place of other fuels. But to many of the poor people of the world, coal or wood are the only feasible energy sources. The best hope for the six billion-plus people on earth is conservation."
- Use of alternative sources, such as wind power, photovoltaic and geothermal, will remain fairly insignificant, mainly because of cost. "In 2001, there are no important breakthroughs in sight," according to Mitchell. Some new nuclear power plants are likely to be built, "but they're never foolproof safe."
- Fuel cells will be widely used. "Their environmental impact could be huge, their impact on energy conservation not much," says Mitchell. But the fuel cell, whose main application will be in transportation, will reduce dependence on oil imports.
- The U. S. will be importing substantial amounts of liquified natural gas. Mitchell says that although the nation currently has 50 to 75 years of

reserves left, by 2020 exploration will necessarily be focused on deeper, more-difficult-to-produce deposits, and imports will be required to supplement domestic supplies.

- With technological advances in seismology, production and drillship design, drilling in 10,000 feet of water or more will be routine, but, Mitchell says, the possibility of spills will be ever present.

Mitchell's success through the years has been based on his ability to look just beyond the horizon and act accordingly. His batting average is definitely big league.

The principals of Christie, Mitchell & Mitchell, a predecessor company to Mitchell Energy & Development Corp. (MEDC), were, from left, George P. Mitchell, H. Merlyn Christie and Johnny Mitchell, shown in new Houston Club Building offices in the mid-1950s.

MEDC's Board of Directors in 1972: from left: Shaker A. Khayatt (seated), J. Leonard Ivins, Walter A. Lubanko, Bernard F. "Budd" Clark, George P. Mitchell, Leland Carter, Morris D. Thompson, Jr., and B. J. Houck.

George Mitchell's mother, Katina Eleftheriou Mitchell, and his father, Mike Mitchell. Photograph was taken in Galveston in about 1915.

Graduation from Texas A&M University in 1940. George Mitchell, 21 at the time, was a major and battalion commander in the Aggie Corps of Cadets.

Mike Mitchell and his family, probably at the dedication of the Texas A&M Mitchell campus, Galveston, in about 1960. Mike Mitchell is seated. Standing from left: George Mitchell, Christie Mitchell, Maria Mitchell Ballantyne and Johnny Mitchell.

George Mitchell (110 pounds) with tarpon (120 pounds) at Bettison's Fishing Pier in Galveston. The year is 1934; George is 15 years old.

Aggie cadets from Galveston in 1940: At left is George Mitchell, 21. Seated center and looking to his right is Buddy Bornefeld, a long-time friend.

In this picture, taken in Galveston in about 1931, George Mitchell was 12 or 13.

The Mitchell brothers in the service: from left, Johnny, Christie and George. The year is about 1942 or 1943.

Mitchell Energy drilled directionally toward Galveston's beachfront from an on-shore warehouse in order to develop the Lafitte's Gold Field.

The noted R. E. "Bob" Smith, right, and George Mitchell after a day's fishing off Galveston. Smith, a major investor in Houston real estate, was an early backer of Mitchell.

Advances in the technology of massive hydraulic fracturing have made the North Personville Field in East Central Texas one of MEDC's most prolific. The array of equipment and manpower gathered above is typical of what's needed for a massive frac.

In the mid 1970s, MEDC found it economical to use portable plants, such as the one above, to extract liquids from natural gas.

A portable natural gas liquids plant is transported from New Mexico to a new gas stream in Texas. The year: 1984.

The MEDC natural gas liquids plant in Bridgeport, Texas, by far the company's largest, went on stream in December 1957. The company consistently ranks among the nation's largest producers of NGLs.

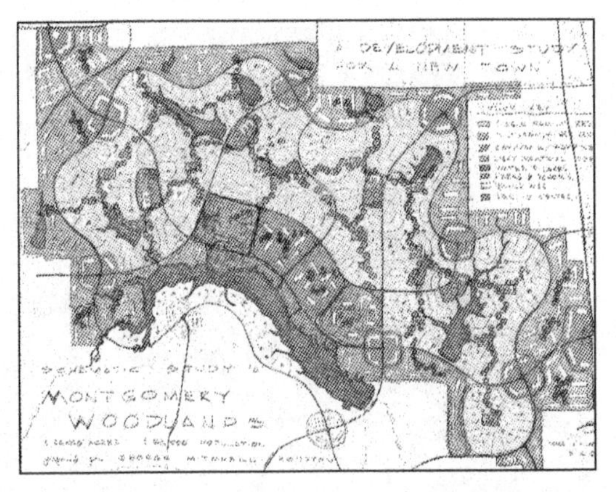

First plan for The Woodlands, prepared in 1966 by Karl Kamrath and Hugh Pickford.

A meeting of planners, in the early 1970s.
From left, Robert Hartsfield, Ian McHarg,
Morris D. Thompson, Jr., James Veltman, B. F. Clark.

Mainstays of the early planning team for The Woodlands:
from left, Ian McHarg, Richard Browne, William Pereira.

Aerial view of The Woodlands Executive Conference Center and Resort, host to international conferences since The Woodlands opened in 1974 and home of The Woodlands Country Club.

Montgomery College, part of the North Harris Montgomery Community College District, opened in 1995. It was followed by the University Center which offers bachelors' and masters' degrees from six universities.

The Houston Advanced Research Center campus, located on 11 of 100 acres presented to it by MEDC. HARC, a scientific and engineering research consortium, is in the Research Forest.

Cynthia and George Mitchell celebrate the opening of the Cynthia Woods Mitchell Pavilion, an outdoor entertainment center, on April 27, 1990.

The Pavilion has been expanded to a capacity of 17,000, from the original 10,000. Its presentations range from the classical arts to rock.

Maestro Christoph Eschenbach of the Houston Symphony conducted the opening night program. The celebratory weekend also included a performance by Frank Sinatra.

The Houston Open golf tournament, held at The Woodlands since 1975, attracts the world's top golfers and huge crowds each year.

Lake Woodlands, a 200-acre man-made lake, provides a recreation center for sunfish sailors, as well as for fishermen and women and people in search of a pleasant place for relaxation.

The Woodlands Athletic Center features an Olympic swimming pool and a 10-meter diving tank. Facilities also include a basketball court, tennis courts and weight equipment.

Thirty congregations representing diverse faiths tend to the spiritual and many human service needs of residents of The Woodlands. Left to right: The Woodlands United Methodist Church, Woodlands Parkway Baptist Church, The Woodlands Christian Church.

Besides serving as a recreation amenity, Lake Woodlands provides a desirable setting for choice residential and possible commercial development. The lake was filled in 1985.

A network of hike 'n bike paths interlaces The Woodlands, providing residents the offstreet means to jog, walk and bike-ride.

A broad range of housing, from apartments to million-dollar-plus custom homes, is available in The Woodlands.

A crowd of 80,000 enthusiastic shoppers took part in the grand opening of The Woodlands regional mall on October 5, 1994.

The lit doves in The Woodlands Town Center are an annual signature of the Christmas season. The spectacular is visible on the nearby interstate highway.

The Hardy Toll Road, opened in 1987, provides an express highway link between The Woodlands, Bush Intercontinental Airport and downtown Houston.

Through George Mitchell, MEDC took strong public stands on the nation's energy issues. Above, an early 1980s press conference with a sizable cross-section of Houston's local and national press.

President Jimmy Carter and George Mitchell at a conference, held in The Woodlands, on Mexico-U.S. business relations, sponsored by the Houston Advanced Research Center.

George Mitchell was a prime mover in reviving Galveston's Mardi Gras, an annual event that attracts hundreds of thousands.

Mitchell Energy & Development Corp. shares were listed on the New York Stock Exchange in 1992 after 20 years on the American Stock Exchange.

Richard P. Browne

Today is Sunday, September 29, 1996. This is Joe Kutchin. I'm with Richard P. Browne in my home in The Woodlands. Richard is in The Woodlands after having spent about a year-and-a-half in Indonesia. Richard, please give me some highlights of your career before you came to The Woodlands. I know you're from New Jersey.

Yes. I came out of New Jersey Institute of Technology in 1950 as a civil engineer. Early in my career I worked at producing tens of thousands of subdivision lots all over New Jersey. Besides planning and designing real estate development, we did highway designs, drainage projects and sewer and water plants for communities. This experience was a good foundation for the things necessary for building new towns later in my career.

You were based where at this time?

I was based in Wayne, New Jersey, and headed up a multi-disciplined firm of engineers, planners and architects that I founded in the late '50s.

Right out of college you had your own company?

No, I worked for about eight years for a small engineering firm in Clifton, New Jersey, then formed my own firm. Within five years we had grown and had offices in five states and we were 186th on *Engineering News Record's* top 500 consulting firms in the United States. We were huffing and puffing with a couple of hundred guys on the payroll in offices in New York, New Jersey, Maryland, Georgia and Louisiana. We were doing a lot of work.

What year are we in now?

This would be 1965. I formed my own company in 1958. I probably should back up. After college, I worked that first eight years for a small consulting engineering firm. Extra-curricular-wise, I got to be a national director in the U.S. Jaycees and because of my exposure to the process of contributing to the community life, I wound up with an invitation to run for council in a town called Wayne, New Jersey. A year later, I was elected the town's mayor. Then the form of government changed to a strong mayor-council form and I was elected mayor again. I spent about eight years in local community work in one of the fastest-growing towns in the United States–3,500 building permits a year. That volume is almost triple the building volumes that we've had in our best year in The Woodlands.

Was that a paying job, mayor?

Yes, it was. Back in 1960, it paid $16,000 a year, which was good pay in those days for an elected public office. And it was almost full time. I'd go to my own consulting business in the morning for a couple of hours, go to town hall and say I'd be back by noon and usually never made it back. Nevertheless, it was a good foundation for the things you need to know for new town community building. I came out of that public office situation and focused my time on building up my consulting firm's practice. I also became interested in the social behavioral sciences and cultural anthropology because I believe that you can't design and build communities just knowing engineering and planning. You must also understand the end users–people–and the things they need to optimize living a good life.

Was there something you did formally to develop your skills and talents in planning and social behavioral sciences?

Yes, there was. Back in the early '60s, we were the engineers for a large regional mall built by the Rouse Company of Baltimore, Maryland. We designed a jughandle overpass, not unlike the one for The Woodlands here (and that's one reason why we have one at The Woodlands). That's another story I want to touch on later. We also designed the utility services and the site plan. The Rouse Company was impressed with our performance and they invited me to come to Columbia, Maryland, where they had just acquired 11,000 acres to build a new town. I spent a long day there, wrote a five-page report on what I had seen and what I thought they should try to do. When I got back to New Jersey, the phone rang and it was Jim Rouse. He wanted me to come back and talk to him. To make a long story short, I wound up in charge of the final planning, engineering, governmental approvals, construction and management necessary to open up all the infrastructure for new town Columbia, Maryland. In today's dollars it was probably about a $70 million effort, and Columbia opened 18 months later–on time, under budget. The Rouse Company was pleased at the result. That experience convinced me that I could tie together the surveying, engineering, planning, architecture and a concern for the lifestyle of the people in a new community. I decided then that I wanted to spend the rest of my career doing new towns–and did. From Columbia, Maryland, I went to St. Charles, a new town also in Maryland, then Flower Mound near Dallas; later came Gananda up at Rochester, New York, The Highlands, The Woodlands,

Peachtree City, etc.

Did you physically move to each of these places?
No, I did not. We went in as consulting engineers-planners-architects under contracts with the sponsors of these projects. We were responsible for putting the master plan packages together so that the projects could be submitted to the United States Government Department of Housing and Urban Development for U.S. federal guarantees for funding debentures. As a result of the federal guarantees, favorable interest rates were obtainable. The whole process began after the success of Columbia, Maryland, and Reston, Virginia. Congress supported the program believing that this country should be building new towns instead of suburban sprawl on the fringes of the urban cities of the nation.

You're talking late '60s now?
Yes, late '60s. A group of us from Columbia were invited to HUD to give our ideas on how the U.S. Department of Housing and Urban Development could set up regulations to accomplish a new town program. Congress passed the New Communities Act and because of my intimate involvement, I had a unique opportunity to participate in a majority of the new town projects.

What you had was a series of consulting jobs, right?
Right. We were the prime consultants providing most of the technical planning, engineering and feasibility studies necessary for them to get the federal guarantee for the debenture sale. Of course, often times we would stay on and roll with ongoing planning and engineering assignments for the development. That was a very intense project period in my career, from the mid-'60s to the mid-'70s. Eventually, the Government, during the Nixon administration, undercut the funding for new towns at a time when the United States was suffering inflation and they decided that it was inflationary to be supporting new towns and that called a halt to the program.

I believe the HUD Act was passed in 1970.
Yes. I'd like to go back to your original question about social behavioral science, Joe. I sat around the table with the design team at Columbia, Maryland. Here we had 11,000 acres and we were going to build a new town for 110,000 people. I told Jim Rouse that if it was just going to be a collection of subdivisions, I wasn't interested. At that time, a good friend of mine,

Mike Spear, an engineer out of Rensselaer Polytech, unfortunately now deceased, gave me a book that had a big impact on me. Mike came on board at Rouse in the middle of my effort at opening Columbia and eventually became its president. He gave me a book, *African Genesis* by Robert Ardrey. I read that and became entranced by the information on genetic propensities that still are inherent in us today. As human beings, as a result of the evolutionary forces on our lifestyle over several million years, we still carry a lot of genetic traits that shape our behavior and responses even today. Robert Ardrey also wrote *The Social Contract*, to record his personal inquiry into the social behavior of our species. Then he published *The Territorial Imperative.* It had to do with our relationship to space and crowding and the arithmetic of the human group. I was really impressed with what this knowledge could add to the design of man-built environments. I came to call this the "fifth dimension of community planning." We had on the Columbia team land surveyors, civil engineers, specialists in utilities and highways, a subdivision planner, architects, marketing experts and economists. I brought the group together at Columbia and said, "Who around this table is an expert on the end user of what we are doing—an expert on people?" We were all technically oriented and the answer, of course, was, no one. I then wrote a blind letter to Robert Ardrey's publisher in London, England, and said, "Please forward this letter to Robert Ardrey, wherever he is." In that letter I spoke of having the Columbia, Maryland, project under way and of having other new town projects coming. I said that the disciplines working on the projects represented the "hardware" of community development but they lacked adequate input from people who really understood the ultimate users, people.

You had not yet met him?

No. I wrote him, however, that I would like an opportunity, if possible, to meet with him. Two weeks later, I got a letter back. My letter had gone to his publisher in London, and they forwarded it to him in old Rome, Italy. He wrote to me saying that he'd heard about our work at Columbia and, indeed, we did have a lot to talk about. He invited me to call him and to arrange to come visit him in Rome. Within a month, after a couple of phone calls, we settled on the dates and I went to Rome to spend a week with him. That was the beginning of a good friendship between us. He opened up a network of people like himself, Conrad Lorenz, Lionel Tiger, Ed Hall and other cultural anthropologists. They all added to that "fifth dimension" in the process of planning and design of new communities.

Give me a year for this.

This was in 1968. Our time together was very fruitful. I would get up in the morning in Rome and go down to his apartment about 10 o'clock and we'd sit and brainstorm for an hour or two and then he'd say, "OK, it's time for lunch," and we'd take this little elevator down to the street. He had a beautiful apartment, by the way, on a top floor, with two balconies. From one you could look down the Appian Way; from the other the view was to the Vatican. And of course, old Rome, if you've been there, is just delightful. If a tile falls off the roof, you can't replace it with anything but a 2,000-year-old tile. We'd walk on our lunch break down the little cobblestone lanes to a square where there'd be a number of outdoor cafes. I'd say, "Oh, let's have lunch," and he'd say, "No, not here." So we'd walk inside where there were also tables and he'd say, "No, not here." Then we'd walk through the kitchen, out through a back door to behind these restaurants. There one would find little walled courtyards with one or two tables. We would go to a different one every day and somebody would bring wine and cheese and we'd have lunch. And in these intimate courtyards the vines were growing up the wall. They were old and thick and I'd sit there and say, "My God, Caesar sat here." You just had that feeling. Anyway, some months later, we met in London, another time in New York. Then he came to Columbia and stayed at my house for a week. During those exchanges we shared a lot. I wound up with a library of a couple of hundred books on social behavioral issues, went to a lot of conferences on this subject and met a lot of other authors besides him and added to the "fifth dimension" of the process. Up to this day this information has not been documented properly, but given an opportunity, I do hope to write that book.

So, how did you and George or you and Mitchell Energy & Development get together?

Well, I was at Columbia and someone called me up one day and said there's a gentleman here from Texas and he'd like to have dinner with you.

Just showed up?

Yes, as far as I know. We were holding a series of new town seminars in order to share our experience with other people in the country who might want to do new communities.

"We" is who, now?

"We" was a group from The Rouse Company and some people from my company. I had by then opened an office of my consulting firm at Columbia and moved my own personal office to Columbia in the American City Building. We joined the American City Corporation, whose role was to take the lesson of Columbia and share it with other communities and other urban cities that had problems and wanted some help. There were also a number of other consulting firms there. We used the Columbia experience as a backdrop to prove that we had some credibility in what we were doing. George Mitchell, I believe, came to one of those seminars in which I was usually a speaker. We went to dinner and he said he had assembled some land north of Houston and would like to talk about it some more. Shortly after this we met again at another meeting held in Reston, Virginia.

Did this meeting in Reston also have to do with new town developments?

Yes, it did. Reston was another example of a new town started in the early 1960s. That was an interesting day for me. George invited me to play some tennis that day, so we went out onto the hard courts in Virginia at Reston and played. It was a long, long set which George finally won. George then retained me and our firm, Richard Browne & Associates. We came to Houston to do the engineering feasibility on drainage and transportation for The Woodlands and then to sit in with Ian McHarg and others to start to brainstorm on how to proceed to develop the new town, which was later named The Woodlands. George had some previous efforts done by a local Houston architect, Karl Kamrath. Kamrath's plan was a more traditional subdivision rather than a new town after the Columbia model. George used to say, "I have the best from the East (and that was my firm) and the best from the West (Bill Pereira's)." He put the two of us together to argue through what the town should be. I always have felt proud that we won the battle. Pereira wanted to do a linear commercial urban center like a main street that would go from east to west and the residential development villages would be behind and off that main street. He used FM 1960 as a model. I opted for continuation of the Columbia concept which was a number of totally self-contained villages off a hierarchy of town roads that would provide access to the villages and separate the villages one from another. It had its own real town center off a major highway, also.

We're in 1972 now? Had you moved to The Woodlands?

In 1971, I think. And no, I was still based in the East, Columbia, Maryland, and doing new towns from there plus still working with The Rouse Company.

Tell me the advantages of your concept compared with what Pereira was saying.

Well, as I said, Pereira was following the traditional Texas development patterns like FM 1960. This is an east-west arterial road. The commercial strip runs along it and behind the commercial strip are residential subdivisions and he was saying, "That's what Texas is used to, that's what Texas would like, and that's what they expect." I tried to argue, and, I guess successfully, for trying to create village centers surrounded by a variety of housing in a series of self-contained village-scale communities where the arithmetic of the village didn't get so large that people would feel lost in a sea of suburban sprawl. I used the arguments that I had formed from the study of the villages and new towns in England and traveling on the continent. Especially neat were the Austrian villages that were so wonderful to experience. There the walking scale of the community was delightful. There was a beginning and an end to each village, not that you were part of an endless morass of housing.

And how was that issue decided? Did George say, "OK, we're going to do it this way?"

I don't really remember but we collaborated on the final plan and that's the way it went. We all recognized that the plan would be a living document and be responsive to the changing marketplace over time and change, and it did. When I finished with that effort—I think it was about 1972—I was offered the job of general manager of The Woodlands. I turned it down, however, because I had my own company to run and at that time the Peachtree City new town came forward and I was invited to go over and do that one. So I left and went to Peachtree and worked on that for a couple of years. I also went to Florida to work on the Inter-American Exposition for the State of Florida. This was an exposition for all the countries of the Western Hemisphere. Anyway, the phone would ring every once in a while and Vern Robbins would be calling and saying, "You know, George said give Browne a call and see if he won't come back."

So Vern was already here, as were Len Ivins and his associates?

Yes, a group of Rouse people did come down from Columbia and helped start The Woodlands.

Did you work with those folks pretty harmoniously?

Yes. Vern Robbins worked for me on the opening of Columbia. He was my construction manager. Len Ivins did work for me but was an assistant to Bill

Finley, Columbia's general manager, to whom I reported. He was the top guy in The Rouse Company for the Columbia project and was the former head of the Washington National Planning Commission. He's still a good friend.

You said you got calls from Vern?

Yes, every once in a while, he'd call, saying, "George says to keep calling you and invite you to come back." Then on one of the phone calls from Vern, I said, "OK, I'll come by for a month and we'll see how you're doing." I came back to Houston in October of 1974. They had at that time been pushing a lot of dirt, building the opening mix of buildings and were getting ready to open.

I didn't know I predated you. I came in January of '74 and thought you were here before me.

I came back at the end of September 1974 as a consultant, for one month. I saw what had been done, and Len Ivins said, "We need you here," and made me an offer to head up community planning and development. At that point I decided to sell my consulting business that I'd run for 20 years. I also decided to try to help make The Woodlands become the best new town of them all.

You can't do it right without putting your heart in it.

Yes, but after three-and-a-half months, however, I thought I'd leave again. I had a good relationship with Budd Clark. He'd also come to Columbia when the thought was maybe Mitchell should buy my firm and bring all our expertise to Texas and make it part of the Mitchell company. Then they chose not to get into that business and decided to use outside consultants instead. Anyway, I talked to Budd and then he took me to Leland Carter. I told him my perspective of the project and that I was going to leave. This was in January of 1975. Both of them said, "Don't resign. Give us two weeks," and I said OK. And believe it or not, within two weeks Len and the whole gang had resigned and we had to regroup and start anew. And that's what happened.

Did Leland continue his responsibility for the company's real estate operations?

Leland stayed on almost a year and then when Len Rogers came in, Leland resigned. I guess it was in June when I said to Leland, "You know, I think we can make this work and I'm going to move to The Woodlands and help do that." He said, "You're crazy. I'm going to put a chain across the entrance road and shut this down." I said, "You'll what? You've got to be kidding." He replied, "Well, George doesn't agree, but I think that's what should be done." So within a matter of a week or two, he resigned because he really believed that. George said, "No," and George won, as we know he did. I stayed on and we did our best to make this a special place.

Once you joined the company, what were your responsibilities?
Early responsibilities were to continue the detailed planning for the expansion of the project.

We've always had about 25,000 acres, is that correct?
No, we had 17,000 then. One of the things I campaigned for was to push the land ownership all the way to FM 2978. Everybody said, "Oh, my God, more land when we already have enough." I said, "No, you're going to be sorry if you don't, because you're going to create values and then we'll be glad that we've got that land." So Charles Lively got to work and he bought all the land out to 2978. Besides the planning, we were in charge of getting all the engineering done, the detailed subdivision engineering: sewer, water, drainage, pavement profiles, lots, plats and so forth. We also were responsible for all residential lot sales to builders. There was no marketing department at that time and we were responsible for doing the economic model for strategic planning. The finance department was not large enough to take on that responsibility.

I thought Jim Rush was in charge of marketing.
This was before Jim Rush's return. Every year, my department also had to redo for HUD the full-term 20-year economic model for the project. So, in summary, we were doing financing, strategic planning, engineering, urban planning and real estate sales.

Name some of the people who, besides you, were on your team.
We had Robert Heineman with us, Bob Hartsfield was working for Ed Dreiss and then he later left and there were Jim Wendt, Matt Swanson and Gloria Howell. Anyway, The Woodlands was opened on October 19, 1974, and I was here for the grand opening.

We had Settlers' Corner, we had the Inn and Country Club and not a lot more than that, did we?

Yes, besides the Country Club and Inn, the golf course was in, as well as the Swim and Athletic Center, the offices on Timberloch Place, many homes and town homes and apartments, too. They were also starting to build housing around Red Ridge Circle. I wanted to get people like Ryland to come and start building houses, too. They did well at Columbia for us. Walter Jolly was working for us at that time as an employee of the company. There was a lot to do, with such big management changes. One big thing was that there was a lot of money spent for the opening. The opening program was a long way from I-45. And that's where the Country Club, Inn and Wharf were built. There was a big concern about how that was structured, too. Steve Harrison was here then and saying it was going to be a rustic marketplace and combined conference inn and country club.

Have you been able to drive by there now? Just about everything is torn down. They're finally getting an Albertson's supermarket there and maybe that will be the anchor to bring some help to that area. Over time, I associate you with many major projects . . . the Pavilion, the canal, providing access to the mall from northbound I-45, that sort of stuff, with you being too stubborn to say, "I give up" when some others preferred to walk away or not get started on some projects. What are some of the things that come to your mind that you feel you've accomplished or that you left behind before you could bring them to life?
Let's go first for the good things that got accomplished. One of the good things that I'm happy that I won the battle on was the purchase of 10 acres in Oak Ridge so that we could have a jughandle overpass mid-way between Research Forest Drive and Woodlands Parkway, to provide access for northbound traffic to the center of the Town Center. Well, I knew from past experience that malls go at the intersection of highways or they have great access for some other reason. Without having a jughandle, like we did for Willowbrook for The Rouse Company 15 years earlier, our Town Center wouldn't be a good mall site. That northbound traffic would wind up at a mall on 1488 and I-45 and not at The Woodlands. I brought up the proposed land purchase and everybody at the management meeting seemed shocked. There was an attitude that "we have enough land," and I said, "No, we'll never have a mall if we don't have those 10 acres." The response was, "It belongs to Couch, he'll never sell it to us." There was a little discussion and at the next management meeting a month later, I brought it up again. Meanwhile, I had been told not to bring up that subject again, but I brought it up

anyway. And George Mitchell, to his credit, said, "Wait a minute, let's hear this again. Tell me again why we need it." I said, "I've examined every site from FM 1960 all the way north to Lake Conroe. There are many better mall sites along I-45 than we have in The Woodlands. We'll never sell it for a mall unless we can say we've got a jughandle overpass into the site." George said, "What's it going to cost and how are we going to do it?" I said, "All I need is clearance and we'll get it done." And there was huffing and moaning going, but I got a green light. So I got with Charles Lively and to his credit we came up with a "Mexican" acquisition. He got a national from Mexico City to call Dean Couch and tell him that he was looking to invest in some land out of Mexico, with money out of Mexico, on the I-45 corridor. Our man from Mexico got off the airplane and was picked up by Dean Couch, drove up the highway, spotted this, spotted that and said, "I want that 10 acres, I like that 10 acres right there." They made the deal and it was after it was closed that Dean Couch found out that it was really a Mitchell acquisition. The next thing was to convince the State Highway Department to provide for the overpass. It's to Barry Goodman's credit that he helped us get that approved and get state highway funds committed for construction. But that could have easily been lost and the Town Center you see today would not be what it is today if that mall had to shift to 1488 or up to 336 or some other spot.

The concept of the Town Center was part of the original planning, is that correct?

Yes. There would be a Town Center on the I-45 corridor. How successful it could be would depend on a lot of things, but one of them was access. We had problems, of course, with drainage to Oak Ridge. We were part of creating the retention lake (now Lake Robbins) at I-45 and Woodlands Parkway to throttle the surge of water into Oak Ridge so that we could develop up there, and that was important. And now some of the urban drainage goes to Oak Ridge, but it also comes down to Lake Woodlands, which is now to be the waterway.

There was a plan at one time to put Mitchell corporate headquarters at I-45 and Lake Woodlands Parkway, in the same general area.

Yes, that building would have been on the shore of that pond. And that's still a great site there for an office, just south of Landry's Seafood Restaurant. Anyway, you talk about my being stubborn: We had to build a major drainage way through Town Center because urban development in a natural forest like this really intensifies the runoff. When you build parking lots, rooftops, concrete and hard surfaces, storm water needs a new outfall. So we had to provide for a major drainage way for that excess water. Exxon did just that down at Greenspoint, and there's a big ditch with cyclone fences behind the Greenspoint Mall. It is behind their major office buildings today and it's just a trash wasteland. My point at The Woodlands was that I didn't want that kind of eyesore in our Town Center, and suggested we turn the drainage way into an amenity. I lobbied a long time that we turn it into a canal like in Amsterdam or the San Antonio Riverwalk. Since then, I've seen a similar treatment done in Singapore along the river there, and in Perth, Australia, too. Wherever you have waterfront, the chance to improve real estate values is very, very high. You get a premium for being on the water. People love to be close to the water. But it was a battle. I remember being told many years ago to "stop planning on that crazy concept, no one is going to buy it." We outlasted them, but there have been legendary fights on the same subject since. We ultimately prevailed but only because of George's support.

I'm trying to get a better feel for the development of the concept. That was not part of the original plan?

No, it was not.

OK, so you came here and you saw these needs?

I told everyone that we had the Town Center and we have frontage along I-45 that will naturally bring good dollars because of its exposure to the highway. We also have frontage along Woodlands Parkway and that's worth dollars, too; we have frontage along Grogan's Mill Road, and that's good. And we have Lake Woodlands Drive and that's worth premium money. Land on busy roads has value. But, when you get into the middle of the development, how are you going to make that worth something? I'll tell you how. You build this needed retention pond at I-45 and you bring a canal down through the middle of Town Center and hook it up with Lake Woodlands. Then you put "Tivoli Woods," an urban park, in the middle of it. Add an urban village on Lake Woodlands and commercial and office and entertainment along it and anchor it into the mall. The result is an urban spine that just sings of real estate value enhancement. That was the argument I kept battling for over 10 years. I hope it does successfully get done. I also wanted the Pavilion there. The story on the start of the Pavilion was I took Roger to the 18th green at the TPC course after lunch and walked him up on the hill and said, "Can you imagine instead of looking down on the golf green, that there was a stage down there? Why don't we do this in the Town Center's urban park?" He said, "OK, how much will that cost?" I said, "We could start a little one for a million dollars." So, I got an OK and I got a little tent structure designed by my friend Horst Berger and planned for a little hill where you would bring your lawn chairs and sit down and watch some performers. The original thought was that we'd have a little blue grass and little concerts. Horst Berger worked for me at InterAma in Florida a decade earlier. We did a big pavilion design down there so I was familiar with his work and liked him and brought him into The Woodlands project. Roger asked what it was going to cost and I said, "Well, the first cost estimate is $1.3 million." He said, "You said a million." I replied, "That was an off-the-wall figure." And he said "Get the price down to a million." Next we scheduled a meeting with Pace Management to see what kind of shows I could bring to this little theater.

Had you had any contact with Pace prior to this?

No. I just knew they were presenters and show producers. They came to a meeting that I set up, and George said he wanted Cynthia Mitchell to come, too. So she came and Bryan Becker of Pace was there, too. I presented the concept and Cynthia said, "Well, that's lovely" and everyone looked at a little model I had made—in fact, it's probably still sitting up in my former office. And she turned to Bryan Becker from Pace and asked, "What do you think?" He said, "That's a nice little design, too bad nobody will come." She said, "What do you mean nobody will come?" He said, "Well, it's too small," and she looked at Roger, he didn't say anything. Then she looked at me, looked at Bryan and said, "Well, that's simple, we'll make it bigger." So, we came up with some ideas on how to make it bigger and Roger said, "How much is that going to cost?" I said, "I don't know, I'll have to figure it out." We came up with $3.75 million. And I was told we couldn't afford that.

Was Allen Becker at that meeting?

No, not Allen, but his son, Bryan. We had a number of meetings where we hollered back and forth about costs but finally it got approved. I guess George said, "Yes, we'll do it for Cynthia." I was walking down the hall and I said to Roger, "Roger, why don't you give me $6 million and we can do it right?" He replied, "Don't push your luck. No!" But, as you know, people would call George and say, "We don't have an orchestra pit, we can't fly scenery, we need air conditioning on the stage, etc." We wound up costing close to $10 million. I don't know how many people know this story. There's one other story that's related to this: I wanted Frank Sinatra to open the Pavilion. By then I was working pretty well with Pace and they were happy that we were able to succeed in all these upgrades of the facility (thanks, again, to George's support). They wanted to make certain it was attractive so that they could look to an alternative to putting all major shows in the Summit in downtown Houston. Pace had experience around the country and I flew to Atlanta and visited some of their other pavilions. I believe all agree that ours is still the best. Anyway, I was a kid when Frank Sinatra hit the big time. Throughout my life, it seems, Frank Sinatra's music was in the background. I loved it. Then the man from Pace came back and said, "We've got good news and bad news. The good news is Frank Sinatra is available on opening night. The bad news is that he wants $200,000, so he's way over our budget. We've got $120,000 or so for the opening show, so that's it–end of story." I said, "No, you book Sinatra." He said, "No, you don't understand." I said, "You don't understand. I want Frank Sinatra." He said, "We can't afford it." I said, "I'll tell you what we'll do. We've got 10,000 seats, right? You can have 9,000 of the seats, you give me the first 1,000 seats next to the stage and I'll guarantee the first $100,000 and you've got the other 9,000 to get the other $100,000." He said, "You must be kidding." I replied, "I am not." He said, "Frank Sinatra is too old, he doesn't sing well any more, nobody is going to come." I said, "Trust me, I'll guarantee the first $100,000, give me those first 1,000 seats." He asked, "Are you sure," I said, "Yes." And we shook hands on it and he gave me the thousand seats and I said, "OK, 100 bucks apiece." They were gone in two weeks, and we had Frank there on opening weekend. It was an event of a lifetime and Frank was great and got six standing ovations.

You were both right because it did not sell out, but your good seats did go. I remember that.
Yes, there are enough people who wanted to see a living legend one more time. And fortunately, I didn't have to go into my own money and we did pay for our part of that show.

Before you leave the Pavilion, which, as you know, is now 13,000 instead of 10,000, tell me something about its design origins.
I went to Atlanta at Pace's invitation to see how their pavilion was going but it didn't have anything to do with design. It was a crowd control study and seating details and things like that. The design of our Pavilion is very similar to what I had done with Horst Berger in Miami, Florida, at InterAma. The

first architect on our Pavilion was a local Houston firm that Coulson brought in but I didn't think their design was right, and George agreed. He said, "I want something inspiring." I said, "Turn me loose. I can bring in a design you're going to love." And I went and got Horst Berger again and we did it with a teflon-coated fiberglass tensile structure. Twenty-seven thousand square feet of fabric over 3,000 seats. The hillside, which seats 7,000, took 18,000 trucks of excavation that Plato was glad to get rid of.

I know that cost control was a consideration throughout the project. You found, for instance, you couldn't afford Chris Jaffe to consult on acoustical design.

Chris is a great acoustician and did well. As we expanded, however, we did switch to another local consultant because of the cost. As far as research is concerned, we did visit other pavilions to experience the size of them, how many seats felt right on the slope and where you have the restrooms and how did you handle the concession outlets for people, the parking, etc.

The Pace people were very helpful, weren't they?

Yes. They told us a lot of things we didn't do, but they were helpful. Of course, they were used to the hard rock shows and not the wide variety of shows that we wanted, so we had to modify their concepts with something to make it more appealing for the other types of shows–opera, symphony and ballet, for instance.

Yes, they said, "You don't need an orchestra shell, the Symphony can play without one," and the Symphony would not have come out here without a shell, period.

Right, Joe. Let's talk about the "Tivoli Woods" urban park for a moment. I don't know how far they've gone since I've left. The concept was to form an urban park around the Pavilion, with a lake and dam with a noisy spillway to make it an exciting place. I know George never liked the name "Tivoli Woods" because there's a Tivoli in Italy and one in Copenhagen.

Yes, actually I thought it was Cynthia who objected.

But the fact is, "Tivoli" is synonymous with fun and entertainment and it's good to use that word. I will always disagree with not using that name because the minute you say it, anybody who's been around knows that's a nice kind of place. Another thing I'm happy with is the Teacup Island and North Shore Park on Woodlands Parkway at the end of Lake Woodlands.

That project was a battle, too. I said that we needed a visual landmark so that people coming in don't see just the water. We needed a focal point, a reason for people to ride their bicycles along that hike-and-bike trail, both from Grogan's Mill and down from Panther Creek. We also needed this little off-shore island with a nice pedestrian bridge and an overview pavilion where you could get up and look down the lake. It is a place where young teenagers and lovers could go. People need a place like that, a destination point. The pedestrian bridge cost $80,000. There was a lot of disagreement about going ahead. I don't know how we won but we did, over some strong opposition, and it's there. It is a visual landmark statement for everybody coming and going on Woodlands Parkway and someone is always there!

Let's come back to the Pavilion. Why did we need a pavilion?
I went through that at the Columbia new town in Maryland, and worked with the team there to bring the Washington Symphony to the Merriweather Post Pavilion. I became convinced that when you're building a new town out in the suburbs, you don't have your mall in your early years. Therefore, you need to create a destination to help people realize that it's fun to be there, and I said that for every hundred people who come out, there's going to be some home sales that spin off because we had a Pavilion. Later, I heard that people from Kingwood were unhappy over the fact that they had chosen Kingwood rather than The Woodlands because now we've got the Pavilion and they don't. I kept arguing that point. It is $8 million or $10 million, but we've got Kingwood out there, and all this competition around Houston. This Pavilion put The Woodlands on the map forever and that was my argument.

It was a very well spent $10 million.
One of the other things that I'm real happy about is the development of the villages and neighborhoods and sub-neighborhoods of The Woodlands. The proliferation of cul-de-sacs that show up in our land plan is unique and it's an outgrowth of my own belief that the arithmetic of the group should be as close to that of the original primate hunting troops of man as possible. We were there into those numbers for so long—50 to 60 people living together on a defensible piece of territory. In community building, culs are friendlier and more secure than living five blocks in, six blocks in, 10 blocks in off some street where there's no beginning or end. We've arranged our subdivision development with that in mind. At last count, I think, over 70 percent of our houses were on cul-de-sacs. You're not subject to through traffic, the

kids are safer in the street, you get to know your neighbor because you don't have to meet 1,000 people, there's just a handful and you can cope with that. And I think post-occupancy evaluations that we did over the years always reinforced that this is the friendliest place, quiet, secure, and people like it. And that, again, had to stand to the criticisms that people get lost and you get in and you can't get out, the fire department can't come in with the big trucks and the garbage truck has to turn around slowly and the moving van has trouble maneuvering. But all of those really are minor things.

The other argument was to do linear streets?
Yes, longer streets. You can't do more than 600 or 800 feet on a cul-de-sac. In most places you have to have a through street. If you've got a through street, traffic comes and goes. But with cul-de-sacs, I think security against burglaries is enhanced because there's only one way in and one way out. They can be trapped in there so easily. I'm happy about that. What am I not happy about? Battles lost? No urban residential program in Town Center. I really wanted to build a "Georgetown" on the Riverwalk on the eastern shore of Lake Woodlands.

Is that lost forever, do you think?
I didn't have any support when I left and I don't know what kind of advocacy is there now. Another thing, we didn't maintain the landscape illumination program that we had at Woodlands Parkway in the beginning.

Tell about that, please.
John Watson came and put up lighting in the trees along Woodlands Parkway from I-45 in and made coming home here like a Walt Disney experience. It broke my heart that they took it down and threw the lamps into a warehouse. But there again, we lost the battle because of dollars—penny wise and pound foolish, I think. Another situation: Down at the Wharf behind the Conference Center, at the eastern end of Lake Harrison, is a great place for some benches to watch the water and to watch sunsets. I could never get the park benches approved for that. There's also a courtyard that was built out there for the Conference Center folk when they take breaks. They'd walk out and have a smoke. The low walls around the courtyard there are designed with little cubicles to put in these benches. But when the budget got cut, so did the benches. There's an article that I read some place that's about how good benches make good communities. Communities that don't have good benches are cold and heartless. Benches encourage people to sit

down and talk to one another, sit down and have social interaction. If there's no place to sit, you just get back in your car and go on.

But in a climate like Houston's, does that work as well? On a day like this it would be perfect to walk, but when it's 95 degrees and with Houston's humidity, would people do that?
Absolutely, just like they'll come to the Pavilion on a hot night and listen to a show on the grass and brave the mosquitoes. When I was doing the Pavilion, a lot of people told me that the word was out to "distance yourselves from Browne, he's going to fall on his face on this one. Nobody is going to come out of Houston to sit on the grass in The Woodlands in the hot and humid weather that we have here. Nobody is going to come. He's going to fall on his face this time." But I insisted that they would come and they did. Anyway, I think for the little money it costs, you can locate benches in the shade at places with delightful views. It encourages people to stop and enjoy their lives a little bit more than if the benches aren't there. It's what I used to call "finishing touches." And, finally, I was disappointed that our village centers are not done better. Starting with Grogan's Mill, then Panther Creek and then Cochran's Crossing. They are all three really strip commercial centers, not unlike anything you'd find along FM 1960 or any other place. When I left, the schematic plan was in place for the next, most western village. It was to be a real village center, one with a real village square and park where you would have a good kind of social interaction, a nice place to walk and to be. We'll see whether it happens.

The reason for not doing it is what, the requirement for land and the cost?
No. There was a feeling that they should build strip centers because that's what has been built in Houston. That's what sells and they just want people to get out of their cars, go into the store, come out, get back in the car and go home. And this business of sitting around and mixing it with recreation and cultural or social activities is contrary to high-pitch commercialism, so it's a different viewpoint. I know that real village centers can work. Columbia has done some nice ones and they still work fine. I'm disappointed, too, that Lake Catamount came on too late in George's career for it to happen. There wasn't any support for that, beyond him and me initially, but that's the way it goes. Somebody, hopefully, will pick it up and do it some day.

Give a minute's worth of description of what Lake Catamount is.

Lake Catamount is a primary location for a major year-around resort in the Colorado Rocky Mountains, and I'm convinced that with cellular phones, fax machines, computers, more and more people can afford and desire to be in a location like that instead of urban centers. You could get high-end people to invest in a unique setting, 10,000-foot vertical drop on a ski mountain with a lake. In the winter you can ski in and ski out of your house and skate out and skate back. In the summer you can golf out and golf back, sail out and sail in, go fishing. It is just a unique and beautiful location. It would be the sister ski mountain to Steamboat Springs, the second or third largest ski area in the United States, which needs a place to grow because it's out of growth capacity. The Steamboat region has the biggest jet port of any ski area in the United States just minutes away so that it conceivably could be an international destination, as well. It's a winner for somebody to do right. It would be the only U.S. ski area that is also on the shore of a lake.

We have partners in that, or semi-partners?
We were partners with Martin Hart, who started out as the owner, with his group, of the Steamboat ski area. But he sold out to the Japanese and wound up still being our partner. I understand from what I read in the papers that MND plans to sell its Colorado property and George is quoted that selling Catamount is consistent with the company strategy to focus on development at The Woodlands and dispose of everything else.

That seems to be exactly what's happening.
Yes, that's the decision. Like I say, if George were 40 years old when this started we could have done it, but he was 70 and so it's too late. It takes somebody with a "can do" drive attitude to help make things happen and from my position I didn't have that power to make a change. It's now "get in and get out, liquidate for the most bucks." I understand all the northwest properties are now pretty much gone and the San Luis, too.

That's gone. Well, that is probably a good business decision because some of that real estate was eating our lunch.
Sure. I'm not saying otherwise. You have to pick and choose what you want to do. Maybe they sold it and made a good sale and maybe the next guy will do it and make a lot of money.

Going back to the very early part, Pereira and McHarg were essentially in place when you hooked up with Mitchell, is that correct?

I don't know the timing because the early maps and reports say Ian McHarg, William Pereira, Richard Browne and Associates in the title box. I think we almost came on concurrently.

Did you have a role in the original submission to HUD?
Yes, absolutely. We provided the engineering feasibility study on how the drainage would work and the highway system and we did presentations on the master plan, as well.

Was that largely under your wing?
I think Hartsfield, who was full time with George, was the general manager of the process, and Pereira, Ian McHarg and I were consultants.

Obviously this was before you came on full time. We got that HUD guarantee in '72 and you were not with us full-time yet. You left the company in April of '95. What have you done after that?
I went immediately to Irian Jaya in Indonesia. There we built the landmark new community which was a commitment to the government of Indonesia by Freeport-McMoRan of New Orleans. They have the world's largest copper and gold mine there. Actually, I didn't work for them, I worked for an Indonesian company because the Indonesian government insisted that Freeport stick to the mining and that real estate development and the trucking and other businesses ancillary to service the mines be done by Indonesian companies. But they introduced me to an Indonesian company and they hired me as executive vice president to oversee the new town's completion. It was under way when I got there, but still in the mud. We polished it up, finished it up, landscaped it, they got it all open and it's running.

In your 24 years at The Woodlands, it has to be the most important project you've had in your life.
Yes, I spent half of my working life at The Woodlands. But I left before the icing was on the cake. I still think the urban center can be a crown jewel if it isn't chopped up and short-changed for expediency or the fast dollar. If done properly, it will cost some money, but it will enhance the real estate values of all the sales to come. It will more than pay for itself.

You've been at arm's length for a year-and-a-half, more or less. What grade would you give to The Woodlands development—A, or B, or C?
I would give it an A, and a B to Columbia, Maryland, and a C, maybe, to

Peachtree City. Those are the three best communities.

If everything you wanted had gone through, would that have been good, or at this distance are you able to say, "Well, in retrospect maybe it was just as well that such and such didn't happen"?
I do have the courage of my convictions regarding the creation of a landmark community. One is needed in the U.S.A. to be a role model for community development in this country for the next century. I think that should have been part of our mission. I resist the American philosophy that earnings per quarter is the only measure of success. I think this society that we live in should say the creation of assets that benefit this nation should also be a measure of success. To short change reaching a pinnacle of greatness just to wring out the last almighty dollar is wrong. One of the things that I found in Indonesia, and they're a centroid of three billion people, is their resentment of American capitalism in its unbridled form. They feel that much of American capitalism just worries about dollars and doesn't give a damn about people. I've spent 21 years plus consulting time before The Woodlands, and have given my life to creating real estate values here and to making this new town. And then maybe three or four years short of my retirement and finishing the job the way it could have been, the company decides that they can save a couple hundred thousand bucks and throw me away when I'm not ready to be thrown away . . . that's cruel. In Indonesia, they preach that the goal in life is harmony and peace and a good life and to make enough money to be comfortable. In this country, it's the almighty dollar and everything else be damned. I don't think that's a goal to be envied or desired in the long run. It's cold and heartless. Is our human species to be treated that way? Why? I don't think that $4,000 or $5,000 a year for benches in The Woodlands would have affected the price of Mitchell stock one bit. But it would make the community a lot better place. To create a mixed-use urban center with housing in our Town Center, where people who want an urban experience can walk to the park, walk to the Pavilion, walk to Lake Woodlands, walk to the mall, walk to the restaurants, to the movie theater—that would be a wonderful thing for young, single professionals before they've got kids and want to move out in the suburbs where their kids can play more. It would be wonderful for senior citizens or even divorced couples. Everybody doesn't have to live in a four-bedroom house with a two-car garage. The Woodlands has a 40-year supply of land for office buildings. Let's take half of that and convert it to urban residential and cut the 40-year supply back to 20 so you can get it done. In the meantime, it's

more profitable to do it the way I'm saying. Provide another dimension of housing that is needed. There are people out there in 5,000-square-foot houses who when their kids leave to go to college and then graduate, they look around an empty barn of a house. They can easily say, "We don't need to live here anymore. Let's move into Town Center." The Woodlands could have some nice luxury condos and things like that for these people. They wouldn't have to mow the grass and instead they could go out and watch the kids and have fun. You can't put everybody in the same box, they don't belong there. We need to take a more holistic view of life and meet the needs of people who come with different propensities, different cultural upbringing and desires, plus their own varieties of likes and dislikes. When you have something for everybody, the pace of development quickens and, therefore, interest costs are reduced and it's then more profitable. If you sit around waiting for another 40 years worth of office buildings in Town Center, you'll wind up losing money. The good ideas I've proposed, I think, are not inconsistent with the long-term economics of the project.

Very good, very thoughtful. Thank you very much, Richard.

This is Joe Kutchin on Thursday, May 30, 1996. I'm in my office in The Woodlands in the MND Building with David Bumgardner. David, tell your full name and then give me four or five minutes on your background, education, experience with the company and so on.
My full name is Gene David Bumgardner. I have gone all my life by the name David, but oddly enough I married a girl named Jean, so it was very convenient not to have us both called Gene. I came to work for Mitchell Energy, which at the time was Christie, Mitchell & Mitchell Company, in 1956. I came here right out of high school as the mail boy. I had graduated, planned on going to college the next fall and work that summer to earn money for college.

You were a Houstonian?
Right. I grew up in Houston. I did not earn enough money and through a friend learned that the mail boy at Christie, Mitchell & Mitchell was leaving that fall to go to school himself. So I applied for and got his job. I started there and moved into the land department after a few months.

You were located where then?
I started out in the Houston Club Building and eventually we moved over to One Shell Plaza and then finally up here to The Woodlands. I moved into the land department after just a few months and I was there for 10 years. I think I held every position in there, up to and including that of manager by the time I left. I started out as a recording clerk. I saw that all of the company documents were recorded and put back into the proper files. I did that for a while, worked on division orders for a number of years. During this time, I went to night school and got 90 hours and then went into South Texas College of Law.

Where did you go to night school?
At that time there was a South Texas Junior College located in the YMCA downtown, which was just a few blocks up the street. I went there and then went to South Texas law school, which is also downtown. That all took about 10 years.

South Texas Law was 10 years?

No, the two together. Undergraduate as well as the law school.

But the law school was night school?
Yes, all of this was at night. About the time I was finishing up school, we had a young manager of the lease records department, a part of the land department, by the name of Bennett Young, who died very suddenly of a heart attack. I was the next senior person in the department and had to take over under what essentially was a crisis situation—tremendous trauma to a lot of people there. In a few months I was heading that department. Fortunately for me, about that time the company decided to establish its own legal department. It had always used Smith & Fulton as outside oil and gas attorneys. They had brought in Paul Wommack as a real estate attorney when we got more active in real estate. About the time I graduated from law school, they brought Paul in as in-house legal counsel and about a year after that I moved in with him.

By this time you had passed the Bar?
Yes. I had finished law school and passed the Bar and then moved into the legal department with Paul. That was 1967 or 1968. I basically continued in that position as the Number 2 man under Paul and eventually under Tom Battle for the rest of my career here. We obviously brought in other people, staffed up both the real estate side and the energy side and then we would contract. I eventually became the general counsel of The Woodlands Corporation after Bob Hinton, who was brought in by Len Ivins during his era. When I retired on April 30, 1995, I was senior vice president and general counsel of The Woodlands Corporation.

* * * * * * *

The editor of *Terra Sol*, the Mitchell Energy & Development Corp. employee magazine, asked David Bumgardner for his recollections of early days in the development of The Woodlands. Bumgardner, the head of the company's real estate legal department at the time and a long-time employee, responded with a memorandum dated September 20, 1994. The main part of that memorandum follows.
These are stories of people whose names, with a few notable exceptions, will not be found on the lists of those who made significant contributions to the development of The Woodlands, and stories of events that will not be listed among the many significant milestones in the history of The Woodlands. But to me, they are significant because without them, The Woodlands might

never have been, or at least not be what it is today and what it will become.

Two of the earliest characters in The Woodlands' colorful history were John Taylor and James Stephens. Some time in the 1800s, John Taylor acquired title to land comprising a part of The Woodlands known as the John Taylor Survey, wherein lies the Village of Grogan's Mill and much of the Metro Center, and James Stephens acquired title to the lands known as the James Stephens Survey, wherein lies part of the Metro Center and the Research Forest. The Stephens Survey adjoins the Taylor Survey on the north and, as originally patented, the south 200 acres of the Stephens Survey overlapped the north 200 acres of the Taylor Survey (surveying not being the exact science we would like it to be; or perhaps, more accurately, it is an art, not a science, according to Plato Pappas). This problem was easily resolved, or so it seemed, by a court ruling in the early 1900s, decreeing that the land properly belonged to the owners of the Taylor Survey, being the older of the two. Through mesne conveyances, The Woodlands Corporation acquired this 200 acres from John Taylor's successors and rightfully ignored any claims by those holding title under James Stephens. At least until a prominent Conroe lawyer pointed out that another surveyor, undoubtedly a practitioner of modern art, had determined that the south 200 acres of the Stephens Survey did not completely overlap the north 200 acres of the Taylor Survey, but instead extended just a little east, thereby creating a small strip of land still belonging to the descendants of James Stephens. Fortunately, these descendants were soon located by Charles Lively and his constant companion, the late Gerson Haesley. (Presumably, Charles and Gerson are no longer constant companions, but with Charles, you never know.)

Also, in the early 1900s, a Mr. Smith Winslow allegedly settled on land in this area and eventually claimed squatter's rights to 160 acres. In 1925, a deed appeared of record in Montgomery County whereby Mr. and Mrs. Winslow purported to sell 160 acres to a Mr. Bell. The only description of the land was 160 acres near a common corner of the (you guessed it) John Taylor Survey and the James Stephens Survey, including a strip of land out of the James Stephens Survey. Mr. Winslow was apparently deceased, and the deed was actually signed by Mrs. Winslow and three of their five children.

Now we had to contend with finding the heirs of two of the Winslow children and all of Mr. Bell's heirs. No problem. If they could claim title by squatter's rights, so could we. A fence was erected around the strip of land,

and we merely had to wait our 10 years. Unfortunately, between the good citizens of Shenandoah using The Woodlands as their back yard and Plato constructing roads and utilities, the fence didn't stay in place two days, much less 10 years.

So, with the legal expertise of Eileen Stilson and Kaye Applewhite, and more sleuth work by Gary Calfee and John Cook, we located all of the Winslow and Bell heirs and purchased their claims. During this process, we heard several versions of the saga of one of Mr. and Mrs. Smith's sons. This son, a rather cantankerous sort, was driving a wagon loaded with logs from this area to the sawmill in Magnolia. Along a dirt trail located approximately where FM 1488 is today, he met a wagon coming from Magnolia. The road where they met was too narrow to allow them to pass, and one driver had to back up. Neither was willing to do so, and, after a short exchange, Winslow drew his pistol and shot and mortally wounded the other man. Unfortunately, the decedent was a nephew of the sheriff, who promptly tracked down and killed Winslow.

We located the remaining Winslow heirs with the help of their attorney, who tried mightily to convince us that Mr. Winslow had in fact established a valid claim to 160 acres and he (the attorney) knew right where it was, more or less. After several trips into the woods with oldtimers in the area who claimed to know something about the old Winslow homestead, the attorney was ready to give up. On one such odyssey, we left him panting beside the trail while we continued across a creek in search of an old woods road that purportedly ran in front of the Winslow place on the way from the Houston-Conroe Road (now I-45) to the place of one of Mr. Winslow's sons on the Conroe-Magnolia Road (now FM 1488).

The attorney finally turned up a niece of the Winslows, who, though 90 years old, blind and in a nursing home, had a remarkable memory of visiting her uncle's place as a child. She even recalled climbing on the old rail fence (fencing being critical to establishing squatter's rights). Unfortunately for the Winslow heirs and their attorney, she also recalled climbing on the fence and looking out at the creek. The problem was, for the 160 acres to extend from a common corner of the Taylor and Stephens Surveys to the nearest creek, it would have had to be in almost a straight line to the west, a rather odd shape for a 160-acre homestead. Moreover, in this area, the creek is on someone else's land.

Nonetheless, if we could locate all of the heirs, it would be worth something to buy their claim and clear the title. Needless to say, the Winslow heirs were very helpful in this, being anxious to get the money. Their dilemma was, the more heirs they turned up, the more they had to share the money with. We had to corroborate all of the heirship information, and, much to the chagrin of the other heirs, one lady remembered one of the deceased heirs' having a daughter, but didn't know what happened to the daughter. This raised the specter of an unknown heir no one else had ever heard of. After much wailing, some of the heirs trooped off to a cemetery in Magnolia and reported back that there was indeed a daughter, but she and the mother had died during her birth and were buried in Magnolia.

Some time in the 1960s, a visionary named George P. Mitchell gazed upon this veritable paradise flowing with milkweed and honeysuckle and determined he could build a new city that would solve all of the urban ills of Houston. He proceeded to buy the stock of a timber company owning some 60,000 acres in this area and surrounding counties, Grogan-Cochran Lumber Company. Mr. Mitchell gave promissory notes for the purchase price of the stock, secured by mortgages on the land. As payments were made on the notes, the mortgage was released on portions of the land, which was promptly mortgaged to an insurance company to secure a loan which paid the notes given to the shareholders of Grogan-Cochran for their stock. Mr. Mitchell then sold the timber to another lumber company under a long-term contract which provided for payments sufficient to pay the mortgage to the insurance company. All of which is to say, Mr. Mitchell didn't pay anything for the land, although he claims to have bought it for $135 per acre. Along with the Grogan-Cochran Lumber Company came some of our most colorful employees, Raymond Johnson, the late Max Newland, the late Charlie Thompson, and Joe Cliff and Charlie Mock. (I am not sure Joe Cliff and Charlie Mock were with Grogan-Cochran, but the only times I ever saw them were at the hunting cabin on FM 1486, and they seemed to have come with the land, just like Raymond, Max and Charlie Thompson.) All of them were good, down-to-earth, honest and hard-working, although a lot of people would be surprised to hear Max described in those terms. Charlie Thompson and Max were foresters, who, like their successor, David Townsend, knew and understood the land and the timber like few others. Charlie was quiet and unassuming, always friendly and helpful. Max was also friendly and helpful if you were on his good side, but anything but quiet and unassuming. He was tough, hard as nails and, I suspect, downright mean if you got on his bad side.

When we bought a 1,000-acre tract from the Catholic Church, it had been leased for hunting to a Houston policeman. Max notified the policeman that we had bought the land and would not renew his hunting lease. The policeman promptly informed Max he didn't give a rat's behind who owned the land (this being before the days of sensitivity training), he had been hunting there for years and intended to continue hunting there. In fact, he would be there Saturday morning. Max said fine, he would meet him at the gate. Max showed up, but the policeman didn't.

Max was obviously well-known and well-respected in both Conroe and Magnolia. A friend and I were driving through Magnolia on our way to hunt at the aforementioned hunting cabin, located between Magnolia and Dobbin. We were stopped for speeding, and while we were talking to one officer, another officer walked around the car and noticed our rifles in the back seat. He motioned to the first officer, who glanced at the guns and asked where we were going. I replied that we were going to "Max's cabin" to hunt. The officer handed my friend his driver's license back, tipped his hat and told us to have a good time.

Raymond had been with Grogan-Cochran since time immemorial. He has always been an invaluable source of information about the land, its title and its uses, as were Max and Charlie Thompson. I recently spent a pleasant afternoon visiting with Raymond and a Mr. Everett Harper, another gentleman who has a wealth of knowledge about the history of this area. Mr. Harper actually worked at the old Grogan-Cochran Lumber Company mill located in the vicinity of what is today the intersection of Woodlands Parkway and I-45. Mr. Harper is healthy enough, physically and mentally, to still be working in that mill. It closed in 1928. (The 1920s seem to have been an important era in the history of The Woodlands.) Mr. Harper and Raymond kept talking about the job of riding the log as it went back and forth through the saw. I never quite figured out how they did this and avoided the saw blade. Perhaps that's why most men who had this job were called names like "Shorty," "Stump" or "Tripod."

Following our acquisition of the land through Grogan-Cochran Lumber Company, Charles Lively, then a landman in Bridgeport, was brought to Houston to help block up additional land for the "new town" project; and

Paul Wommack, then an associate with our oil and gas lawyers, Smith and Fulton, was brought in as general counsel to handle the legal work.

Charlie Magan was a draftsman and surveyor for Mitchell Energy & Development Corp., and transferred to real estate to assist with the acquisition and development of The Woodlands. Charlie knew the land like few others, not only as a surveyor but also as an amateur archeologist and historian. Charlie was likewise a valuable source of information before his untimely death shortly after he retired. Early in the development of The Woodlands, Charlie and I stood on a dirt road, known as Golden Road, one hot, dusty afternoon, staring down at a surveyor's stake in the middle of the road. It seems we had platted one of the first subdivisions in The Woodlands and, for some reason, one of the lot lines ran off of our property and into Golden Road. Since we had already sold some of the lots, we weren't quite sure what to do. So we simply kicked dirt over the surveyor's stake and assured ourselves no one would ever know the difference. This particular lot was sold soon after, and the surveyor for the buyer dutifully surveyed the lot line, found the stake in Golden Road, and promptly called it to the attention of the title company, who promptly called it to the attention of Paul Wommack, who was not amused at our solution. The problem was resolved by a correction plat, and an important lesson was learned—if you have a problem, deal with it. Don't kick dirt over it and hope no one will notice.

Golden Road is the access to an out-tract in The Woodlands affectionately known as "Dogpatch." I am not sure if it got that name before or after Charles Lively visited one of the residents there and returned to his car covered with fleas; however, that may explain why we never bought Dogpatch.

Charles Lively, with the invaluable help of Frank Karnaky, Gerson Haesley and, later, Tom Ledwell, was very successful in buying hundreds of other tracts that make up the present day Woodlands, some of them twice. Much of the John Taylor Survey had been subdivided into 10-acre lots with roads that literally went nowhere, leaving the lots with no access. These subdivisions were created in that mystical era of the 1920s by some people named MacDonald, and lots were sold, presumably sight unseen, to people on the East Coast. Shortly thereafter, a Mr. J. H. Rose moved in and established squatter's rights on many of these lots. Mr. Rose sold his claim to a Mr. George Parrish and a Mr. R. J. Carter. Charles bought their title, then proceeded to locate the purchasers of the lots from the MacDonalds and bought their title, as well. Some we either could not locate or they would not sell, so we had to file suit and gained title by proving that Mr. Parrish and Mr. Carter had indeed used and occupied the land long enough to establish superior title by limitations (i.e., squatter's rights). Ernest Coker, Sr., a longtime district judge in Montgomery County, had two sons, Ernest, Jr. (Bo), an attorney practicing in Conroe, and Lynn, who was himself then a district judge. Ernest, Jr., was our lawyer in these cases and Lynn was the presiding judge. It was a tough fight, but we prevailed. Imagine how tough it would have been if the other side had shown up and contested our claim. Ernest, Jr., is still practicing law in Conroe. Lynn has since passed away. (Somehow, writing this is making me feel very old.)

Like Mr. Winslow, Mr. Rose did not have a deed to his land. He acquired title simply by fencing, using and occupying the land for 10 years. Other fine, upstanding citizens figured out that if you had a deed to the land, you could acquire title by fencing, using and occupying the land for only five years. The fact that the deed came from someone having no interest in the land made no difference. Knowing the record owners were not around, and couldn't locate their property if they were, one gentleman deeded a number of the lots to another, who promptly fenced the lot and waited for the five years to run. We were fortunate enough to retain the services of another renowned Conroe attorney, the late Tommy Green, who so intimidated these gentlemen that they promptly quitclaimed their interest to us.

Two rather large purchases helped complete the original 17,000 acres of The Woodlands. One was the acquisition of the Sutton and Mann interests, the principals in a former timber company who had come to a parting of the ways. The Sutton group was a Mr. Sutton and his three sons, all very nice gentlemen. Their former associate, Robert Mann, was a banker from Woodville.

During the negotiations with the Suttons, Charlie Magan, Charles Lively and I went on a tour of the property with the Suttons to check boundary disputes. One of the Sutton boys had a new four-wheel drive Bronco with a winch on the front, which he assured us would get us anywhere we wanted to go. Like a kid with new rubber boots looking for mud puddles, he set out across the worst possible terrain and through the most dense underbrush he could find. We blazed a trail through yaupon and young pine saplings until we met a sizable tree that would not budge, whereupon the young Mr. Sutton meekly drove back to the road to finish our tour.

Mr. Mitchell rarely got involved in direct negotiations, but Robert Mann was an exception. Mann was as tough as chewed leather, and we simply couldn't reach an agreement with him. Mr. Mitchell went with us to meet Mann one Sunday afternoon at the Tenneco facility at Intercontinental Airport, from whence Mr. Mann was scheduled to depart shortly, a fact of which he constantly reminded us throughout the meeting. Several hours of intense negotiations ensued, during which Mann got so animated at one point he literally fell out of his chair. Mr. Mann finally gave Mr. Mitchell an ultimatum, either make the deal the way he (Mann) wanted it or he was walking out to get on his plane and leave. Mr. Mitchell said no, and Mann and his entourage stomped out of the room, only to return a few minutes later and sign the deal on terms acceptable to Mr. Mitchell.

The other large transaction was a land exchange with Champion Paper Corporation. Champion wanted timber lands further away from Houston so taxes would be cheaper, and we wanted their lands in this area for The Woodlands. A young attorney representing Champion was Randy Hendricks, now a noted sports attorney and agent for several of the Astros, Oilers and other players.

Randy is also the brother of Dave Hendricks, who, along with Jim McAlister, formed the original Woodlands management team (or perhaps the aboriginal team, since the Len Ivins group is generally thought of as the original team). We started out in One Shell Plaza, where all of the offices were defined by movable panels, and prestige or rank was determined by the amount of space within your panels. Dave and Jim officed next to each other, and a tremendous power struggle developed over who had the larger office. Jim started each morning in a fit of rage as he struggled to move the panel dividing them back towards Dave's office from where it had somehow mysteriously moved during the night.

Jim is, of course, now a prominent real estate broker in Houston. Not to be outdone, Dave and Randy developed Mostyn Manor, a subdivision on FM 1488.

The Ivins management era came (and went) bringing with it some of the most interesting characters you will ever meet. My favorite was probably Sam Calleri. Sam often either started out the day with a breakfast meeting with his staff, or had dinner with them in the evening. He commented to me that they were like a bunch of hogs and the only way he could get any work out of them was to feed them. (Mr. Bell, who you will recall acquired the phantom 160 acres from the Winslows, was reputed to have trapped wild hogs in this area. David Townsend still does, although as far as I know, neither David nor Mr. Bell ever trapped any of Sam's employees.) I always said I hoped nothing ever happened to Sam so I would never be the skinniest person alive. Alas, the last time I saw Sam he had gotten fat.

Ivins had a management meeting every Monday night. Since we were on a crash schedule to get the Inn and Country Club opened, roads and utilities in and houses constructed for our grand opening in October 1974, each manager had to account for his progress (or lack thereof) at each weekly meeting. Invariably, each Monday afternoon brought a flurry of requests to the legal department for legal services. And that night, each manager could dutifully report that he was doing everything he could, but he was "waiting on legal." Ivins accepted no excuses, least of all "waiting on legal." On one occasion, he instructed his manager to take his lawyer to lunch and do whatever he had to do to get the job done. Since the management meetings themselves included a rather elaborate feast, we tended to eat well during this period.

Ivins also brought in his divorce lawyer to serve as our chief legal counsel—Bob Hinton, a former law school classmate of mine whom I came to respect and admire as much as any other lawyer I have ever known (with the notable exception of Tom Battle—after all, I still have to work here). Bob brought in as his chief deputy Dale Ossip Johnson from Austin, also a divorce lawyer, who specialized in obtaining child custody for fathers. Dale had his own plane and spent quite a bit of time commuting to and from Austin (one other manager during the Ivins era commuted to and from Aspen, Colorado).

Every communication to the legal department went through Dale's office, or at least into Dale's office. Dale had a huge rolltop desk which was filled with a stack of files and papers three feet high. Since Dale was seldom there, we had to go through the stacks several times a day and eventually knew where to find everything. At one point, Dale shoved everything onto the floor, where a huge pile again sat for months. We soon figured out where everything was located in this pile as well.

In the early days of The Woodlands, finding a place to eat lunch was a problem. There was Hyden's, the Falls and Bill's or Brad's Barbecue (I forget which, but it was named after the former University of Texas football star Bill Bradley). We often brought a sack lunch and went to the deserted ice skating rink at the equally deserted Wharf Shopping Center to eat. This was mainly out of sympathy for Gwen Bert, a former secretary who got caught up in the hype of all of the wives' coming with their husbands to conventions at The Woodlands Inn and spending all day shopping in the Wharf. Gwen opened up an antiques store there. Regrettably, hers was the only store in the entire center, and I think we were the only living souls she saw all day.

One day, just to break the monotony, we drove to the Conroe Airport where Dale kept his plane, with the intention of flying to Hooks Airport for lunch. A fierce storm had come through the night before, and when we got to Conroe several of the planes were standing upright on their propellers. Dale's plane was OK, but we decided not to fly that day.

The Ivins era also was noted for its parties. (I think Len, like Alan Vitale, lived by the creed, "Life is uncertain, eat dessert first.") When we were still at One Shell Plaza, the partying started in the Plaza Club sometime in the afternoon and continued well into the evening. If something you were working on came up, you were summoned upstairs to discuss it. Sometimes you were invited to stay for a drink, but more often not, which was fine since we were very busy. During this period, we entered into the Project Agreement and Indenture with HUD, created the HUD tract system that is the basis for our current property control map, created companies for and got into the businesses of construction, insurance, title insurance, telephone, gas, CATV and a mortgage company, negotiated a long-term agreement with Gulf States Utilities and continued purchasing property for The Woodlands.

It is probably safe to say that Budd Clark was not enamored with the Ivins management style. Nonetheless, Budd remained philosophical and gave me probably the most sage advice I have ever received. After listening to one of my complaints, Budd calmly told me not to worry about it. I asked why, were things going to change? Budd said no, I would get used to it.

During this time, we also had our first encounters with our neighbors. The same engineer designed the drainage systems for Vicksburg and Shenandoah, in each case collecting all of the water from the subdivisions and discharging it in one spot, on their neighbors' land. Vicksburg's neighbor was a Mr. Kraft, who promptly obtained an injunction to prevent the discharge of water onto his property. Fortunately, The Woodlands had constructed a large drainage ditch (affectionately known as the 9x3 drainage ditch) along its east boundary, to carry drainage water resulting from the development of The Woodlands to Spring Creek, rather than discharge it on our neighbors. Even though many Vicksburg residents had protested the construction of this ditch, The Woodlands allowed the developer of Vicksburg to buy rights to discharge Vicksburg's water into the 9x3 ditch. In spite of this, Vicksburg still had internal drainage problems, and the residents still blamed The Woodlands.

The Woodlands also was the neighbor receiving Shenandoah's water. Unlike Kraft, The Woodlands did not seek an injunction but instead negotiated an easement to allow the discharge to continue. During the negotiations, Shenandoah kept expressing some urgency to conclude this matter, which we did not understand since we were not threatening to stop the discharge. Unbeknownst to us, Shenandoah needed to get the drainage easement resolved because they were planning to incorporate as a city and include The Woodlands Metro Center as their downtown. Fortunately, our old nemesis, the surveyor's error, reared its ugly head, but this time in our favor. After brief litigation, the Metro Center was excluded from the City of Shenandoah. This time, we were represented by a very prominent Houston attorney, Bob Burns, an urbanized counterpart to Max Newland. Bob didn't win you many friends, but you got a lot of respect.

Like the Ivins management team, HUD also came—and went. The Title VII Program was, like so many other governmental programs, a well-meaning but shortsighted attempt to address America's urban problems. HUD did not foresee the amount of money it would take to sustain developers of large communities over long periods of time, with all of the down cycles in the real estate market (getting in bed with real estate developers was probably shortsighted to begin with). Unlike other government programs gone awry, however, HUD chose to extract itself from the program rather than continue to throw money at it. Closing the program down should have been easy. Twelve of the thirteen projects were in default and could be foreclosed on. Most were divided up and sold. One was sold back to the original developer for 20 cents on the dollar (at least according to the developer). The

Woodlands was the lone exception. HUD, searching for some reason to declare us in default, came up with a ridiculous appraisal that seriously undervalued our property, a condition of default. We promptly commissioned a legitimate appraisal, which determined that our property had ample value to secure our bonds which had been guaranteed by HUD. When HUD finally told us they needed to put us in default so they could close down the program, we voluntarily went into default to accommodate them.

During one of the down periods in the real estate cycle, probably following one of the oil embargoes by the Middle East, Mitchell Energy's primary lender advised Mr. Mitchell to get rid of his real estate holdings. Mr. Mitchell got rid of the bank instead.

Our drainage problems with our neighbors continued, but this time in reverse. A Mexican gentleman who owned a tract of land on Spring Creek complained that the 9x3 drainage ditch was discharging debris on his land. He and his lawyer accompanied Plato Pappas and me to the site of the damage. The attorney was not feeling well (obviously feeling the ill effects of an exhausting day of taking advantage of widows and orphans and filing spurious law suits against deep pocket companies), and we stayed behind while Plato and the Mexican gentleman walked into the woods to the site of the alleged damage. The debris was obviously old and had been on the site for many years, long before construction of the 9x3 ditch. Although I could not see them, I could hear Plato reading the riot act to the Mexican gentleman. I did not hear the reply, but he must have responded with the commonly used retort, "Just remember, my friend, what goes around comes around." I assume this because Plato, in reporting this at a management meeting several days later, quoted the gentleman as saying, "Just remember, my friend, the world is round and I am coming back to you." (Plato, of course, was translating from English to Greek and back to English again. I think Greeks use word pictures rather than phonetics. Another "Platoism" is his translation of the expression "taking something under your wing." Plato's version, complete with demonstration, was that the District Supervisor of the Texas Department of Transportation was "putting our project under his armpit.") While we're on the subject of "isms" and drainage, another gentleman who will never get the credit he deserves for his role in the development of The Woodlands is Mike Page, our MUD attorney (MUD being the acronym for Municipal Utility District, not a reflection on Mike's legal work, which is impeccable). Mike came up with the concept of a large MUD which would initially cover a good portion of the watershed within The Woodlands in advance of development. As MUDs were created to serve portions of the area for development, the large MUD would detach the area to be included in the new, smaller MUD. Thus the larger MUD was titled by Mike as the "incredible shrinking MUD," or "ISM," and became known as the Mike Pageism. (For some reason, this proposal never took place. I don't think it had anything to do with our then state representative, but I do recall Mike describing him as "not the brightest bulb in the chandelier.")

I am not sure that the incredible shrinking MUD would have solved our drainage problems with our neighbors in any event. Long before The Woodlands came into existence, homes in Timber Lakes/Timber Ridge were built, some deep into the flood plain. These homes tended to flood in a heavy dew and really had problems when there was a heavy rain generally throughout the watershed of Panther Creek and Bear Branch to the north, or the watershed of Spring Creek to the west. Several days after a storm in the upper reaches of the watershed, a huge flood would find its way down the creeks, using the flood plain for which it was intended. Unfortunately, this often included the flood plain in Timber Lakes/Timber Ridge. To avoid exacerbating this problem, The Woodlands' drainage was designed to retain the flood waters in some areas, and in other areas expedite the flow so that the drainage would reach the creeks and be well downstream before the flood waters from upstream arrived. In order to accomplish this, we constructed a number of ditches from The Woodlands into both Panther Branch and Spring Creek, in the area of Timber Lakes/Timber Ridge. Unfortunately, the people who lived in Timber Lakes/Timber Ridge seized upon these ditches as the cause of their flooding.

Unbeknownst to us, and to the Texas Water Commission, a statute had been recently passed requiring the Water Commission to issue permits for all drainage ditches throughout the State, a rather formidable task. After several of our ditches had been constructed, the good citizens of Timber Lakes/Timber Ridge called this statute to the attention of the Water Commission and demanded that our ditches be filled in. This demand was made in spite of the fact that one of the ditches (commonly referred to as the "little pink ditch," because it was colored pink on an exhibit used at the hearing before the Water Commission) was constructed to divert water away from Timber Lakes/Timber Ridge which was generated by the recent opening of Glenloch Drive from Timber Lakes/Timber Ridge to McCullough High School, at the request of the residents of Timber Lakes/Timber Ridge.

Several hearings before the Water Commission followed, with an entourage from Timber Lakes/Timber Ridge arriving by bus. In one instance, they brought along their state representative, who obviously knew votes when he saw them.

At one hearing, Plato testified the first day, and the following day we had Olaf Asgeirsson on the stand as our witness. Olaf, of course, is from Iceland and, after a few minutes of his testimony, the opposing attorney asked if he was Plato's interpreter.

One of the issues in these hearings was a large amount of dirt that had been stockpiled adjacent to Woodlands Parkway for construction of the golf course. Although the dirt pile was quite high, it was located at a lower elevation below The Woodlands Parkway bridge across Panther Branch. An engineer testifying for Timber Lakes/Timber Ridge, in attempting to calculate the volume of dirt, made an error which, Plato realized, would have had the pile of dirt extending 30 feet above the bridge. When I asked the engineer if the dirt pile in fact extended 30 feet above the bridge, the Timber Lakes/Timber Ridge entourage cried out en masse, "yes".

Needless to say, all of the permits were eventually obtained, and the Texas Water Commission convinced the next legislature to repeal this statute, although I can't imagine why.

At some point during the course of these proceedings, we had to determine the history of the flooding in this area before any development. I interviewed people throughout the watershed who had lived in this area for many years and would be familiar with the extent of the flooding. One lady, in describing a particular storm, stated that the water had gotten "all the way up there," pointing to a spot about 20 feet up the trunk of a tree. When asked how she knew the water got that high, she replied that when they returned after the water went down, that was how high up in the trees they found the cows. (Life on the farm must have been rough in those days, particularly if you were a cow during the rainy season.)

Another gentleman who probably will not get the credit he deserves is Dick

Browne. Dick, like George Mitchell, is a visionary; but unlike Mr. Mitchell, he is not the boss, so his visions are sometimes subject to mockery, derision and ridicule. Early in the development process, I attended more than one meeting at which Dick would explain his concept of a riverwalk, which was met with derogatory comments. Dick, however, again like Mr. Mitchell, is a man of some perseverance. The riverwalk is now included in our plans and the first part of it is currently under construction. I recently helped Dick on another of his visions which, like the riverwalk, was turned down because it was not economically viable. Knowing Dick, it, too, will become a reality some day.

Probably the best quote regarding the history of the development of The Woodlands belongs to Skip Christie (I don't know if he was the original author of this line, but he took credit for it). In filling out our report to HUD, Skip responded to a question inquiring as to the innovative measures we had taken in planning and developing The Woodlands with the comment, "We have landscaped a forest and irrigated a swamp."

One of the problems in mentioning certain people is the risk of offending those you do not mention. Without question, hundreds of people have made significant contributions to The Woodlands over the years which somehow escaped my attention, for which they will undoubtedly be eternally grateful.

That, as best I can recall, is the history of The Woodlands at least from my limited viewpoint, which is not unlike the view of the outside world as seen by a rat living inside a beer can.

B.F. 'Budd' Clark

Today is Tuesday, January 16, 1996, and this is Joe Kutchin in the office of Bernard F. Clark in Texas Commerce Tower. We're touching up the edges of what had been lost in transcribing an earlier interview. Tell me about your early background.

I was born on September 3, 1921, in New York City and grew up in the city and went to Fordham Preparatory School, which is in the Bronx. From there I went to Fordham University. I graduated with a bachelor of science with honors in chemistry, and since I was in the ROTC, I was commissioned a second lieutenant.

What year did you graduate?

I went to preparatory school from 1934 to 1938 and to college from 1938 to 1942, getting out just months after Pearl Harbor. At the time of my commissioning, the ROTC unit at Fordham was in the coast artillery, but since I and three others had taken the courses that earned us acceptance as Professional Chemists in the American Chemical Society, we were commissioned in the chemical warfare service. I went to Edgewood Arsenal, and from there the four of us split up. I was transferred into the Air Force (the Air Corps at that time) as a chemical officer. My first station was at McDill Field, Tampa, Florida. From there I went to Barksdale Field in Shreveport, Louisiana, which was a major peacetime air base. I was there from 1942 to some time in 1944, at which time I was transferred to Eglin Air Force Base in Florida where all of the research was carried on into bombing and bombs. My job focused on use of Napalm (gasoline and detergent) in wing tanks dropped on enemy positions. I was returned to inactive status in February of 1946 as a major in the U.S. Air Corps, and was later separated from the then U.S. Air Force. I went to New York after my discharge with my wife, whom I met at Barksdale Field and married in July of 1944. Since my family was in New York, I went around looking for a position as a chemist and contacted General Electric, Celanese Corp. and Pan American Refining and Transport. I received offers from all three but accepted the Pan American offer since it was in Texas and my wife was from New Orleans and used to the South. We located at Texas City, in the Galveston area, in April of 1946 and I went to work at the Pan American refinery there as an analytical chemist. After about a year, I realized I had lost too much time during the Army, that chemistry had progressed significantly. The idea of getting a Ph.D. did not seem feasible due to the four years in the Army.

You thought you were too old to undertake that?
Yes, to catch up. It would take too long. I became aware of the Harvard Business School through my wife. I applied for admission and was accepted for the September 1947-49 class.

You were an employee of Pan American?
Yes. I took a leave of absence from Pan American and came back and worked there during the summer of 1948. I felt that from reading the literature, there was no question but the future was in petrochemicals. That's why I wanted to retain my ties to Pan American, which was a subsidiary of Standard Oil of Indiana, now Amoco. Pan American was just getting started in the petrochemical business. Upon graduating from Harvard Business School with a master's degree in administration, with distinction, I took a position with J.T. Baker Chemical Company, Phillipsburg, New Jersey, a subsidiary of Vick Chemical Company. I worked there from 1949 to 1951, at which time I heard about Standard of Indiana setting up a new subsidiary in New Orleans. I contacted the company and joined them in 1951 in the economics department. Subsequently, the company was down-sized out of existence.

When you say the company, you mean the operation in New Orleans.
Yes, the Pan Am Southern Corp., which was only in New Orleans and six states, was wiped out. I would have gone to New York as an assistant to the head of research, who originally hired me back in 1946, but who was now president of the company. However, the idea of going north with five children did not appeal to my wife, particularly since during our first winter in the North, in 1946-47, we saw 27 inches of snow in New York City in one storm. She thought that was just normal for a northern winter. Actually, there were 144 inches in Boston that year when I was at Harvard Business School. (The next year was quite warm.) So I sent a resumé to the Harvard Business School Alumni Association in Boston and to the Harvard Business School alumni group in Houston, Texas. How this got to Merlyn Christie, I have two stories. One is Merlyn's, but he was never constrained by details. He stated that he was at a bar in the Ambassador Hotel in New York, where he was a director, and he asked this man next to him, How do you find a Harvard Business School graduate? The man, according to Merlyn, told him, Well, contact the Harvard Business School in Boston. The other ver-

sion is the résume I sent over here went to the vice president of First City Bank who sent it either to George or Merlyn—I think it was more than likely Merlyn. So this was a quirk of fate. Wherever the résume went, it ended up with a phone call from Merlyn Christie asking me to come over to Houston. I told him I was coming over in a couple of weeks to see Conoco at their invitation, at which point he said, "You want the job or don't you?" I said, "OK, I'll come on." So I met Merlyn and George Mitchell and Johnny Mitchell. I joined them on September 15, 1956. At that time, Christie, Mitchell & Mitchell was a Texas oil wildcat exploration company with investors. They had been drilling in south Texas and the first hit was at the Vienna Field. The arrangement was that Merlyn had the connections with the Texas Commerce Bank and Johnny had the connections with the Jewish businessmen in town and George was the geologist and was considered one of the best on the Gulf Coast. He had graduated first in geology and petroleum engineering at Texas A&M in 1940. The Jewish group, which was called the Big Nine, were the leading merchants in the city.

Were they all merchants?

Yes, the two Weingartens were grocery, Lack was auto parts, Oshman was sporting goods. Mr. Zinn, I'm not sure. Irving Alexander, he might have been involved in real estate, but he was also connected with the Weingartens. Anyway, there were nine of them. The big change occurred when a deal was brought to George covering Wise County in north Texas. This area had been drilled over many times since the 1920s, and it was no longer of any significance. The thing that caught George's attention was that the Boonsville Field was a stratigraphic trap (actually the deltas of many small streams) in the Fort Worth Basin. That meant you could have wells with as many as six possible horizons that produce oil and gas, each one having been deposited at a different time. The course of the streams varied, so while you might not hit what you were drilling for, you would hit something. George agreed to a proposal from John Jackson and Ellison Miles and they started leasing up the county, plus part of Jack County and part of Parker County, until they had about 300,000 acres under lease.

Can you put a time on that?

This was 1953 to 1958. They drilled many wells, and near the end of 1957 they started delivering gas to the gas plant. As a matter of fact, the pressure for capital got so intense that the Big Nine said, "Thank you, George, we've had enough."

Pressure in the sense of financial pressure?

Yes, everything was going great and it was profitable, but the need for capital was just limitless. For them, this was a hobby, not a primary business, so they just said, "We'll hold what we have and we won't claim any right to be in what comes in the future." They never did. How they got in touch with new investors, I'm not sure. George would know. It could well have been Raybourne Thompson, who was one of the top attorneys at Vinson Elkins and a friend of Merlyn Christie and George Mitchell. He had worked and had contacts with the FPC, now known as FERC. So whether they knew him before they retained him for that or as a result of retaining him learned about his contacts, I don't know. Raybourne was close with Graham Mattison, who represented Waterford Oil Company, which was in the estate of Barbara Hutton. As a result of Waterford's joining in the drilling, the estate of Stephen C. Clark, which was the Singer Sewing Machine estate, also joined. George contacted R. E. "Bob" Smith to join and I believe through Raybourne, John Riddell came in. As it turned out, Riddell took 25 percent, Waterford took 37 percent, and I'm going to end up with too many percents, and R. E. "Bob" Smith took 37 1/2 percent, and Stephen C. Clark took 12 percent. Oil Drilling and the Johnny Mitchell partnership took positions. Riddell decided not to enter the gas plant deal. He did not take a position in the gas plant, but Warren Petroleum had 20 percent in the gas plant although none in the drilling. Basically, they agreed to a percentage and then they took all deals in that percentage.

Just to wrap that up. That was the early financing for drilling Boonsville, is that correct?

Yes, that's where the money came from. This happened before I came to the company. The big key for Christie and the Mitchells was that the head of the energy department of the Bank of the Southwest had been George Mitchell's engineering professor at Texas A&M. He might have been head of the department. He did something that was done very rarely, if at all, in that he lent George money on the basis of the logs showing what reserves were behind each well, even though the well wasn't producing. In addition to Oil Drilling, he lent money to the three partners. Further, George and Raybourne worked to have Natural Gas Pipeline Company advance money in the form of production payments to be paid out after the wells were on production. This amounted to $7.5 million for all of the investors. Initially, the Big Nine was in the first or second financings, but then even that wasn't enough help. So that's about it.

* * * * * * *

This conversation with B.F. Clark took place on July 12, 1995, at his office in The Woodlands. He was asked about early investors in the company and begins his response with comments about the estate of Stephen C. Clark.

The Clark estate actually was the Singer Sewing Machine estate, and they followed behind Waterford which was Graham Mattison at Dominick and Dominick investment brokers. So when you got one, you got them all. I think it could have been Raybourne Thompson at Vinson Elkins, one of their top lawyers, who dealt with Merlyn and George, who got them in as investors. After they were in, George got R. E. "Bob" Smith into the drilling syndicate, where they committed to a percentage of the cost of every deal that came up.

We're talking specifically about Boonsville?

Well, no, not only Boonsville. Once they had this group, they were more active in South Texas, too. So with that, the Big Nine just held their position, and these new investors came in. Among them, Smith, Riddell Petroleum, Clark, and Waterford took a combined total of 75 percent, and the remaining 25 percent was split, 15 percent to Johnny Mitchell, Trustee, and 10 percent to Oil Drilling, Inc. The background on those two companies is that Oil Drilling, Inc., was a corporation owned 50 percent by Merlyn Christie, 37 1/2 percent by George Mitchell, and 12 1/2 percent by Johnny Mitchell. Well, this was an unequal distribution, so they formed a partnership called Johnny Mitchell, Trustee, in which they each owned one third. Johnny Mitchell, Trustee, took 15 percent of each deal, Oil Drilling took 10 percent. (And as a matter of fact, it was not long after I was here, about mid-1957, that even these new deep-pocket investors were crying uncle and came over to have George slow down the process.)

You mean the wells were not connected?

Yes, up to the end of 1957. George had decided he wasn't going to sell gas for 10 cents per thousand cubic feet to Lone Star. He built his philosophy on two facts, or to him they were facts: First, there would be inflation as far as the eye could see, and second, the price of gas had to approach equivalency with oil, which it wasn't. With those two ideas, his main thrust was to accumulate as much reserves and land holdings as possible. And that paid off finally in 1973 when the price of oil exploded and all these properties that were hardly worth running were there and we had them and could then drill additional wells on them, and we were in a very good position. So his basic long-term strategy was, you might say, brilliant because it worked. If it doesn't work, it isn't brilliant. Anyhow, it might have been through Raybourne Thompson, who handled negotiations with Natural Gas Pipeline from a legal standpoint, that George negotiated a deal with Natural for 14 cents per Mcf between 1953 and 1956.

We're talking specifically Wise County?
Wise and Jack and Parker Counties. In 1953 the Supreme Court ruled that the Natural Gas Act of 1933 applied all the way back to the wellhead. The result was that the deal with Natural had to go to the Federal Power Commission, the forerunner of the Federal Energy Regulatory Commission. And they came back and said no, 14 cents was too high, lower it. So George made a deal with Natural for 13 cents, but they had to put in the pipelines, the gathering system. Now this was very important, because that system cost $35 million, which George and his partners would have had a hard time raising. It went right through the FPC, though. They had gotten their penny off, and they didn't know what a gathering system was anyway. One of the requirements was that all gas that they took out of this area had to go through a processing plant. You had to have a way of taking the liquids out because otherwise the liquids would just block the line going to Chicago, and in the winter it would freeze, so you had to have a processing plant. We went ahead and built the plant, with some help from Bob Smith. His main bank was First National Bank of Chicago (as also was Warren's), and he went up and told them this was a good project, and those were the days when that was about all the collateral you needed. They agreed to lend $4.3 million to build the plant.

Prior to that there had been no plant?
No, nothing.

So the gas was being produced, liquids and all?
There was no production. The gas was just sitting down there. That's what was killing the investors. One of the first things I was involved in at the company was working with the bank on the loan for the plant. George and Raybourne Thompson had the most to do with it. We borrowed $4.3 million from the bank.

Clarify what the Mitchell corporate entity was °at that time.

Johnny Mitchell, Trustee, and Oil Drilling, Inc., were the investment companies, and Christie, Mitchell & Mitchell Company was the operating company for all the investors. So, they—Smith or Waterford, for instance—had a position in Christie, Mitchell & Mitchell Company as an operator because it operated for everybody. And that company was a shell . . . well, it was more than a shell because all the employees were in it and the furniture and fixtures, but other than that it had no interest in the properties. It just operated on a fixed fee. Drill at cost and operate on a fixed fee.

But it would be Christie, Mitchell & Mitchell that is the ancestor to Mitchell Energy & Development?

No, Oil Drilling is. Christie, Mitchell & Mitchell Company is the ancestor to George Mitchell & Associates. That was just a change in name and we had George Mitchell & Associates up till when we collapsed it into MEC. Oil Drilling, Inc., bought out the partners' interests in Johnny Mitchell, Trustee, and at that time, Christie, Mitchell & Mitchell Company became George Mitchell & Associates, once again keeping an operating company between the principals and the investors. They were allowed promotion. None of this applied in Wise County. Wise County was pretty straightforward. So Christie, Mitchell & Mitchell became George Mitchell & Associates, and Oil Drilling, Inc., became Mitchell & Mitchell Gas & Oil. Now that's a little bit down the line. We're still in '56, when they started working on the plant. It was completed in December of '57. At that point it was Warren Number 53 plant. Warren Petroleum had 20 percent of the company. Of the gas plant, 20 percent went to Warren because they were going to run it and the old man Warren was in with the First National Bank of Chicago, so with Bob Smith's recommendation and Warren's recommendation, there was no problem on the financing. So on December 27, 1957, the first gas flowed to Chicago under the contract with NGPL, which was to expire on January 1, 1997—it was originally a 20-year deal, and there was a 20-year extension. Shortly after, the first gas flowed through the plant for processing. There had been three original partners in Oil Drilling, Inc. George was definitely the brains and did call the shots on all the operations. Johnny was supposed to be Mr. Outside to investors in the Houston area, whereas Merlyn Christie was supposed to be the financial partner, the partner who does the financing. Actually George did the financing. I know because I worked with him on it. Merlyn also handled the investors like Waterford, Clark, Riddell. Nobody handled Smith but George Mitchell because Smith wouldn't deal with anybody except him.

Would he deal with you?
No, never. George Mitchell was the only one. Merlyn had done some work for him in the '30s up in East Texas picking up leases, but that has nothing to do with this company. Johnny, of course, got the publicity and he brought in the Big Nine.

I have the impression that back then George was the inside man, that he was not at all as aggressive as he is today.
Oh, yes. George avoided all publicity. He was just work, tennis, work, tennis, work, tennis. He did all the geology. We had geologists, but basically he had the final say on getting leases, setting the terms and conditions. Of course, the other partners could give their opinion. It'd be a joint decision. Most of the time they'd agree.

You were not in on any lease acquisitions?
There was no general manager prior to the time I joined the company. I believe there was a vice president. He left as I came in. I stayed out of operations, and that is true to this day. I worked with the accounting–the audit reports had been coming out six months late, and the investors were being billed three months late, so we put in an estimated billing to get the money first. They weren't happy about that at times. So it was a matter of bringing the accounting up, although I wasn't an accountant. I changed the people around to the point where one man, Jim Harrison, was chief accountant. Armintrout was the man I replaced. A Mr. Ploog was the treasurer, but I replaced him in time. I worked on financing with the Bank of the Southwest, which was very important because you couldn't get a loan on a shut-in well unless you had other collateral. Where's the income? What's going to pay the loan? George had a professor at A&M, Harold Vance, and he became senior vice president for energy at Bank of the Southwest. He saw George's strategy and the sense it made and he authorized lending George the money. Then George, with help from Raybourne Thompson, got advances from Natural Gas Pipeline Company of America. George would tell them, look, we can't hold this deal together while we get the negotiations done and the plant built. So they gave them production payments of, I think, $6 to $7 million.

Let me be clear on something. You're saying production out of North Texas started 40 years ago, or it would be 39 years ago?
Well, it would be 39.

And something else. All the leases in North Texas had been substantially acquired by the time you joined the company?
No, a fair amount had. The Hughes Ranch was the key. Ellison Miles and John Jackson had that lease and brought the deal in to George. They had tried to shop it elsewhere. Now the story goes that a gambler in Chicago brought the deal to George, a bookie.

It wasn't the gamblers from Galveston?
No, in my time with the company George dealt directly with Jackson and Miles—there was no third party. However, that all happened over three years before I joined the company. You will have to ask George about the facts. In any event, it boiled down to George seeing the merit in looking at all of Wise County. Wise Country had been drilled and was a big coal producing place in the '20s. I've heard that they kept finding gas, but they just capped the wells. It wasn't worth completing them. And there were oil wells there, too. Wise County had been drilled for 30 years, but the returns were small, and in those days if you didn't bring in a gusher or even find a big field, you'd just throw the fish back. But if you found enough oil or a big enough fish, it was a pretty good deal. So, George saw that the Fort Worth Basin underlay this whole area, which was really (and I didn't learn this until recently) a collection of rivers, creeks, where the sands built up. But George saw that he wasn't just dealing with a well here or a well there.

When would that have been?
This would have been '52. It was '53 when they started drilling, so they had to be negotiating. Possibly early '53. And Ellison and Jackson already had a deal on the Hughes Ranch and he had four drilling rigs. Jackson, a geologist, had been a high school science teacher. Ellison got an exclusive on the drilling, so they went out and drilled Hughes Number 1. They eventually drilled eight Hughes wells, and those came in. Those were gangbusters. They were the best wells, I think, that ever hit in Wise County. So when Jackson and Miles came in to lay out the idea, showing all this area, Jackson had done a lot of grunge work, you might say, which all geologists have to do to find a lead. George saw immediately that this was a big deal. Ellison and Jackson each got an override on all leases starting at one-sixteenth. It gradually reduced as the payout grew bigger.

But did George own or did the company own some acreage in North Texas prior to this?

No. North Texas came specifically through Jackson and Miles.

Then the deal with the bookie, or the gambler, would that have come later? I'm trying to get this straight.
It would have been at the same time.

Well, George, anyway, had the insight that here's this big Fort Worth Basin, and he had the attitude that things are going to appreciate, gas is not going to stay at 10 cents forever, oil is going to go up and gas is going to go with it.
There's going to be inflation, so what money you borrow today you'll pay with cheaper money when you pay back tomorrow.

He built this huge land position in North Texas and then indeed the things he saw came to pass, increased by the deal on the liquids.
Yes.

Enhanced by the deal on the liquids because that was just kind of gravy on top of the whole package.
Yes, but he knew in this case that the gravy was the best part and the gas didn't matter. He made more on the liquids than on the gas. That may not be exactly true. I'd have to do some calculations. At that time, gas was 10 cents and propane 3 cents–2 cents in summer and 5 cents in a cold winter. Oil was about $2.80 and condensate was $2.50.

I think when I came to the company, it was around three dollars, and George said, If we ever get oil up to five dollars, it would be so wonderful.
Just to recap a little, they had a lease position, obviously, when I joined the company, or they wouldn't need 250 people. I think the difference may be that Jackson saw something in terms of alluvial, or river-like, deposits, maybe just around the Hughes Ranch. George looked at it in terms of three counties.

* * * * * * *

Today is Wednesday, July 19, 1995, and this is Joe Kutchin again. We resume the interview of Budd Clark in his office in The Woodlands. We were talking about North Texas when we finished our first session.
Yes, but first a couple of things that came to mind after we talked. One, the

Harvard Business School alumnus who got my résume here was named Grover Ellis and he was with First City National Bank. And as for ownership of the plant, Warren Petroleum got 250,000 shares of stock at $1 a share and $500,000 in notes on a ratio of 2 to 1, versus the stock that the investors took.

2 to 1 what?

Two dollars of debt to one dollar of equity. And this in a way reduced future taxes. Warren had 20 percent of the plant, I think Signal had 7,500 shares, and Jackson and Miles each had about 3,000 shares. Now, one of the big problems with the Wise County operation was that it consumed capital at an alarming rate.

When you say operation, are you talking abut the whole thing?

Overall, yes. Taking the leases, drilling, operating the plant, etc. We dealt with the demand on capital by doing farmouts. Capital was the reason that the bigger investors almost followed the route of the Big Nine. It got too heavy for them. We did find another investor in Northern Illinois Gas Company, or NIGAS. One outstanding farmout was to Whitehall Corporation, which was basically a syndicate put together by the Empire State Bank of New York. They were putting $2 million into drilling for which, in those days, we were able to drill 50 wells. They came in October 1958, I think. We started drilling (on the same terms as below) in November, and had all 50 wells down by the end of December, which was necessary from a tax standpoint, and then completed them in the subsequent year. Another way of relieving the strain was cash flow from the gasoline plant, the Bridgeport Plant, which was now making a substantial profit. With reference to Whitehall spending only $2 million for 50 wells, a million dollars went a long way. The plant program was in groups of 10 or 15 wells. While the plant was getting its money back, it would pay a 15 percent override to the investors, and after it got its money back it would pay 85 percent of the net profits to the investors.

Summarize who the investors in drilling North Texas were.

At this point it would be the Big Nine, if some of the acreage in which they still had an interest was included. That acreage would be acreage inside a gas unit which was drilled at the time they were active. They didn't take any new deals once they said to stop, but in the meantime they'd acquired acreage. They also had these wells which were on 360 acres and eventually would go down to 180 acres. Then there were also Smith, Clark, Waterford and Riddell.

And what was our corporate entity?
Oil Drilling and Johnny Mitchell, Trustee.

All these together were the investors?
Right, so each package of these wells had a different ownership which to this day still exists in the subsequent drilling and re-drilling of these wells. It's gotten to be quite involved, but somehow or other, the land department has kept up with the ownership. These were followed by I don't know how many farmouts, but they went on until 1968. In the meantime, Waterford had quit, which meant Clark had a lot of undeveloped acreage. Basically, we told them that if they couldn't drill it they had to forego it. So, as well-drilling deadlines would come up, they would give up the acreage, and NIGAS, who we brought in, got a very ground-floor deal because we just assigned the acreage to them with whatever Waterford's costs were in calculating our net profits.

Come back to the Big Nine. Did they act as a syndicate?
They were individuals, but they all talked to each other, all very close friends.

So actually they were investing as individuals?
They were investing as individuals but acting as a syndicate because when the lead horse would go, all the others would follow. In the late '50s, when the Waterford, Riddell, Clark group had become unhappy with the heavy pace of drilling and a feeling that the program wasn't being operated properly, they brought in an audit firm, Touche, Livin, Smart and Bailey. Smart is still there. They spent a year and finally they could find nothing that was out of line, and at that point, this group talked about buying out Christie and the two Mitchells. I think it was strictly by chance that a loan officer from the Chase Manhattan Bank who knew Johnny Mitchell was just dropping in to see what was going on, and was told about their efforts to buy us out. It turned out that Waterford had a production payment from the Chase which was in default. By making a production payment to Oil Drilling and Johnny Mitchell, Trustee, the Chase could retire the Waterford payment which was on other properties. With that, George, Johnny and Merlyn bought out effectively the oil interest of Riddell, Clark, and Waterford. Bob Smith did not sell at that time. What this meant was that 85 percent of the

cash flow would go into retiring the production payment. So it didn't create any new funds in that sense. This was, more or less, the start of George's overall next strategy, which was to take over all the partners. He did not like to work with partners.

You're talking about liquids?

I'm talking about anything, anything he was in, if he could do it. So he went after the Waterford, Clark and Riddell group and bought them out through the Chase.

We're still in the late '50s?

This was 1959. Riddell stayed in on the Gulf Coast; he sold out in North Texas. Smith did not sell out anything, and Waterford and Clark sold out across the board. Riddell kept certain properties in the Gulf Coast. The next logical group to be taken over consisted of the other two partners. Beginning in 1962, Merlyn Christie wanted out—he was older and did not want to risk what he had acquired over the years by remaining active any longer in an increasing program. So the procedure that was followed was this: The properties involved in Johnny Mitchell, Trustee, were put into Oil Drilling. In effect, Oil Drilling purchased the three partners' interests, except that wasn't completely true in Johnny Mitchell's case. The gist of the whole deal was that Merlyn was paid roughly $4.2 million cash plus assumption of his share of debt on company properties. That was the beginning of 1962. The company became Mitchell & Mitchell Gas & Oil Corporation. George changed the name. I had moved from general manager in '56 to vice president in '58, and George made me executive vice president when the corporate name was changed. George owned 70 percent of the corporation and Johnny owned 30 percent. The reason for that was, Johnny retained his interest (I'm uncertain on the acreage) in the plant farmouts, which at that point would have been the first and maybe the second or third such farmouts.

I don't understand how you have a plant farmout. I know about an acreage farmout.

I meant a farmout to the plant. A plant farmout is a farmout to the plant.

All right.

George and Johnny then created M & M Investments. It was in subsequent plant packages. M & M Investments took farmouts.

The creation of M & M Investments—was that Mitchell & Mitchell Investments?
Yes.

Did that get rid of a lot of the other corporate structures?
There were no corporate structures. (Johnny had an interest in the acreage [see above]). But the main reason for the differential in the two properties was that Johnny did not put all of his interest in Johnny Mitchell, Trustee, so the ratio which would have been 46-54 became 30-70 because he didn't put as much in. Johnny had 30, and George had 70. Then George started after the Big Nine, and over a space of four or five years, they fell one by one, which brings us up to 1970. George had been after Bob Smith and at this point, Bob was willing to talk about selling. One of the problems was that if Merlyn sold out, he insisted on a press release indicating that he was paid $16 million. It was $4.2 million in cash and the rest was in production payments and debt taken on by the corporation, not only in his interest, but Johnny and George's, too. So the value for the Smith property was about $7 million, and Smith said there was no way he was going to sell for less than Merlyn. So, in recalculating it over a 12-year period, if you took the gross amount without discounting for present worth, you come to a $12 million number. This was satisfactory to Smith, and he was bought out of his oil and gas properties and his plant shares, which brought the ownership close to 80 percent. To bring it where the plant income would shelter the tax write-offs of Mitchell & Mitchell Gas & Oil, you needed 80 percent ownership of the plant. At that point, we started working on going public, and we started on buying out Warren Petroleum, which we did. We bought out Warren for some cash and also a contract tying the products to Warren, which had subsequently been made a division of Gulf Oil. Warren had originally been an independent back in the early '50s, but Gulf bought it and kept the old name up to 1970, or somewhere in there.

Let's go back on one thing. Before we started talking last week, you were remembering back to Roxoil.
Roxoil was formed by Roxie Wright, Jimmy Gray and Merlyn Christie. Roxie Wright was the driller. He had what they called ironworks. I'm not sure whether Gray and Merlyn might have bought into Roxoil and then the name went from Roxoil to Oil Drilling or Roxie had Roxoil and then he joined with the other two and formed Oil Drilling, as a new corporation. I vaguely remember seeing the minutes, the book with the minutes of Roxoil,

20 years ago. Anyway, you end up in 1947 with Oil Drilling, which is owned one-third each by the three partners. Then Gray sells out his ironworks to American Machine and Foundry—Roxie didn't want to buy. Merlyn went to buy out Jimmy and took half of Jimmy Gray's one-third, which would give him one-sixth of 50 percent. He let George buy the other half. Up to this point George was doing geological and consulting work. Then in about 1952, Roxie sold out. I think he must have died soon after, because I remember dealing with his son, Dick Wright. After Roxie's death, George and Johnny bought out his share from his wife. In the end, Merlyn had 50 percent, George had 37 1/2 percent, Johnny had 12 1/2 percent. They then formed Johnny Mitchell, Trustee, which was owned, one-third each, by the partners. They also formed Christie, Mitchell & Mitchell Company, which was the operating company and which was created for one reason—that any liability from the operations wouldn't automatically fall on the partners.

What has happened in the past hour-and-a-half of tape, more or less, is that you talked about the status of the company and the different steps along the way and the corporate organization. Summarize a bit for me. Give me the high points that come to mind.
Well, you've formed the company and then George joins it. Then Johnny joins it and they have a side company, Johnny Mitchell, Trustee, one-third each. Oil Drilling takes 10 percent of the deals and Johnny Mitchell, Trustee, takes 15 percent. Christie, Mitchell & Mitchell Company is an operating company which handles all the operations for the partners and their investors. At this point you're talking about drilling down here and South Texas, all over.

But we're not talking Wise County yet, right?
No, Wise County hits about 1953.

All right, continue.
And Miles and Jackson come in with this deal and George sees the merit and also the potential in it and signs on, and in the meantime has acquired investors, pretty much on the basis that they will take everything that's proposed as long as the partners are in it. And that's the Big Nine, which is the local Jewish merchants in Houston. Pretty soon George is gobbling up this acreage, up to 300,000 acres which he reached, I'd say, by 1958 or so. And this was too much for them, so they pulled back and sat on what they had, plus any drilling that affected what they had. Everybody has that right. Then

these bigger investors, Waterford, Riddell, Clark and Smith came in. Then they were run sort of into the ground by the magnitude of George's scheme of drilling, as evidenced by one farmout to the Whitehall Group of 50 wells which were all drilled in two months' time at the end of 1958. Plus other farmouts which helped relieve the pressure. Then, finally, George bought out the three, Clark, Riddell and Waterford.

Say something here about the Supreme Court decision.
After the formation of Christie, Mitchell & Mitchell Company, George saw the value in Wise County and started drilling but had to shut in the wells. Fortunately, there was an energy banker at Bank of the Southwest who had been George's professor at Texas A&M. He saw the logs and the fact that the wells were good, so he had his department lend against them which was highly unusual. Also, George then started talking with Natural, since Lone Star was only paying 10 cents and was taking more or less on a demand basis. Otherwise they'd take a minimum amount.

George was looking to get 11 cents?
He was looking to get more than that. He made a deal with Natural for 14 cents, but this was 1953 and the Supreme Court ruled in Phillips against the State of Wisconsin, I believe, that natural gas was controlled all the way back to the wellhead. At that point, any negotiations had to be approved by the Federal Power Commission, which had been in existence to control the pipelines. Well, Mitchell had a contract for 14 cents and the Federal Power Commission said that was too high, so George cut it to 13 cents with the provisions that Natural would do all the gathering plus all the gas they bought would be processed at one gas plant. So two things were extremely important: One, having the pipeline and gathering system put in by Natural; second, all the gas Natural bought would have to go through. Eventually, it was necessary to go public to raise cash in order to meet the limits set by the banks. Eastman Dillon had shown an interest in working with us. Bill Houck knew one of the top partners in corporate finance there, who had been the chief energy officer at Chase Manhattan in earlier years. We worked on a registration statement with the local representatives of Eastman Dillon. It was especially difficult since there were so many related-party transactions between the partners back and forth in the corporation, and that made people leery, but we convinced the attorneys there was no problem in it and we were ready to go. This is '71, I'd say June. A man named King in New York was the top guy working with us. He went into the cor-

porate finance committee and they voted him down 24 to 1. I think those were the numbers. They suggested that the company sell assets and pare down debt, so the two young representatives here came over and saw George and me and told us what their committee had said and we said we didn't have to go to New York to be told that we should reduce debt. There was no unpleasantness, but the meeting suddenly ended. Previously, a man named Shaker Khayatt came by here in 1968. At that time he was with Eberstadt and Company and he talked about doing a private placement with Travelers. I actually went up with him and talked to Travelers. Then Eastman Dillon came in. So when they came back with this advice, I called Khayatt and I said, "OK, you've got your deal back." At the time, he was going to the opera with some undersecretary of the Navy. I told him we had all the papers, we had everything done, and then shipped them up, sent a young fellow by the name of Clark, no relation, to New York and he met Khayatt at the theater and handed over the briefcase to him. The following Monday, Khayatt said he had three offers of interest. One was from Eberstadt, one was Wertheimer, and I don't remember the third. Of the three, Eberstadt was the most likely since it was the most flexible, so we told him, "OK, we'll go with them." We were asking $20 per share at this point, and then Eastman Dillon came back Tuesday or Wednesday and said they would do a preferred stock offering. We said, "It's too late, we've already committed to this other firm." So this goes on, and in September we're ready to go public. When I went on the due diligence road trip, the market was at 970. In November, when the syndicate was put together and it was priced and put out, the market was 720. Obviously not a good time, so the syndicate said, "No way, forget it." So it was still $20 at that point. Then, in December, Walter Lubanko of Eberstadt and Khayatt came up with some deal of warrants, debentures and stock, and I said, "Forget it." Then in January, the underwriters wanted to wait until we had a new audit, and I went up to New York with Houck and I said, "We can assure you there is nothing new going to come up in an audit," and finally I said, "Do you want the damned deal or don't you?" They said all right, they'd go, but then they cut the price. Khayatt was pushing for 13 1/2 or 14 and the Eberstadt crowd was pushing for 12, and it was settled at 13 1/4, 770,000 shares. So that morning they cleared the Blue Sky laws in all the necessary states and they put it out. Members of the syndicate were carefully asked, were they firm on what they wanted, did they want additional shares? Bear Stearns swore they had it right. I remember having lunch at the Wall Street Club. A phone call came that the stock had gone public at 11 a.m., and here at 12:20 Bear Stearns dumped 25,000 shares, their allotment, on the market. At which point all hell broke loose. And I think Khayatt and Lubanko went to Loeb Rhoades and convinced them it was a good deal. I think Loeb Rhoades was in the deal, but not at the top. Word got out on the street that Loeb Rhoades was supporting us, and the stock ended at 13 1/4. It had been down in the 12s earlier.

To wind things up for today: Operationally, we've been in the oil and gas and liquids business, mainly in North Texas. Have we entered the real estate business yet?
We entered the real estate business in 1962, it seems to me, when George started Pirates' Beach. That's when he bought the land.

Thank you.

* * * * * * *

Today is Wednesday, July 26, 1995. This is Joe Kutchin with a continuation of the interview of B. F. Clark in his office in The Woodlands.
The first of the corporate transactions in real estate was done in 1962 and was the purchase of land on west Galveston Island which would be the location of Pirates' Beach and Pirates' Cove (the bay side) of a recreational second home development.

We had been a successful energy company?
Yes.

Why go into real estate?
Fortunes were being made in real estate. This is my opinion. George may give you another story, Oh, well, we're an oil and gas company and we have the lease. Why not buy the land? Back in the late '50s, the three partners bought 160 acres, or maybe 300, in Montgomery County. By strange coincidence this land became the industrial park which is now a part of The Woodlands. As I said, this was surely a coincidence, because at that time, I don't believe George was thinking of The Woodlands. So I would say, Bob Smith made money in oil and gas and then George saw that he made a ton of money in real estate by buying up almost all of Westheimer Road from Hillcroft practically out to Fondren. Inasmuch as George dealt personally with Smith on his oil and gas investments in the company, he would be familiar with how well Smith's real estate went and could have been influenced by it. That's my answer.

Another version I've heard is that the price of oil was then $1 or $2 a barrel, something like that.
It was up from $1.25 in the '40s to $1.30 in the '60s.

So George was looking for diversification. You wouldn't give credence to that?
No. The price of oil was frozen during World War II. Then after the price controls of World War II were removed, the price more or less leveled out at $1.25 and moved up over $2. At the time of Pirates' Beach, I believe we were getting $3.10 for oil and $2.85 for condensate, which is what drops out of the gas stream. So there was no need there to get into real estate. Plus, there has to be some other reason for the real estate because real estate and energy are both capital intensive investments and this was a capital deficit company from day one. It was always like this. So getting into another one where the returns are far off just didn't diversify anything. As I say, it's just my opinion.

So we bought Pirates' Beach.
Soon thereafter, George found out that the company owning Pelican Island, a national company–something like DeVoe Paint–had bought the island and put in the bridge to the mainland (the mainland being the island of Galveston), and in all had put $24 million in it. This might have been '63. The company owning it didn't proceed further with it since, I imagine, it was over budget and George learned that they had made a transaction where they had a large gain. So he approached them on selling the island and offsetting the loss against the gain to minimize their tax liability. The company paid $7 million for it, all borrowed. There was a road as far as Todd Shipyard. He gave acreage to the Texas A&M maritime department and extended the road on down to what was called the Quarantine Station which hadn't been in operation for decades. And he sold off tracts to industrial customers needing a water location. In the meantime, unbeknownst to anyone in the company, George was working on a purchase of the Grogan-Cochran lands in Montgomery County.

The company no longer owns Pelican, is that right?
We still own it. After the initial flurry of sales for about two or three years, interest died out. It's just still sitting there. It's used by the Corps of Engineers as a spoils dump, which is important because such areas are becoming scarcer and George has hopes of trading it to the Corps of Engineers for

land on Galveston Island itself.

Let's go back to Montgomery County.
George, with an attorney and a forester who was with Grogan-Cochran, put together the entire deal on the 50,000 acres. No one in the company had anything to do with it. I didn't even know of it until it was almost consummated. I believe it was—what is the number he gives out about how many transactions?

300, I believe.
Something like that. As I say, he worked it entirely out of the company.

50,000 acres includes part of what is now The Woodlands, plus timber properties?
Right. He paid $6 million, which was $125 an acre, and there was $85 of timber on each acre. So essentially he was paying about $40 an acre for the land. He bought it on notes from the company. There were four family groups. These notes varied at 10 years, 12 years, and some were even 15 years. We borrowed $250,000 at Bank of the Southwest. Actually, collateral was the company's interest in a plant package in Wise County. Then to get the financing simplified and also have a strong partner, we tried in New York to see the Whitney Foundation and New York Life (or it might have been Mutual of New York) and Wall Street investment bankers, to see if we could get someone who would meet the obligations of all these notes over the succeeding 10-plus years for a 50 percent interest. We were not successful, but in coming back to Houston, George approached the management of a local insurance company and actually banked the total amount of the notes. They were familiar with the land and realized its value. In other words, he was able to borrow the face amount of the notes from Great Southern Life Insurance and pay 6 1/2 percent interest for the first 10 years with no principal payment and 6 percent interest in the second 10 years with principal payments due over a 10-year period. You can't do much better.

Put this in time again.
We closed it in '64, and then we went hustling. I'd say 1965. About a year after closing.

At that time he was not planning a new town development, as far as you know?

You don't know. I remember reading something that might reflect his intent before George told me what he was doing and it went back to the early '60s. There was a study put out that the population of the U.S. would grow by 30 or 50 million people over the next 30-40 years. And the question was, where do we put them? Do we put them in the same old tired cities? Or do we put them out on the farm or extend the suburbs even further? And the answer seemed to be the planned city, which had been done in England and Finland. I do believe two lines of thinking were there: 1) he would have bought Grogan-Cochran whether he had the idea of the city or not, and 2) because he had the idea of the city, he bought Grogan-Cochran. I think the two are intertwined. He immediately started on blocking out the land. He walked the entire acreage, I believe, over a period of time. He then decided there were about 4,000 acres that seemed to him to be the best location. Next we traded 12,000 acres in Grimes County for 4,000 more next to the Grogan-Cochran acreage, and then George started picking up large blocks that either were owned by some of the leading individuals in Houston or had been given, in one case, to a Catholic church. In time he was able to build up 25,000 acres.

The total land holdings of the company by that time had gone to what?

About 65,000 acres. We started dealing with the new town department of HUD. There were about eight or nine others being done at the same time. A lot of those others just had options. They were a gleam in the eye of the developer. But here, the land was under contract and was owned, although there was debt on it. HUD wanted a guarantee that we would go to an auction on bonds to finance the land purchase, rather than negotiate the deal. HUD felt we could get a lower rate, and we did auction them off at 7.1 percent interest, 20-year bonds, with no sinking fund until about the 15th year. The unfortunate part was HUD insisted that all mortgages be paid off on the land, that they would have to have first claim on it. So we did away with the 4, 5, and 6 percent interest we had been paying and instead had to pay 7.1 percent. The deal also denied the use of much of that money for development, so the HUD guarantee had a double effect.

I never did understand that.

The government was backing those bonds and they wanted to be sure their investment was safe.

I just assumed that in 1972 we had $50 million to develop the place.

You have to remember that the land involved wasn't just only Grogan-Cochran. It was all these little tracts. I forget how much was left out of the $50 million after paying off the first mortgages. I doubt it was more than $10 million, plus or minus $5 million. So about this time, George brought in a group of men who'd worked in Columbia, which was a successful new city. Private, not HUD-backed. A James Rouse development.

This was about 1970,'71,'72?

1972. There were five, six or seven men, maybe as many as 10 or 15, who formed the nucleus down here. We had no one fully experienced in a project of this size whatsoever. In the course of time, the legal department of the company was split between energy and real estate. Real estate got their own financial staff, their own legal department. The people who had come from Columbia had relationships with financial people on the East Coast. They went to Chase Manhattan Bank, which had been our energy bank, along with First National of Chicago, for 10 years. Unbeknownst to the energy people at the bank, our real estate department was borrowing from them. The real estate department was on the 17th floor. Energy was on the third. Unfortunately about this time—'73 or '74—Chase had its own problems. Banks were overloaded with oil tankers that they couldn't peddle. Then Chase had gotten involved in real estate development in Puerto Rico which demanded quite a bit of funds. A typical thing with real estate is that it overruns budget. Then there was a hotel in Athens, Greece, where they advanced $25 or $23 million to build, and through an accident they guaranteed an overrun of $23 million. At this point, the real estate department of Chase was not in great favor with top management. So, our people went up and explained that we needed additional money (that was our real estate people) and they went up to that group and explained the need, at which point, the real estate department of Chase suddenly realized they needed help. They went running down to energy, which was energy's first notice that the other half of the company was borrowing upstairs. That, of course, brought all kinds of ill will from the bank.

This would have been '74?

Yes, the big opening of The Woodlands was in October 1974. The blowup with the Chase came in December of '74. At the time we had planned to borrow $10 million from First National of Chicago and $10 million from Chase, but after this blowup we had to cut back to $5 million from each at a time. Eventually, we got $20,000,000 total—the amount we had originally requested. First National of Chicago had no problem with the company since their collateral was the interest in the wells. Chase realized their loan to us had suddenly doubled overnight (by real estate loans), although actually it hadn't been overnight. They lent $17 million on the conference center and it cost $29 million. This might have been the trigger that sent our boys up to talk to them. Of course, our Board got involved. The bank made lots of noises but we still had one advantage in that we had two banks. The year just rocked along with the capital budget in calendar '74 set at $100 million. In 1975 the budget was $21 million. Everything was cut back. We were placed more or less under the control of the bank. With the plant company it was First of Chicago. We managed to continue working during that period. When the boys went up to New York, I believe the conference center had run over budget. I went up and got $2 million from the bank. I forget what basis I used except that I knew the new officer who had been put in charge of the department after the other problems they had. We were actually building The Woodlands all this time. We were going to sell municipal utility bonds, but our financial statement wouldn't support it. So basically it had to be a private sale. The Chase said to get the local banks to do it. I talked to Bank of the Southwest and National Bank of Commerce. I believe we were talking about $2 million again. And they said they would buy the bonds if they could be taken out at the end of the year. I then went to Chase and got them verbally to agree to take out the local banks. Somewhere in the next week, I imagine, the local banks called Chase to confirm it and Chase said no, they wouldn't take them out. So that started putting a little more pressure on the deal. Then we went to New York and got $5 million from Chase. We needed $20 million, $5 million of which we had gotten out of First Chicago without too much bother. There was always a large amount of grumbling and maneuvering out of Chase, but we did get it. We later got another $5 million out of each.

The price of oil went up to $12 in 1975?
It went from $2.50 or $2.60 to $10 to $12.

I've always understood that was our salvation. I think it came in the spring of '75 or so and suddenly we were worth a hell of a lot more than we had been the week before.
The banks didn't realize it. The employees didn't realize it. The directors didn't realize it. The real estate people didn't know it because of their prob-

lems. They kept coming in saying, What's with George? He isn't upset. I didn't realize what he had done. He had taken our reserves and just multiplied them by the higher price in his mind. No problem on value. The only problem was there was no cash.

I also had the impression that the peculiarities of the people imported from Columbia kind of compounded our problems.
They were just plain old real estate people. Blue suede shoe.

But there was also the huge front end investment with no cash flow. Isn't that in the nature of trying to do a development as large as The Woodlands?
Yes, but if you're Exxon, it's not a problem.

We weren't.
Anyway, '75 was an unpleasant year. Towards the end of it, or early '76, this young man from Goldman Sachs came by and wanted to know if we'd be interested in having his firm finance us. So Tom Walker, who was in charge of their Dallas office but who was one of their top officials, came by to meet George and me. He said he felt sure they could make a private placement. In the meantime, I had known one of the top partners in Kuhn Loeb in New York, which doesn't exist anymore. He came down and looked over everything and said, "We can sell $40 million of public bonds." I called Walker and said, "Look, are you going to do a private placement? We've got this deal with Kuhn Loeb and we're going to have to take it." Well, don't worry. He sent one of his young men down to check out the company. Then Loeb comes up with a deal. In the meantime Goldman had sent someone down to look at us. I hadn't heard anything back. So I called and said, "Look, if you want it, let me know. Let's go." And they looked at it and said, "Well, we could do $75 million." Then they decided to do $100 million. In the meantime, Manufacturers Hanover Trust Co. in '75 had banked 10 portable processing plants. In the meantime, I hired Bill Tonery and I put him to working with Manufacturers. I called the top oil officer of Security Pacific, a Los Angeles Bank, who I had met at previous conventions. He said they would be interested. So Tonery talked to him about a line of credit. We had always borrowed on collateral, which is a weaker position.

Was the deal now with Goldman?
This was going on simultaneously.

But you hadn't closed anything yet?
No. Goldman lined up Metropolitan Life and Equitable Insurance. Our director, Shaker Khayatt, brought in Travelers. At that point, they jumped to $100 million and in the meantime Tonery was dealing with Manufacturers and Security had told them about the insurance company interest. He got a $75 million line of credit out of them. So all this was going on parallel tracks. Insurance companies came down, looked at the Bridgeport plant, looked at this, that and the other, went back, and sometime in late October or November they approved it. Of course, it was important that we not let our old two banks know what we were doing because we had approached them on all this and they wouldn't do it. We told them about the insurance companies. Chicago would do it, Chase wouldn't, or they wouldn't do it on terms that we considered acceptable. Chicago wanted something extra, but I felt that once you have a question mark on a relationship, you're never going to get rid of it. It will always be there in the background someplace. I felt it was better just to get with two new banks and three insurance companies, which were Metropolitan, Equitable and Travelers. In the course of wrapping everything up in November '76, I had a detour–I had a heart attack.

Going back to the very difficult period with the Chase, do I remember correctly that George kind of lost control of the company? There was a story that some young man from the Chase was dressing George down in a private meeting that I'm sure you were a part of.
Yes, he did. They pretty much dictated to us what we could and couldn't do. We had to get their approval because if we didn't, then they wouldn't give us any money and then we would drown.

Well, we'll try to pick up loose ends another day. Thank you, Budd.

* * * * * * *

Today is Wednesday, August 9, 1995, and this will be the concluding interview with Budd Clark. This is Joe Kutchin; I'm with Budd in his office in The Woodlands.
One thing I might clarify: Before my heart attack in 1976, I was handling the insurance company loan. I turned the banks over to Bill Tonery, whom I made chief financial officer, to negotiate, primarily because we had two banks and I didn't want to antagonize them by having them find out that I was dealing with other banks, since I had been the sole contact outside of

the real estate borrowing. I had him do all the negotiating. After the heart attack, I turned it all over to Tonery. Then we closed in January, at which point I was back at work.

How long were you away from the office?
About five or six weeks, I would say. On going public, to get back to what I said earlier, you had two capital intensive segments of the company and the company was capital deficient. For the longest period, from, say, 1953 through 1970, we had managed to work through and with the banks for our capital requirement plus the interest we would earn (that's what is known in the industry as promotion fee). In other words, with our investors we're entitled to 25 percent net profits after they got their money back. In later years, that share of the profits started coming in and helped. But once again, it was banked. It appeared that we were about to run out the banks' patience. They felt there should be equity.

We're talking about 1971?
1970. So we started talking to Eastman Dillon since Bill Houck, our chief financial officer, had known the corporate finance officer for energy over at that firm from the time when he was at another company. I can't remember the individual's name. Eastman Dillon showed an interest. We started working on our S-1, getting reserve reports, etc., for a filing. We finally got it all together in June '71, at which point the corporate finance guy from Eastman Dillon took it to their committee and the committee overwhelmingly voted down doing a stock deal and gave advice to us recommending that we should sell assets, reduce debt, and then we might qualify. Well, since we knew that answer anyway, I called Shaker Khayatt who had shown an interest in financing through a company he was with. It was called Coggeshall & Hicks at that time. Khayatt felt when we started work with Eastman Dillon that they had taken his deal away, so I told him we weren't going with Eastman.

Lubanko was with Eberstadt. How did you know Khayatt?
Khayatt was sent by Norman Olansen, who was our contact at Chase Bank.

You said last week that the offering was 700,000 shares plus another 70,000. But wasn't a big chunk of that a secondary offering?
No, there was no secondary offering. Without a secondary, you could sell it better to the public, meaning the other brokerage firms. An unknown com-

pany doing a secondary to start off can be done, but it sure indicates that you don't have any confidence that the things are going to have value in time.

You might say something about the idea of listing the stock on the American and then a number of years later going to the New York Stock Exchange.

We decided to list on the American because we couldn't get on the NYSE. In those days NASDAQ was OTC, or over the counter. We had a very good relationship with the management of the American, and, as a matter of fact, the president of Amex, Arthur Leavitt (who is now head of the SEC), at that time was very innovative and put on an overseas junket and set up American Stock Exchange clubs in major cities where you could go and present your company's information. Then the NYSE started coming in and pointing out that they would like to have us on board. By that time, Arthur Leavitt had left the American and Jones, a former congressman from Oklahoma, had taken over and it had lost its spark. We felt we would have a better image by being on NYSE, so we switched to it in 1992.

Do you remember the year of the big capital budget?

1981. The budget ended up at $530 million. Early in that process we were borrowing $20 million a month on top of our regular cash flow. In June it was necessary to slow the operation down or we were going to be in serious trouble if we didn't.

What year are you talking about?

This is '81, which would be fiscal '82. The budget had been set up at $375 million, then raised to $425 million, and then in the process of stopping the freight train, we ended up at about $525 million and very little to show for it. Soon thereafter, the price of oil started down. One thing that helped us along was natural gas liquids. Of course, we had the price controls on gas, which actually worked in our favor because they allowed the price to go up each year. In fact, three or four years ago, the price of gas was $8 under the price controls. Of course, once the price controls were lifted, the price crashed. How do you explain that one? Price ceilings usually become a price floor, if you want a generic explanation.

The plan then was that there was no end in sight and prices were forever going to go up?

Right.

That's really a huge budget for a company our size.
It was totally out of proportion. We eventually settled down with budgets of about $275 million. I think the year after that, the budget was still $300-something million. But then after that, it dropped below $300 million and has stayed there pretty much ever since.

Thank you, Budd.

Roger L. Galatas

Today is Wednesday, August 9, 1995. This is Joe Kutchin speaking with Roger L. Galatas, president of The Woodlands Corporation, in his office in The Woodlands. Tell me something about your background.
I have a degree in geology, worked as a geologist for Humble Oil & Refining Company which later became the Exxon Corporation, in exploration and production. I also had an opportunity to work in a management development role with the board of directors of Exxon USA, which gave me a wonderful opportunity to understand the major corporation and see how it worked. I moved into the real estate area of Exxon, which is called Friendswood Development Company, because I thought it would be a wonderful opportunity to have a job and not move around the country, having been a geologist and moved every year. I thought, "Well, they only own property in the City of Houston, so I couldn't go anywhere." For that very fundamental reason—well, other reasons, too—I got into real estate with them and worked at places called Kingwood, which is jointly developed by Friendswood and the King Ranch, and Wood Lake in Houston out on Westheimer and Gessner Road. One day I was called by a head hunter and apparently Mitchell Energy was looking for someone to join the company and help run some portion of The Woodlands. Ed Lee was president of The Woodlands at that time and things seemed to work out favorably and I joined Mitchell Energy. It was not called The Woodlands Corporation back in those days, it was just the real estate division of Mitchell Energy. That was in 1979, and I've been with the company continuously since that time. I worked as senior vice president in charge of residential development in the beginning, became the executive vice president of the company and later its president in 1986.

Let's go back a moment. Your degree is from where?
Louisiana Tech.

Do you have a degree in meteorology?
Well, I spent a year in graduate school in meteorology. In the Air Force I was the weather forecaster at a famous Air Force base, Homestead Air Force Base.

And where are you from?
The little town of Benton, Louisiana, close to Shreveport. Grew up there,

didn't know there were other places in the world until I went to college. Didn't know I had a Greek name until I graduated from college and went into the Air Force. One day I was filling out a weather form at the Homestead Air Force Base for a pilot and he started speaking in a strange language which turned out to be Greek and he was asking me if I was a Greek and I said, "Gee, I don't know." He said, "I think you are, your name means 'milkman.' There's a book, named Elani, and it describes activities of the milkman in Greece. The milkman there was a German fighter pilot who would fly over Greece in the early morning and drop bombs and they called him 'milkman' because he'd come early in the day."

Please describe the circumstances, the state of the company when you came in 1979.
Well, things were kind of chaotic. The Woodlands was just coming out of a severe economic downturn. It hadn't really recovered fully when I came, as I recall. A lot of things were not complete, and the residents in the community were ill at ease about the progress being made . . . simple things like installing in a timely way the street lights in the neighborhoods. There was an uprising we had about building an Exxon Service Station and clearing trees. People here had, and still have, a possessive sense, a strong feeling for The Woodlands. I would say it was just coming out of some rather chaotic times. George Mitchell was personally and deeply involved in day-to-day operations in The Woodlands in those days. Ed Lee had joined the company not too long before that and he was learning what he should be doing and I came along and I was trying to learn what I was supposed to be doing, and I guess it was a period where a lot of people were learning.

Do you remember the population about that time?
Not precisely, but I'd say about 5,000 people, really not a very big place.

And today it's about 43,000. What was your immediate responsibility?
I was responsible for all of the residential development in The Woodlands. That included land development, the planning that preceded land development, getting the homes built, lots sold to the builders, increasing the volume of home sales. I think the first year I was here, the home sale volume was maybe 300 homes that year, something like that. We are now up to almost 1,000 homes a year.

What kind of things did you do?

Well, address some of the complaints of the community, like installing street lights in a timely way, you know, the fundamentals. Getting it a little better organized, working with builders. No one had been here before who really worked with builders and coordinated their activities and helped with their marketing, and we got some of that organized. We put together a team that consisted of our people in our company, the builders, the sales people, marketing group, advertising and PR, and then really built a system that started with planning and went all the way through marketing, sales and follow-up with customers, in dealing with our customers, the builders. And also during that period, Houston was recovering, too. This was in the late '70s, early '80s, when the economy was doing very, very well. All the way through 1982, '83, things were really going well in the Houston economy. So the combined effort of getting a better organizational focus in The Woodlands among ourselves, superimposed on economic improvement, led The Woodlands in the direction of increased sales, increased growth. We also were able to attract a few companies and create some jobs.

Name one or two, please.

Betz Process Chemicals was one of the first that was here. It was here when I came, and then Pennzoil built a products testing laboratory here. Asea, a company from Switzerland, I believe, had been here for a short period of time. Not too long after that, George Mitchell began to focus on research activity, and HARC was conceived and funded initially and gave the initial thrust to the Research Forest, and some companies moved in out there. This was just a formative period when we were getting the community off and running.

Did you find The Woodlands different from where you had been before?

Yes it was. The Woodlands is more comprehensive in nature. By way of comparison, Kingwood is a quality community on the north side of Houston, developed by Friendswood and King Ranch, 13,000 acres of land, big community. But its primary focus is on residential homes, and people who live there commute to Houston to work. By way of comparison, the objective in The Woodlands is to create jobs so people can live and work in the same community. I think that's the biggest distinction between The Woodlands and any other community you look at in Houston today–more comprehensive program, much more focused on creating jobs and economic growth within the community. So people could live close to where they work. I think another distinguishing characteristic is the focus on community amenities here—and some of these came along later—but things like the Cynthia Woods Mitchell Pavilion, you don't duplicate that anywhere in Houston. It's here, it's a unique thing. The focus on the hospital in the early days: We were the only free-standing community who used our own resources to build a hospital, which is now operated by the Memorial Hospital system. It's there because we took the risk and guaranteed $12 million in financing to sell bonds and cause it to happen. And the John Cooper private school. George Mitchell and John Cooper and others of us worked hard to develop it and we supported it financially.

You would include, wouldn't you, The Woodlands Athletic Center?

Yes. Other communities have things like that, but ours is more substantial, with the Olympic diving tank, the only one in Houston that has an Olympic free-style swimming pool, both inside and outside pool, heated pools. The things that were done here were just a little more extensive than they were in other places. I think another distinguishing characteristic of The Woodlands is that, clearly, the focus has been on working with nature, developing in harmony with nature, saving the trees. It's been a hallmark of The Woodlands, still is; it's a very strong marketing image now. Other people are emulating that, which is good. I'm glad they are. The control of signage, attention to detail, agonizing over things that other people may not pay much attention to.

Tell me about your early days here.

I remember the first two or three meetings with Ed Lee and George Mitchell. I wondered why I had come. I thought, "Gee, boy, this thing is moving quickly here." It was an exciting time, but for most people who joined The Woodlands back in those days, there was a period of adjustment. It was a place not operated exactly like other places you might have been, but, as you got acquainted with what was happening and could see what the vision really was and understand the people, it made quite a bit of sense. And clearly those are the things that made us different and made us successful, but it was an adjustment period, too.

It sounds very intense.

I've said this to George, so I guess I can say it on tape. I was just hopeful that if I could do half the things that George thought I was doing, I'd be judged as successful.

OK, you came in '79 into a high velocity situation. Did things change dramatically with Ed Lee's illness and his death?

As you know, Ed passed away in 1986 and I guess he was ill for a couple of years really, so over that two-year period, we all went through some very troubling periods. I wouldn't say there was a drastic, abrupt change . . . things evolved over that period. When I was responsible for the home building activity, or residential activity, we created a home building company. We built homes, we built the first company-owned apartments in The Woodlands–Village Square Apartments were the first to be built, and we built them with a general contractor. We designed them and got the financing in place and we used a general contractor named Bill Aydam to build those apartments. They turned out pretty good. Aydam was a personal friend of mine. I had known him before I came here and encouraged him to come and help us understand how to build apartments and he did that. He and a gentleman by the name of Dick Kilday, who later came to work for us; he worked here a number of years. After we built Village Square, we built the Holly Creek Apartments and the Wood Glen Apartments. That was an experience. Ben Barnes, the former lieutenant governor of the State of Texas, was a business partner with John Connally. They had a construction company, Connally, Barnes or Barnes, Connally, I can't remember which it was, but anyway, we contracted with them as a general contractor to build Wood Glen Apartments and the first phase of Holly Creek Apartments. Barnes is an interesting person, too. He's the only person I've ever met who could come into your office and within five minutes be sitting behind your desk with you sitting across from him and you were just pleased to be there. He was a kind of person who could really take control of the situation and he did a good job for us.

We had nothing before Village Square?

Before Village Square was the Grogan's Landing Apartments built by an investor who did not do a good job. Construction quality was not good–the investor/owner got into financial difficulty and we acquired the property from him. Spent a lot of money on renovation. This led us to the conclusion that we should be more involved in apartment design and construction to ensure a better quality product. This led to development of Village Square and later, Parkside Apartments.

Do we still want to do that?

We still do that. In the last year we just built about 240 units on Sawdust

Road through a general contractor and we own them and we lease them and operate them ourselves. We are in the midst right now of trying to secure financing to build 250 additional units. But we also formed a home-building company and developed the capability of building apartments ourselves—we built the Creekwood Condominiums, we built Phase 2 of Holly Creek, and we built Parkside Apartments, the first phase of that, then we built maybe a couple hundred single family homes a year. We did that for several years till the economy collapsed in 1986-87, and we elected to get out of the home building business. We had the choice of being either the dominant and only builder in The Woodlands and selling lots to ourselves, or we could get out of the business and make room for third party builders and share the risk, and we elected to do the latter. The volume of business in '86-'87 was just not enough to support our own home building activity and two or three other builders in production housing, so we elected to step aside and close down the home building company. I think Larry White, who ran that for us, moved away. That was the first cutback in employees I had to deal with here.

How big a cutback was that?
I think it was, overall, probably 25-30 people, but it was directly related not to a reduction in force but a change in direction that brought about the closing down of part of our business operation.

Has that turned out to be a good decision?
I think so. If you'll look around the country, not too many companies are successful developers and home builders. It seems that separating those two activities is good for both entities. I think we made the right decision.

So it was in '86 that you became president of The Woodlands Corporation? That wasn't a great economic time.
No, that was not the best of times. It was right at the bottom of the economic downturn in Houston. I guess recovery started slowly in mid-'87, so '86, '87 was really the bottom of the market.

You took the presidency of an organization that was a going concern but certainly less than prosperous, is that correct?
Well, it was not as prosperous as it is now. We had to cut back on staffing, we had to conserve our capital, we had to increase our sale of non-Woodlands assets, like timber and things of that nature, to pay the bills. It was not the best of times, but it was not only related to our company but to the economy

of Houston, as well. Oil and gas prices fell off the cliff in Houston, and there was an overall loss of 200,000-250,000 jobs over a very short period of time. So we were just part of what was happening, but the interesting thing about the downturn, I think, was what happened in The Woodlands and what happened in other parts of Houston. For the first time, there was a good measure of whether or not The Woodlands was really marketable. Our home sales held up better than sales in other places. There were five large-scale communities around Houston at that time–Clear Lake City, Kingwood, The Woodlands, First Colony and Copperfield. Those five communities became dominant in the residential marketplace. They captured almost 50 percent of new home sales. So that told me that quality was what people were looking for. There was customer satisfaction in large scale communities where there was a quality developer who was financially stable and could deliver on the promises he made in the marketplace. Those five communities became the dominant places for people to seek housing.

All but The Woodlands are built out, aren't they?
Most of them are reaching completion. Copperfield is completely finished. I guess Friendswood Development Company is in the process of selling their remaining properties in Kingwood and Clear Lake City. But regardless, they probably only have a few more years remaining in their development life. First Colony is nearing completion, but probably has several more years to go. So out of those four major communities that were dominant in the mid '80s, only The Woodlands continues to be very active in the marketplace.

And with the building out of those communities, is it fair to assume that your opportunities increase?
On the north side of Houston that's true. Other communities have been started in other parts of Houston, particularly in the southwestern quadrant. Cinco Ranch, Greatwood, New Territory are very substantial communities, but as you look on the north side of town, there are very few competitors to The Woodlands. As time goes by, I think we will have a more dominant position in the north Houston marketplace. A lot of small developments that were competitive in the early-mid '80s in the FM 1960 area are complete. They were small communities to begin with, and not many are coming along to replace them. Congestion on FM 1960 is bad, the unsightly nature of 1960 doesn't encourage people to move there, the improvements of the Hardy Toll Road, which opened in 1986, gave a direct link from

Houston to The Woodlands and from The Woodlands to Intercontinental Airport, which relieves some of the pressure on Interstate 45. So that, coupled with the improvements put in place in The Woodlands—such as the regional mall, the restaurants, the shopping, the growing job base and the transportation and mobility issues we've addressed over a long period of time—I think those advantages are putting The Woodlands in a very dominant position.

Have you or your team done anything in terms of planning or strategies that you feel have been extremely important in the development of The Woodlands?

Yes. I think the general plan that was developed in the late 1960s, early '70s, was a very good general plan. We've been fairly faithful to the overall concept.

This is after the HUD plan, is that right?

It's part of the HUD plan. As you know, we succeeded in spite of HUD, not because of it. I think The Woodlands probably wouldn't be here today if there had not been a $50 million loan guarantee provided by HUD. That enabled George Mitchell to have the courage and financial ability to go forward with The Woodlands, and had that financing not been available—and it was a loan guarantee and not a gift—I'm not sure George would have followed the course he followed. But I can tell you that HUD did nothing after that to make us succeed. They were a stumbling block. Dealing with them was an administrative nightmare. It added to our cost and performed no good function. I was delighted when the federal government decided to discontinue that program and offered us the opportunity to get out of it. I will say, though, that we were the only one of 13 large-scale communities under the Housing and Urban Development Act of 1970 that succeeded. We were the only one who repaid all of its financial obligations, we were never in default, we made all the payments in a timely way, we didn't default on any operating obligations or requirements. We were faithful to the contract, even though from time to time I thought it was burdensome.

Despite all the problems associated with the HUD deal, isn't it correct that the development plan done for them remains the basic plan, even 25 years later?

Yes, the general plan that was evolved during the late 1960s, early '70s was a very good master plan and we've been fairly faithful to that. There have been modifications, scheduling changes, improvements, I would like to think, but it's been a good framework and that framework was not only a land plan, but it was a program that suggested that the community needed to be complete, needed to have jobs, churches and schools, and provide a whole array of amenities and quality of life that would be better than anywhere else in Houston. That's what we've tried to adhere to.

You talked about some refinements or modifications. Those would have come about when, Roger?

They evolved over time. The general plan had a town center but we didn't know where a regional mall would be on the original plan. We didn't know what the waterway would look like. We didn't know there would be a Cynthia Woods Mitchell Pavilion.

But a waterway was part of it?

A waterway was part of it. We just didn't know what it would look like, how it would be shaped or how it would be implemented. We knew that entertainment would be an important element, but we didn't know there would be a pavilion that looks like the one we have now. We knew there would be regional shopping, we didn't know where it would be or who would be our partners or what its architecture would be. So all of those things have evolved, but they remain fairly faithful to the general plan.

Tell me more about what's happened in the nine years that you've been president.

I think we work pretty much as a team here. It would be hard to assign credit to one individual, but the things that have happened in The Woodlands over its life that make it different are what I call defining events. For instance, the creation of Research Forest was a defining event. George Mitchell's vision was that we could do something here like the Research Triangle in North Carolina that would have an anchor in academic institutions through the Houston Advanced Research Center. The idea was that it would engage in contract research programs for business, industry and government. I think that was a defining event that made us different than other business locations in Houston. I think the formation of the hospital was a defining event. It provided medical services and established this community as a substantial community. The moves of one or two companies here have been defining events. The world headquarters of Hughes Tools moved here in the early 1990s. They moved from Houston where they'd been for 88 years. They became the first world headquarters here. And that created a lot of awareness in the commercial brokerage community in Houston. They said, "Gee, if Hughes is going there, maybe it's a good place, maybe my clients would like to go there." I think that was a big step in our successful commercial business development operations. I think a very substantial event was the activity we undertook to include all The Woodlands property in a single school district. We had the original 17,000 acres which now has been grown to 25,000 acres. The land was in two different school districts: Conroe Independent School District encompassed the portion of land we developed first along I-45, and about midway through the property going westward was the Magnolia Independent School District. Magnolia is a good school district but didn't have much tax base, didn't have the ability to fund rapid growth of schools. In a special session of the legislature, the law was changed to allow a land owner to petition school districts to modify the boundaries of school districts.

Explain the advantage of that.

Well, before that change in law, only residents could petition to change a school district's boundaries. That meant you had to first have people move into the district and then petition to solve the problem. It made a lot more sense to us to allow those boundaries to be adjusted to accommodate growth in the future. The boundaries of school districts in The Woodlands, when they were first formed, were just lines drawn through the woods; they just divided land between districts. It makes much better sense from an educational standpoint if there could be a common school system. Instead of dividing the community into several school districts, we would have a more systematic program, consistencies, transportation efficiencies, educational efficiencies. Had we not done that, I think The Woodlands would have stopped growing dramatically in 1992.

Is it fair to say that there was a perception here, or otherwise, that the Conroe district was of higher quality?

Yes, the perception was there. The fact was that Conroe was able to build schools in neighborhoods in The Woodlands, and Magnolia did not have that capacity. So Woodlands kids would have had to get on the bus and ride 25 minutes to an hour one way to elementary schools and intermediate schools and high schools. Had that change not been done, The Woodlands would have curtailed its growth significantly, and we would not have been able to develop the commercial base either, because the residential population is what supports the mall and the shopping centers and things of that nature, so it's all tied together.

Was that realized in the early years?

No, not really. There was an assumption that somehow when we got to that point, we'd solve the problem, and sure enough we did. But we didn't plan ahead very far. We were right up to the deadline, and it cost us $12 million to solve that problem. We agreed to pay the Magnolia School District $12 million over, I think, 20 years, or something like that, in annual payments and they agreed to allow the Conroe School District to annex all the lands that lie within The Woodlands which were originally in the Magnolia School District. I think that was a major, major event. So the HUD funding, the loan guarantee, the school district boundary issue, the creation of HARC, the opening of the mall, and the hospital and the pavilion, those were things that made us stand out as being a significant place.

None of these was accidental in the planning?

No, none of it was accidental. All were intentional. Another thing that made us different, too, has been George's visionary leadership. As I said, when I first came, it took me a while to get used to where it was all leading, but after looking at it for 17 years now, The Woodlands has been under a single common ownership since its beginning. There has been a consistency of plan and vision and there has been a strong interest in implementing the plan as originally conceived. All of that goes back to George Mitchell's personality and his vision and also his financial capability to make things happen. He has been committed, and I think his commitment has really made The Woodlands happen the way it's happened.

I know you well enough to know that you don't sit and wait for the phone to ring. I'm trying to get to the forces that direct where we've been going. Did somebody sit down one day and say, "Hey, it would be keen to have a pavilion because it might accomplish certain things"?

I can tell you how the pavilion got there; it has to do with a hot dog. Dick Browne and I went to lunch one day, at the TPC restaurant, and we got a hot dog. We had just rebuilt the 18th green on the TPC course. I had been down to see Sawgrass. Duke Butler, who was the executive director of the Houston Golf Association then, had presented an approach to remodeling and improving the TPC course in The Woodlands. It involved the creation of stadium seating around the greens the way it is on the Sawgrass course in Jacksonville. We had just redone the embankment around the 18th green and rebuilt the green, and Dick and I walked up on the mound out there and we had lunch, hot dogs, and I said, "You know, Dick, if that green just had a tent structure over it, we could have a small theater, just the kind of thing you've been talking about." He had been talking about it for a long, long time. He said, "We could probably do that." We thought that maybe if we could get George to spend $1 million, we could build something that would seat maybe 1,000-1,500 people, and that would be the initial outdoor amphitheater. We started talking to Pace and we started talking to people who knew about outdoor amphitheaters and we got Cynthia Mitchell involved in it and she got enthusiastic. As we talked to Pace, they said, "You've got to build it bigger because you'll never get any performing arts groups to come. If it's too small it won't be commercially viable. You've got to seat at least 10,000 people." So the hot dogs that Dick and I had grew from a $1 million proposal to almost $10 million before we got through building it. But George and Cynthia were the guiding lights in that and insisted on doing it right, and I give a lot of credit to Pace and their leadership for helping us understand what we were doing from a commercial viewpoint.

Phil Hoffman has said that he was one of a group that included Ben Woodson and John Cater who put the bug in George's ear at lunch one day. It's a great success and everybody claims parentage.

In '81, '82, there was a proposal to build it and some design work had been done. It was just put on the shelf when the economy went to pot, and through 1986, 1987, nobody talked about anything in that regard. But as we were coming out of that recession, we had those hot dogs, it wasn't an original thought–somebody already had had it, but that's what reinitiated the activity.

What has the pavilion accomplished for The Woodlands? What would you say the benefits have been?

It's like the Astrodome or the Summit, the theater district, the Alley Theater. It's a place where people go to be entertained and they travel from all over the region to come here. There are about 50 performances a year now, 250,000-300,000 people a year come here. Clearly, it does more than support the entertainment needs of the community. It attracts people to come to The Woodlands for another purpose, and I think that gave The Woodlands a bit of definition in the Houston marketplace that it didn't have before. Now it's benefiting things like the mall and restaurants around the mall because people come not only for performances, they come to dine and shop, and with the movie theater that's being constructed, a million people will go to the movies every year. So all of these things have a way of working together.

Do you know how I became president of the pavilion? I got a call from Tom Battle one day saying, "Did anybody ever tell you you're supposed to be president of the pavilion?" I probably said something like, "Why not?"

I remember the conversations in the early days. There was a great sensitivity, especially from the tax standpoint. Neither The Woodlands Corporation nor Mitchell Energy could control the pavilion, or we'd lose tax advantages. So, it was organized as a not-for-profit with an independent board, separate and apart from our management team.

What else has been important?

The Mitchell headquarters building completion and the corporate move here in 1980. I think that was a defining event. That gave people knowledge that Mitchell Energy was firmly behind this. It was the first major company to move here.

About 150,000 square feet?

Yes. It broke ground about the week I came to work, as I remember. It was the first groundbreaking I went to. Shortly after that, the second building to be built here of any substance was the Woodlands National Bank building, which became the Bank of the Southwest, which became M Bank, which became Bridge Bank, which became BankOne. We've outlasted all those banks. Then the Houston Advanced Research Center started in '82.

Did the people in The Woodlands Corporation have any special role in HARC?

I didn't have a personal involvement in the early days, but I was aware of what was happening. George had the notion of creating something like the Research Triangle, but there was really no formal plan, and Ed Lee took the initiative to get the planning done, with George's concurrence.

Was that Arthur D. Little?

I think that's right. He got Arthur D. Little to do a feasibility study and a plan of how to do it. Actually the plan wasn't followed precisely, but at least it was a starting point. I think the thing that really made HARC happen was the commitment by George Mitchell to provide funding through his own resources and through the company to build a building. I went over to the Research Triangle in North Carolina and took a look at it, and we involved the leadership of Texas A&M and Rice and the University of Houston in the early planning stages, and Baylor College of Medicine was involved to some extent. It didn't get off to an easy start, but I think that once again, George's strong desire to see it happen caused it to happen.

The growth issues program that George had initiated in the mid '70s was one of the things that got wrapped into HARC.

And David Gottlieb was with the University of Houston then. When I first met David he was with the University of Houston, on their staff, and he was working on the growth program for George as part of his responsibility at the university. That was one of the formative things that helped HARC. The opening of the Hardy Toll Road in 1987 really was an event that helped The Woodlands, too. That was one where George worked with County Judge Jon Lindsay and Harris County. I don't know to what extent we were able to influence that, but the road was built, it's there, it's a tremendous asset for people who live here and for the community at large. It's helped the company considerably in its development. George was the one who had first discussed this with Kathy Whitmire. His real focus was getting some access to Houston Intercontinental Airport and downtown Houston. Of course, the Hardy Toll Road serves that purpose but also comes to the county line, and to The Woodlands. I think the John Cooper School is a defining event that separates this community from others. People have moved here because they wanted their kids to be in the school. I think one or two companies, small companies, have moved here because the owners wanted their kids to be in the school. I think it's been easier to recruit some very talented people for companies in the Research Forest because of the alternative of a private education that's available here. Although CISD is a very good public system, some people just want the alternative available.

The school was an initiative from George, or did it come out of the Ballantynes?

Actually, we used the model for the hospital. For the hospital, we created a not-for-profit corporation. We agreed to donate the land to this institution.

We did all the planning and we got all the permits to build the hospital. We contracted with Methodist to do the programming and help us with the application. Methodist Hospital was the first manager of it. But the model was a not-for-profit organization where the company provided some land and seed money and we got people with talent to run it. George got John Cooper to move here to help us think about education. We used the model of the not-for-profit organization, with the company donating the land and providing some seed money, having a Board of Directors of capable, interested people in the community, the community being all of Houston and some from The Woodlands, and we went about doing the planning and programming and design of the building. We got that off and running, and Marina Ballantyne came along in that process and made a tremendous contribution.

Talking about the school?

Yes. Physically where it would be and how we would organize it and when it should be available. John Cooper and Joel Deretchin and I, I guess, were the three people who did the first scratching on paper. The first stop John Cooper and I made was at the superintendent's office in the Conroe Independent School District to make sure he understood we were not trying to be competitive and not suggesting the public schools were not good, and we gained his support. Private schools and public schools work together very well here.

In what year did the Cooper open?

It opened in 1988, right after the downturn, as we were coming out of it. We agreed to donate 40 acres of land to the school for its campus. First donation, I think, was 10 acres, and we also provided $2 million to build the first school. That is, we guaranteed the loan.

While we're talking about acres, how many did you give to HARC? And the Cooper School—did it have all grades at the outset?

The campus for HARC is approximately 100 acres. That will be donated over a period of time also. I think now they have maybe 10-15 acres under their ownership, and as they grow and expand, more land would be made available to them. Same pattern as for the Cooper School. First the elementary school opened and then it went from K through the fourth grade, I believe, and each year an additional grade was added to the school, and now it's up through the 12th grade. There are 800-900 students there and probably it has reached its maximum size. I was very impressed at the first commencement exercise, which had as its speaker President Bush, just after he left office. John Cooper had Bush's children in the Kinkaid School and John Cooper had extracted a promise from the President that he would be there for the first graduating class at the John Cooper School. By that time, John Cooper had passed away, but President Bush remained faithful to his promise and he came to the graduating class that included 25 happy seniors. It was a really wonderful event, very personal event. Also, the school originally was not named the John Cooper School. John Cooper was a very modest person, and he didn't really want his name attached to it and it was called The Woodlands School, as I recall. We thought it appropriate that it should be the John Cooper School because he had made such contributions to it and also we felt that having his name attached to it would remove the "stigma" of its being a developer school, so we did not want to call it The Woodlands School. The John Cooper name would help in fund raising and would help people understand that it was for a broader community than just The Woodlands. So anyway, the name was changed. John Cooper reluctantly let us do that, which I was pleased about. Of course, The Woodlands Mall was a big event in 1993. It changed the complexion of The Woodlands significantly. About 10 million transactions are conducted in the regional mall annually. That's a lot of retail business. Of course, it created an environment that attracted other retailers, Target and Service Merchandise and Toys 'R' Us and Blockbuster Music and a host of restaurants, because people who come to shop also come to dine. And now with the theater going in adjacent to the mall, all of that complex is becoming a destination for the north Houston region. And the concept of our Town Center is not just to be the downtown for The Woodlands, but to be, I like to say, "the urban center of north Houston," the focal point for all of north Houston where you can shop, work and be entertained. It's the downtown for a million people within 20 miles.

This concept of the downtown for north Houston dates back to when?

I'm not sure when it was articulated, but clearly the downtown that was planned for The Woodlands would have to be supported by a bigger population than that of The Woodlands community itself, so I think it became better defined as we looked at the service area for the mall. It takes 300,000 to support a mall. That's the population of the area along I-45 from 1960 through Conroe and then over to the airport. If you look at north Houston, there are no real downtowns. If you look at the rest of Houston, there are a number of downtowns. There's the downtown central business district, there's the Space Center in Clear Lake, the Galleria Post Oak area is a downtown, Greenway Plaza is a downtown. But the north side is just a hodgepodge of small businesses and scattered residential subdivisions along FM 1960, which is really not cohesive in any way. It doesn't cause you to want to go there. So our mission is to be the downtown of all of the north Houston region with its population of a million.

Was it DeBartolo who was talking about a mall in Conroe?
Yes, there were competing sites. The one in Conroe jointly owned by Friendswood Development Company and DeBartolo would have been really very competitive. We were concerned that they would be the successful entry into the marketplace. We attached our hopes to Homart, which is the development subsidiary of Sears, and I'm glad we did because as it turned out, they were financially the strongest one in the end. I remember a trip that was a turning point. We were having a tough time getting things under way. We had been talking about it for 10 years and built an overpass on I-45 to serve our site, but it seemed we could never get everything in place. One anchor store would say yes, and one would say, well, not yet, and so we decided we'd go see Ed Brennan, the chairman of Sears in Chicago. George Mitchell, Ben Love, Mike Richmond and I got on an airplane and met with him and his chief staff people, and I think it was really the help we got from Ben Love and George Mitchell personally who were willing to have lunch with Ed Brennan in his office and get the commitment that led to the development of the mall in a more timely way. Had we not taken that action, I think it would have been much longer in coming.

Actually the possibility of the Conroe mall accelerated what you folks were doing, is that a fair statement?
Yes, but by the time we made the trip to Ed Brennan's office, DeBartolo and Friendswood were having a little trouble getting theirs off the ground. It was a matter of the commitment by Sears and Homart that (1) Sears would be there, and (2) that they would be willing to build a mall if we could get a total of three anchor stores. Dillard's had already said yes, so that was two, and the next one that we tried to get was Penney's, but they were not willing to make the commitment. We got Mervyn's, which is not one of the top end anchor stores in a mall but it was the very top end for us because it made three. Then when we announced we were going to build it and actually

started building, Foley's came along immediately. So we got four; three very strong ones and Mervyn's, and we have room for two others. We are now talking to Penney's and they're likely to be the next one. We're saving one of the remaining anchor sites for a very upscale retailer for the future.

Anything else come to mind?

One of the things that happened in the early days that made a difference was something that would probably not seem significant today. It was the opening of the first grocery store in The Woodlands by a gentleman named Jamail. People didn't believe this was going to be a community, because there was no grocery store here. The grocery store building had been built, but it had never been occupied and Jamail made the commitment. We loaned him the money to put the tenant improvements in the store. He came and became part of the community. I think that gave a lot of credibility to the first village, Grogan's Mill.

Jamail's was a community center for a long time.

Jamail sold the store to Randall's and Randall's ran it for a while. They vacated it and the building once again is vacant. But we have great hopes for the future and the next time we have an interview, we'll talk about how wonderful that's going to be. But the fact was that Jamail's, a very upscale, recognized retail store in Houston that also served the River Oaks community, made the commitment, a very strong one. That made a tremendous difference in the attitude of the people who already lived here.

Did Jamail's do well?

Yes, he did very well. He still has a store, I believe, in Kingwood. He opened this one initially and then he opened one in Kingwood. He was running both of them, and I think Jamail did quite well when he sold the store to Randall's. I'm not fully aware of all the financial implications, but I think he did pretty well. He continued to run the one in Kingwood for some period of time.

* * * * * * *

Today is Tuesday, November 21, 1995. This is Joe Kutchin in the office of Roger Galatas in The Woodlands and we are concluding the interview started a couple of months ago. When we finished talking last time, you mentioned the possibility that something good is going to happen in Grogan's Mill Village, in the shopping center.

Grogan's Mill Village Center was the first retail center built in The Woodlands. It was built as part of the first effort in constructing the Wharf, then a small shopping mall. It was attached to the conference center. It really didn't work very well. Jamail's grocery store was the first thing that came along that gave it some light. After Jamail's sold out and went away, the center just collapsed. There was no anchor, no reason to go there. Architecturally it doesn't rate very well, wood-frame construction, not state-of-the art. But two events are coming along that will enable us to engage in our first urban renewal program in The Woodlands, which I think is interesting. After 21 years, we get to rebuild something. The Grogan's Mill Center was ill-conceived when it was initially planned and if there were mistakes made, and clearly we made mistakes, that would stand out as being one of the major architectural and planning mistakes we made in The Woodlands. It was not visible, not easily accessible, had no major tenants, no major anchors. The first attempt at internalizing shops along the hallway just didn't work. People didn't respond to that very well.

Was it too small?

It was too small. I've never been able to identify the person who planned it. There are people who claim planning a lot of things here, but nobody has ever claimed planning the Grogan's Mill retail center. But whoever did it made some mistakes. It was on a smaller scale than it should have been to be successful. It was envisioned that there would be small shops and a small grocer and a small butcher and everybody would need those retail stores, but it just didn't work that way. Major grocery stores were growing in the area along Sawdust Road on land we didn't own. They were 50,000-60,000 square feet of full-service, high-quality facilities and small grocery stores just couldn't compete, so it was destined to fail from the outset. Anyway, we're in serious discussion now with a major retailer, a grocery store operator, Albertson's. We think that deal will be announced in a week or two. They will build a very substantial grocery store sufficient to serve the entire Village of Grogan's Mill and the area outside. That will allow for the renewal and rejuvenation of the other smaller shops over there. A portion of the center will be demolished and the major grocery store will be built. The old grocery store space, which was occupied by Jamail's, is in the process of being converted to a learning center for Shell. Shell Oil Company is going to establish a state-of-the-art learning center for all of their senior management team to be located here in The Woodlands. It will be a real plus for our community and our company and we hope a good arrangement for Shell.

We will convert the old grocery store space into a 30,000-square-foot learning center with conference rooms, meeting rooms and facilities of that nature. They will also have use of the hotel rooms in the conference center. Shell has recruited the director for that operation, so it's going to be a world-class operation for Shell.

That's a done deal?

That's a done deal, it will be announced by Shell. We're waiting on them to do that. We'll tag on to whatever announcement they make, but they're going to take the lead in terms of timing and how to position that announcement. Those two events, Albertson's grocery store plus the Shell Learning Center, are going to be two major anchors. One is retail, one is not, but it gives us the opportunity to renovate the thing architecturally, structurally, functionally and put some life into it. We'll tear down two of the old buildings. This week we'll also announce a new information center. The information center which has been in the Village of Grogan's Mill since the early days has served us well. It is now in a location that's not appropriate for marketing. It would need substantial renovation, so rather than renovate the old building in that location, which is in the wrong place, we're going to build a new information center on Woodlands Parkway near the Mitchell headquarters building, near The Woodlands Corporation office building, and we hope to get that opened in June or July. That will enable us to use a portion of the site where the old information center is for an office building. We think there will be other retail facilities, perhaps a Shell service station near the Exxon station, some restaurants. We hope that within a 12-month period of time, the renewal of the Grogan's Mill Village Center will be substantially complete and will be an asset for not only that village, but all The Woodlands.

As a Grogan's Mill Village resident, I'm very pleased.

Well, I hope all will be. I remember the fight we had when we built the first service station there. Now we may add a second service station, so hopefully, we have built a little more credibility over the years with people in terms of quality. I think that's going to be a big plus. It will certainly be a big plus for the retail operation.

The Woodlands has been a major part of your life for the past 15 years, more or less. I think according to the original plan it would have been built out by this time. Generally, are things going the way they should?

I think so, Joe, particularly over the last three or four years. I think The Woodlands has gotten to a new level of activity with the opening of the regional mall, the Cynthia Woods Mitchell Pavilion, the new 17-screen cinema which is under construction, the hospital, some of the major corporations that have established operations here—Hughes Tool has moved to The Woodlands. Companies in the Research Forest are beginning to grow and develop and make substantial additions to jobs. The residential housing is moving ahead at a pace which is, I think, more appropriate, about 900 to 1,000 new homes a year. We have good relationships between the company and the community, which I think is a fundamental reason for succeeding. I spend a fair amount of my time trying to talk to people in the community, either directly at meetings and presentations or through writings or personal contact, and other people in the company do the same thing. We need to make sure we preserve the right to continue to develop and that's going to be a community relations issue over time. We see around the country a no-grow sentiment; people who are there don't want any more people to move in. So, I think as we get bigger in The Woodlands, the challenge for the future is going to be this relationship within the community between the developer and the residents who live here. I think we have a good foundation for having a productive relationship. I believe, clearly, that this is a better community because of the quality of people who have moved here and the interest those people have in making it a better place. A developer can just do so many things, we can build streets and utilities, we can build buildings, but really the heart, the fabric of the community comes from the people who live here. I'm encouraged by the attitude of the community, the interest of the community, the support of the community, and I think as we go forward to the next 20 years, the challenge is going to be to keep that relationship as we have done in the past, on a personal basis, the relationship between employees in our company and people who live in this community. Most of the people who work in our company live in the community. I think that's a real plus. George Mitchell is a real plus—he's seen in the community, he's active in the community and people have great respect for him. I think we have a good relationship with elected officials, we work with civic groups and chair a lot of activities. We participate. I think that's just part of community development, it's just as important as making sure the streets and roadways and utilities work the way they're designed to work.

Is there some point in the future where The Woodlands will be structured as a city?

Under existing state law, the City of Houston really has control of the destiny of how The Woodlands will be organized. All of the land within The Woodlands, 25,000 acres, is within the ETJ of the City of Houston. Under state law, Houston has a right to annex The Woodlands at some point in the future and I think the City of Houston will preserve that right, I don't think they'll give up that right. So I guess absent a change in state law, or absent a change in the attitude of leadership in the City of Houston, we can look forward to one day being part of the City of Houston. This is a very controversial thing and it's something that we've been forthright about. If we took a survey of the people in the community today, residents who live here, I'd say 95 percent of the people would be opposed to being incorporated into the City of Houston. George Mitchell has been forthright about it. He holds the view that the health of the economic region that encompasses Houston and the surrounding countryside needs to be preserved, and if you separately incorporate small towns and cities, you fractionalize the economic strength of the region. There are good examples of that, if you look at the cities of the Northeast which George likes to focus on. Boston, Baltimore are surrounded by small incorporated towns and the city can't grow. Productive, tax-paying people move out of the central business district into the suburban area. The people who are left behind are less fortunate and things kind of go downhill. Where it's going to go, I'm not sure but right now I don't see the initiative by Houston to annex The Woodlands over the next five, ten, fifteen years.

You don't see that?

No, I do not, absent something to force them to do it. Each time the state legislature meets, a bill is introduced by someone that would in some way limit the power of the cities to annex lands within their ETJ. Each time this happens, there is an initiative by the people of The Woodlands and other suburban locations around the city to get the state legislature to change the law and, in essence, allow people in suburban areas to separate themselves from the city government and separately incorporate. That's when the cities take the aggressive action of annexing to protect the rights they already have. Each two years when the state legislature meets, we see that interest peak and then after the legislature goes away it dies down somewhat. As we get bigger and more people are here and there is more unrest in cities—and there is some unrest in the City of Houston—you're going to see more interest by people who live in The Woodlands to be somehow separately incorporated and become a city in their own right. I really don't know how that's going to play out over the next 20 years, but I hope that we as a development company and we as a community can preserve the opportunity to continue to develop The Woodlands comprehensively as we've done in the past, and not have small city incorporations that fragment the government within The Woodlands and splinter the community so that it becomes impossible to implement the comprehensive plan. We've had freedom and flexibility within the broad framework of the general plan to go forward and implement it without confusing governmental regulations that would likely be imposed by smaller incorporations or even the City of Houston. It's worked well for our company, it's worked well for the people who live here, and I think it will continue to work well. The community associations impose an assessment and collect the fees from people who live here. Those fees are used, I think, very productively to provide municipal services, very cost effective. The level of services is good, people seem to appreciate that, the system that we have is working well. I just hope the system is not set aside in favor of something else before we are substantially complete.

In the original deal with HUD, a number of goals were established, many of them social, some of them developmental. Multi-use of parking lots by churches, for instance. Does anybody ever go back and measure where we are against those goals?

I have thought of an organized attempt to do that routinely at scheduled times, but on reflection, I think those general goals we were trying to accomplish have been matched pretty well. For example, one of the goals was to create jobs in the local community, one job per household, and we're pretty much on target. We've got about 16,000 jobs and we've got about 16,000-17,000 homes. So that general goal is being maintained. The general goal of addressing religious opportunity is being matched pretty well. We have, I think, 20 to 25 churches in The Woodlands today and probably 30 different congregations. So, addressing that aspect of community life has been done pretty well.

The notion of sharing facilities has not worked?

We do have shared facilities. For example, we've just finished planning and are now constructing a parking garage in Town Center, adjacent to the theater. It is designed so we can build an office building adjacent to the garage on one end and the theater on the other end and much like the Summit uses parking for office buildings at night, so, too, the theater will use the garage in off-peak times. People typically don't go to the movies in great numbers during the working day. When the office buildings are not using the facility as a parking garage, it can be used by the theater.

Affordability of housing?

Affordability of housing is a great challenge. The original goal was to have a wide range of housing for all income levels in The Woodlands, to fully utilize all the subsidized housing programs in HUD. We still do that but there are just not many subsidized housing programs left in HUD. In the past three years, we built probably 500 rental apartment units that are subsidized, tax credit units, high-quality, a good program. We are applying now for an additional 200-250 units. Whether we get them or not, time will tell. It's competitive, the environment for funding those at the state and national level. But we have stayed with the goal of using all the programs that are available. The programs have disappeared by and large, so it makes it very difficult to provide a wide range of housing opportunities that have subsidy or support. We have maintained a reasonable price range of homes from the high 80s. Town homes—we have a few in the 80s. Single family detached homes, today probably $110,000 is the least expensive new home you can buy, more likely it's $120,000-$125,000 bottom end. Of course, at the upper end that price continues to grow over time, so there's no upper limit other than the marketplace. We still have the objective of providing a range of housing as affordable as we can make it in the marketplace, using all the federal programs that are available for housing support.

When it comes to stand-alone housing, essentially, the product is market responsive?

Single family detached homes? There are no federal programs that I'm aware of. There was the FHA-235 housing program which enabled people who qualified by virtue of income to buy single family homes. That program doesn't exist anymore. So everything is market-ranked in terms of single family detached housing. Rental apartments are the only place where there is a limited subsidy remaining and we've got to avail ourselves of that.

The toughest one has always seemed to me to be the commitment to seek a mirror image of the socio-economic makeup of Houston in The Woodlands. Any hope for that ever coming to pass?

It's going to be very difficult. If you look at suburban planned communities around Houston, The Woodlands probably has a higher percent of minority resident population. We're the only planned community around Houston that has subsidized housing. I don't think any other planned community includes any subsidized housing at all. We've made the best effort we can and continue to make that effort. We aggressively seek out minority realtors and try to get racial diversity in the community. We helped form an African-American Baptist church here in The Woodlands that has now built its church building and is successfully operating. It's a good congregation and is growing. We've done these kinds of things very actively and aggressively, encouraging all who choose to come here to feel at home and welcome.

You say we had the highest minority representation of any suburb? What minority, is that black and brown?
Black, Hispanic, Asian and others. I said we probably have the highest percent minority of any planned community in a suburban location, not just the general suburban locations but where a single developer planned the community and created an environment that encourages people to come live there.

Increasing that representation is a company commitment, right?
We made the commitment that we would aggressively do things that would attract minorities to move here, feel welcome here, and I think that opportunity increases as we create jobs at all levels, expand the community college and implement the outreach programs we have in place. All of that tends to move in a good direction, but there's still some reluctance. For whatever reasons, people make those choices themselves. I think our challenge is to create an environment where people feel welcome and comfortable and find their own home and build their own life. I think that's a stronger approach. I believe this is a very tolerant, embracing community.

Do you have any comments about other things—traffic, drilling?
I talked about maintaining this right to develop in the future and the no-growth attitude that people develop. That no-growth attitude arises out of a lot of things, being too crowded, too many people around you, traffic, personal safety and security, decline in the quality of education. So those are the things you have to work on to make sure people who come later feel as good about the place as the people who came early, and the people who came early still feel good about the place they live in. Traffic—I think we addressed that pretty well, and continue to address it. A lot of major thoroughfares have been built. I think we've been aggressive and innovative in attracting public funding for public infrastructures. I believe to date we've

gotten about $270 million in public funding for public facilities, which is appropriate. We create roadways and overpasses to benefit not just The Woodlands, but the broader population in the region, so attracting public funding to do that is important to us. We've gotten support from the county government, the State Highway Department, the federal government. I think traffic works pretty well in The Woodlands. It's a little more congested now than it was 20 years ago, but compared to other places in Houston, it flows quite well.

Traffic is everywhere—it's going to kill us all. What about the future?
Well, as I say, the future looks good for The Woodlands as a community. We have a lot of infrastructure in place to support future growth that does not have to be duplicated. It can be expanded incrementally in a cost-effective way to provide growth opportunities. The challenge is going to be the human side of this business, dealing with people, dealing with services for people. Also, the challenge, if we're going to grow effectively, is to continue to attract jobs. Attracting jobs is based on rents and location and cost of providing facilities, but it's also related to the work environment, the living environment and the quality of life and the attitude people have about a community. It's got to be cost effective. Businesses today are looking at the lower-cost solution. It's a real challenge to build new buildings and attract companies here when there's an overhang in the supply of vacant office buildings throughout Houston. You just can't build a new building and lease it at the same rental rate as existing buildings on the market today. We've got to sell The Woodlands as a job location on something besides rental rates and the cost of the building. It's got to be the quality of life and the related services, like the relationship between the community college and Tenneco Business Services, Inc. The college has an excellent training program to upgrade the skills of employees and train new employees here in The Woodlands. That's a cost savings for Tenneco, which will not have to do that in their own facility. So those are the kinds of partnerships between the educational community and businesses which will attract people and jobs here in the future.

That will also permit you to attract a higher quality of jobs, rather than a minimum-pay job.
Right. Of course there are a lot of minimum-pay jobs, too, as we get more retail in place. We added 4,000 new jobs in 1994, which was a big year, and I guess 3,000 of those were related to retail development of the mall and other facilities around the mall. It's a full range of job opportunities.

What about the completion and effect of the waterway?

Well, we have a lot of construction work underway as we speak about it today. The excavation of the significant part of that is being conducted to provide fill material to build an overpass which helps the transportation issue we talked about. If we're successful, we should get about a $4 million federal grant that will enable the substantial completion of a section of the waterway which will lead from the Pavilion easterly towards the mall. When I say completion, that would be final excavation and grading, bulkheading, pathways and roadways on the banks of the waterway. We will still need to add some additional elements such as street lighting and signage, but I would think that by mid-next year we will have a substantial section of that waterway complete so we can illustrate how it is going to look ultimately.

It's going to have a sizable impact on retail?

Yes, that will attract other businesses. That's part of the environmental development of The Woodlands, the quality of life.

What will the population be at build-out, and when will build-out occur?

If we stay faithful to the plan, the population at build-out in The Woodlands will be about 150,000 people. We've got 44,000-45,000 now, so we're 25 percent there if that's correct arithmetic. It will be substantially complete by the year 2025, so we've got another 30 years to go. I think we will see an acceleration of growth as we go through time, too.

Actually that's occurred, hasn't it?

Yes, back in the '80s we were selling 500 homes per year, now we're up to 900-1,000 per year. I think we'll probably peak out at about 1,200-1,300 homes a year. That's about as fast as the infrastructure can keep up with growth—public schools and other services that people need. One of the things we're seeing, more and more people who are "empty nesters," whose children have left home, they may be retired, or may be nearing retirement, or looking for a place to spend the rest of their lives, and as those people find The Woodlands, they discover it's a good place to live for personal safety, shopping, medical services, recreation. All the other things that you read about Sun City, you can find them right here in The Woodlands. You don't have to identify yourself as a Sun City resident when you can live here in a more traditional community and still enjoy all the other benefits. I think this is a strong marketing opportunity for us. Perhaps we can accelerate the pace of home sales to 1,600 per year because we would not have the limitation of school capacity to deal with. The superintendent of schools likes that, too. People move in, buy a house, pay taxes, and don't have any children, so it's a good deal for the school district.

What about future economic success?

I think the real opportunities for earnings in the future will be related to an annuity that we can create by taking advantage of this franchise that we've created here. We can own an interest in office buildings, own an interest in shopping centers, own an interest in apartments and we can manage those and get property management fees and leasing fees and we can participate in future revenues of The Woodlands by virtue of ownership of income producing property. When we get through developing the land, we don't just shut it down one day and walk away, we have an ongoing business operation that we created and the value of what we've done has manifested itself in this earning opportunity for the future. I don't look at this as just a land development operation, although that's been the substantial contributor to revenue and profit in the early years. I think we will continue to get that profit and revenue out of land development in the next 25 years, but we will add to that the income from ownership and management of commercial properties.

That was part of the original concept, wasn't it?

Yes, we're defining it better as we go along. We've created some strategic alliances, a word that we coined but it sounds pretty good. It's really joint venture ownership of commercial properties, income producing properties. We've attracted some very substantial investors in that regard. Stanford University is a joint venture partner in developing neighborhood retail centers. We own them 50-50, we manage them, we operate them, we share the income, we share the risk. It's a source of capital funding to develop additional neighborhood retail centers as we go forward. We've done the same thing with office buildings and a REIT owned by Richard Rainwater. It's Crescent REIT, and we just sold an interest in 10 office and technology buildings to that group. We retained a 25 percent ownership and we manage and lease them so we get the fee income. We've done the same thing with apartments with American National Insurance Company and Sumitomo. Of course, the regional mall is a joint venture with Homart (now General Growth), so we've tried to align ourselves with financially capable institutions with good reputations who are leaders in the industry, whatever that industry happens to be, office, retail, regional mall or apartments. We take advantage of the franchise value we created in The Woodlands.

And with each success the value of that franchise increases.

Yes, the advantage of large scale development is that if you insist on quality and if you insist that dollars be spent in a quality way, it doesn't matter who spent the dollar, the value of your remaining land is enhanced. So, if we spend a dollar wisely or we cause somebody else to spend a dollar wisely, we still have 15,000 acres of land to develop and sell, and the value of that remaining asset is enhanced by what's being done by ourselves and others. If you look at an office building developer, say Tishman Speyer or Gerald Hines, when they build an office building on an urban site, the only earnings they get is from the office building itself. They don't benefit from the value of the surrounding blocks of land in the city. Here, if we build an office building, we create value in the remaining land that we own. That's been particularly true with the regional mall. I think we added probably $100 million in value to The Woodlands by building the mall because the retail values and the destination aspect of The Woodlands are enhanced by having a mall here. That will create additional value beyond the mall itself.

What have we failed to talk about?

The Woodlands is getting to be nationally known. It's nice to go across the country and have on a name tag or something that says "The Woodlands" and people say, "I've visited The Woodlands." Usually I ask them, "What did you visit?" And it's amazing how many different things they come up with. The conference center has created a lot of national recognition because people from all across the country come here to meetings and seminars and conferences and things of that nature, and we're doing a pretty good job, so that impression is good. People have seen The Woodlands on ABC Sports when the Shell-Houston Open is broadcast each April, so that gives us national recognition. I think generally people in the environmental community recognize The Woodlands as a place where careful thought has been given to working with nature. It's become something of a symbol of how development can be done in a way that's environmentally sensitive. That's good. That's a good marketing thing for us and the way development should work. A lot of people live here now who have relatives in other parts of the country who have become acquainted with The Woodlands, so for a host of reasons people are now aware of what The Woodlands is and I think that will be a plus for us in the future, too.

Well, with 44,000 more or less, we are also getting to be one of the larger cities in the state.

A study was done by Barton Smith at the University of Houston and based on his view of our projections, they seem to be realistic. So upon completion of The Woodlands, South Montgomery County about the year 2020 would have a regional population and economic strength of probably Austin or El Paso. If you look at the economic benefits that are coming from The Woodlands, it's going to be a sizable economy compared to many other places in the State of Texas.

Thank you, Roger.

(Roger Galatas in October 1997 became chief executive officer of The Woodlands Operating Company, L.P., which was known as The Woodlands Corporation before it was acquired by Crescent Real Estate Equities, Ltd., and Morgan Stanley Real Estate Fund II, L.P., on July 31, 1997. He resigned in October 1998 to open his own consulting firm.)

Today is Wednesday, July 17, 1996, and this is Joe Kutchin. I'm in my office in The Woodlands with Joyce Gay, an old friend and veteran of the company. Joyce, when did you join Mitchell?
1957 or 1958.

Give a little about your background. I remember you're from Louisiana.
Yes, New Iberia, Louisiana. I'm married and have three children and I had the three children when I went to work. I went to work for Merlyn Christie first. He was chairman of the Board at that time.

Of what company?
Christie, Mitchell & Mitchell, and this was 1957. I worked for him for 18 months. After a while, I decided to leave. It turned out that George's secretary was leaving as she was pregnant and so it was decided that I would take that job. George handled it with Merlyn Christie and that was the way I went to work for George. Everybody joked about my lateral transfer a lot, but I always thought it was one of the smartest moves I've ever made in the company. Merlyn was very inactive by then, so the job wasn't as interesting as working for George.

George was pretty much running the company?
Yes, he was the president. They had made Johnny president first, but that was before I was with the company. When I came, George was the president and Johnny kind of went out and made deals but George was running the company and Merlyn was very, very inactive.

You were in the Houston Club Building?
Yes. We were there until we went to One Shell Plaza which was just maybe a year or two before I left the company.

You've known George now for about 40 years. What was he like then?
I remember he drove an old pink Cadillac when I first went to work for the company which sounds more outlandish now than it was then because pink was popular, but he still lived pretty modestly. He lived on Memorial Drive but it was not a home they had built and it was not a real grandiose home by

any means, not by Memorial Drive standards, and he had all of his children except one when I went to work for him.

Nine out of 10, you're saying.

I remember all of their pictures were lined up on one wall in his office and a man came in one day and said, "Aren't you girls around here afraid of that track record?" and I said, "We don't even look him straight in the eye." And I used to say, "He won't have another one because the next one would cover the light switch." The offices then were very modest. George took a long time in really going to a higher standard of living. He lived real well, but huge homes and great big fancy cars and jewelry were not ever that important to him.

Still aren't, as far as I can see.

Right, and they were both very much into rearing their children. They had help but they really reared their children themselves and I think were very good parents. He bought Merlyn Christie out while I worked for him for, I think, $13 million and that was headlines in the paper. I remember the mountains of legal documents that were drawn up for that sale and how exciting we thought it was. This was a very nice time in our lives. I remember when I got the job—my twins were almost two years old—and I was so excited that I got home and realized I hadn't even asked how much money they'd pay me. George was extremely active. The first thing he did the whole time I've known him is come in, throw his coat down on the chair in the secretary's office and go from department to department checking with the head of the department and the petroleum engineers and the geologists. Of course, that was his background, so they were kind of his pets. That was the first stop he made until he went into real estate and then he got involved in that, too.

In the early years the company was not significantly involved in real estate?

No, we weren't much involved in real estate until we started The Woodlands. George was always very, very nice to work for. The people who knew him personally all liked him a lot. George is not status conscious. I think he treats everybody alike. I think he's as nice to the janitor as he is to anybody else, so people liked him a lot.

Have you an idea of how many people were in the company when you

came?
In the central office, I doubt if there were even 100 of us. I'd say more like 50. And then you had the people in Wise County. No, I don't remember how many there were.

With George, Johnny, Merlyn Christie and by that time Budd, you were working with a lot of rugged individualists.
I learned a management style that has been to my detriment the rest of my life. They used to joke and say it was management by decibel and they would scream in the offices, in the halls, everywhere. George and Budd would scream in the hall until Budd had arrived at the four-letter word and then George would yank them all back in the office and shut the door. At one time, they even put soundproof taping around the doors between my office and George's office so that the people outside couldn't hear them. I never thought it was vicious. It was just their style–they'd scream and yell.

I know Budd and George did that—Merlyn and Johnny, too?
No, no, but they weren't really as active either. The decisions weren't made that much with those two. It was Budd and George the most and then other people had to learn, including me, that if you didn't yell you never got to say anything. They'd interrupt you in mid-sentence and if you let that stop you, you'd never get to tell them what you wanted to tell them. You pretty much had to jump right in. I remember one time when they were on something really important, one of the lead attorneys with Vinson Elkins came out and sat in my office and shut the door between George's office and mine and said, "I'm not going back in until they quit yelling." It was a good five or 10 minutes before the rest of them realized he wasn't there.

You did not have at that time a role in public relations, did you?
No, I finally had a series of promotions. I was still at my secretarial desk. First I was secretary, then I was administrative assistant and then I got the title of assistant vice president.

At that time were you at One Shell?
No. I was still doing some secretarial work, but I also got involved in writing an in-house newsletter and I got involved in news releases but still doing my secretarial work. Then, George and I shared a secretary and then I became a full-fledged vice president and we each had our own secretary. By then we were at One Shell Plaza. And then I didn't do any secretarial work any more,

I just did public relations but I always did it with you as my outside consultant. There was a lot of background I didn't have. I had a lot of company background, but I didn't have that much public relations background and so I worked with you on the things that I didn't know how to do on my own.

That was about 1972, I think, when Harshe-Rotman first got involved and by that time the company was at One Shell. The story is told that One Shell was so desirable at the time that an offer was made to the company if it would relinquish its space.

Yes, but we didn't. I was the point man that the decorators worked with, but I didn't actually do the decorating. However, I had gotten to know George real well and could almost always anticipate his preferences. They would show me several things and say, "Which one do you think he'd like best?" And I would select one and I was right so many times that it got to where we didn't bother him very much any more because I really had kind of learned what kind of decor he liked. He was always a very contemporary man and liked very contemporary things. If there was a wall covering he would definitely go for the one that was more minimalistic than the others. He liked clean-line furniture and things of that nature. So, I did work with the decorators a lot, not just on George's area but on all of the areas, not actually doing the decorating myself, but I was their clearinghouse.

I believe that you migrated into being the person responsible for the art program but I get mixed up as to whether that was here at The Woodlands or at One Shell.

That was at One Shell. When I went to work at the Houston Club Building, everything was already in place and we didn't do any decorating there, not during my time. It had been done and they weren't redoing. When we went to One Shell Plaza I worked naturally into the art program, because we were selecting art when we were decorating those new quarters. Art was something that I didn't have a formal background in at the time. I later went back to school and had a minor in art history, but I had worked with the visual arts a lot, I was just always interested in it and always went to a lot of art galleries and museums and things and so I was sort of self-taught. I worked with the designers in selecting the art.

Who were the people you worked with?

Tom Jackson or Bruce Monical. Monical owned the company, but Tom Jackson was the person who handled our account. In fact, when I opened up an art gallery of my own, Tom Jackson went with me, but that was later. We selected the paintings together; we weren't into sculpture then.

Let's not leave One Shell yet. On what basis did you select the paintings? What kind of suppliers, what were your standards, what were you aiming to do?

The first criterion was that we selected well-established regional artists. That was one guideline.

Were you working alone or did you have a committee?

No, there was no committee then. Whatever Tom Jackson and I thought was pretty much where we went. Now the executives, like the senior officers, very much had a say on what went into their offices. We would make a presentation to them that we thought was acceptable, and from there they would select what they wanted. We had already weeded out anything that we thought wasn't acceptable. We'd go to the leading galleries in town and even though we didn't have exactly a formal budget, we definitely knew we weren't going to be spending the kind of money we'd need to spend for a national name. So we bought well-established regional artists; it was real good art. Of course, we selected what went into contemporary decor. You always have to take into consideration the environment, the colors, the carpet, everything has to harmonize and so those were basically our guidelines—the environment in which the art would find itself and selecting from well-established regional artists. I don't know if there were any guidelines other than that. Mostly we selected paintings for the walls; we were not into sculpture then. We only had the offices, we didn't have The Woodlands yet, so there wasn't any great outdoors to place sculpture in. We did select a few table-top pieces. I definitely remember a marble piece by Charles Pebworth that went on a pedestal in the lobby. So there were a few pieces like that.

At One Shell Plaza you had at least two full floors.

We started off with two, and then we needed more almost immediately. The more important pieces went in the executive suites and the public areas. We always selected things that were, of course, in very good taste, I think. We went into prints. They weren't always signed and numbered, but a lot of them were.

Do you feel that you were kind of on the ground floor with anyone who later developed into a bigger name?

I can't remember if we ever bought any Donald Roller Wilson for the company. I did for myself, and financially, it's the smartest move I ever made in the art world because he's really big now. His paintings were a little bit on the bizarre side. I don't know if the Mitchells bought him. Everybody we selected already had a name. I didn't take any chances with this collection and so if you're asking whether we bought someone who went from being a well-established regional artist to a well-established national artist, right off the top of my head I can't remember. In the sculpture program, we have works by some people with bigger names than they had when we started out. But when we bought them, they were already established, so I don't know if anybody made a great big quantum leap.

At the same time you were also responsible for PR?

Yes, we did the annual report, the house organ, news releases, press conferences. I later went on securities analysts meetings around the country, setting up the dog-and-pony shows. We produced the literature that would be taken to various financial seminars in New York and elsewhere.

Did you do the one for Galveston with all the controversy about drilling offshore?

Yes. That was quite an experience. We had a public hearing in the Moody Center that's on the seawall. The Corps of Engineers was there and George's people were there and the public was invited and the room was packed. It was a very emotional hearing and I remember that Jon Conlon, one of the men who worked for me, was in the middle of one of the aisles setting up a mike so that people from the audience could leave their seats and come and ask a question directly of our people and while we were standing there setting up the mike, one woman said to another woman, "She's one of them." I thought we were going to get tarred and feathered. It was a rather hostile group. There was some emotional testimony. One woman got up and cried and cried because she said tar had gotten on her little girl's bathing suit and of course tar has been in Galveston waters forever. I felt that she cried because she was nervous, but there wasn't anything to cry about. For a little girl from southern Louisiana, a lot of this was very exciting stuff. I'll have to say I rather enjoyed it. I loved it. One of the big benefits of working for this company was the people I met. George was important enough that there were people of note who came to the office. Senators came to the office. I always get tickled at one. He was known as being a big glad-hander. He walked into my office and said, "My, you people always keep these offices so beautiful. Everything is always pristine." This is when we were in the old offices and he walked out and I looked at my door jamb that was full of old, dirty fingerprints and I thought, "What's he talking about?" But he was like that, he was always saying the right things. Now, if he'd come back a couple of years later, we did have a beautiful office.

I was knocked out when I first came here. It was really beautiful.

The offices in the beginning were not that outstanding. You walked into a little bitty reception area that had a little glassed-in box where the two switchboard operators were and they were also like the receptionist. We had a little receptionist's desk but it was always empty. The switchboard operators did double duty. We had a little room that had four tables in it and that was the coffee bar and my office was so small you couldn't have put another folding chair in it if you wanted to. George had an office that had been real nice, I'm sure, when it had been done but it was passé by the time I went to work for him, and they had never, ever touched it. But the three main people had nice-sized offices. Budd's office was a postage stamp office, real, real small. There wasn't anything flashy about those offices. They just didn't think that way in the beginning.

I first became involved with the company, of course, in 1972 right after the public stock offering and before the HUD loan guarantee came through. Those were very exciting and very stressful times.

Well, we had always had a volatile company. I told you about the screaming and everything, but I don't think any of it was vicious. I really and truly think people were enjoying themselves when they were screaming, and it was just a management style. The level of volume went up but everyone was working towards one goal. I don't think I'm remembering that with rose-colored glasses. I think that's the way it truly was. We were a house divided after that. The real estate people came in, they were a different breed. All of a sudden there seemed to be a double standard and, of course, then the people on the oil and gas side were not happy. It was really a bad time and it was extremely uncomfortable for me because I was supposed to straddle both sides. I had to work with both, but I was never able to work very well with the real estate people.

And finally in 1974, spring, I think it was, you left?

Yes. There were any number of reasons why. The most important was that my husband had grown in his company and was doing a number of things that were exciting. He was on some boards in Washington and he was taking some trips I couldn't take with him. I felt the need to become more involved in his business.

Let's go to the sculpture part of the program. Did that get underway while you were still at One Shell?

Yes. In fact, I was totally responsible for everything that's out here up until about a year or two ago with only two exceptions. I did not commission the piece that is in front of the hospital. That was done by Mr. Cooper, who started the private school. The reason he did it was, he had access to some people who were willing to contribute for that piece and he wanted to do it and so he did. Then the other piece is the serpent in the lake that Cynthia Mitchell commissioned. She knew the artist.

She actually, besides that one, bought some that are at the Pavilion or right next to the Pavilion.

If there's anything in front of the Pavilion, I didn't do that. But there are a number of pieces in The Woodlands now and I commissioned all of those.

Those are all commissioned?

Yes, some of them were made especially for us and sometimes I bought something that had already been created but that I thought was really good. And the way we worked in the early days, I was the one in charge of the program. I did the research and I brought the pieces for consideration to The Woodlands. George and Coulson Tough were the only two people on my committee and it stayed that way for a long time. Then, Ed Lee wanted on and he stayed on until he died and then Roger Galatas took his place and then Roger wanted Dick Browne on. Actually, Dick Browne had a good background in art. George just automatically goes to what's good. I've never seen George make a bad decision in art. I'd see him walk through an art gallery and automatically go to something that was very, very good. No one directed him to it, but he just automatically knew. And Coulson Tough had done this kind of work a little bit for the University of Houston and he brought sensitivity to the job, he didn't have an art background but I felt like he understood what he was doing. The larger committee, I felt, didn't work as well. I generally would bring about three or four or five pieces for their consideration, all of which I could live with. I didn't bring them anything that I thought was wrong for the environment. I always had a preference as to what I would like. Cook came with a good reputation. I knew I wasn't taking any chances. Cook's pieces have an energy that I find very intense. In some areas maybe that's what you want, but in a lot of environments I just knew Cook wasn't right because his are always very high-energy pieces and that's not what we'd be looking for. You don't just think about the artist and his background and how much talent he brings to the possible commission, but you also have to consider the environment within which this is going to be set. Is it going to be viewed from a moving car, is it going to be viewed by a lot of people or just a few people, what kind of background, what are you trying to say? A lot of things have to be considered before you place a piece in a certain area.

It seems that you did a fair amount of Pebworth.

The Family piece at the entrance to The Woodlands. There again, Pebworth is one of your better regional artists and he got the commission for the front entrance. It was supposed to be cast bronze, but a cast bronze piece that large would have been beyond any budget and so it was fabricated. I was not real happy with that decision. I didn't think the piece translated as well in fabrication as it did in cast bronze. You went to flat planes and it gives it a contemporary look which George likes, but it also did some things to the piece that I didn't like. But it's a good piece, and yes, Pebworth also did metal reliefs. He doesn't paint but he does metal reliefs for the wall. And he did table sculptures, mostly in Italian marble.

Do we have those? Would they be at the Inn?

One piece I kept seeing being moved around. It's sort of a black dome-shaped piece, very abstract, black marble. He doesn't work that way any more. The Woodlands probably has four or five pieces of his, which is more than of any other artist.

Again, the same question that I asked about the paintings and I'm guessing that the answer is the same, any noteworthy acquisitions?

What I need to do is send you a list and sort of evaluate them, because right off the top of my head it's not going to all come back to me. Some of these pieces are awfully old and I'm not able to remember. One of my favorite pieces is a very representational piece which is in front of the back entrance to the Inn. It's a boy and a deer. The sculptor's also the one who did the boy and the seagulls for the Mitchell corporation in Galveston, next to Harbor House and Willy G's. He comes close to having a big national name; in fact, in art circles he does. The buying public doesn't readily know him as much as the art world but he does have a national reputation in the art world and he's very, very good. He has two sons who followed in his footsteps and one of them did the Elvis Presley piece in Memphis. And there's a Corbin Bennett piece in front of the MND Building. It's been nicknamed the "Pringle." Cook, who I mentioned a while ago, comes close to being a national figure. But none of these are known like Picasso or somebody like that.

Let's go back to the time that you had the Ars Longa art gallery. When was that?
I must have started in 1973, and I think we closed in about 1975, maybe 1976.

And during the time you were there you were the art and sculpture consultant?
Yes. The gallery didn't belong to the Mitchell company in any way, but they were my biggest client. We had a complete corporate department and we had two sales people who did nothing but call on corporations. So they weren't our only account by any means but they were our biggest account, mostly because by then they had decided to put one-half of one percent of all their building costs into outdoor sculptures.

The outside builders as well, not the residential builders but the commercial-industrial builders? One-half of one percent of the total cost of the building?
Actually, I migrated to handling nothing but outdoor sculpture. I didn't do paintings anymore. These offices were already designed and it wasn't like you were opening tons of Mitchell offices here at The Woodlands. More were opened and some art probably was bought for them. I didn't do it. I decided to stick strictly with the sculpture. You work less and make more money.

Do you feel that The Woodlands needs a centerpiece?
It needs an artist who is a household name. It needs one piece to give it the authenticity and the weight that a piece like that could bring to this collection. It would totally change the complexion of what they've done here, but we're talking probably a half-million dollars, maybe even a million dollars, and they haven't wanted to do that yet.

Let's talk for a moment about the Joyce Gay Report. What were the

years that you had that going?

I did it for eight years and I've been gone about four or five. The program idea started without me. There was a little program that Randy Woods hosted on The Woodlands cable. There was a man on Louetta who had a little cable company and they did some broadcasting from there. They had started something and that was when I happened to mention to George that I had gone with my husband and some of his executives to do media crisis management training, and they liked the way I handled myself, and I mentioned this to George and he said, "Well, we've got a little program at The Woodlands, why don't you do that one?" He didn't even know Woods had started it. They talked to me and it was decided I would take Randy's place, and Randy is the one who came up with that name. At the beginning, mostly what it was, was almost a calendar of events for The Woodlands. You kind of talked about some of the nice things that were happening here and you interviewed some of the people and we expanded from that.

At that time did you have your own equipment, your own production capabilities?

No, I was still working with that company on Louetta. I went to work over there thinking, "I'm going to be surrounded by professionals who are going to show me how to do television," because I didn't know the first thing about television. I'd stood up with a mike and talked to a room of people, but I hadn't done television. It is a different technique and I did have to learn. I always said I was surrounded by a troop of Boy Scouts because they were a poor, small, new company that couldn't afford to hire anybody but kids right out of college. They had a lot of vim and vigor but not any experience. Finally, I went to the Mitchell company and said, "I don't want to stand in front of the camera making a fool of myself. This is a lousy program. I will do this: If you will give me a contract, I will buy my own equipment and I will hire better technicians. You don't have to do anything you weren't going to do anyway, but any video work you are going to do is mine." And I built from that. That contract allowed me to go to the bank and borrow the money I needed and then I hired a salesman to call on other corporations and we did ads, promos for other companies and we started doing longer promos for the Mitchell company.

Your facilities were in The Woodlands?

Yes, JG Productions. What we did at first was, we did the Joyce Gay Report that was aired on that little cable company on Louetta. Also, there was another smaller program that was now originated from The Woodlands. They had a smaller budget than mine. We produced both of them.

The other one was a separate program entirely?

Yes. We had nothing to do with content, we just showed up with the cameraman and they did hardly any editing. So it was a very raw, inexpensive little program, and it truly was kind of a calendar of events for The Woodlands. It never progressed beyond that or went to PBS when I did. Finally, and I think George is the one who saw this, it was decided that we should try to be aired on PBS, locally anyway. George saw it as a way into a bigger audience. I tried to get on PBS for two or three years before I made it. They don't just accept every program that comes along. They told me some of the things that were wrong with the program before they could consider it and so we worked real hard at changing what was wrong and we obviously made it because not only did we get on PBS but we won a lot of awards. Right now at home I have three awards that are called "The Telly." They are the highest award you can win for non-network broadcasting, and, of course, from the Houston Festival, I could paper my walls with all the awards we've won there. Some other things happened that I was real proud of. We did a program on the oil crisis which was picked up by many, many, many PBS stations around the country and it was shown to all of Congress. Lloyd Bentsen arranged for all of them to see it. Lloyds of London said they thought it was the most informative piece that they had seen done on the problem and they bought hundreds of copies and shipped them to their people all over. So, we were the proudest of that piece.

Just for the record, the Joyce Gay Report program was essentially a series of interviews?

Not entirely. It was a news magazine and it was sort of on the order of "60 Minutes" in one way. It certainly wasn't "ambush" interviews or anything of that nature. We didn't conduct hostile interviews of any kind, but it was like "60 Minutes" in that we had segments, we went on location, we were not just talking heads. We shot a lot of B-roll, in other words, visuals–and so we would edit in a lot of action with the talking heads. The greatest part of the job for me was the people I met. We interviewed senators, we interviewed former President Carter, we interviewed Nobel Prize winners. We did Al Gore, we did Governor White and we did Governor Clements and we did John Connally. We have never been turned down by anybody we wanted to interview except one and that's a funny story: Phil Gramm turned us down when we were doing the oil and gas show and then when he found out that Lloyd Bentsen was showing it to Congress he asked them not to show it until they could do a little trailer—a little few feet of video—and they tacked it on to the tail end of our show so that he would be in it. But the reason he stands out in my mind is that he is the only person in the eight-year history of that show who said no. A lot of well-known senators are on the show. Now at the Medical Center, we had all of the well-known doctors there. We had Denton Cooley, Charles LeMaistre, who is head of M. D. Anderson Hospital. In fact, there's a funny story: I did a show for the 50th anniversary of M. D. Anderson Hospital that won a really important award, and I was at M. D. Anderson myself once for a little exploratory operation and the doctors found out (I didn't tell them, but someone found out) who I was and that certainly explained the wonderful treatment I was getting. One doctor came and sat on the foot of the bed and said, "I understand you're doing this show about M. D. Anderson Hospital and I want to be in it." We went around the country when we were doing the new town here; I did ask special permission from PBS to do a new-town show because I knew how they might look upon that as being self-serving.

You did a number of specials, didn't you?
We did some documentaries.

The new town would be one of them?
The new town was a documentary, the oil and gas show was a documentary. We did a documentary on George Mitchell that won some awards. The documentaries were picked up around the country. Mostly the shows appeared on Channel 8 here in Houston. Then we got to where some other markets, PBS markets in Texas, were picking us up on a regular basis but the documentaries were picked up all over the country. We did a Christmas show once that not only was picked up, but was run around the country three years in a row. We kept getting requests for permission to show it again.

And then towards the end, the budget got tight and you had done as many as you wanted?
Actually they never canceled me. I canceled because I moved out of town and it got harder and harder to do. In fact, when I decided enough was enough, they asked me to do three documentaries a year, and I said that would be more work then just doing the regulars. Three documentaries a year is a lot of work. So we thought, well, let's try for one and actually I never

even got the first one off the ground. I truly was finished. I had gone to Bastrop. I was spending more and more time there. I was enjoying it. I felt that even a good show runs its course. If I hadn't been doing anything else, one documentary a year would have sounded great.

You've given me some good material. I asked you about your personal background early on, and I want to get on record that in your retirement you completed work on a college degree.

Yes, when I started here I hadn't gotten my degree. I actually got that after I left. I got a double major and a double minor, a major in political science and English literature and minors in art history and French. I graduated with honors from Houston Baptist University and, of course, I've been very active in the business world even after leaving here. I'll have to say I certainly learned so much here, mostly by observation. I don't think there's a chair in the house for learning about a company better than the chair of the boss's secretary–you hear everything that goes on. In fact, that was truly the way I worked into public relations. I had such a knowledge of not only what the company was doing but how it felt, what philosophy it operated under. I had watched the executives of this company operate for so long, I really felt that I knew how they thought and what they brought to the table and I definitely felt that my biggest advantage was that I knew the kind of image George wanted to project.

Joyce, thank you very much.

David Gottlieb

Today is Wednesday, August 23, 1995. This is Joe Kutchin. I'm in the office of David Gottlieb at the Cynthia Woods Mitchell Pavilion in The Woodlands. First, David, tell me what your exact title is, please, then give a few minutes worth of your personal history.
I'm currently the president of the Cynthia Woods Mitchell Pavilion. The Woodlands Center for the Performing Arts. On May 1, 1995, I took on the additional role of executive director/manager on a full-time basis. Prior to that time I was on the Board of the Pavilion for about two years, and during the past 11 years, I have been vice president for institutional development with The Woodlands Corporation.

That was starting what year?
I began in 1984 and retired at the end of April 1995.

Go all the way back.
Prior to that time, I was with the University of Houston where I had been the dean of the college of social sciences, a professor of sociology, and also special assistant to the president of the University of Houston system for a period of time. Very early in my career at the University of Houston, I would say 20 years ago, I was first introduced to George Mitchell and became involved in the series of bi-annual conferences that George Mitchell wanted to see launched, and was in fact funding, that dealt with issues of limits to growth in sustainable societies.

I'm going to ask more detailed questions about that, but now I'm just trying to hit the high points of your career. Let's go back to prior to your coming to Houston.
I came to Houston about 20 years ago, 1974, in fact. I came to Houston to be in the department of sociology and soon after that was appointed to be the dean of the college of social sciences. I did that for five years, then was in different academic positions, including an administrative position as a special assistant to the then head of the system, Dr. Bishop. During my career at the University of Houston, I was involved as a principal investigator in a number of research studies funded primarily by the United States Department of Labor and National Science Foundation dealing with subjects mostly related to youth and labor markets. An exception was a study of

Texans' attitudes towards energy—the knowledge and awareness of energy issues. I was also one of the three founding fathers of the Forum Club of Houston at that time, with Fran Steckmeist of Shell and Ben Woodson. I served on a number of local committees. One was the Greater Houston Partnership committee dealing with education in the City of Houston. I was on a National Science Foundation Advisory Board and at one time was a member of the President's Commission on Science, dealing with the behavioral sciences.

You're from Detroit, I know.

I'm originally from Detroit, did my undergraduate work at Wayne State University in Detroit, my doctorate at the University of Chicago. After I graduated from the University of Chicago, I went to Michigan State University. That was my first academic appointment. I was there several years, took a leave of absence, went to work in Washington in the Office of Economic Opportunity and then later in the Department of Labor. After completing my Washington stint, I went to Penn State University where I chaired the Department of Community Development.

What year did you go to Penn State? How long were you in Washington?

I was in Washington for a total of three years, then went to Penn State University. While at Penn State University, I took a leave and went back to Washington where I was director of research for the 1970 White House Conference on Children and Youth. I went back to Penn State and stayed there until I came to the University of Houston.

Your first contact with Mitchell Energy & Development and George Mitchell was in connection with the series of conferences on growth issues?

Right.

And what was your role?

Well, the way that came about, I received a phone call from the then president of the University of Houston, Phil Hoffman, who I later found out was a friend of George Mitchell. He told me that George Mitchell, really at that point I had no idea who he was, was interested in doing a series of bi-annual conferences dealing with issues of sustainable societies, limits to growth, etc. Phil Hoffman said that he wanted me to represent the university because evidently they wanted the university involved in coordinating the programs.

This would have been 1974?
Right. It wasn't something that I particularly was interested in intellectually. I thought it was pretty much futurism which did not have a very good name in the behavioral sciences at that time. But I did get involved.

Others from the university had roles, as well?
At that stage Allen Commander was involved, but I can't remember to what extent. My feeling is that it was a short period of time and he had a pretty limited role. I think soon after that he moved into some other administrative responsibilities.

I think he did go through the conference. I remember a very heated conversation between him and John Naisbitt after the conference. My recollection is pretty clear on that. *(Later note: My clear recollection seems unreliable; most likely the conversation in question took place prior to the conference.)*
In fact, my feeling is that for the first conference, my role was quite marginal. I was much more active in the second conference, where Dennis Meadows was involved, and I think the third one, but far less in the first.

Tell me then about the second one. As I remember that was run by Willem Boichel of the Society for International Development in Washington.
Right, and also involved in that one at that time was a guy who worked for the University of Houston who was quite prominent at that time, Roger Singleton. He was working out of Phil Hoffman's office and it wasn't really clear to me what his role was other than I know that he controlled the money that the university had in regard to this project.

I believe he was supposed to control the company's money, as well.
At the same time, of course, George Mitchell was giving money to the University of Houston in his effort to create a University of Houston freestanding campus at The Woodlands.

And you had a sizable role in that?
Over time, I had a sizable role.

Let's come back to the growth program.
At the first conference my role was quite marginal. I can only remember one meeting that I attended with Naisbitt. My involvement with the second conference was greater, and I remember visiting Boichel's office in Washington and wondering how, given what needed to be done in the international aspect of this and getting out all kinds of invitations, planning a program, how this guy was possibly going to do it because it was like a little "sweat shop." I remember also we had some international meetings. Was it the second one where we came up with this very elaborate process for evaluating papers to judge who the winners were going to be? I remember there were different categories, and I helped sort out and stamp papers for judging. Dennis was involved and played a pretty prominent role.

He was prominent in the first, too, with Naisbitt. They shaped that program, and the first time it was done it was gangbusters. It was a tremendous success, but it kind of slid downhill after that. The novelty wore off. Why don't you talk about the third one that you managed with Harlan Cleveland?
Harlan Cleveland and Jim Coomer from the University of Houston at Clear Lake. What I remember about that was that Harlan had his own agenda and he had his own collection of friends and he wanted very much for it to be tied in with pieces of Aspen, because he was involved with the Aspen group. I don't think he really was all that committed to the same theme George was. I think Harlan was very much more interested in this whole business of leadership and where would future leaders come from and would we ever have the Harry Trumans and the Franklin Delano Roosevelts and the Winston Churchills again.

Was it your feeling that he pretty much set the course for that conference?
Yes. I think we really tried to stay within the bounds of what we thought George was interested in, but I feel that wasn't always too clear either. We weren't sure precisely what George had in mind other than calling this issue to the attention of people because he had been heavily influenced by the Club of Rome and the *Limits to Growth* book. And I think, also, there was the hope that some of these papers would get international attention and lead to policy changes. I never really saw how that would happen. I never saw the route that the conference results would take to be turned into national policy issues, or international policy. There was an attempt to involve people

from the United Nations. In fact, Harlan had the head of the environmental issues section, I forget his name, attend the conference. We got involved with some people in Washington, D.C., too.

If we're thinking about the same thing, you and I visited the National Academy of Sciences, but I think this was at a time that you were acting president of HARC.
We met with Gus Speth of the World Resources Institute. In between there also, we held a conference . . . we did one on Mexico.

That's the one that you and Tony Lentini worked on?
Yes, Jimmy Carter came. That was done under the auspices of HARC. Skip Porter was then on board. I remember he did one of the opening speeches. That would have been about 1984.

Didn't a book come out of the conference with Harlan Cleveland?
Yes, by Jim Coomer and Harlan Cleveland. Harlan had a lot of contacts, and I think that's how we got involved with the UN people. And I also remember going to Washington and meeting with Peterson and then William Ruckelshaus, who was then head of EPA. You mentioned the point that the first conference probably was the best one and I don't question that. I was very impressed with the press attendance. There were people from all over. I was surprised that we got that kind of coverage.

From the viewpoint of press, it kind of went downhill after that. At the first one, we had immense coverage and we really had an all-star program. That was really largely thanks to John Naisbitt and to Dennis Meadows, whose *Limits to Growth*, the book, was still fairly current. The second conference was not bad as far as press was concerned, the third a little less so and then gradually the press found that there was nothing there for them. In that respect, were you especially pleased with any of those conferences where you were closely involved?
It's such a jumble to me now. I was trying to think about it because I knew it would come up. The thing I remember being most involved in, or devoting most of my attention to, was the Mexico-U.S.

That was knock-out for press.
It was something that George was particularly interested in because he was

always indicating that Mexico–you know, the whole business of when we sneeze they get a cough, or vice versa. I thought that from my point of view, it was a pretty good conference. It got a lot of good media, but as is the case with all of them, in retrospect you say, "Well, what came of it, what can you point to and say that this has somehow made a difference or had an impact somewhere?" But on the other hand, I can say that of every conference I've been a part of. I remember working in fairly high leadership roles in two White House conferences on children and youth and I'm hard pressed to find a recommendation that came out of those that made any difference. I guess one of the things that I concluded from that is that there were a lot of people out there who were quite willing to take advantage of George's money and were going in the wrong direction toward their own ends, with their own agendas. If asked if it was worth all the time and effort, I guess the answer would be no. That doesn't include the first conference because I don't really remember that much about it.

I've been close to at least all the conferences from the first and on through the mid-'80s—the first conference was a big assignment when I came to the company in 1974. I think you hit it on the nose: When you look back in 1995 and say that with all the effort and all the expenditure and all the high-powered intellect that we got together here, what came out of it? Very little. I'm not sure what they accomplished.

I agree. By the way, I also have the sense that it's less of a topic of importance to George than it was before. That may be reflective of the fact that it's not a terribly popular topic right now.

And as time goes on, most of the basic thinking that was put forth then has been pretty much shown to be wrong.

In fact, now that I think of it, you asked me what do I think is one of the highlights that came out of all this. That would be the wager between Julian Simon and Paul Ehrlich. And Simon won. I remember that–that was pretty exciting–that was one of the results of this conference.

There was an article in the *New York Times* magazine which covered that bet pretty well. This must have been three or four years ago, maybe longer. I made sure George Mitchell got a copy and he responded to the effect, "Very interesting, but not enough time has passed."

I know that Julian Simon came and spoke to the Forum Club of Houston. *The Journal of Social Science* (published at UT Austin) devoted practically one entire issue to this, Simon and Ehrlich. Now, somewhere along the line, I got involved with HARC.

What do you prefer? Do you want to talk about HARC first or do you want to go back and cover the U of H campus? First of all, you have absolutely no relationship any longer with the growth program that happened for 10-15 years?
No, I haven't since the time that it was taken over by HARC. I think I was involved in the work on the first issue of the newsletter that the Mitchell communications department prepared for them. Is that still published?

That's dead. OK, go back then to the university campus.
I was aware early in the game that this guy George Mitchell wanted to do a campus of the University of Houston and I became more actively involved just immediately following my first work with the growth conferences.

This was in your position as assistant to Bishop?
No, first it was while I was dean of social sciences. George was asking me questions about the campus and you know how he'll go to anyone and say, "Go do this." And so I started to try to learn about the history of it, etc., and got directly involved under Bishop in making presentations to the Board of the University of Houston, to the various presidents of the other campuses, trying to generate their support for going back to the Coordinating Board and getting legislative approval to start the planning process again.

Put this in some kind of time frame if you can.
This was in the early '80s and Coulson Tough had been asked by George to monitor what was happening to the money that he had given the University of Houston to plan and to advocate this campus. We finally did get a plan together, we had a proposal and it did go to the Coordinating Board. The plan responded to three of the objections raised earlier by Kenneth Ashworth, who was and is commissioner of the Coordinating Board of Higher Education for the State of Texas. The first time, this thing had just about gone through the Coordinating Board and gotten its approval–this was in the late '70s–a member of the House of Representatives who was from the Huntsville area, I can't remember his name but he was with Sam Houston State University, came in and pretty much said, "If this goes through, it will

be the demise of Sam Houston State University." So there were three restrictions that were placed on the proposal, and the Coordinating Board was saying, "Until you meet these three requirements, don't come back." One was that Sam Houston State University's enrollment reach at least 10,000, another was that 750,000 people be residents within a 15-mile area, and the third that The Woodlands campus commit to offering only the junior and senior levels so as not to conflict with a freshman and sophomore community college.

The Higher Education Coordinating Board is a state organization that kind of acts as traffic cop? Their imprimatur was necessary before money could be obtained from the legislature, is that correct?
Before you could go to the legislature and introduce legislation to establish a new campus and then get funding, you had to have their approval. They're the gatekeeper for everything. Any state institution that wants to introduce a new degree program has to have their approval. They set limits on how many courses can be taught off campus, etc. In any event, we came back in 1982 having met all of those requirements.

Who was the "we"? Was the company represented in that when you say we came back?
No, the company was indirectly represented by me.

You were still with U of H, weren't you?
Yes, and my assignment as special assistant to the president was to work on this. So it first went to a committee of the legislature and a lot of people were working on the lobbying part of this. For the company it was John Watson. The governor then was Mark White and Mark White set aside money that the state would allocate for a planning effort to get this new campus started. It was clear with the Coordinating Board that they wouldn't do anything to stop the planning. It would not be like saying, "There will be a campus," but it would be an important first step. That money was the only money that was removed later from Mark White's recommendations on higher education, and it was done not by the people from Sam Houston State University, but by several black legislators who thought it would have a negative impact on Prairie View A&M.

Did it involve a lady from Dallas?
Yes, very influential, DeLoreto or something like that.

Actually she's got a very good reputation.
Yes, she's not a flake. But that was totally unexpected because Texas A&M itself had been an advocate of having a campus here. So it died and it was very depressing, demoralizing because it had come so close. That was it. Nothing more happened on that until several years ago, maybe six years ago, when we started talking about a new community college campus out here of Montgomery North Harris Community College which now is open. It's the new Montgomery College campus of the community college system. John Pickelman came in as chancellor of that system, and he and I had a meeting with George, and Pickelman said he wanted to explore the possibility of creating a university center that would be on the campus of this new college, and while they would offer the freshmen and sophomore years, they'd get a consortium of universities that would be willing to offer the junior and senior years and then eventually some graduate courses. Knowing that having a campus out here, a free-standing college, was not going to happen, we got pretty excited about that. What they asked for is the same thing George had already offered to the University of Houston, comparable to what U of H did in west Houston—that is, put up a building. Everybody agreed to that, and we got the Coordinating Board's approval to go ahead with the planning and the concept. That was primarily the result of the woman from Houston who happened to be the president of the Coordinating Board. She's a lawyer, I'll try to remember her name. Now that position is going to Leonard Rauch, who used to be on the Board of the University of Houston. He is involved with Compass Bank. The concept has been approved, and for the last three years, they've been working with the various universities, have identified what courses. The university center is going to be a go. They've allocated the land.

Going back to early history which I know precedes you, but you may have knowledge of it. In the original land plan, was there a place for the University of Houston campus?
That idea was first broached 18-20 years ago—300 acres were set aside. As The Woodlands developed and nothing happened with the U of H, that was changed. Now, thank God, they never took it because if they had, we'd probably have no Town Center.

Through all this time, you started representing the university. Yet, in a way at that time, you were also representing the company's interests. Then you came to the company.

I went through a period of time where people at the University of Houston wondered who I worked for. I went through a period of time when I wondered, too. I was becoming much more of an advocate for Mitchell, not The Woodlands because that didn't really mean anything to me, but for George and what he was doing, and I really did believe it would be a great thing for the university. My sense was that if they would have gotten on board when they should have, they could have had their first and only residential campus, and a really beautiful one. We talked about the things they could have done here. This could have been their center for performing arts, this could have been the site for their programs for the future, architectural design, urban planning and so forth. They never got it.

OK, so at present, 20-plus years after The Woodlands opened, there really is no profound and continuing relationship with the University of Houston that I know of. Is there one that you can think of?
There is the Texas Music Festival at the Pavilion, there is a professorship named for Cynthia in the drama department, there is the HARC involvement, but I know of nothing else.

* * * * * * *

Today is Wednesday, August 30, 1995. This is Joe Kutchin and I'm in the Cynthia Woods Mitchell Pavilion office of David Gottlieb. This is the second of our conversations. Last week, we talked about what you had done before, about the growth conferences, your work in establishing a U of H campus here. To a large extent, last week's conversation brought us up to the point where you had left the University of Houston and joined the company. But before going into that, I don't recall that you talked a lot about the last growth conference that you and Tony Lentini co-managed where Jimmy Carter was the headliner.
George always expressed a real interest in Mexico, the sneezing, the coughing, who gets sick. Also, there were some connections with Carter because Carter wanted to do this center in Atlanta. George helped arrange for Carter to come to Houston and give a talk.

And George gave considerable money towards the Carter Center.
Yes. Duncan in Houston, who had been secretary of energy while Carter was president, and George co-hosted a breakfast downtown where Carter could talk about the center. But in any event, for the conference that we had organized, we tried to get "X" number of American business people and "X"

number of Mexican business people and some academics and government people, trying to match interests from both sides, and have them come together for a conference dealing with economic and social implications and how do you enhance trade, etc. Mostly under the rubric of sustainable societies. We got a pretty good turnout, and I was impressed with some of the American companies that participated, and Mexican companies, as well as some of the other people from different disciplines and professions. It got quite a bit of coverage and I thought it was a good conference. That was supposed to have been taken over by HARC after that and was to be the foundation of trying to do similar meetings where you could really focus in upon some very specific policy issues instead of grand global notions. Let's take two political, national, geographical entities, particularly at the border relationship, and see what we can come up with. But I soon dropped out of the involvement that I had with HARC. I remember that about that time Skip Porter had been announced as president of HARC and he was at the Mexico conference. I remember he said a few words, and I had transitioned out of HARC. So after that conference, it was pretty much in their hands. I think it was soon after that also that the MEDC communications department initiated the newsletter. In fact, the newsletter did have some things about the Mexico conference.

In my recollection, the very first conference, as I said on the first tape, was extremely successful. And then the Mexico conference, which was not designed to attract as many bodies, was the next most successful.
Yes, it was on a much smaller scale. There were no rewards, no papers, it didn't have all of the dynamics. But still it worked pretty well.

Let's go back to the HARC situation. I don't think we went over that other than superficially. You were the U of H representative?
I think it was when I was dean of the college of social sciences at U of H, and George expressed an interest in developing a consortium of universities. He had in mind U of H, Rice, A&M and UT, and he visited with the presidents of A&M, U of H and Rice, but not UT at that point.

Can you put a year on that, approximately?
It was in the early '80s, very early '80s. The president of each of the institutions expressed an interest because George had pretty much said, "What I want to do is create this and I will give the land and I will give start-up money" and made it pretty clear that there would be no cost to them. It was

a very loose idea for some sort of consortium. This was even before it was defined to the point where people were saying, "This is not going to be competitive with each individual university because this is going to be big-ticket major research which calls for interdisciplinary collaboration, etc." The consortium would get into things that weren't defined yet, but they would be such that no one institution could take it on itself. Instead, a unified effort would be required. Each of the institutions was invited to have two representatives for these planning efforts. I was designated as the chair.

You were one of two from U of H?
Yes.

Who was the other one?
Abe Dukler, who was the dean of engineering. From Rice, I remember, one was John Margrave, I can't remember the other, and A&M. I don't know who it was from A&M, but it wasn't Porter at that point. In any event, this group got together and quickly decided that we had neither the resources, the time or whatever to really do a systematic analysis that would allow us to come up with a plan. So the decision was made to hire an outside consulting firm to do that and be advisors to this group. The firm that was selected was A. D. Little, out of Cambridge. A. D. Little had a lot of experience nationally in looking at new technologies, what's emerging, how do you match government with universities and so forth. I think they were given $50,000 and they eventually came up with a report. The report said such a consortium is a very good idea and could be very beneficial, both economically and from an educational point of view, and it pretty much said everything that George wanted to hear. They set priorities as to what they thought the focus should be in terms of what are the areas to get into first. With very few exceptions, everything they recommended as far as 1-2-3 and 4 never happened. The reason is that all of these things were based on their assessment of the faculty and where science was going and so forth, but did not take into consideration what was going on in the heads of some people at the universities at that point or to put it very simply: They were not aware and we were not aware of Peter McIntyre and Russ Huson at A&M, who were knowledgeable about the Department of Energy's interest in something called the Superconducting Super Collider, but that's jumping ahead, so let me back up. The recommendations were made that they create this center, Houston Area Research Center was the name that it was given, and that we go ahead and have two representatives from each of the universities,

plus some people that George had talked to in the private sector, and to look at funding. The most active member from the non-university community was Paul Howell, who had been on the Board of Rice and had his own corporation. And he's a good friend of George. He was the most active of the non-university people. In any event, one of the first things that happened is George wanted to get somebody to come in and be president. So the committee selected a head-hunting firm from Houston and I remember we gave them $10,000. It was Heidrick & Struggles. We talked about what we thought the characteristics were, the educational background of that person. While that was going on, George had been talking to George Kozmetsky at UT. There was nothing to really run at that point. Some things were going on with the A&M people vis-a-vis the Department of Energy, but that wasn't really on the agenda yet. In the very early stages, we were located in the building where the WCA is now. Somewhere around that time, there were some attempts made to get some projects started. There were three that I remember best. One was out of Rice University and it was somebody who was an associate at that time of John Margraves, who turned out to be the son-in-law of Hackerman. Do you remember Professor Barry? He was doing some things with lasers. George gave him some money to encourage him to do this work. He was a distinguished professor at Rice. The other was, there was this operation in Conroe where they had a facility to test the impact of freight trains banging against each other. It was never clear to me how that got into the picture but it was somebody at A&M, in their Department of Transportation, who knew about this. George was eager to see things begin to happen in spite of having a committee which now became a search committee and was trying to do the search in a professional manner by soliciting vitae and working with the search firms. George Mitchell had some kind of discussion with Kozmetsky and all of a sudden Harvey McMains appeared. He had a place to live here, in the Tree House near the Inn, and he was now not the interim but the president. It happened so quickly, he just moved in and then he had the checkbook. It was one of those large-size checkbooks that has three checks on each page. He was negotiating with Barry, he was going to do this and that and he was going to renovate the building. I'm chairing this committee and I remember Abe Dukler particularly saying, "What's going on here, who is this guy, does anybody know anything about him?" Yes, he worked for Kozmetsky, he was with UT, he was a scientist with Bell Labs, he was this and he had a degree here and there.

He claimed to be an architect.

Yes, that's right, that's what led to his downfall—people thought he was an architect. But you know him because you were very much involved at that point. It was ludicrous because I remember Paul Howell's saying, he just looked at George and said, "George, look, we each have one vote, you have two. Why don't you just tell us what you want to have happen, it will save a lot of time and a lot of money." So meanwhile from the search point of view, the guy from Heidrick & Struggles came to me and he said, "Look, I feel guilty about taking your money, because it is very clear to me that George has already made up his mind who he wants to do this, and bringing this guy in from Tennessee and this one from Chicago is not fair to them and it's not fair to you." Harvey was still there and George was negotiating with Skip.

Harvey McMains?

Yes, he was hired on an interim basis, but he never used interim himself. George made it very clear to all that Harvey was an interim, that he just wanted somebody there who knew what he's doing, that George Kozmetsky recommended him, he's a good scientist, he'll get things going but you guys go ahead and finish this thing. And we said OK. The search process was increasingly leading to Porter. It was obvious that's what was going to happen. Every time Heidrick & Struggles would say, "We've got a candidate here," George would find a reason to put the guy down, like, "If he's so good why does he want to leave?" Well, he hasn't said he wants to leave but a certain position has been described for him and he's interested in looking at it. That's what the search process is about. This guy had worked for the Atomic Energy Commission, "Well that's not going to work because we're not going to do that kind of stuff." Then the newspaper story came out about Harvey.

He was interim director at the time and there was still a search going on for a permanent director? I didn't realize that, I didn't think that search started until Harvey left.

No, that's one of the reasons why everybody on the committee was puzzled. Why do you bring somebody in while we're doing a search?

What period of time was covered by the search?

Six months maximum. Meanwhile Harvey, of course, is not behaving as if he's interim. Harvey is behaving as if he in fact is the president of this new organization and he's not particularly involved with or concerned with this committee, like he knows something we don't know. Which was always the case in all of this, that no one ever really felt that we had all the information, and then it became increasingly clear that we were just going through this ritual. Then the article appeared in the paper that Harvey's credentials, for the most part, didn't really exist. It was on the front page of the *Post*. In a very short period of time after that, I got a call from Mitchell, like he knew it was going to be in the paper. He said, "You've got to handle this, you take over and you tell Harvey he has to leave and make sure you get his key," and then Clyde Black called and said, "Make sure you get his credit card," and all that stuff. That's before I really felt that I knew Mitchell. And I said, "I'm a University of Houston employee, what do I tell him?" "Just tell him that he can't do that any more, he has to resign, you tell him that." And I said, "Mr. Mitchell, you hired him, don't you think . . ." "No, I don't want to do it, but don't be mean to him and we'll try to help him out." I remember this so vividly, "We've got to help him out." So early the next morning I walk in there and Harvey is there and Harvey is already starting to take stuff down off his wall. And I said, "Harvey, I'm really embarrassed, I hate to do this." And he said, "No, actually I understand." I couldn't help but ask, "Harvey, why did you do this, how did this happen?" And he simply said, "For the longest time people had assumed that I have a Ph.D.," and he had on his wall these honors he'd received and they all said Doctor Harvey McMains, or Ph.D., and I said "What about the law degree?" He said, "Well, you know, that's something I was really interested in and I knew a lot about law and so forth . . ." So he turned everything over and I learned soon after that George had given him a chunk of money and helped him. Harvey went to Oklahoma. Off he went and the next thing that happens is that UT announces, "We ain't coming near that HARC organization for anything, we have been embarrassed, humiliated."

So you then got appointed?

That's the other part that was funny. George said, "You be the interim president." I said, "Well, there's a committee." He said, "That's all right. You have a degree (but he didn't say he wanted to see it), I think you should hang it up somewhere." I remember I had not looked at my Ph.D. from the University of Chicago since the day it came in the envelope, but I knew where it might be, and that night I was rummaging through all this, and I did find it and I brought it in. He never asked to see it, no one ever asked to see it.

I got asked by some reporter whether George had looked and I lied, and said, "Oh, yes."

He said, "Just have it around, or hang it up or something." So, anyway, I went to the committee and I said, "This is the situation" and they were having a ball with this. All of a sudden, nobody had trusted the guy, everybody was suspicious of him and boy, that'll teach George. This academic world is a very sacred kind of thing, it's not a good idea screw around with it unless you know the business.

But actually Dukler was the one guy who spoke out?
He was the most outspoken, yes. In any event, I took over as interim for 11 months.

I didn't realize it was that long.
I didn't either until I just read that, and then Porter came on and it became very apparent to me that it would be best to move on.

But you were still an employee of the University of Houston?
Yes, but while I was interim president of HARC, George was paying a portion of my university salary, reimbursing the university, I think it was 25 percent. He did give me a raise, and when I came in, he gave me some compensating money. The first thing that we got into seriously, the train-smashing thing, we looked into that and there was really nothing there. The money had been spent, but that effort wasn't going anywhere, so I worked with Mike Richmond and we closed it down. The organization itself was owned by the Pritzker family in Chicago. Meanwhile, McIntyre and Huson were pushing the Super Collider, and we went to Chicago to meet with the Department of Energy and signed an agreement. So the Texas Accelerator Center became the first entity within HARC. George provided them a facility and some housing, and he was excited about it. Then about the same time came Barry from Rice, and he and Margrave set up a laser project, related particularly to the health field. Then there was a guy who was associated with the energy business who was working on a project where electronically you would be able to clean the sludge out of oil pipes, and he had a little lab there.

As I recall, you were headquartered then on Grogan's Mill?
Yes. Harvey had also hired several people to work there. It wasn't really clear what they were doing, and I came in and started looking at that. So we made some changes there. Those people left, or were asked to leave. This went on and on and Skip, meanwhile, had announced that he was going to be the president and was going to come in at a certain time, and was doing things

on an interim basis. He brought in Dan Davis from A&M and began to hire some other people, and I worked with him. Finally, I left and everything was settled.

So you did your 11 months as interim director and was it after that when you left the university and came to The Woodlands Corporation?

Yes, a few months after that. George saw the Texas Medical Center as a potential partner at The Woodlands. Let's do A. D. Little again and so A. D. Little again went to work. This is when Phil Hoffman was the president of the Texas Medical Center.

Phil Hoffman, after he left the university?

After he left the University of Houston, Phil took over the job that Dick Wainerdi has now. That's one of the reasons why George said, "The Medical Center is my next target." So they did a study and concluded, yes, biotechnology is a big thing and it has good potential for The Woodlands. And then Hoffman left and Wainerdi came in and assigned someone from the Medical Center to serve as an interim coordinator out here.

We're not talking in the context of HARC now, are we? We're talking about the Research Forest?

We're talking about what became eventually the Research Forest. But the idea initially was that there would be HARC with its focus, and there would be a Medical Center component, and hopefully they could work together. But it became very apparent that it was a mistake to think you could create such an entity by going through the Texas Medical Center administration.

By this time you're a Mitchell employee?

No, at that stage I was still on some Mitchell time but really was still at the university. I know that I was not an employee of Mitchell Energy or The Woodlands Corporation full time when we started discussions with Dominic Lam and John Lanier. Soon after that I did come over to The Woodlands Corporation, and I began to work with Michael Richmond to develop what eventually became The Woodlands Research Forest, where HARC was a component of it and technology transfer of Baylor Medical School was the other. Nobody at that point had negotiated an agreement with the university about their taking their research projects and moving them into commercial products. Between Michael and George they created

The Woodlands Venture Capital Company. Michael designated Marty Sutter to lead it because he had known him, and this entity was created. One of the things you could give a Medical Center organization, in addition to space and a contribution towards equipment, was to come up with venture capital, and George invested, at that point, $15 million. That was another message he was now sending to the city, that we've got to diversify, we're losing all of these things to the East and to the West and the Medical Center is, in a sense, a gold mine and we ought to take advantage of it. So Marty set up his organization, The Woodlands Venture Capital Company, and we began very early to work with Dominic Lam and John Lanier. Dominic Lam was from Baylor, Baylor is a private institution, and you didn't have to go through all the state approvals that we did with Lanier at UT Health Science Center. Dominic at that point was considered a star by everyone. Baylor didn't want to do anything that might offend him, George was charmed by him. He was not only considered an established scientist but an artist, raconteur and whatever else; very persuasive guy. I remember hours that we worked on getting this deal done so that Baylor would put a building up here and it would be Houston Biotech, and Dominic would have the building, and we would also put money into Houston Biotechnology. Venture Tech I was the first building that was put up, and Houston Biotech was on one side and on the other side we had the University of Texas Health Science Center, with John Lanier.

Where are we, about 1985 or so?

Closer to '84. John Lanier and Steve Livesay together moved from the basement at UT Health Science Center out here, and they were working on cryobiology. They were the first two. Then the Center for Biotechnology was negotiated, where Baylor was given the land. They put up a building and along with that, where the WCA is now, we did an agreement where Baylor was going to set up the Magnetic Resonance Imaging Center there. George had given them about $750,000 worth of equipment to get them there. That was to be basic research and eventually would have a clinical component. That was in a sense the beginning. In the meanwhile, Marty Sutter was going out and looking at venture capital investment prospects. The ground rules at the beginning had been that anything that he invested in would have to come to The Woodlands, but there were a couple that did not. But usually we were able to use that as the leverage, and one of the earliest, even before Dominic Lam and John Lanier, was the University of Texas Center for Preventive Medicine, part of the public health school. The dean wanted very much to set up a center out here where he could do longitudinal studies and saw this as a wonderful population. George got involved with that first–it was the first thing he invested in. The University of Texas Public Health Center has been doing this longitudinal study on cholesterol and cardiovascular conditions for kids. That was even before we officially designated the area as The Woodlands Research Forest.

Are they still out here?
UT Health Science Center is still out here. They're probably going to leave because they're not going to get NIH funding anymore. The next one we went after, which was not related to the university, was Surgimedics, and that represented a very significant relocation from Houston to here. After that, the Research Forest just continued to grow.

You were a part of The Woodlands Corporation?
Yes, vice president for institutional development, but actually when I first came in, I didn't really have a title and Budd Clark said he was making me a vice president of Mitchell Energy & Development. Then there was a question of whose payroll I should be on. So they put me under The Woodlands Corporation to work with Michael Richmond part of the time and part of the time for George. If Ed Lee had any problem with that, he never said anything. Roger had a bit of a problem with it.

You mean Roger, after he became president?
Yes, we had a problem when unilaterally Budd Clark said, "You're vice president of The Woodlands Corporation," which I didn't realize at the time created a few problems for Roger. My major assignment at that point was to work on development of the Research Forest. That led to opportunities to do things with the Japanese because they became potential partners, and I had this opportunity to go to Japan for 30 days as a guest of the Japanese government at their expense and that led to Sumitomo's becoming involved in financing and some other things that we did as joint ventures with Japanese companies. So my heaviest involvement was with the Research Forest and with George's unrelenting desire to have some kind of university campus here. It's been a wonderful tenure.

Talk about what's happened with the Cynthia Woods Mitchell Pavilion. You took over as chief executive officer two or three years ago?
Yes, about two years ago, with 40 percent of my time spent on the Pavilion.

The Pavilion reimburses The Woodlands Corporation for that 40 percent. The big part of it was trying to learn and to understand, to try to do it on a part-time basis.

At the start you had Kirk Metzger as general manager?
Yes, Kirk was general manager and Steve Baker was here for marketing and development. One thing led to another, and we made some changes, the major one being Kirk. The idea initially was that we would get somebody to come in to manage the Pavilion . . . I was going to do that part on an interim basis, but at that point we started talking about expansion. We were going through a period where we really weren't sure what the outcome would be. My feeling was that this is not the right time to try to bring somebody in because we were in a state of some upheaval. So we went through the expansion and negotiating contracts, trying to learn and understand. I have real concerns about the future of the Pavilion; George's thoughts about the future have never been really articulated. Anyway, then The Woodlands Corporation and Mitchell Energy last spring announced this voluntary retirement program where you could voluntarily retire and reap some benefits or take a chance that you wouldn't be fired. I had a talk with Roger Galatas, and based on that discussion I felt there was a high probability that I would be one of those who would fall into the "you're out" category. At the same time, there was this expressed desire to "keep me in the family," and I think it was Roger who first suggested that I devote full time to the Pavilion. He had mentioned the idea to George. Then George mentioned something to me and I added everything up and decided this is a good idea, what with being newly married and living out here, I thought it was the thing to do. I have a three-year agreement that's renewable. We went through the expansion . . .

That's from capacity of 10,000 to 13,000?
Yes. It included expansion of the plazas and adding some 16 new points of sale, new bathrooms, a 15-year agreement with Pace, a new 10-year agreement with Ogden, and a seven-year agreement with EMI. So I came on full time in May of this year, 1995, and developed a much greater appreciation of what people like you must have gone through as I see more and more of the dynamics of this thing.

For accuracy, I was once removed from what people went through. I did my aggravation with and through Kirk. I was never general

manager or involved to the extent that you are.
It's been an incredible learning experience, been very exciting because it's something new and different and I'm glad to be doing it.

The theater, where does that stand?
There are two theaters. One theater I'm involved in at George's request, God bless him. I made the mistake of crossing the line and entering into Galveston politics. Big mistake. That's the Strand Street theater, which now has a general manager and a reorganized board and, hopefully, has a very good chance of success. The theater at The Woodlands: I think this is something he really wants to do for Cynthia. The thing we're looking into now, given the money that George says he's willing to put up, is, can we get a theater of at least 500 seats plus some classroom space? With 500 seats we'd be in a much better position. We already have some programmatic commitments to it. Theater Under the Stars would do something there, Pace would, the University of Houston, the Humphrey School, so there are some possibilities there.

Humphrey School? What's that?
Humphrey School is part of Theater Under the Stars, it's a huge educational program. There have been so many false starts with a theater, and I don't know if we're on one now. Cynthia had more or less commissioned Michael Graves to do the theater. The financial constraints are tight. In addition to other costs, Graves would receive a standard 8 percent fee. He would oversee it from Princeton. I think we'd take a terrible beating. That's aside from the fact there's very little in his accomplishments that are theater-related. It doesn't take away from the genius of the guy, but it does raise a question. Still, we have now identified two potential sites where there could be a theater. George's preference is right adjacent to the Pavilion, on the Pavilion side of the waterway. The question is, would that work? Coulson and I are going to San Antonio next week to meet with the architects there to see what we can do, and we'll go from there. Another thing is, there are serious discussions now going on about The Woodlands Corporation's forgiving the debt that the Pavilion has with the company, because it's an opportune time for them in terms of tax deductions. That would be very helpful; George's idea has always been that if there's any deficit with regard to the theater, it would be handled by the Pavilion. And he seems to think there's a management here that could manage both the Pavilion and the theater, and that's a pipe dream. If they forgive the debt, I'm sure that George would then increasingly

take the position, "Now you can afford it." So that's where we are on that.

You've had an association of 20-plus years with the company, some of those years closer than others. Over that time, what are the things that you've been involved with that you're proudest of, that you're most pleased about?
I think there are three things. One is the whole emergence of the Research Forest and the fact that we were able to do some things that no one had ever really done before in terms of uniting universities with companies and venture capital. Two would be the fact that after all these years there will be a viable academic and university presence here. Not anywhere close to what had been planned in terms of what finally got here, but the fact is that it's here. I'm very proud of that.

Did you play a substantial role in that?
Yes, I think so. The third is the opportunity to be with the Pavilion and the idea of acting as an advocate for the performing arts, because I think that's terribly important. I'm really pleased and privileged—it sounds corny—to be able to do this. I would say those are the major positives.

Thank you David.

ROBERT HARTSFIELD

It's Tuesday, August 1, 1995. This is Joe Kutchin. I'm in my office in The Woodlands, in the MND Building, and I'm talking with Robert Hartsfield, who left the company in 1980. He came in 1969 and was one of the pivotal persons in the development of The Woodlands. To start with, Bob, tell a little bit about your personal history.
I studied architecture in college–graduate of Rice and the University of Texas in architecture, degrees from both. Then I spent a couple of years in the service.

What year?
This was '58 through '60. Army Corps of Engineers, a lot of fun but no battles. Just a lot of the headaches. Typical ROTC lieutenant. After that I went to graduate school at the University of Pennsylvania, which was an excellent move on my part because they had a super faculty there, just top-notch professors. I was in a curriculum that was a blend of architecture and city planning. One of the important connections there was one of the professors I had, Ian McHarg, who is a renowned environmental planner. I was later able to bring him in as one of the prime consultants to help us do the environmental planning for The Woodlands.

What years were you at Pennsylvania?
'60 through '62

Your degree from Pennsylvania is a master's degree?
Master's degree in city planning. But I took a program that involved architecture, as well. I already had degrees in architecture, so I wanted a planning degree. Then there were several years of movement. I came back to Texas and worked for an architect named O'Neil Ford and a planner named Sam Zisman. Ford is dead now, but he was one of the better Southwestern architects. We did a research park in North Dallas which is still operating. It's now owned by the University of Texas. Sam has long since departed, also. These two guys were very close and Sam was doing planning as a consultant around the country, and it was actually Sam who interviewed me and hired me to come down and be a designer in the Dallas office of an associate of O'Neil Ford named Arch Swank. We worked on that project there for almost two years, designing and laying out the campus of the Graduate

Research Center of the Southwest.

What years were you there?

From, I guess, the summer of '62 to early '65 or somewhere in that area. Everything kind of worked in two-year increments. I think that about every two years I got the itch to move on. About that time I had an opportunity to go to Little Rock, Arkansas, and take a senior position with a planning affiliate started by a major firm there called Wittenberg, Delony and Davidson. They're probably still operating in Little Rock. A very good colleague of mine named Jack Mitchell, whom I met at Penn, was a junior partner there, and he asked if I would come and be sort of the second in command and main planner/designer for one of their people who was going to head up a firm called Community Planning Associates. His name is Jim Hatcher, and he's more of a businessman than a planner. So I spent a couple of years in Little Rock. Very good experience. We did a lot of rural planning around the mid-Southwest. Jim and I would get in a plane, he was a pilot, and would fly from one city to the next selling government-funded community plans to small towns—that's what it boiled down to. Still it got me into the business of interpreting contracts, working with people, doing planning and understanding the business. About the time I decided that wasn't going to go far enough for me, I had an offer from a firm in Tulsa, Oklahoma, that was the subsidiary, or the branch office, of a larger firm in Oklahoma City and they paid my way over to Tulsa. But I wanted to go talk to another person named Bob Jones while I was there, which I did. Bob was a partner in an architecture firm called Murray-Jones-Murray. And Bob Jones and I really hit it off and I ended up taking a position with them and became the director of a joint venture between Murray-Jones-Murray and a firm called Braisch Engineering. Both Bill Braisch and Bob Jones remain close colleagues and good friends today. And I spent some two to two-and-a-half years in Tulsa, and we were doing the same things, except I was running the show. We were out promoting planning and engineering contracts in communities around Oklahoma and Tulsa. Everything from small master plans to major site plans, park plans, you name it. Plant planning at every scale, almost. But again I got restless and felt that in order to grow professionally, I needed to move on. By then I was ready to come back to Houston. I am a native Houstonian. So I ended up contacting and talking with some people at Caudill Rowlett Scott, CRS, which later became CRSS after they joined forces with Sirrine Company, but that's way after my time. At any rate, the upshot of that contact was that I left Tulsa in about '67 and came to Houston and became an associate in what they called the development section, which was people who were supposed to be sort of flag carriers for different aspects of architectural planning, and I became the urban design specialist. It was my job to consult with other project teams where urban design issues were involved but also to write articles and make speaking engagements and this sort of thing, and to generate visibility and business for the firm. That was quite a business. We got several contracts, and I was in and out of Washington a fair amount. We had some contracts with HUD. I worked with teams that were planning campuses around the country because that was the big thing that CRSS did. And many times there were urban design issues, either in adding to the campus or in relating the campus to the community that surrounded it. This is where the story of The Woodlands comes in. It so happened that one day when I was in Washington and about to leave later that day, I had a call at the hotel while I was getting my things together, from Tom Bullock, who was the head of my section, one of the senior partners of CRSS, in August or September of 1969, saying that we had a meeting. I think this was on a Thursday or Friday. We had a meeting the following Tuesday with a fellow named George Mitchell who had bought some major amount of land north of Houston and wanted to plan a new community. I was to, if necessary, stay an extra day and go to HUD where he had a preliminary application in. George had early on blocked out the land. First he tried to do it on 1960 and I-45 but couldn't assemble enough land, so he moved up to the present site of The Woodlands. At any rate, the process was that George, through somebody, had hired a local architect named Cerf Ross. I have no idea whatever happened to him. Last I heard he was trying to buy some mansion in the midst of litigation in Florida, but old Cerf Ross had submitted some drawings and some information to HUD in George's behalf as part of their Title IV New Communities Program. This was a preliminary submission, as part of an application for the grants and assistance available under Title IV, for what was then just called Woodlands Village as a working title. Well, I talked to a fellow named Paul Brace at HUD, whom I had known through other work there, and he said, "Goodness, I'm glad you asked. We have this application. We think Houston has a great potential for a new community. We think the site that Mitchell has is great. We think Mr. Mitchell and his company, basically an oil and gas company that seems well-managed and has good assets, is very good," he says, "but the stuff they submitted is God-awful." He says, "You guys need some help. I hope you can go down and give them some better help because we can't respond to what has been submitted."

Was this before or after Kamrath?

This was after Kamrath. Karl Kamrath remained a good friend of George for a great many years, but I think after Karl wrote a paper and did some original sort of dream-thinking with George about what a new community ought to be and how it should be done well, for reasons I don't know he was not involved in this HUD application.

I know he did submit plans and George kind of considers them the granddaddy of The Woodlands.

At any rate, we came back to Houston and did the interview and I had several people from CRS with me, met with George Mitchell, Plato Pappas and Morris Thompson. It was a very cordial get-acquainted meeting. I really don't remember very much about the meeting. They were officed in the Houston Club Building at the time, and we gave them our spiel that we were a team leader type firm, that we had a broad philosophical understanding, we had the talent, that we could bring in specialists to do all kinds of things. Our objective was to become the team leader for planning The Woodlands and to bring in whatever talent he needed to round out a multi-disciplinary team.

You remember how big the land package was at that time?

It was about 16,000 acres, and by the time we actually got started it was a bit over 17,000. I understand it's over 20,000 now.

About 25,000.

This was in the fall of 1969. Things happened fairly rapidly. We were going to have a follow-up meeting–you know, you write the follow-up letter and you leave them your brochure and all of that. He was going to interview several firms and we were one of the first ones, I guess. So we were not expecting anything to happen too soon. But it so happened that there was a conference to be held that was sponsored jointly by HUD and, I believe, the University of North Carolina. The conference was on new communities planning, and it was held at the country club in Reston, Virginia. George Mitchell and Jim McAlister were on the program. Jim was going to present a discussion about organizing the financing, Lively, who was principally the soldier in the field, was going to explain and talk about assembling the land. George was going to talk more about his process of buying the land and his thoughts about developing a new community and how he felt it would be compatible as an addition to his oil and gas business.

I know that Lively was with the company then. Jim McAlister was, too?

Oh yes. In fact, McAlister may have been at that first interview. I'm a little vague about that. At any rate, I had an engagement to speak in Chicago on the same dates, but Tom Bullock, who was very resourceful and a very pushy guy, said, "I want you to cancel your date in Chicago and I'm going to get you on that program and you talk about team leadership for developing a new community and you get to know George Mitchell a little bit better." So that was my job, to go and promote and to do this.

At that time, the experience in new communities amounted to Columbia, and that was just getting under way. Is that right?

Columbia had just barely got started, really. At this time Columbia was more on paper than built. I think they had dammed up and built the first lake and had one office building, like OB 1. Reston was the only real new town in the United States. That's why they held the meeting at the Reston Country Club.

I didn't realize Reston was that old.

Reston pre-dates all the rest of them by several years. It was privately done between Robert E. Simon and Gulf Oil. But the reason they held it there was because of its history and its being the first one and because people coming from all over the country could fly in and it was just a few minutes' drive from Dulles Airport. Well, I resolved to get friendly with George but not do any direct promoting. I told him why I was there; we shared breakfast and rides. We used my rental car and I drove him and Jim back and forth from the airport to where we stayed at the airport motel, and we drove back and forth to the country club at Reston. It was about a three- or four-day conference, so we had a little time to visit and socialize. And the whole time, I never talked about CRS, I just talked about the idea of doing real estate in a new community in ways that were socially responsible but still profitable and environmentally correct. Jim McAlister and I bought a pair of handball gloves and I played handball with McAlister and he beat me thoroughly. He was very good. At any rate, we had good conversations and pretty good congeniality built up. So they made their presentations and I made notes, and I made my presentation and I was reasonably satisfied. It was a good conference. About three days after that conference, Jim McAlister called me and said George wanted to talk to me and I thought, "Hot damn, I think we've got us a contract." I called Tom Bullock and said, "Tom, I just got a call from

McAlister who works with Mitchell. George wants me to talk to him. So I'm going down tomorrow afternoon and meet with him. I think we've got a good shot at this." That was like on a Tuesday and I went in on Wednesday afternoon and met with George. This was still at the Houston Club building. And George's first comments were, "Well, you know, I've got a really major, interesting project to do here and I realize that I'm going to have to have help inside in order to manage the consultants that we need. I don't have anybody other than Plato, who can handle the engineers. Nobody seems to have the understanding you do of the other professions and how to make them work together. What I would like for you to consider would be to come to work for me and head up the planning team. I don't mean to put you under any great pressure, but Jim McAlister and I are scheduled to leave day after tomorrow." He said, "We're going to Los Angeles to interview Pereira and Victor Gruen and some engineering firms." And he said, "It's important if you take the job that you come along because you need to satisfy yourself and you're going to have to help us make choices."

These were all city planners?

These were all different firms. Pereira was a world-renowned architect and urban planner. Gruen was an urban planner and architect, also of great renown. The others were firms that did large engineering projects and major real estate developments in California. It hit me like a ton of bricks. I was really floored. I told him I was very pleased and flattered and would certainly think about it and he told me to just tell him whatever salary I needed and he would appreciate it if I could let him know and if so I was to go with them. I had to tell them in time so they could make the arrangements. I said, "Well, I'll try to let you know," but before I hit the elevator I had the decision made. I figured, however, that I would wait at least until tomorrow to tell him. I went back and sat down with Tom and told him what the situation was. I said, "Tom, if I take this job I'm going to have to be very objective. There might be a role for CRS, but I can't just turn around and hire you guys. I've got to go through with George and others and some major interviews, and you guys will still have to earn it. Your leg up is that I know you and believe that CRS could do a hell of a job, but the guy that I would want to head it would be me, except that I would be working for me so there is a problem. There's also a protocol problem. Let's wait and see." Tom was gracious. He said, "Hey, it's a hell of an opportunity, I wish you well, keep us in mind and do what you can do, but go for it." And I did have a little soul-searching because if you're in the profession and you're working, you're kind of a purist. You feel like you're the consultant and you know things. You have the attitude that morality is on your side and the client is always looking at the bottom line and cutting corners and scrimping on professional principles in order to do things. And developers, particularly in land development, are the black-hat guys. And here I was going into the valley of the whales, so to speak. I thought, "Gee, where else can you make a mark and what kind of a project can you have that much influence on?" Well, I did say "yes," and I did fly there, within a day or so, with Jim and George, and we did interviews for a couple of days.

How old were you at the time?

That would have been '69 and I was born in '34, what's that? 35-36 years? My birthday is in April, so it's 35. I was not that young, I had had some good experience behind me and I had a damn good pedigree, but I hadn't done anything this big. I never had any doubt that I could do it. I immediately foresaw exactly how to do it and I was used to writing contracts and doing all kinds of things and working with people and so, suddenly having major other firms instead of being competitors, being a client to them, having them work for me, was a nice feeling. And I managed to get along very well with all of the firms.

You reported to George?

Yes.

Please describe your responsibilities.

They gave me a little office. I said, "What is my title?" George said, "Well, make it whatever you want it to be" and I said, "OK, it'll be Director of Planning and Environmental Design." I said "Do you want me to explain that?" He said, "No, I understand it." I wanted to get not only planning but environmental design in there because I had been very oriented towards respecting and dealing gracefully with the environment in my work. And this goes back to, as I've mentioned, my studies under Professor McHarg at Penn, who really opened my eyes to this whole thing. And he's probably the father of the environmental planning movement in not only the United States but really the world.

What kind of organization did you walk into? How many people worked there?

George Mitchell & Associates. There were four or five people. My role was to be director of physical planning with the exception of engineering, civil engineering. All land planning, all urban design, all transportation planning, all of that sort of thing were mine. Before me they'd hired a man to head up social planning. And Jim McAlister was going to handle all the economic and financial stuff, and Plato, of course, was to deal with the civil engineering and construction. So the idea was that in addition to me directly supervising and directing the physical planning, I was to make sure that everything properly fit into a master plan and program that we would submit to HUD with our application for the grants and benefits available under the New Communities Program at HUD. At some point in there it became Title VII. It had been Title IV and it became Title VII. I already had good relationships with HUD and so I was able to just say, "I've got the job, fellows, and we're going to work together and get you a good project." And so we had to work hard and they had a very demanding set of criteria that had been set up. What I did then as Director of Planning and overall coordinator to produce this package was, first, lay out a scope of services that I felt were needed in order to carry out the original program and master plan for The Woodlands. Then I talked to George about that and my best recollection was that we were in complete agreement. He had in mind that we needed an engineer, a transportation planner, somebody to do urban design, and so on. I gave him a firm that I recommended to fill each of the slots. The team that we were going to hire was our interdisciplinary consulting team. And the real estate organization was really me, Dave Hendricks, Jim McAlister and, of course, Plato, who was already involved. Plato had been carrying the ball for everything. He'd been involved in all the Galveston and the Point Venture work and maybe some of the earliest of the subdivisions. But we didn't have any separate organizations. We didn't have offices all together, we were scattered around. This was just before we moved into One Shell. I had a little windowless office that was Godawful. I mean it was a come-down, but I thought, "Well, I've got a great job and I'm with some great people and I'm making a lot more money . . . well, not a hell of a lot." I'll just tell you that I was making a little over $16,000 in '69 at CRS. That wasn't too bad in those days. I asked for $18,000, and George agreed immediately, and about two months after they hired me, he said, "Well, Bob, we're going to give you a raise because you deserve to make at least as much as the other people who are working here." McAlister made more and I'm sure Plato was making much more at that time. I think Dave Hendricks was hired at $20,000 and they thought, "We've got to bring him up, he's got a more responsible job here."

Who were the main consultants?

My best recollection is that George and I agreed on all the main consultants. He wanted Pereira of Los Angeles, and I agreed Pereira would be good. Pereira's main role was going to be overall master planning and urban design. We used Richard Browne & Associates, an engineering firm that was deeply involved in the planning of Columbia. They were going to do some of the social programming, some of the community structure analysis, some of the early transportation planning, and possibly a little bit of the engineering. We hired other people to work just on housing programming and some social things. There were some research people out of the University of North Carolina who were going to work for Dave Hendricks.

McHarg, I know, was one of the consultants.

McHarg is interesting because, you see, I think George had always intuitively sensed that he wanted to protect the natural setting. He wanted to call it Woodlands Village or Woodlands something. By the way, I'm the one who finally said, "George, forget all these other names and just call it The Woodlands." We used to sit on airplanes writing names, trying to figure it out.

Another version of the story is that Cynthia Mitchell had a role in that name.

Well, she may have. I can't tell you anything except that I remember saying, "The Woodlands will be distinctive. It's simple, it's direct and it has distinction." I recommended it and it was agreed to. Another important thing I said was that what is going to set this apart from other real estate is the way that we protect and emphasize the natural environment as part of the way we develop. The man who's best in the world to do this and to give us the foundational guidelines that everybody else's work will relate to is Ian McHarg. That's a slot that George hadn't yet identified. And I just called it environmental planning, and I said, "I really believe, George, that it's very important and it's one of the first things we need to get started on." I said that I wanted to bring McHarg down here and have him meet George and have George consider having him on the team. He was then with Wallace, McHarg, Roberts and Todd of Philadelphia. He taught at the University of Pennsylvania and headed the landscape architecture department but he was also a partner in Wallace, McHarg.

Had he written his book yet?

The book, *Design with Nature*, I think, had come out but just barely. In fact,

I think it was pre-publication when I was in school there.

George did read it, is that true?

I probably gave him a copy. McHarg himself gave me one of his last copies about four years ago. It was re-published just a few years ago, but before that he gave me an original. Charles Magan drove a jeep, and we picked up McHarg at the airport. George and Charles and I and Ian in this jeep. We drove straight to the site and began driving logging trails. I was in the front and George and Ian were in the back. These two guys really hit it off. They got along very well, but they argued. They kind of jostled back and forth and argued and fought and just had a grand time.

We were in 1970 by now?

No, this was still 1969, I think. It was not really cold, but it was winter weather. It was probably November or December, in '69. At any rate, they got along well and George immediately agreed to go ahead and add him to the team. I had outlined to George the general scope for these major players and he agreed. I wrote each of them that they were going to be retained and I gave them their general roles and responsibilities and said I would follow it up with a definitive contract. I was used to doing this sort of thing. I proceeded to do that and my best recollection is that I asked George if he wanted to read any of these things, each of them several pages long, a certain amount of normal boilerplate, the schedule and the normal things that go into a consulting contract. My recollection is that George said, "No, that's all right, go ahead." He just gave me free rein. I reported to nobody but him and he wanted to know what was going on but he basically seemed to respect and trust what I did. So I went ahead and sent these contracts and went through a process that took probably 30 days of negotiating with each of these firms. I had purposely put a little bit of overlap in each of them because it's hard to find a dividing line between all the disciplines. Anyway, there's a natural overlap and if you structure it right, the overlap can be creative because it gives you more than one point of view about a critical issue. So I structured them that way and I had some good arguments. Everybody wanted to do more. They wanted to do it all, each of them. But the authority George gave me just enabled me to say, "No, look, this is the way it's going to be." And so we got everybody signed up. I can remember I had the whole package sitting on my desk one day and we had to schedule a big first get-together, sort of an orientation meeting. It hit me like a ton of bricks: "You know, I'm in a corporation now, I've got lawyers here, I signed these things on behalf of the company. George is giving me carte blanche authority, and he'll stand by it, but I should have run these through the legal department here before I did this." I just wasn't used to having to do that sort of thing. So I called Paul Wommack and said, "Paul, I need to see you." And I went up there with these contracts and I said, "Paul, I really have to apologize, but it's my lack of experience, I guess. I've been in charge of this kind of thing in my own work for a number of years and George has just given me free rein and I just took the bull by the horns and went ahead and did all this stuff and I've got about six contracts here that total over $400,000 in consulting fees and they're all signed. Do you want to look at them?" And he kind of thumbed through it, and he said, "Well, all I can tell you is, good luck." Paul didn't get mad and he didn't complain to George, and the truth is I didn't have any trouble at all. The one incident that came up, I don't mind telling about it. Dick Browne's firm ran $20,000 over and I wouldn't pay him. And George backed me up. And finally when Len Ivins came in after several months, he went ahead and paid off Dick Browne.

They were buddies?
Yes. Dick Browne was brought in by Len Ivins.

I knew that Browne had done work for Columbia, but I didn't realize the two of them were close.
Browne had had some kind of hiatus with his firm. They retained his name but he wasn't involved with them. He was kind of floating around doing projects, did something down in Florida. He was kind of looking for a home and Ivins said, "Hey, come on." I was being pushed off to the side . . . that's another war story of The Woodlands, when Ivins came in with what we called the "Columbia Mafia." Dick Browne was part of that group that came in and took over positions.

I remember that period too well.
At any rate, we went through a very interesting four to six months of planning because we wanted to try to open up in the spring of '71. We had everybody under contract in the early spring of '70 and my recollection was that we finished the initial package for HUD before the end of 1970.

This was basically the application that your whole team was working to prepare, is that right? And tell me about the size of the grant.
Yes. The end result was a package that followed a prescription from HUD as

to what they wanted in the way of all the bases that they wanted touched for the program and the plan to qualify for the new community grants. The main grant was the guarantee on some bonds, which I understand was worth at that time about a point and one-half in the market. The maximum guarantee you could get was $50 million, and we got the $50 million. I have to say that I put together a really good package which was in preliminary review with HUD at the time that Len Ivins came in. And the final negotiations at the top levels took place between Ivins and HUD and an attorney whose name I forget now. I can't say that the fact that we got the largest grant and all the other negotiated aspects of the contract with HUD was due to my work alone. There was a lot of tough negotiating and a lot of good work done. At that time it was very good work done by Ivins and some of the legal help that he brought in.

Your main work from the time you joined the company until the application was filed was to get the application ready, is that correct?
It was to coordinate development of a master plan for The Woodlands, which would be documented as a fundamental application to HUD.

Were the Columbia people in at that time?
We basically had the master plan virtually done. We had already submitted the preliminary materials to HUD, it was in review and ready for the final submission about the time that Len Ivins came in.

Which would be about when?
Early '71, possibly. Late '70 or early '71.

I have the idea that when I came in, which would have been in 1972, from the agency in Chicago, that Ivins was still in Columbia. I think some of his people at least were still there.
I can remember that we submitted the application from One Shell and Ivins's first office was in One Shell. I think that in many ways, Ivins did a lot of good. George needed somebody at Ivins's level, with his experience. Clearly, I was not capable of doing it, and at one point George was considering somebody from California. I remember telling him after a visit with Ivins in Columbia, "You really need to talk to Len Ivins. I think he might be a good candidate and I would like you to talk to him."

So at that time you were really looking for a guy who would head the

project?

George was looking for a vice president to head real estate, and I was like one of the vice presidents of real estate, but I wasn't general manager. He wanted somebody who had been manager—or in Ivins's case, assistant general manager of Columbia, and that was his main qualification. George wanted somebody who had that experience, who knew how to run a new town, how to build a new town. I knew how to do the planning, but the business of running it and building it was a whole new ball game.

OK, so under your wing the application for HUD was in, at least the first early version?

The main thing that Ivins and his people impacted on was not the plan at all. It was just the financing and the bond negotiations and all the mincy stuff in the contract about what we had to do to qualify for this, that and the other. He got a lawyer, whatever his name was I can't remember, out of Washington, I think. We came out of HUD with almost everything we asked for. One other little incident there that's important. After we completed our studies, it occurred to me that although I was very strongly involved with the environmental ethic, I didn't have the technical know-how or the time to really put into effect the work that Ian McHarg had done in the master planning process. We needed to take the general guidelines, and the general plan, and then begin to focus on the first village and the first neighborhoods before we started turning dirt. And in order to do that, I needed a chief environmental planner to work with me. I was able to hire Jim Veltman, who was the project manager for the team McHarg had sent to work with us on The Woodlands. You might remember Jim.

I remember Jim as being a really decent guy.

Yes, Jim was one of the good guys. At any rate, Jim came and immediately started working with me to develop the village plan, and we had some follow-up contracts with McHarg to translate the plan into more detail for village and neighborhood development. And finally, development standards, you know, things that we use today as guidelines for the way we do drainage ditches, the way we cut trees and clear lots.

But that was part of the HUD submission, wasn't it?

No, that was after HUD, after we got our loan guarantee and were ready to go into planning for the first development. But the reason I brought this up is that one of the last things we had to submit, almost as an afterthought because HUD was on a learning curve, too, came when HUD told us, "You know, one thing we've got to have is an environmental impact statement." And we said, "Fine, we're better equipped to do that than anybody, what do you want?" and they said, "Well, we'll tell you in about a week." In a week we said, "What do you want?" They said, "Well, we're working on it." In about a month, they still didn't know what they wanted. Veltman and I went and met with them. And it was apparent in the meeting in Washington that these guys were floundering around looking at reports and trying to figure out, they just didn't know what the hell to do. So Jim and I came back. I said, "Jim, you figure what should be in an environmental impact statement and let's submit that as a recommended content and then if they like it, we'll do it." A guy named Tony DeVito was the guy at HUD who was over all the review process. And so we got that to Tony, and he was tickled. He said, "Yes, this is what we had in mind." He liked it, so I really believe that we gave HUD the first criteria for an environmental impact statement they ever had.

Is that what they looked for from the other 12 to 15 towns?
Yes, they used us as a guide and to everybody else that applied, they said, "This is what you've got to have," which kind of didn't sit well with some people because we offered a huge amount of detail. The other thing that was interesting is that I did a little PERT diagram showing the progress of the work. The chart showed different disciplines blending and coming together and different circles and dots where you'd have reports and send preliminary reports. I gave that to DeVito. He had it on his wall and was telling everybody, "This is the way you do planning for a new community." So we really kind of broke the ground to show them how to do it.

All right, the guys came in from Columbia and you and I are both familiar with the strange working relationships that developed, but Bob, they came in '70 or '71, but you say you stayed until '80?
I stayed until '80, but the later years were difficult for me. There was little harmony between the existing staff and the new people from Columbia. It was a very, very painful time, and in the early '70s, I had to leave.

At the end of '74, early '75, a great number of the people from Columbia were dismissed—that job, I remember, fell to Leland Carter. This would have to be before 1974 was out, I would think, because by the end of the year the company had deep financial problems.
I struggled around doing a few things on my own. About a year later Leland

Carter called and wanted to debrief me. He had a whole list of names, and he said, "I just want to get your personal experience with or opinion about these people." We spent about two hours and he went through everybody. It was shortly after that, I think, that there was great concern by some people who held Mitchell debt. It was my understanding it was mainly Chase Bank in New York. I had met Leland on several occasions and we had a reasonable rapport, so he called me and we talked and I gave him all the stuff. And he said, "Look, I know George would like to bring you back aboard but we can't quite yet. But if you can hang on . . . " I was doing a little free lance work. "Hang on and we'll see if we can do it." Some time passed. What they did finally was hire me as consultant and I was to do work for the non-Woodlands stuff and to help plan and do the planning as a consultant. Dick Browne was, by then, entrenched as the director of planning for The Woodlands. George wasn't going to bring me back in and displace Dick. Dick had Bob Heineman, whom I hired. I worked with Bob Heineman at CRS and when he graduated from Harvard I hired him to come aboard for The Woodlands.

Browne's not there any more, you realize that?
No, I didn't know that.

Well, Dick left as a result of the recent Voluntary Incentive Retirement Program.
At any rate, I came back as a consultant to work on non-Woodlands projects, and for several months or a year I was a consultant. I was provided a car and an office and it was borderline whether I was an employee or a consultant.

Who was your boss then? Whom did you deal with?
It wasn't Browne. There was a succession of different people, and finally Ed Dreiss. At some point, it cleared up and they said, "OK, we can just make you officially part of the company now, and you're the non-Woodlands planner," and then I was under Ed Dreiss and that lasted for about three years or so. Dreiss and I got frayed a little bit with each other. It was arranged then that I would be transferred to report to Dick Browne. I had an office and I was in charge of the non-Woodlands planning for Dick Browne. I basically did my own thing and Dick took a cursory overview. Then, in '80 I left the company again. And then my relationship was as a consultant again for another three to four years. I went on to do some free-lancing and some investing and some other work on my own, beginning about '83 or '84, and no longer did consulting work for The Woodlands.

Let's try to summarize some of these things. Under your wing from about 1969 until the Columbia people came in two or three years later, some important things happened. Say again what those were.
The most important thing for The Woodlands was this: I think George intuitively had a feel for nature and for what was then beginning to be called environmental planning. It was intuitive with him and not explicit. What I did was make it explicit and make it work by bringing in the right people and understanding what was required to get those site analyses as a foundation for the rest of the work. I still have in my slide collection the presentations we made. They used to just impress the hell out of people. We'd go from vegetation maps and soil maps and the drainage maps to land use overlays and show how land use and transportation were blended and shaped and organized. So that, with me bringing in Ian and us working together and then hiring Jim Veltman and making all that work, seeing all that put together, that was my biggest contribution to The Woodlands.

Another major thing would have been tying all these loose ends and disciplines together and creating a successful HUD application. Is that a fair statement?
Basically, I was involved in a lot of coordination, but mostly I would sit in on meetings and listen to what was going on with these other guys and consultants they had for doing marketing and economic analysis. There were other esoteric types that Dave Hendricks hired; I would try to sit in but not always. I reviewed all their reports and asked questions and I was always dove-tailing and coordinating, and we all got along. I took parts of their work and I did do a final editing to put this book together.

The book is the application?
The book was the application, and I still have a rain-soaked copy. The cover letter that went with that was signed by me, not by George Mitchell, for the submission to HUD. I've got copies of most of the slide shows that we had. And I've got one or two boxes of reports and memos, I've got a pretty good archive myself. It's not in very good shape because my garage leaked. But those were two of the main things, besides carrying the flag for the right ideas and the right principles.

What are the right ideas and the right principles?
The main idea, I think, was that a community that succeeds in implementing these principles and respects its natural environment is going to have a

better market value than one that doesn't. And we came up with some pretty innovative things.

Name some of those.

We have what we call vegetation buffers along the major roadways. What we did was create an extra setback of 50 to 75 feet, depending on the width of the roadway, and you can see it now as you drive along Woodlands Parkway. By the way, I insisted that Woodlands Parkway be the name. I put it on the maps and they said, "This is Robinson Road" and I said "No, it's Woodlands Parkway." Finally everybody bought that. We were able to prove that by having treelined major streets and having the setbacks and separating the buildings and the backs of the houses, you maintained the quality of The Woodlands.

If it's any comfort to you, when I pulled into the parking lot the day before yesterday, I had to stop for a deer in the road.

Yes, in the midst of this, that's what's so striking about it.

What are some of the others?

Another thing that we tried to do, when you're trying to fit land into nature, sometimes automobiles just chew up so much land, and we came up with the idea of shared driveways that were really sort of easements.

That's the first I've heard of that.

We saved a lot of land and were able to get houses nestled into small pieces of land relating to the drainage and vegetation that way. The whole concept of natural vegetation was based on keeping the water table stable, because normally when you build on ground, you put in a lot of hard surface. Rainfall hits the hard surface and runs off, runs into pipes straight to sewer systems and out to the Gulf. And this wet forest has what is called a "perched water table," which means that there's water two to three feet down or sometimes right at the surface. There's a lot of clay lenses under this that hold that water up. If you poke holes in that clay, or if you pave over too much, that water dries up and there goes your trees. So we said, "We're going to use natural drainage, we're not going to have concrete ditches, we're going to have natural swales." Natural drainage permits the recharge of water and maintains the wetness of the soil. We pioneered that and now it's copied. Many smaller communities claim they do the same thing, but they don't have the woods to work with. They're doing it on the prairies and haven't done very good jobs of them. But all these things are so logical once you think about them. And the main idea was, let's have good design standards. I found out to my dismay early on that there was no way we were going to sell lots to builders and be able to insist on modern architecture because they probably wouldn't stand for it, but at least we said that we would have some design standards for housing, but most of the standards would deal with landscaping and clearing for the site and the siting of the house. Basically the design standards that we set up were to implement the major environmental influence on all the infrastructure for the community, which Plato pretty well went along with. I assume that much of the design standards have held up over time.

Anything else?

The thing that set The Woodlands apart was, of course, all the things that we did—good, sound planning and good environmental planning, and the other thing was that financially, we had oil and gas to fall back on during hard times. Most of these other places were real estate operations that had no other resources and were in trouble. That was the genius of George and his other businesses that sustained The Woodlands.

Thank you, Bob.

(Bob Hartsfield died on September 2, 1995, after a long illness.)

Today is Thursday, September 21, 1995. This is Joe Kutchin. I'm in The Woodlands office of Robert Heineman. Is it Bob or Robert?
I've always been Robert, except for the short time when I was with the company with Bob Hartsfield and Mr. Mitchell called me Bob Heineman, and I never corrected him.

Tell me about your background, where you are from, your education, what you did before coming to Mitchell.
I was born and raised in Lubbock, graduated from Rice University, in the architectural school, in 1969 and worked for four different architectural firms during the next year. I decided at the end of that span that I wasn't interested in working for an architectural firm. I went back to Harvard to graduate school in urban design, and graduated there in 1972, with a master's degree in architecture in urban design. In the summer of 1971, in between school years, I was looking for a summer job and read about The Woodlands. Bob Hartsfield, who was head of The Woodlands planning at that time, had been at CRS when I was there. I wrote him a letter to see if there was employment during that summer and he hired me. I worked for Mitchell that summer of '71, and when I graduated in 1972, I inquired again to see whether there was full-time employment and was offered a job working on The Woodlands by Bob Hartsfield. So I've been with The Woodlands ever since, 1972 through now, 23 years.

Give me an idea of your responsibilities when you started.
Initially the planning department was very small. It consisted of Bob Hartsfield, Jim Veltman and myself, and shortly thereafter, we added James Marshall. Scott Mitchell was with us for about a year or two, maybe a year or so after I came to the company. Initially, I'm not sure what my title was. Everything was in start-up stages and everyone got involved with everything, from general planning to detailed planning. At one point, I was the residential design coordinator, but I think that was further along in my career. I'm not sure what my job title was, other than planner.

You started before the HUD guarantee came in?
Yes, in 1971 when I was here in the summer, I was involved in the planning, in fact, the preparation of many of the exhibits for the HUD grant.

Talk about some of those.

The major exhibit that I remember back then was a map that Mr. Mitchell especially liked, an environmental map that combined different environmental factors from vegetation to soils—an environmental synthesis map. The map had different colored dots on it which indicated whether the land should be left as open space or developed low intensity or high intensity. Green was open space, yellow-tan was low intensity development, and red was high intensity development. On top of that map, with a sheet of acetate, we overlaid the plan that we submitted to HUD which again had colors for different land use intensities. If you looked at the environmental synthesis map with the development map, you could see a harmony between the two which showed that the development was in response to the environmental factors.

What are some of the environmental factors that you looked at?

The two major factors were soil permeability and vegetation type. Ian McHarg's office had evaluated the entire area in terms of those factors and the suitability of the land for different types of development. Then you overlay development on it. For example, a soil with high permeability was good for low intensity development because the water recharge capability would be maintained. From a vegetation standpoint, the answers were a little bit more subjective. We dealt with questions such as whether pine trees were better than oak trees, big trees better than small trees and that sort of thing. Assuming that agreement was finally possible, you could end up with vegetation graded with respect to quality. We determined that the higher quality vegetation should be matched with lower intensity development, the lower quality vegetation with higher intensity development. The development was graded from single family-low densities to higher densities in commercial development. So you could take these two factors and try to mesh them at a general plan level, and that's really what that dot matrix map did.

Hartsfield was your boss then?

Well, actually Veltman was. Hartsfield was my boss in 1971 when I joined the company, through that summer session. When I came back in 1972, Veltman had been hired from Ian McHarg's office and Veltman was actually my boss. Hartsfield was in planning but seemed to be on a parallel plane of responsibility with Veltman. The reason is that there was the beginning of some in-house political conflicts at that time. Within the organization, Hartsfield was on the "outs," so to speak, with the new group from Columbia which had recently joined the company.

They were in place by then?
They came very shortly after I joined the company in the summer of 1972.

What I was driving at is, who was essentially the quarterback in preparing the HUD application?
Hartsfield was, as far as the HUD application. During the HUD application, as I recollect, Hartsfield was the only planner employed full time by The Woodlands. James Marshall hadn't been hired and I was just there for the summer. I wasn't there during the actual filing of the application.

In my mind, the HUD application and granting of the HUD deal were pivotal in the creation of The Woodlands, so that a lot of attention on everybody's part had to be paid to that.
That was my understanding. If the HUD application hadn't been made and hadn't been approved, The Woodlands certainly wouldn't have developed as it has. It may not have developed at all by the Mitchell company.

The map that you were just talking about is not the same as the environmental impact statement that Mitchell feels is unique.
No, the map I'm talking about was done in 1971, prior to the environmental impact statement in 1972.

Tell me a little about the situation when the Columbia group came in.
I have a difficult time keeping track of dates. It seems like they were involved for two-and-a-half years but it seems like 10 years looking back on it, and the last 10 years seem like two. There was a small local contingent within The Woodlands development group, a small West Coast contingent, and a larger group that came from Columbia with Len Ivins. It seemed obvious that the Columbia group meant to replace people in the organization that weren't part of their cast. In fact, within the first year, I was to be laid off on a Friday. This was back in 1973. We were still back in One Shell Plaza. Bob Hartsfield had hired me. I was working for Jim Veltman. They were both out of town and neither one of them had been informed as to my impending layoff. The story was that Len Ivins decided that someone in planning must be laid off. There were three people in planning who worked for Hartsfield or Veltman—there was myself, there was James Marshall, who was a 6-foot-7-inch black architect, and there was Scott Mitchell, the son of the owner of

the company. You can imagine who drew the short stick. As the story goes, everyone in the company, including all of the secretaries, knew that I was going to be laid off that Friday. Everyone except me. Vern Robbins and maybe one or two others salvaged my job. The pink slip never came. In fact, I didn't hear about it until some time after that. In retrospect, I remember walking around the company on that Friday and people giving me strange looks. I didn't sign up for sharing in the company stock program for several years after I joined the company simply because I had no reason to believe that I would be with the organization for a period of time long enough to be vested. A lot of this stems from the political upheaval within real estate back then. It turns out, of course, that I've stayed with the company for quite a few years.

Twenty years later, can you look back on it and figure that that period was productive, counterproductive, laid the foundation or what?
It was some of both. The ideas and concepts behind The Woodlands were so new at that time that they required a lot of time and effort to develop. However, there were also wasted steps and effort. I feel we had some consultants who were hired to give us their best expertise on how to develop a new community like The Woodlands, but, frankly, what we got was not what we paid for. The reason why is because they didn't have it organized either. Everything was so new that it just was not neatly packaged and what you got were bits and pieces, probably 99 percent of which were worthless. It was the 1 percent that you were searching for. In hindsight, we spent a lot of time and money back then that was wasted motion and wasted effort. Whether you could have gotten through that period without going through that, I couldn't say. Even in the Houston area, what we were proposing was completely new for Houston development, from the hierarchical road system to saving trees as you developed, to the natural drainage system. There were so many things that were different that some of the trial and error process was probably necessary at that point.

And ground down humans in the process, unnecessarily.
Yes, there's no doubt about that. The politics within the company, that sort of thing, was completely unwarranted. What I was talking about were the planning exercises and studies, etc., that were conducted back then. But as far as the politics within the company, the human element, that to me was just totally wasteful, had no productive end result at all.

Name two or three of the important innovative things that were incorporated in the original plan?

One main one was saving the forest after development. If you're going to name the project The Woodlands, don't name it for the forest that was there prior to development but which was destroyed in the development process. I think we've been successful in that.

Please tell more about that.

There were several components to the concept of saving the forest. One objective was setting a natural forest preserve along major roadways where trees and understory could not be cleared. The end result is that as you drive through The Woodlands on the major thoroughfares, you see trees. In fact, the houses and buildings are often hidden. It gives the visitor or the resident the distinct impression that the forest is still there after development. Another objective was to maintain natural forest within a parcel of land to be sold and developed by a third party. We did this by imposing a clearance/coverage ratio to each parcel of land when it is sold. The result is that there are trees in parking lots, trees next to buildings, etc., and significant amounts of the forest have been saved on individual parcels. Those ideas were pretty much unthought of in Houston development at that time. In fact, we had some pretty adverse reaction to these concepts initially, because people expected manicured landscaping which required clearing all of the understory. The feeling was that commercial buildings were meant to be seen rather than hidden. So it took a critical mass of development of several years for these ideas to be accepted in the marketplace. Another innovative idea was probably the hierarchical road system. In The Woodlands we have a major thoroughfare such as Woodlands Parkway, which on a Houston area map is the same type of roadway as Richmond Avenue or Westheimer—but yet, when you drive down Woodlands Parkway, it's completely different from Westheimer. The difference is that we did not allow buildings to front directly on the major thoroughfares with corresponding private curb cut access. We also looked at having a mixture of land uses along the thoroughfares. As you drive along Woodlands Parkway you'll see residential homes, golf courses, commercial buildings—a mixture of uses as opposed to a continuous strip of commercial and retail uses along Westheimer. In Houston, the capacity and speed of a major roadway such as Westheimer, which was also a farm-to-market road at one time and used for long distance and high-speed travel, breaks down over time as development occurs along the corridor and you have many individual businesses with their corresponding curb cuts. The speed limit goes from 45 or 50 miles an hour down to 20 miles an hour, and congestion results. In The Woodlands, we really could not afford to do that. As you look across the 25,000 acres of The Woodlands, from I-45, the furthest west point at FM 2978 is about eight miles from I-45. If you were at the Galleria at Loop 610 on Westheimer and you were to drive out eight miles on Westheimer, you would be past Beltway 8 and almost to Dairy Ashford. So you can imagine how hard it would be to market houses out at Dairy Ashford, having to drive on a street like Westheimer with its traffic and congestion to the West Loop, and at that point you're just on the freeway ready to go to downtown Houston. You can imagine the problem we would be having right now. So the hierarchical road system really set up a system whereby the capacity and speeds of major roadways would be maintained over time rather than reduced. It involved more up-front dollars from the developer because if you're building Woodlands Parkway that provides no access to individual buildings, then you have to build a separate street, like Timberloch Place, to access those buildings. In total, you're building probably 10 to 15 percent more infrastructure than with the conventional system.

You're also losing land that could be developed, aren't you?

That's right. The reason you're doing it for the long term is it will allow you to develop six or eight miles west of the freeway with minimum traffic congestion 20 years down the line, while you would not be able to do that with the typical system. The other thing that was different was in residential areas. Rather than a grid system that was popular in Houston at the time, we went to more of a loop/cul-de-sac type system in order to minimize the through traffic that would penetrate residential neighborhoods. It forces the longer-distance traffic on to major collectors, such as East or West Panther Creek Drive or a major thoroughfare, such as Woodlands Parkway. It really reduces the cut-through traffic. What we're seeing now in the City of Houston is that in those older residential areas that have a grid system, they are having huge problems because of cut-through traffic and crime problems, and many neighborhoods have requested closings of streets. The problem is that if you have a street system like the City of Houston's, which does not have a hierarchical-type framework, if you close one residential street and force the traffic onto an adjacent street which is not planned to handle the additional traffic, residents on the second street are going to complain because they've got twice as much traffic.

Anything to say about natural drainage?

Yes, drainage probably was a third major innovation. Initially the plan for the natural drainage system was more widespread than what has been implemented. What has been maintained is the natural drainage system in the major streams as well as open channels which flow from neighborhoods into those streams. Initially, we experimented with open-ditch drainage along residential streets within neighborhoods. The market acceptance of open-ditch drainage is very poor compared to curb-and-gutter drainage. For example, for a typical neighborhood lot 50 feet wide, with a two-car driveway 20 feet wide, the result is 20 feet of culvert with two head walls and 30 feet of open-ditch drainage. The maintenance factor and the visual impact are very negative to the consumer. After Grogan's Mill and parts of Panther Creek were developed, we went to a curb-and-gutter system for intract parcels, but maintained the open-ditch drainage system along collector streets and major thoroughfares and from residential neighborhoods to major streams.

You talked about three or four features of The Woodlands as it developed. Are these things unique?
The hierarchical road system was new to Houston, but it was similar to what had been developed in Columbia. The maintenance of the natural forest, including the understory with the natural drainage system, probably was new to most real estate developments at that time. A more current innovation that we're still working on is the Town Center development and the waterway corridor.

That was in the original plan, wasn't it?
No, the idea of having a waterway/pedestrian/transit corridor developed over time. I think if you go back and look at the different drawings and plans for The Woodlands, most current concepts evolved over time. We've had so many planners and consultants. Virtually every one of them can say, "I was involved in the idea of the riverwalk." And they would be right, because if you review the original plan there was a drainage/green space corridor in the Town Center where the riverwalk is. It just didn't have water in it on those plans, but plans evolved over time and one person's idea was expanded by someone else. It's really been an evolutionary process, and we're not there yet, either. We've still got a ways to go and are in need of more ideas and concepts.

You came to the company in the early '70s when you were quite a young man—I would guess in your early twenties.
Actually, in my late twenties . . . it took me a long time to get out of school.

Over the years, obviously, you've taken over increasing responsibilities and you've been given increasing responsibilities. Would it be fair to say that in the early years you were doing what you were told to do?
Yes, that would be very fair. I was not very far up the corporate ladder. I remember a meeting to plan the Town Center. We had a huge room with Ian McHarg's people, William Pereira's people, Hartsfield, Veltman and Coulson Tough. Bob McGee kicked the session off. He said, "I have the vision for the Center (Metro Center was the name back then), I know what it should be," and then he walked out of the room. So here we have a group of planners, who were supposed to come up with the vision of the Town Center and, of course, the obvious reaction was, "If he already has the vision, why not share it with us and we'll save some consultant time." Of course, the reaction was that he really didn't have that magic vision after all. Other projects such as the Grogan's Mill Village Center had very little input from the staff planners within the company and were planned by a special group formed by management at that time.

Are you suggesting that with more careful planning, results would have been better?
I think they could have been better.

Have we as a company learned to do things like that?
Yes. If you were to replan, with 20/20 hindsight, you would still have a commercial or village center. It probably would not be in that location, and if it was in that location, it would not have been in that architectural configuration, and would have more visibility. We've gotten better. If you look at Panther Creek or Cochran's Crossing, they're on major intersections. In retrospect, the village center should probably have been located at the intersection of Woodlands Parkway and Grogan's Mill Road. Other things happened in the retail business which affected the center. One was the size of the grocery store—a 20,000-square-foot grocery store can't compete with a 60,000-square-foot store. So we've learned from that.

Your responsibilities increased and over time, rather than being the young guy on the staff who was told what to do, you began to take on increasing responsibility in determining how this place is going to develop. What kinds of things do you feel you were involved in where the main job has fallen on your shoulders?

One area has been transportation planning. In Town Center, rather than just planning roadway infrastructure for development as it occurs, not knowing whether you'll have the ultimate capacity for what's built, we tried to look ahead. You can examine the Galleria area and see the problem they've got right now. They've got a huge problem on the West Loop. There is not a solution to the West Loop traffic problem. It's primarily caused by the Galleria development. In Town Center, we examined all of the traffic and transportation development potentials, including road widenings, grade separations, turning lanes, transit, and quantified all of those in terms of costs. Then we said if all of those transportation improvements were done, how much development could they support, still maintaining an acceptable level of service. As we got to those outer limits, we quantified the cost benefit of specific improvements. For example, improvement "A" might cost "X" and from it you could generate an additional 500,000 square feet of office-equivalent space. From this, a benefit ratio would be generated that would suggest which transportation improvements were cost beneficial and which ones weren't. As a result of that, we ended up with a plan for Town Center that would accommodate 18 million office-equivalent square feet of development. Office equivalent considers different land uses–residential, commercial, retail and so forth–and their traffic generation factors and converts all of it into a single base number, which is office equivalence. That's something that few developments do. Usually, you build when the market's there and the customer suffers if congestion occurs and there's no timely solution to the traffic problem. Part of the evolution for us was the pedestrian/transit corridor in the Town Center. Town Center, with its 18 million square feet at buildout, is larger than Greenway Plaza, smaller than the Galleria area and about the size of downtown Kansas City. The question is, how can we improve on those "downtowns." One question involves transportation and land use. The lack of integration of transportation with land use has hurt the Galleria area. It's got all the elements, high rise residential towers, commercial and retail space and recreation (the Transco Park and fountain), but yet they're not integrated. The area is too auto-dependent. You have to get in your car to go from one side of the street to the other or to lunch. The end result is that you end up with traffic congestion even at lunch time. Our feeling was that in a successful downtown, a transit/pedestrian element is necessary; you should not be completely dependent on the car. That's where this pedestrian/transit corridor, coupled with the waterway, evolved from.

Does that mean a place to walk?

Yes, you might walk from an office building to a restaurant, or to a transit stop, which would then take you to the restaurant. You'd have an alternative way to travel—transit, water shuttle or walking—fairly short distances within the Town Center as opposed to relying exclusively on the automobile. And the trip would be fun and eventful, since it would be along the waterway—a major amenity within the Town Center. The end result would be similar to the San Antonio Riverwalk but yet different. In San Antonio, where the Riverwalk is basically U-shaped, you can walk a mile and you're back where you started. In The Woodlands, you can walk a mile linearly along the waterway and you're a mile from your car. That's where the need for transit comes in.

When do you see that coming to completion?

It will evolve over a long period of time as did the San Antonio Riverwalk. But, the beginnings of it have already begun with the construction of the mall ring road bridge and the Six Pines bridge. These bridges have a unique architectural character and are designed to span a transit corridor and pedestrian path system below. If those two bridges, which were required to access the mall, had been built following conventional thinking, with only a drainage pipe underneath, the corridor could not develop. Now there is the distinct possibility that within the next year to year-and-a-half the waterway corridor from Grogan's Mill Road to the mall ring road bridge will be developed. Up until now, we've used a pretty picture to convince the people that the ditch will become a riverwalk and to design their buildings so that they will orient to a future riverwalk. At some point, the pretty picture must begin to be implemented for people to believe it's really going to happen. That time has come. Completion of the two bridges and the beginning of the waterway within the next year or year-and-a-half will ensure the ultimate development of the corridor.

I believe that Browne was head of this department for a long period of time.

Dick was head of the department for the entire time that he was with the company, which was about 20 years. As I remember, there was a very short period when Rob Lapp was head of planning.

At least initially, Dick identified with the group from Columbia?

Dick was hired by that group, and he was from Columbia. I think after he had been here for some period of time, there were differences of opinion between Dick and some of that group. Dick and Vern Robbins were from the Columbia group. They were probably two of the people from Columbia who didn't necessarily fit with the rest of them.

Let's go back to your experience, Robert. Clearly you have had a major role in transportation planning, and you've talked about several things that are relatively unusual if not unique in The Woodlands. In addition to transportation, is there anything where you have been basically the ball carrier?

For approximately the last 10 years, I've been the chairman of the architectural review committee, Development Standards Committee, for The Woodlands. This began when the Water Resources Building was submitted to the DSC. The committee rejected the design. It was what was considered back then a post-modern building design. I was not on the committee then and wrote a couple of memos to the members of the committee and complained about their actions. While the committee should prevent poor design from occurring within The Woodlands, it should not put the clamps on architectural creativity. This building went on to be featured in *Time* Magazine as one of the 10 most significant buildings of the year within the U.S. and is still the only building in The Woodlands that's ever been in a similar type of a national publication. The end result of that was that I was named chair of the committee. I've also represented the company in court as a planning/architecture expert on several different occasions when lawsuits have arisen as a result of planning-related decisions. I've been on the executive board of the Chamber and chaired committees with other organizations. The Chamber, through their transportation committee which I've chaired for several years now, was instrumental in taking the first step for The Woodlands to secure outside state and federal funds for major roadways. Up until that time, The Woodlands had been unsuccessful in securing outside funding, which is really essential for construction of major freeway ramps, interchanges and major thoroughfares. The Chamber committee developed a mobility plan for south Montgomery County. The plan was presented to the Highway Commission in Austin and was the basis for forming Road Utility District Number 1, which floated a $10 million bond issue. That $10 million was the local share required to leverage the construction of the ramps on I-45 at the front door to The Woodlands, The Woodlands "jughandle" interchange that serves the Mall, the Kuykendahl Bridge across to Harris County, and also provided seed money for the Sawdust Road-I-45 interchange and State Highway 242. That was really the first local effort which got the community involved from a financial participation standpoint and was needed at that time in order to leverage significant outside funds for those improvements. It was really the thing that began the extensive transportation improvements in South County. If you compare south Montgomery County with north Houston, you'll see that I-45 is just now under construction in the 1960 area, and the I-45-1960 interchange has not been built. Some of the direct freeway ramps at I-45 and Beltway 8 are under construction and two of them are not yet budgeted. That initial cooperative with the Highway Department resulted in millions of dollars of outside funds being spent in South County. That's been continued recently with the securing of funds for the Woodlands Parkway overpass, Gosling Road extension into Harris County, the widening of Research Forest Drive, funds for a new Park & Ride lot and obtaining funds for the Town Center pedestrian/transit corridor.

The past 20-plus years of your life have been involved in the creation and development of The Woodlands. What are the things that you're most pleased about?

The pluses have been maintaining the forest after development, sticking to our guns as far as maintaining landscape, signage and architectural controls for retail buildings, such as gas stations and McDonald's. Now it can be shown that a retail business, such as McDonald's, which is buried in The Woodlands (it's not on the freeway, has very few signs and no tall signs, and has no entrance from a major thoroughfare) can be successful. McDonald's is one of the five or 10 top performers in the Houston area. The regional mall has been another resounding success. The larger success yet to be achieved (as the regional mall expands to the south, with one or two more anchor stores, and as the waterway develops) is the integration of the mall to the waterway. I think the transportation system is something to be proud of. Most of the problems that we have experienced to date have been related to funding needs to implement the plan. The difference between this and a West Loop or a Westheimer traffic problem is that there is a solution here. On Westheimer there's not a solution yet to that problem. Overall, The Woodlands has really developed and is continuing to develop into a city with a real sense of community, something that I did not detect in Irvine or some of the other communities I have visited. I think people who move to The Woodlands really feel that community spirit. Disappointments? One is the Grogan's Mill Village Center. Another, the failure to acquire some land at our southern entrance at Grogan's Mill Road and Sawdust Road.

Is the Sawdust problem an aesthetic or a business consideration?
Both. It's a business consideration in that it affects the Grogan's Mill Village Center, with the construction of the Randall's Center there. It's aesthetic also in that Sawdust Road is not an inviting entrance corridor into The Woodlands. A disappointment in the Town Center was the lack of grouping buildings together cohesively to produce a sense of synergism early on. We had a plan, for example, for the post office to be located near the Water Resources Building to form a "government center," but that was rejected by management. Also, several of the office buildings are spread out on Timberloch Place. At that time, management followed the "dumbbell theory" that if you develop both ends, the middle will develop more quickly than if you start in the center and spread gradually outward. Probably the opposite is true. If we had six buildings grouped together, they might generate the need for a restaurant; six buildings scattered might not result in a restaurant. Within the residential areas, I think one disappointment has been the creation in Panther Creek as well as in Cochran's Crossing of very successful retail shopping centers, but not village centers. The community elements are missing. We've got a plan to create a real village center for Alden Bridge, where we'll have not only the shopping element, but also a village square or village green, and some community elements. I think there are future opportunities in The Woodlands relating to some of the neo-traditional planning elements that have been written up within the last three or four years in *Newsweek*. Some of these have to do with the benefits of integration of different types of residential units into a single neighborhood and a closer integration of residential with retail/commercial.

You can't have your cake and eat it too, can you? If you're going to have the cul-de-sacs that you described, how are you going to do much commercial/retail?
I think you can combine some of the concepts of the neo-traditional "grid" and The Woodlands cul-de-sac/loop system and be very successful. This could result in a little bit more urban character within a neighborhood—street trees, sidewalks, possibly front fences, front porches, that sort of thing. But yet the grid system does not extend outside the residential neighborhood, so you control cut-through traffic and still maintain the sense of privacy that exists in neighborhoods in The Woodlands. We're looking at a bungalow neighborhood now in which all of the houses have front porches

and alleys, all in a modified grid system within the neighborhood. The plan still calls for retention of natural vegetation. Some other elements of the neo-traditional planning make sense. One is that the employment situation within the American world is changing; more and more people are working out of their houses, mainly because of the communication revolution. I was at a luncheon today where the editor of *Futurist* Magazine says by the year 2010, 25 percent of our work force will be working out of their houses. Right now it's 3 percent, so that's a huge increase. Over the last few years, the number of jobs has increased, but office space hasn't been filled. One of the reasons is because people aren't working in offices, they're working in houses. We must look at what types of houses people want to live in, what types of neighborhoods do people want to live in and what types of communities do people want to live in. If you work out of your house 100 percent of the time, where do you interact with your neighbors in the community? You don't do that at work anymore. You may not want to live in downtown Houston—why live in that environment? But on the other hand, you may feel somewhat isolated living in a residential suburban community on a cul-de-sac street. That may have been fine when you commuted to downtown Houston and you interacted with people there during the day, and after work you looked forward to rest and relaxation. If you're working out of your house, your interaction is going to be local, within your neighborhood. So these ideas of the front porch, more emphasis on the walking environment near the home, more emphasis on a close connection from the home to retail, to shopping and to other services may be needed in the future. That's really what The Woodlands has to think about over the next 20 years, because we've got 20 more years of residential development. If, in fact, this does happen, how does that affect the design of our neighborhoods and the housing product that we put in them?

That's kind of off the track that's been established over the first 20 years. How does George react to that kind of thing?

I don't think George has been exposed to the bungalow neighborhood yet. We're still in early planning. We have plans for neighborhoods, we have some plans for houses and elevations, and within the next month, we'll be conducting a market focus group to see whether they would prefer this different type house and neighborhood versus what's currently being offered in The Woodlands. For The Woodlands, this can be treated as a niche market, of possibly 100 sales out of 1,000 to 1,200 units a year. If it appears promising, we'll be able to develop it more extensively over time and increase our sales pace and our market penetration.

Does your department have any significant role in social institutional planning—everything from entertainment to education to who knows what?

We do from the standpoint of location of land uses and buildings: location of schools, location of parks. Also, we have a significant input into design. As the community associations have grown and taken on more responsibility, they have much more input into the design of parks than they did initially. One thing that has been lacking is studied feedback on what we have designed and built. We don't have public reaction to a number of projects (such as parks), on how is a space used, how we could change things to improve things, what works and what doesn't work. From a social planning standpoint, there hasn't been nearly the evaluation and feedback on things that we've built that I think we should have gotten over the years. A lot of it just probably had to do with budget.

Robert, thank you very much.

Earl Higgins

Today is Tuesday, February 13, 1996. This is Joe Kutchin and I'm in the office of Earl Higgins, who is vice president of property management of The Woodlands Corporation. Give me a quick idea of where you're from, your education, what you did before.

My college degree is in accounting from Sam Houston University. I grew up in my early years in Tomball, near Treeline Golf Course, and completed high school in Jacksonville, Texas, where my dad was in the oil business. I worked my way through school and got my accounting degree in 1965. Growing up in the oil patch, working on oil rigs and so forth, I wanted to have a career in accounting in the gas and oil business. In 1965 when I got out of school, in the middle of the Viet Nam situation, I had an opportunity to go to work for Skelly Oil or for Standard of Texas, which was a division of Chevron. As it turned out, I went to work for Chevron and worked for them about a year-and-a-half. But I was in the reserves and went off to basic training in California. When I came back, Mitchell Energy was looking for a gas and oil accountant.

When was this?

I went to work at Mitchell the day after Christmas in 1967 as a gas and oil revenue accountant. I just answered an ad and that's when Mitchell was pretty aggressive in growth. I can remember that was a really exciting time for me with the Mitchell company. We were packed into the old Houston Club building like sardines, desktop to desktop. George never had a set time to be sitting in the Houston Club building coffee shop. It didn't matter which door you came through, there were two doors, he was going to pop you with a question because he knew what everybody did, what the status was, and he always had questions.

Who was your boss then?

Back then, Benny Hruzek had left. I guess it was Chuck Koehn, who was controller back then. During that time, Mr. Mitchell through trusteeship was acquiring land primarily for The Woodlands project.

He had specifically The Woodlands project in mind?

He was acquiring all this land, as I recall, during '65, '64. And I remember that within 10 months of my joining Mitchell is when he decided to get involved in real estate development. One of the partnerships was with Norman Dobbins back then. And that's how we got involved in what's now known as The Esplanade, which was a spinoff of Dobbins and Weingarten. Mr. Mitchell took the north side of FM 525, which is acreage we still own.

Approximate to the North Belt?

Yes. Then there was Pelican Island, and the Pirates' Beach project was under way. One of the first things I got involved with was the Port au Prince Apartments.

So you were migrating away from oil and gas.

I became the first accounting supervisor for real estate. I can remember Mr. Mitchell's knowing every situation. If we didn't send out water bills on time at Pirates' Beach, he knew about it. Those were fun days. We then went through the process of the company's going public, I believe it was '71 or '72. Then immediately following that, Len Ivins was hired to come in and set up a real estate division.

You're entirely out of oil and gas by now?

Yes. When Ivins came on board, he had full control of the real estate division. He was often in disagreement with Bill Houck, who was senior financial officer in our company. At that particular time, Chuck Koehn took me to lunch and said, "You've got a new boss. I don't know if you have a job, but . . ." The first individual that Len Ivins hired was a man by the name of Tom Cole. He was brought in as controller. I went to talk to Tom Cole. At that particular time I had five individuals in the accounting department, and we started reporting directly to Tom Cole. Of those five, Bob Hibbetts is still with the company; Maggie Bilski is with the credit union, and Tom Jaeger is still with the company. Boy, we worked during that time, night and day, seven days a week applying for the HUD bonds. I made several trips to Washington, D.C., to develop an understanding of the rules and regulations. I've been involved since day one.

When did you take up golf?

I grew up in golf. My grandparents built Treeline Golf Course in Tomball. I played in the HGA Junior Golf Program through high school. But as far as the company is concerned, the Houston Golf Association tournament in 1974 was at Quail Valley. The HGA was struggling financially, the course conditions were not the best, and at that particular time Larry Kash, who was one of our senior officers in the real estate division who was a good golfer, took the lead and helped negotiate a contract with the Houston Golf Association.

You were part of that negotiation?
I didn't participate at that time. The reason the tournament came out from Quail Valley to here is because Mr. Mitchell, in essence, gave them $50,000 to bail them out of their problems at Quail Valley in exchange for a commitment to move the tournament to The Woodlands in 1975, even though the course hadn't been finished. At the time the contract was executed, the course was still under construction and there was a lot of concern by the PGA whether we'd be able to host the tournament in 1975.

Was it George's initiative or Larry's?
I'm not sure who made the initial contact, but the president of the Houston Golf Association at that time dealt with Larry Kash to get a contract consummated. And that was basically a $50,000 check to bail them out financially from Quail Valley in order to move here in 1975 to a course that was not even open yet.

Had there been a long history at Quail Valley?
No. As a matter of fact, this year is the 50th year of the Houston Golf Association. And it's had its tournaments at Memorial Park and Champions and various other places. It was to the developer's advantage to have as a marketing tool the hosting of a PGA event. The agreement would normally last one, two or three years and then they'd move on.

And the deficit had been accumulating?
No, that was just that year.

Would the HGA have folded if it hadn't gotten that $50,000?
I'd say they were on the brink of being financially insecure, being a nonprofit entity. And, as it all worked out, we finished the golf course which was the original 18 at The Woodlands conference facility.

That's now part of the country club course?
Yes, the original 18 was split into what is now called the West Course and the North Course. We added nine holes on each one of them. The HGA tournament started in 1975 and it did a lot for The Woodlands, from a mar-

keting standpoint. It's been very beneficial to us. And I joined the Golf Association in 1975, the year it moved to The Woodlands.

By joining the Golf Association, what do you mean?
It's an all-volunteer force which exists primarily to raise money for children's charities. At that time we had just one charity group, Boys' Harbor down at LaPorte.

So you became one of the volunteers?
Yes. Being that I grew up in the junior golf program, I'm interested also in that aspect of the Houston Golf Association.

Did you play golf competitively?
As a 15- or 16-year-old competitor in the City of Houston.

How good are you, Earl?
Today I'm probably a 6 handicap. I've played with a 6 handicap but I've been a scratch player in my lifetime. In college, I played on the golf team.

Was management of The Woodlands Corporation and/or George aware of your close ties to golf?
Yes, early on. Over the years, I sort of represented the company, as one of their employees, with the HGA. I handle all the inter-company sales and coordinate things for George Mitchell and Phil Smith's office for getting the credentials for the tournaments. I sort of serve as liaison. I participated in the negotiation of the first contract with the PGA Tour for a TPC license agreement on the existing course.

That was '75, too, wasn't it?
No, let's see. I became an officer of the HGA, got on the Board, then was president of the Golf Association in 1984, which was the last year the tournament was held on the old course. In 1985 we moved over to what is now called the TPC. The year I was president probably would have been 1984, which is when we negotiated that contract–Len Rogers, Mike Richmond and I. The HGA executive director was Duke Butler at that time.

Going back to 1974, were there other developers who wanted the tournament?
Yes, there was a lot of development activity around the city, but at that par-

ticular moment we were willing to put up $50,000 to get the option.

Early on there was the involvement, I don't remember exactly what it was, of Doug Sanders. What was his role?
When we opened the original course in 1974 or '75, Doug Sanders was brought in as part owner of the pro shop operations. He was sort of the director, so to speak. He actually had a percent of the profit of the pro shop operation and I would say that was very important. As a matter of fact, the original HGA Pro/Am had been the Doug Sanders Pro/Am. Doug's contacts and influence back in those days were with the Bob Hopes and all the California group. Evil Kneivel was a hot item at that particular time. Doug helped the HGA and The Woodlands from a standpoint of exposure to that element of people. I can't remember how long it lasted. Today we've got Jeff Maggert and John Mahaffey as touring representatives.

Was he still competitive at that time?
Yes. Very colorful. I want to say that lasted for about three or four years and then he decided to move on.

I have the impression that it was a disagreement over terms.
As I recall, that might have been. I suspect there was a contract and that contract was up for renewal. What I heard was that he was asking more than we were willing to pay.

I imagine you can probably find an oil and gas man anywhere, but a good golfer . . .
A good golfer with all these contacts he had on the West Coast, that's rare.

But would you say that in the early years he played an instrumental role in establishing the tournament here?
There's no doubt in my mind that, at that time, especially coming from a class B event at Quail Valley, to elevate it and make a new statement at a project 30 miles out of town, where nobody in the marketplace knew what The Woodlands was even about, to open the project in October 1974 with the conference facility and the old wharf retail area and an 18-hole golf course way out in the woods, I'd say Doug should get credit for helping establish The Woodlands at that time as a golf community.

OK, so here we are, 20 years later. What has been the role of the

Houston Open and golf in general in the development of The Woodlands?

There's no doubt that the PGA event in the first five years, at least, helped accelerate our marketing capability for selling homes.

How would it do that?

Just by being a class A golf facility. The recognition that that brings in the golf community as a place to live and enjoy excellent facilities definitely accelerates the sales and marketing pace. I'll give you another way to demonstrate that. The reason the PGA Tour started building stadium golf is because the PGA said, "Hey, with a tournament event we can go into a new developer's parcel and we can guarantee you, the developer, that you can market your lots at a much faster pace with our name recognition, and in exchange for that we want half the project. You contribute the land, you put up all the capital improvement money and for that we'll operate it and sell memberships and so forth as our participation as a partner." A golf course will do that.

The Houston area is a really intense golf area.

There are a lot of golf courses in Houston now, and to this day, more are being built and especially in developments because they're a strong selling tool. There's obviously different levels of pricing and quality. From being involved in the Houston Golf Association, I'd say in today's times most of these clubs at real estate developments are doing very well. Golf has really changed in the last four or five years. Everybody is maxed out on the number of rounds they're booking. There's just a big demand, as demonstrated in our project where we've got four courses now plus one course that has 27 holes and we need more.

We do need more? Is that on the drawing board?

Yes, sir. We're looking at municipal golf courses. As a matter of fact, we're under contract with the PGA Tour to build another TPC facility, west of Kuykendahl.

Does that mean that somewhere in the future the present TPC course will take second place to the new one?

That's the plan right now, yes. Basically, what it is, there's a 75-acre lake that's part of the water retention system that already exists out there and this future TPC facility will be around that lake area. As the housing demand goes west of Kuykendahl, then that would be obviously the time to start construction of the golf course, which is an amenity that adds value to the lots.

It seems that in the creation and development of The Woodlands, a number of strategic decisions have been made. One of the key ones has been to provide plenty of playing opportunities for the weekend golfer.

As of right now, we have geared it more to the premium values of a country club or a high-end TPC daily fee type of course. But there is certainly demand for meeting the need of a municipal, affordable course.

What is the daily fee of the TPC?

I'm not sure. It's a seasonal deal and I guess in the peak season it's probably $60 to $70 per round.

It gets a lot of use, I know that.

It gets too much use. For hosting a tournament event, you really need to decrease the number of rounds in order to prep the course for the tournament week. As a matter of fact, I know that The Woodlands Corporation and the HGA are negotiating right now about possibly shutting down the course after this year's event and completely rebuilding all the greens. It's just like everything else, there's a life cycle of that asset. You may recall two years ago we experienced in our tournament very bad greens conditions, and there was a lot of concern about the pros coming back, and so forth. Last year, because of the way the weather in the winter transitioned into the spring, we had excellent greens. There's a challenge this year for Carlton Gipson, who has been hired as consultant, to be sure that this year we get through the golf tournament with as good greens as we possibly can. I think the plan is to immediately shut it down for approximately nine months and rebuild the greens.

The present deal with Shell, is that a long-term thing?

The Shell contract, I believe, runs out in '97. The parties involved are PGA Tour, which enters into a contract with HGA year-by-year; HGA as a sponsor that provides all the volunteer labor force for the event, and which has an agreement with The Woodlands to use their facilities; and Shell, which underwrites the tournament as the primary sponsor. All of those parties are currently in the process of negotiating and possibly going ahead and entering into a contract that would cover an extended period of time. And part of that equation is commitment by the company to go ahead and recondition the course. Further, with us as a partner with the PGA Tour, we're negotiating the new TPC course.

You say you're on the Board of HGA?
Yes, on the executive committee and on the Board of the Houston Golf Association.

And at one time you were the president?
I was president in 1984, the last year of the Coca-Cola-Houston Open.

Another advantage of The Woodlands' involvement in the Houston Open is great publicity.
It's on national television, ABC coverage. Shell is their commercial sponsor. Last year was, I'd say, our most successful year with the Houston Golf Association from a marketing standpoint and financial results to the charities. We contributed $1.6 million to charity. We had a record 200,000 people come out here and park on the streets and attend the event. We had one solid week of perfect weather–and that's rare. We've always had at least one day of rainout or interruption, and we've had one year after moving over to the TPC when we only got in two days of play versus four. That event was canceled and we hosted it again later at Champions.

Is there something else that needs to be talked about?
Well, I take a lot of pride in my position and as one of the original employees. I've gone through the cycle of being the first accountant on the front end. I've been here now going on 30 years. Roughly it's been 10 years of accounting. I was the first real estate budget director through the first 10 years of the development of The Woodlands and then the last 10 years I've been vice president of property management, overseeing the day-to-day operations of all our office, industrial, retail and apartment portfolios. That job has been very rewarding, not only maintaining our buildings to the quality of what we say The Woodlands is, but also insofar as the interaction with third-party tenants and residents and community and country club members who are tenants of ours. It's been fun.

To some extent, at least in that job, the more successful you are, the smaller your empire grows because we keep trying to sell those assets, don't we?

It doesn't affect our area because as we sell assets we roll them into partnerships, but our investment in partnerships still requires more and more management. The focus of the real estate division from bottom line earnings is switching to where fee management is becoming a big part of our business. We're still general partners, but we're also fee managers.

Name two or three or four of the main ones, as examples.
Since 1995, the new partnership arrangement with Crescent, where we sold five office buildings and five of the research buildings into a portfolio consisting of 10 buildings. We're the managing general partner, plus in fee management, like no other company, we get a 5 percent management fee from revenues. For 1996, that revenue string could consist of $40-plus million. I'm responsible for in excess of $10 to $12 million a year in expenses. My capital budget consists of about a million-and-a-half. It's a big operation. I've got 75 employees who basically work for me, and these are mostly employees who are on the front line dealing with the customer, day in and day out, all the way from maid service in the apartments to certified engineers maintaining buildings.

Was a part of the original thinking back in the early '70s, mid-'70s, that we would retain a partnership interest and the management role in these things that we sold?
In the old days the strategy was, it's George's company and George's master planning of this community and the conversation at the coffee table was, "We'll never sell anything." We'd always buy more than we'd sell and we would never even think about partnerships. It's just that when you are a pioneer in a project of this size, and have to put the seed money into so many start-up projects, be it mortgage companies, title companies, whatever, until we prove to the market that we're just not all hot air talking about our projects, we have to demonstrate that we're long-term investors, long-term developers and that we'll stand behind this project for a long period of time. Consequently, we did business unlike a typical developer, who would get in and get out. And I think the proof of our success today is that we haven't deviated from that philosophy of quality, long term. In the real estate business, you roll assets over in certain life cycles of, say, every 10 years, you roll it into a partnership, pull your equity out, and keep control of the management and ownership. Then you invest that money in the next opportunity.

Because actually you've been nicely profitable.

I haven't seen year-end numbers now, but I think real estate this year is going to be very proud of its contribution to the parent company. The challenge ahead of us is to continue this growth. I think that for the first time you're going to see, maybe the whole company is going to see, what role property management and the management of this asset portfolio is yielding as far as profitability.

I would guess that's not widely understood.
I don't think it's an area that we have concentrated on. There's the development side of the business and that's broken down between commercial/residential. That's our bread and butter. That's really what our company has been about. This last year we sold the remaining half interest in the cable company. It's obvious what that did for our company—it's paid off. It's a good investment. We took the risk and started it and we're now rolling those kinds of investments into other partnerships.

I have the impression that one of the original concepts was, let's build, finance it and then sell it for a profit.
As long as we're going to be in business, I guess our company, just like all other corporations right now, is going through corporate re-engineering and head counts and all that kind of stuff. But in our part of the world, we can't be that concerned about head counts because if you're in this business, you've got to service the customer. When we add an apartment project, yes, that's seven more people and you can't outsource it. So, on our side, we're in a mode of growing opportunity, and with that, to service the customers, you've got to add head count. We're concerned with the overall company philosophy and head count controls, but at the same time I think management's been fair in recognizing that, yes, the commitment is still quality and we need to service the customer. I can remember back, probably in the mid-'80s, when the oil crunch hit and I can remember having to reduce staff. I had 70 people back then and I eliminated 10 people, for example, in our building operations—painters, carpenters and engineers—and here we are 10 years later and we've added lots and lots of buildings and I'm still back where I was 10 years ago. But obviously, technology and a lot of things have allowed managers like myself to be more efficient.

Let's go back to the significance of golfing in the development of The Woodlands. Is there more to tell?
The real reason I got involved in the Houston Golf Association is not

because of the company but because I played in the Junior Golf Program growing up and I know what that discipline, that exposure did for me personally. The first time my wife and I went to a charity supported by HGA, it was Boys' Harbor. You sit down at the table with twenty 4-, 5- and 6-year old kids. It's real easy to get involved in trying to do everything you can to raise funds.

Is there more about the role of different sports in the development of The Woodlands—everything from the diving and swim program to golf and tennis?

You've got to offer all of these things to make it complete. The first day we opened for business we had an 18-hole country club facility with many tennis courts. That was a statement to Houston primarily: We are a real master planned community, which at that time were new buzz words. A real hometown, as you might remember, was our marketing theme.

The Woodlands Athletic Center, I think, came a year later.

Yes. Dick Smith, the diving coach—he was just as important as Doug Sanders in a smaller scale, in a smaller sporting event. But even today, being in the fee management business, talking about commercial billing rates versus apartment rates: We say we're getting a 12 to 15 percent premium over competition because of the amenities you get when you live in The Woodlands—numerous parks, play areas and tennis facilities, swimming, pavilion, golf, playgrounds. It's a complete package. It's a quality of lifestyle that any individual or family can take advantage of.

You use the term fee management. Define that, please.

In real estate, there are asset managers. Like in our particular case, I report to Eric Wojner. Eric Wojner is vice president of commercial assets investments. He plays the role of the owner, while I play the role of a fee manager running his day-to-day operations. We're under Mike Richmond's area of responsibility.

I don't understand what fee management is.

Fee management is, let's say the Crescent Partnership is a 50-50 partnership between The Woodlands and Crescent. As fee manager, I get a 5 percent cut off the top of revenue to staff and oversee all of the day-to-day responsibilities of running an office building.

OK, so fee management means you get a fee for managing?
And administering repairs and negotiating contracts for services within the buildings, like an agent.

Yes, it's a little bit of trade talk. Have we covered it?
I went from startup to the grand opening. The year we opened, we were also broke and had to go through the process of laying off people. I can recall back then being budget director. Right here in OB 1, our overhead budget was $7 million and consisted of approximately 300 employees. We met the deadline and opened this project in October 1974. It was time to do some layoffs and get back to normal operations. There are not many of us who came through that transition.

There were 300, pretty quickly dropped to less than 200?
As I recall, the target was about 150. We actually went through a three-phase layoff to get to that number. I can remember Mike Richmond and Larry Kash at that time with contractors lining up in the lobby, having to tell them, "Sorry, we don't have any money, but if you'll ride with us we'll make it somehow." Luckily, most of those contractors took the risk.

Earl, thank you for your time.

SHAKER A. KHAYATT

Today is Tuesday, April 9, 1996. This is Joe Kutchin and I'm in the Texas Commerce Tower offices of Mitchell Energy & Development Corp. with Shaker A. Khayatt, who has been very much involved with the company since what year?

I've been a Director since the company went public. I'm going to say that I got involved with MND in the very late '60s. The company was still domiciled in the old Houston Club Building. Merlyn Christie also had an office there.

Give me a little bit of your personal history—education, what you do now, and so on.

I was an engineer, went to MIT. I had left Egypt and, of course, had no money.

What year would that have been?

1957. I held down several jobs while at MIT, then I got a job doing research. It supported me through my master's degree.

How could you afford MIT?

Well, I worked three jobs to start off with and then, as I said, the research job paid a part.

So when did you get your bachelor's from MIT?

No, I got my bachelor's from Egypt and got my master's from MIT.

When was that?

'59. Then I went to Harvard Business School. There I borrowed money, because I couldn't afford it, and got my MBA in '63. I did well and decided to leave research and engineering and went to work on Wall Street for a firm called F. Eberstadt & Company, and got to know the people at Chase Manhattan. The loan officer for Mitchell was a gentleman called Norman Olansen, who became a friend. One day, I got a phone call from Norman who said, "We have a client in Texas. I wish you'd go visit him." I said, "Could you tell me why?" And he replied, "Well, we'd like to get some of your ideas." So, I came down to Texas and met Budd Clark, Bill Houck, the chief financial officer, and George Mitchell.

Could you say what year this was—late '60s?
Yes, it must have been probably '68 or '69.

Did Olansen provide an introduction?
Yes. I was visiting at Chase's recommendation. I got interested. Two more disparate characters than Mitchell and Clark were hard to find. Add Houck and I must say the group was unique. The first thing that impressed me was Mitchell's vision. Once you got over the novelty of a man who thinks in decades instead of six months, you were fascinated. Here was a man who, with a straight face, told me what the world would be like 10 to 20 years in the future. The world according to Mitchell. What Houston was going to be like at the end of the century. He was talking 20 to 40 years ahead. Mitchell is unique. He started with nothing, and he used to rent, borrow, or whatever, the seismic from Shell at 8 o'clock at night, spend the whole night up, and the next morning, take it back. Listening, you start by being somewhat suspicious–is this a fairy tale?–and then very quickly you come to the conclusion that it is real and you really have met somebody who is extraordinary. He also has a sense of morality that these days is hard to find. He's a family man. I remember when I went into his office when I first knew him and there were 10 pictures of children on his wall. I said, "George, that's very interesting; I never thought of taking pictures of my children at different ages." He replied, "What are you talking about?" "Well," I said, "you have them young here and then about five years later." "Wake up, they're all different." "Ten kids?" "Yes." I thought I was courageous having four, but 10!

What did you expect to find when you came down here?
I didn't know, Mitchell didn't know, and I'm not quite sure Olansen knew. I think Olansen's idea was that sooner or later, Mitchell was going to need the services of an investment banker. And because of the uniqueness of the man and the company, I think Olansen figured out that he couldn't send Morgan Stanley, he had to send somebody who was willing to do the job . . . let's face it, the company was not Exxon. When we first started to think about taking the company public, Houck put on my desk a stack of printouts of debts.

This would have been the '70s?
Very late '60s, or maybe '70. And I looked at the debt schedule and there were people who were owed under $1,000. I asked, "Houck, why don't you pay them off?" Then there was the mouse company. This sort of thing made the company interesting. However, I told George: "Please George, no mice.

I've got a lot of friends on Wall Street, a lot of credibility, but it doesn't extend to mice." George was mad because nobody would want to pay for the mice which were critical for cancer research. And he felt this was something that should be done by somebody. However, we prevailed. One of the things, of course, that was unique in the company was the vision of The Woodlands. I went out with Morris Thompson when we first started working on the offering, and we stopped on the way at a store that sold Western garb. He said, "Come on, I'm going to get you some boots." "Boots, why?" He said because where we are going there are snakes. I said, "All right, I suppose I'm game, maybe." After that, we drove for miles, and finally, we stopped at a big gate with a chain and a padlock. This was The Woodlands. Joe, there was nothing for miles. I had enlisted Eberstadt as the lead manager for the offering, as my firm was much smaller, and I needed muscle. I was sure we had a diamond here and that everybody would make money. Walter Lubanko represented Eberstadt for the offering. I remember flying over The Woodlands with him in a helicopter. And Walter's comment was, "Khayatt, we've got lots of nothing." And there was nothing.

Was Lamar School there yet?

There was nothing. There was nothing for miles, forget The Woodlands. All the development that you see today on I-45 didn't exist. Even from downtown to the airport was spotty, hadn't filled in. All the stuff that you see now, George forecast it, saw it, imagined it. The Exxon development on the corner of I-45 and the airport road wasn't there, there was nothing. We were out in the wilderness. But George had this huge map of Houston and explained the different quadrants. The southeast quadrant was close to the petroleum cracking industry and it smelled bad. The southwest was pretty much filled up, you could only get tracts of maybe 10 or 15 acres. The northeast was swampy, I think–or there was something that was wrong with it. And the northwest was the only quadrant that had everything going for it. Therefore, it was in the path of what George forecast would be the settlement of 3 to 4 million people. Now, if you look up Houston at the time George was selling me this bill of goods, I think Houston was just over a million. An entirely different city from now. To give you an idea, there was one good restaurant downtown and that was Maxim's and there were two or three clubs. I needed to do some checking. I had a classmate named John Maher. His father was a partner of Mecom's. Through John I met his father, who was an oilman, and we talked about Mitchell. I also went to talk to Sarofim. I wanted to get a better understanding of Mitchell both as a person and as a company. I also met Johnny Mitchell, who gave me a fascinating book he had written. He's a character and I am very sorry that he's so unwell. I also met Christie Mitchell in Galveston, also a character. And last but not least, I met Maria and her husband. I've met all the Mitchells. I also went out to see George's house. I was interested to see how you house 10 kids.

Why go public?

For a whole host of reasons. One, it was a good idea to raise equity for the company. Second, for George personally I thought it would be a good idea to put a price on his holdings that the IRS couldn't contest. But primarily this was a way of raising money that in the long run would be less expensive than raising it either through private placements or through more debt. I also think George wanted a vehicle for his long-trusted employees to make some money. We also had to think in terms of The Woodlands. They were going to need major financing and sooner or later we had to have more equity in the company. The opening of The Woodlands was a major achievement in spite of several obstacles—for example, the waterfall next to The Woodlands Inn was torn out, I think, three times. Every time they put in new stone, Mitchell would come and look at it and say, "I don't like it, take it out." "Wait a minute, George, easy now. Every time we take it out . . . " "Take it out." So out it went. At the start of The Woodlands, when they started to clear the land for houses, believe it or not, Mitchell sometimes would stand there to ensure that the big trees weren't cut down. He pre-dated ecology's being fashionable. He came up with novel ideas. The concept that a child could walk from his or her house to the town or village center without crossing a highway. Having buttons in homes that can call in an emergency for somebody with medical problems. Not having homes fronting the street but having a buffer of trees between the house and the street. Some of his ideas make living a pleasure—having office buildings that were not higher than the treetops, with mirrored windows. You ask him why the mirrors? George would say, "So the deer do not see the office buildings." Well what happened was quite interesting. The first office building occupied . . .

That would be OB 1?

Yes. I've seen animals come up and bump their heads against the mirrored window, thinking it was another animal. Or it would sit there and snarl at its reflection. Because of the mirrors in the middle of the woods, the building blends in. So Mitchell was not concerned only about ecology for humans, he was also worried about ecology for animals. I wonder if he thought of anything for snakes. In those days he was, and I still think he is, a complete nut on physical activity. I must have had many lectures saying, "You're too fat, you don't do any exercise." He used to play tennis every day. At 4 o'clock, come hell or high water, Mitchell was playing tennis. But then–and even today–Mitchell marched to his own tune. In those days, Mitchell owned 100 percent of the company.

I think Johnny owned part.

Well, the Mitchells owned 100 percent. In fact, one year Johnny got mad at us and voted against all the directors. We knew it was Johnny because the number of shares voting against us was exactly the number of shares that Johnny owned. In fact, I called him up and said, "Why did you vote against me?" Anyway, say the Mitchells owned 100 percent, so a dollar a share meant a lot of money. Don't forget it was 1972, and that dollar would be worth better than $10 today. But George is a unique man. Even though he made all this money, he is not really motivated by greed. He is a funny fellow. He can be the most generous of men and yet the toughest in business negotiations.

Let's go back to the going-public process. Tell some of the details along the road there.

We aborted it at least once.

The original package was going to be units consisting of debt and equity, as I recall.

It was going to be debt, equity and warrants.

Why was it aborted? Do you recall that?

Yes. Two reasons–one, the market in terms of timing was bad, and, two, it was too damned complicated. And I have to take the blame for that because I was one of the architects of this very complex package. It was a very good package for the company, a very good package for the prospective investor. But it wasn't right, it didn't work out. And it was, as I said, too complicated. We finally went public with a very simple common stock offering. No bells and whistles. But that morning up in New York, I got a call from my floor partner saying, "You'd better be careful, the bears are going to maul you." Now, as lead underwriters we had to hold the price. That means when people dumped or shorted stock, we had to buy it.

Right now we're talking February, 1972?

That's right. The actual offering. We went effective around 10:30, 10:45 in the morning and at 12 noon we went to lunch with Budd Clark and a couple of officers. Walter and I and I forget who else. We both got a phone call, pretty much at the same time. It was our floor partners saying we were being mauled. They were long in the stock. There were no buyers in sight. And that was the time to call in some chits from friends. I called a good friend at Loeb Rhoades. As I was picking up the phone to make other calls, one of our traders said, "Hey, look, look." Across the broad tape came Loeb Rhoades had entered the sheets for Mitchell Energy & Development one way only—buy. In other words, they were saying, "OK, guys, we will buy anything you can sell." And of course it was Loeb Rhoades' capital that really panicked the shorts.

I've never heard the story before.

Loeb Rhoades really bailed us out. They knew that we didn't have the kind of capital to stand up to this onslaught. It was interesting. Walter and I went to see Goldman Sachs to invite them to come into the underwriting, but they turned us down and a lot of people turned us down. Of course, after the 1973 war in the Middle East and the embargo on American oil, we were heroes. We had the opening of The Woodlands and invited everybody and his brother, including the treasurer of the Ford Foundation, Bob Salomon, who had bought stock for his small cap fund. It was rather amusing because Salomon bought the stock. When the fund managers met Budd Clark after the offering, they said to Budd, "Can you tell us a little bit about the company?" Budd said, "You own it." They replied, "We know; the boss knows a lot about it because Khayatt sat down and explained it to him but he never explained it to us. We have the prospectus, but we need to hear the story." And they were good stockholders in Mitchell for a very, very long while, as were firms like John W. Bristol, who managed the Princeton money. I could go on and on. For example, American Capital Management owns a slug of it. Some other friends also own it. We were asked to put together a fairly nice group of institutional buyers and also some wealthy individuals. We're talking about a fairly small offering.

Finally it was 700,000 shares, then went to 770,000.

Yes, we're talking $10 million in 1972 dollars, or about 10 times that much today. It was a respectable offering. We defended the stock, held the line and a few months later we were bailed out. One of the problems with issues like

Mitchell is that the underwriters make the offering and then forget about it. We did not. I remember going over to Europe with Budd Clark several times for Mitchell because we had sold stock to people in Switzerland, London, Scotland and Holland. The reason for that was that Europeans have a much longer investment view than Americans. And we wanted to have people in Europe who would buy the stock when somebody else was dumping it. We all wanted ownership spread as widely as possible. Then we got a listing on the American Stock Exchange and by the time we opened The Woodlands, the oil embargo caused a rise in the price of oil that positively influenced the stock price.

It sure came at the right time, didn't it?

It helped. The interesting thing is, even in those days George, justifiably, was saying, "We are not an oil company, we are a gas company and gas is the fuel of the future." Oil is good, but oil pollutes. Then he started to do some things which were different. He spent a lot of money developing cryogenic techniques to get liquids from gas. Because, again, he forecast that liquids would be a prime raw material for the chemical industry. If you look at the offering prospectus, we tried to explain in a fair amount of detail George's vision, without calling it that. I remember a friend at Travelers saying, "This is a unique situation, but how do you marry oil and gas into real estate?" It's a very big hurdle, by the way. The hurdle was because we could never get analysts that follow both industries at the same time. Thus, the company always fell between the cracks. You talk to a real estate analyst who looks at Mitchell as a real estate play and then questions why he needs the oil and gas. In talking primarily to oil and gas analysts, they ask why do you need real estate, it's only a drag. Mitchell believes, "OK, when we run out of gas, real estate will still be there."

Talk about the highlights and lowlights of your 25-plus years with the company.

I think one of the lowlights is that we have suffered from the tremendous debt servicing costs. Interest rates went up and oil and gas prices came down. Incidentally, when that happened, Chase, the company's major lender, began to get cold feet. This was not helped by the death of Norman Olansen. The company then made the decision that it had better start talking to other lenders. This was a pretty low point. Where are we going today, where are we going tomorrow? I think that for the stockholders' benefit we should be planning to separate the companies. That may take a while to do. We certainly have to get this travesty of a lawsuit behind us. It boggles the mind of any reasonable person. Less than $400,000 worth of damages resulting in a $204 million verdict is unbelievable.

You say that in the future the two companies will split?

Maybe. But that is strictly my thinking. It makes sense for the stockholders and it also makes more sense for the employees. I think there is, in organizations such as Mitchell, people who have a tremendous pride in their jobs. Here we have two disparate businesses and one side of the business says, "Why am I supporting the other side?"

That's been a long-standing problem.

Now real estate is beginning to turn. But don't forget, Mitchell never said it was going to work in less than 20 years. He always said it's a 20- to 50-year plan, not that anybody really believed him. Now it's beginning to become profitable. I think maybe it is a little early to split the two companies, but we are talking about something that could happen in our lifetime, and it makes sense. Personally, I was uncomfortable with real estate. Don't forget, I don't have the vision of George Mitchell. All I saw was a lot of empty land that was never going to fill up in my lifetime. It was an act of faith to open The Woodlands.

Then you compound the question because of the original management team, which made a lot of people nervous.

I suppose it takes people of this caliber to do something as successful as The Woodlands. It is successful. We have seen that people like it. I know that my wife, who came here for the opening, has said, "Let's go live in The Woodlands if we ever move to Houston." It's a wonderful place and it appeals to people in terms of scenic beauty, the practicality and the amenities that you have for children, etc. Now with the mall you also have the ability to shop. And I think 20 or 30 years from now, people are going to say, "But did you have any doubt it was going to work?" However, if you had stopped with Morris Thompson on the way up there to buy boots because there were snakes, you understand my doubts. The next financing hurdle for the company was to get some institutional debt. And that was the first Goldman Sachs private placement. The partner in charge was good. He had some clients and we picked three—Travelers, Equitable and Metropolitan. Our great strength was Travelers and Metropolitan, his was Equitable. A few days later over the phone with Travelers, we said, "We have honored our promise to offer you the first chance at this placement. Do you have any comments?" "Yes, thank you," was the answer, "please circle 25 percent." Sacerdote, the Goldman Sachs partner who was also on the plan, said, "Did I hear you right?" "Yes, Mr. Sacerdote, you heard us, circle 25 percent." And then we went to Stu Kennedy. "Stu, don't tell us you cannot top Travelers." He circled 50 percent.

He was with whom?

Metropolitan. Then we said, "Sacerdote, you go get the rest." The interesting thing about this, as you know, is that we could not get the debt placed at the parent level. So Mitchell made history, it went downstream. It created a subsidiary of a subsidiary which held the properties that it borrowed against.

What year are we talking about now? Tonery was still with the company, I think. I want to say '78, '79.

Yes. I believe you are right. For the first time we had institutions of this caliber lending the company money. It was like getting the Good Housekeeping Seal of Approval. We were delighted with Goldman Sachs and we wanted Travelers and Metropolitan because they are the big boys. If you do not screw up, you can always go back and borrow more.

What have I failed to ask you, what else is there?

If I have to leave you with my contribution to the company's history, it is: "MND–One Man's Vision, One Man's Dream," and this is really what made this company. And if George were 15 years younger he would be clawing, fighting and scrapping to double and triple the size of the company. Yes, he is a brilliant geologist, but, more importantly, he has the ability to dream and concentrate on realizing it for both businesses, not just one. I have had more fun on this Board than in any other relationship that I've had in all 33 years of my working experience. First of all, as I have emphasized, George is a unique person. The people he surrounds himself with, with few, if any, exceptions, are unique. I've never seen anything like it. Joe, one of the things you can highlight in the company's history is the Board of Directors of Mitchell Energy and just list the people who have served. Outstanding people. Why? Is the company paying them more? Is it giving them special perks? No. Thus, George must be doing something right. You admire George. You respect George. We live vicariously through George, his vision, his dreams, his accomplishments. Seeing the birth of The Woodlands, its evolution, it's a unique experience. Seeing things like some of the things he's done in the oil and gas business is very rewarding.

And I think the fact is that he's the best geologist the company has ever had.

Forget just the company, he is that by any standard. You listen to Wanda Jablonsky. Her father has to have been one of the world's greatest geologists. You listen to her talk about her father, and the image that she paints is the image of a dedicated, brilliant technician. But in George's case, you have something that's much wider, broader than a dedicated, brilliant technician. And I think that most people who have been associated with the evolution of this company have had a wonderful, exciting, rewarding experience. Most of them, I think, could have gone to work for Exxon or any other company but it was the man, not the company, that attracted them.

Shaker, thank you for a very informative interview.

Howard Kiatta

Today is Tuesday, September 26, 1995. This is Joe Kutchin. I'm in the offices of Howard Kiatta in downtown Houston. Howard, please give me some information about your background.

I was born in Houston in 1935, went to parochial schools here, finished St. Thomas High School, went to the University of Texas, got a bachelor's degree in geology and then went to Texas Tech, took a master's degree from Tech in geology and went to work as a geologist for Texaco in New Orleans in 1960. My father had been with Texaco for 45 years and I had worked for them during the summers. If it hadn't been for that connection, I probably wouldn't have gotten a job at all. I was with Texaco for seven years, four years in New Orleans, a year-and-a-half in Houma, and two years in Lafayette. When I was in Lafayette in early 1967, I was contacted by a gentleman by the name of Tom Purcell, who was a geologist with Mitchell–George Mitchell & Associates, at that time. We had a mutual friend, Ross Dawson, who had given Tom my name and Tom called me. I was in Lafayette working for Texaco, and he called me and asked if I would consider talking to him about a job with George Mitchell & Associates in Houston. One thing led to another and I went to work for George Mitchell & Associates in May of 1967, in the Houston Club Building in downtown Houston. I had never heard of George Mitchell when they contacted me. Being from Houston, I had some Houston contacts and I asked around, "Who's George Mitchell?" The replies I got were, "That's Johnny Mitchell's brother." Everybody unanimously said, "If you have an opportunity to go to work for George Mitchell, take it because he's probably the smartest oil man in town, very aggressive, he's building an empire and you'll have an opportunity to be a part of it, you ought to do it." I was ready to leave a major oil company. I had decided that I was not cut out to be a major oil company employee for my entire career. I was very grateful to Texaco for training me and giving me a job when I needed it. I worked hard for them and found them a lot of oil and gas and we parted as friends and I wouldn't take anything for having worked for Texaco for seven years, but I was very grateful to have the opportunity to come to Houston and work for, at that time, George Mitchell & Associates.

What did you encounter when you started here?

It's kind of interesting. When I came over here I began to realize that the company had a reputation for slow pay in bills. They had some financial problems and I remember my first experience in trying to order a geological base map to work an area in south Louisiana. I called a company in New Orleans and they said, "Who do you work for?" When I told them, they said, "Send us a check and then we'll send you the map." I replied, "No, just send me the invoice and we'll pay it." They said, "No, you work for George Mitchell & Associates, you send us the check first, if the check clears then we'll send you the map." I said, "Why?" They said, "Because he doesn't pay his bills." Then, right after that, Jade Oil Company went bankrupt. That was Johnny's company. I began to wonder if I had done the right thing, by leaving mother Texaco to go to work for an upstart.

You had no connection with Jade, and George Mitchell & Associates had no connection with Jade?

No, I had no connection with Jade. George, to his credit, put out a memo that day to all the employees clarifying the situation that Jade belonged to his brother Johnny and had no connection whatsoever with George Mitchell & Associates and none of his employees needed to worry about it. I didn't know what the situation was until he put out this memo and it was very timely and it made us all sleep a lot better. When you work for a major company, you are very much in a vacuum, you're in a protected environment. You don't know what goes on in the outside world, and it was a tremendous cultural shock to go from a company like Texaco to a company like George Mitchell & Associates. There were only seven geologists who worked for the company when I came; I was the seventh geologist. Most of those were working in the North Texas area. Tom Purcell, Bob Schrock and I were the only three geologists who were working the Gulf Coast area. Everybody else was working North Texas.

Who was your boss?

Tom Purcell, he hired me. Tom had come from Tenneco and had been with the company a couple of years when they decided to expand their efforts along the coast, so he hired two geologists. He hired Bob Schrock from Cities Service and he hired me from Texaco and put us to work on the Gulf Coast. Bob was working South Texas and I was working the Upper Texas Coast and South Louisiana. Shortly after that, Tom quit. He was there a couple of years, I guess, and then he quit. They had started some things about that time that really became landmark foundation builders for the company. I got to be involved in that.

Talk about those.

OK, but let me finish up on my thought process about my impressions when I first got there. In addition to the slow pay reputation, I found out that George was very vocal in his discussions with Budd Clark out in the hall, and I'm sure you've heard other people talk about the screaming arguments they've had. Again, coming from a sheltered environment, this was a shock to me. I'd hear these guys screaming at each other out in the hall. I thought they were going to kill each other, and it turned out that's just the way they discussed things. They'd just stomp off and you'd think they were mad and would never speak to each other again, but they were obviously very close. It became obvious to me pretty early that George was smarter than anybody around him. He was a visionary, very creative, he expected a lot from his employees. He worked hard and he expected everybody else to. His thinking was way out in front of all the rest of us, and the one thing that stood out among his various traits was his persistence. Once he gets an idea he thinks ought to be done, then it's going to be done, sooner or later. It was a valuable lesson I learned—that is, be tenacious, be persistent, be patient and some day you'll get it done.

You talk about creativity and smartness. Do any examples come to mind?

Yes, a lot. When I came to work for the company, Tom Purcell, who was the only geologist working the Gulf Coast at that time, and Sam Meadows, a geophysicist, had acquired some seismic data in the Galveston Bay/Pelican Island area north of Galveston Island. Gulf Oil Company had made some wells over on the Bolivar Peninsula and they had shot some seismic lines in that area to define this producing area. George—the company—owned Pelican Island as a result of a swap that had been made with the Corps of Engineers. In order for Gulf to shoot this seismic data on Pelican Island, they had to get permission from Mitchell. As part of the permit process, they gave Mitchell the seismic lines across his property on Pelican Island. As the story was told to me, Sam Meadows and Tom Purcell were reviewing this seismic data with George and they all noticed that these seismic lines indicated a huge geological structure that appeared to be underneath the main part of Galveston Island. That was the beginning, the prospect, that led to LaFitte's Gold Field. They were working on that when I came to work for the company. The discovery well was drilled shortly after I came to work and was kept very confidential. That was onshore. That was on Galveston Island, back in the warehouse district, north of Broadway. It was a straight hole, a vertical well, and they kept it very confidential. I was working for Tom Purcell. Tom was actually the lead geologist on the project, and George's idea was to get as much acreage as possible, buy the easy leases that were available. But they didn't want to start a town leasing program until they knew whether they were going to have a field down there because there weren't any geological controls where we were drilling. There were no wells, no seismic data, they were just keying off these lines that Gulf had shot that were quite a distance from the well location. When told how expensive it would be to acquire a large block of leases, George was quoted as saying, "Well, let's just get a few leases because it's probably going to be a dry hole, so we don't need to spend a lot of money getting a lot of acreage." Well, of course, it wasn't a dry hole, it was a nice discovery. Shortly after the discovery was drilled, Tom Purcell went off to Harvard and later quit, and I took over the responsibility for LaFitte's Gold Field. I got to develop that field and that was a lot of fun. We drilled a number of wells from the warehouse. We kicked the roof out of a warehouse, set up a rig and drilled numerous locations from the warehouses north of Broadway. Those were mostly directional holes. We reached as far south as we could, down toward the beach area with those slant holes, a mile, mile-and-a-half, maybe 7,000 feet. We were drilling slant holes that reached 5,000 or 6,000 feet horizontally in a vertical depth of 10,000 feet or so. They were pretty high-angle holes . . . not by today's standards but back then, that was pushing the technology. We were very successful, found a really nice field and developed it very quietly. Once they found out they had a discovery, they started buying town lot leases and it took about two years to put the units together. Most people in Galveston own the minerals under their lot. Each lease was maybe a quarter of an acre and we had thousands of leases down there. Jack Yovanovich led that program and I'm sure he's talked about that. George's tenacity and courage in putting that thing together and developing it astounded me at the time, and still does. I was running the logs on those wells, the main geologist on it, and those were high-angle holes. Every time you ran a logging tool in the hole, you stood the risk of sticking it. On almost every well we stuck a logging tool.

That means that the tool got stuck in the hole?

That means you're running in the hole with a wire line, with a Schlumberger tool on the end of the wire line and the open hole can collapse on you and you stick the tool. Then you have to go in with drill pipe and some kind of fishing tools. Usually, you can get it out. Logging tools are one thing and they're not as apt to stick as what is called the formation tester, a sampling tool that Schlumberger runs in on a wire line and you go down to your sand or your zone of interest and you open the tool up and you take a sample of the fluid in the formation. Those things are very apt to stick in a slant hole like this. I remember George being out on location one time with me and we'd already stuck the logging tool, and we'd already stuck a dip meter tool, I think. We had several potentially productive zones and we needed the information to prove they were productive for our reserve report so we could borrow the money, take the reserves to the bank. But they had to be proven reserves, so you had to actually have a sample of them. You had to prove they were productive by taking a formation test sample, and I knew we were going to stick that tool. I told George, "George, I don't want to take formation tests because we'll stick that tool." He said, "But we'll need the information." I said, "Well, you're going to stick the tool." He said, "We've got to have the information. Go ahead and run it in there and get the samples. If we stick it, we'll just go in there and fish it out." That's exactly what we did. It sounds easy, but you stand the risk of losing your entire hole if you get that tool stuck and you can't get it out . . . you can't push it down and it won't come up. Then you can lose your hole and you have to drill it again. He had a lot of guts in putting this whole thing together and developing it and he didn't bat an eye at some of these things that just scared me to death. The man's got an incredible amount of courage.

Does LaFitte's Gold include the offshore Galveston wells?

Yes. We drilled about four or five wells on the northerly part of the field. That is, the portion of the field that extends from the Galveston ship channel, which separates Galveston Island from Pelican Island, that's the northerly limit of the field. After we drilled some wells up there and kicked some wells southerly under Broadway toward the middle of town, that was about as far as we could reach from our warehouse locations. Our seismic data that we had shot, which we acquired after the discovery well, indicated that the structure was oriented in a northwest-southeasterly direction. It looked like a great big football, with the north end at Pelican Island and the south end about three or four miles offshore. We had to figure out a way to develop the portion that was offshore. Sidetracking from the northerly location wasn't feasible to reach the beach, so we had to figure out a way to get our locations under the beach and under the offshore portion. Incidentally, the seismic data were shot at night, going up and down Broadway and some of the key streets, 61st Street and some of the others. They were shot at night using Vibroseis tools. It's a surface energy source that is non-dynamite. Obviously we couldn't use dynamite in the city. So we used these trucks that had vibrators on them that created a sufficient shock wave to acquire seismic images from the subsurface. We shot those at night, up and down Broadway, with a police escort, and we shot them at two or three o'clock in the morning and acquired fairly good data for that point in time, a fairly good imaging of the structure down under the island. Incidentally, during the development of this field, the Galveston City Council decided that they needed a drilling ordinance; they didn't have a drilling ordinance. They needed a drilling ordinance for the city if the city was going to overlay an oil and gas field. So we helped them write their drilling ordinance and I think they appreciated it. It was mutually beneficial to both Mitchell Energy and the City of Galveston. Actually, it was our engineers and landmen. I was involved to a limited degree, Jack Yovanovich was involved to a large degree and our lawyers and our regulatory people worked with the City Council in developing a drilling ordinance for the City of Galveston. It did not allow any rigs to be located offshore within a mile-and-a-half from the beach, so we faced a situation of setting rigs a mile-and-a-half offshore and kicking back towards the beach. You reach the limit of your directional drilling capacity that way and so we actually ended up drilling three wells, setting them up a mile-and-a-half offshore and kicking them back toward the beach. None of them quite reached the beach and we were trying to develop a method of developing a portion of the field that underlay the beach area. We looked around for drill sites where we could set up some rigs on the island. We came up with the idea of purchasing the Fort Crockett property from the government, actually it was from somebody who had bought it from the government, and setting up rigs there. Geographically, it was a perfect place to locate a rig, but it had lots of problems in getting the product out, laying pipelines down some of the lower-numbered streets that ran near Fort Crockett, trying to get the product out if we had production, and we were sure we would have production. We met quite a bit of opposition from the general public in trying to set up a rig in Fort Crockett, but we bought it anyway and I remember George's comment. We paid a million dollars for the property and George said, "Well, it will be valuable real estate some day, so it doesn't matter whether they give us permission to drill from it or not." It was, of course, later developed into commercial property, but we never used it for a drillsite. We ended up developing the field sufficiently from our drill sites offshore and our warehouse drill sites. We never did have to locate any other wells down there near the beach. It turned out to be a heck of a good field.

Were you part of the campaign to get the offshore sites? You went from geology to politics pretty quick.

Yes, I did. That's one of the things that you learn pretty quickly working for George, that you didn't operate in a vacuum, you didn't just stay in your little cubbyhole and work on geology all the time. You did whatever was necessary to get your wells drilled and get the overall company goals met. One day George came in and said, "I want you and Jack Yovanovich and George Roberts (who was our drilling engineer at the time), and Joe (the landman who worked for Jack, I'll think of his name in a minute), I want you all to go down there and I want you to talk to all the merchants on Seawall Boulevard and explain to them what we're going to do and tell them it's going to be OK. Just do some politicking and see if any of them would have some objections if we put a rig out there, and that way we could work towards public hearings." Public hearings were going to be held relative to our permit request from the Corps of Engineers to set up rigs out there and drill our wells. I said, "George, I'm a geologist, I'm not a politician, I'm not a PR man. Let somebody else do that." He said, "No, you know more about it than anybody else. I want you down there and explain it to them." So, Jack and I and Joe Ramsey and George Roberts went down there and spent one or two whole days going up and down the beach talking to filling station attendants, restaurant owners, shell shop owners, souvenir shop owners, motel clerks, just anybody we could find to talk to and explain to them who we were, what we were going to do, and would they have any objections. It's amazing, I expected a great deal of hostility, but that's not what we encountered. Everybody was very friendly, a few of them were opposed to it but with one or two exceptions, everybody was very nice. Mike Gaido, who owned Gaido's Restaurant at the time, was an ally. He said no, it didn't bother him a bit, and if it didn't bother Mike Gaido then it didn't bother a lot of other people. Obviously, he was a very well thought-of civic leader down there. You can still see those wellheads out there when you're sitting in Gaido's having dinner. It worked out well, we didn't have any pollution problems. The first well we drilled out there, though, we did encounter some pressure. Anywhere in the Gulf Coast, if you drill deep enough you're going to encounter high geologic pressures, called subsurface pressures or abnormal pressures, or geopressures, that are greater than normal hydrostatic pressures, so you've got to be very careful. The geologists and engineers have to be prepared to handle these pressures when you drill into them. Since this was the first well anywhere in the area offshore there, we didn't know exactly where to expect our geopressures and we drilled off into some pressures before we expected to and took a kick, lost the hole and had to redrill it. Jim Stewart, who was the engineer on the job, and I took quite a bit of heat about that.

Heat from whom?

From George. The last thing we wanted was to have any surprises down there. We didn't have a blow-out, we didn't have pollution, it was not a public threat or anything like that, but we lost the hole and it cost us a lot of money because we had to redrill it. But it turned out to be a very successful well and as far as I know it's still producing.

This was from what year to what year, approximately?

They started acquiring the data in about 1968. No, before that, because I came to work in 1967 and they were getting ready to drill the discovery well then. So, the discovery well was drilled about 1968 and the offshore wells were drilled in the early '70s.

Did you have any particular title when you started?

No, I was just a geologist, just Howard Kiatta, geologist. Later on, when Tom Purcell left, they named me district geologist for the Gulf Coast. That was in about 1970 and then as our success continued in these areas, and I was going to name a few of the other areas that we worked on that were a success . . . anyway, as our success grew, my responsibilities grew. I became exploration manager for the Gulf Coast division and then eventually vice president of exploration for our southern division. That's what I was when I left the company.

The vice presidency must have come before 1974, because I think you were a vice president when I came here, and I came in '74.

Yes, it must have been around 1974, somewhere in there. That was a lot of fun in those days. We were very aggressive and didn't have the budget to be doing what we were doing, perhaps, so we had to be very creative in finding partners and developing things. We were, of course, competing with the real estate company which was being developed about that time. They were developing Cape Royale, they were developing Pirates' Beach and Pirates' Cove and they were acquiring property for The Woodlands and starting to develop The Woodlands and that turned out, of course, to be a very expensive operation. We were competing with the real estate division constantly for dollars and we resented each other because we felt like the cash flow we were generating with our successful oil and gas operations was helping to support The Woodlands. There was a different mindset altogether between what we were trying to do in the oil business on a shoe string and trying to build a company with oil and gas reserves and the things they were trying to do–some of the people they hired from outside of Houston to come in and develop the property like they had done it either on the East Coast or the West Coast. Totally different personalities, and we were resentful of some of the things they were doing. Despite the rivalry between real estate and the oil and gas division, we just kind of hunkered down and did our job and in all modesty I think we did a hell of a job. Along with LaFitte's Gold, we had other successes at that time.

Do you have any numbers that can help quantify what's happened with LaFitte's Gold?
I left the company in December of 1976 and they drilled a few more wells after I left. But up until the time I left, we had pretty well developed the field. We drilled some deep wells to test the Frio. Let me back up. The production was from Miocene sands, from a depth of about 7,000 feet down to a depth of about 9,500 feet, roughly. The Frio in that area is the next deepest producing formation, and it's in that area down around 12,000 to 13,000 feet. We drilled a couple of deep wells to test the Frio and the Frio sands were not present on that structure, so the Frio did not turn out to be productive on this LaFitte's Gold structure, which was a shame because it was a huge structure, geologically. Houston Oil & Minerals developed a field called North Bolivar Point about that time and it produces from the Miocene and Frio, but most of their reserves are in the Frio. In LaFitte's Gold, we were a little bit outside of that Frio sand trend and it turned out that we were too far south to encounter the Frio sands that were present at North Point Bolivar. That's an interesting story, if I can digress a minute, about North Point Bolivar. Back in the late '60's when I was working for Tom Purcell, we knew about the LaFitte's Gold structure. I don't remember if we had drilled the discovery well or not, but Tom and I were doing some work in there and he had developed an idea of a structure that stretched from the north end of Pelican Island behind Bolivar Point. It's right where the North Point Bolivar Field is and we believed we saw a structure there. We were interested in buying the leases in the state lease sale. The tracts were located in Galveston Bay and they were state leases. We developed the geological idea, nominated the leases and were going to go to the state lease sale to buy these leases. Tom had arrived at some prices to bid on these leases,

but we were so short of money that right before the sale, our financial people withdrew the bids just to save a few dollars. I'm talking about bidding $25 to $35 per acre on these tracts. We would have been the high bidder. The leases were bought by Sun and Allied Chemical (which was Union Texas) and Occidental. We knew the geologists who worked for those companies. They drilled a well and made a Miocene discovery at 8,000 feet or so. We talked to the geologists at those companies and asked them if we could join them, if they were looking for partners in developing that field or drilling exploratory wells. We were interested in that area because we had a prospect there but were not successful in buying it at the sale when they bought the leases. They said they didn't know what they were going to do with that thing, they may sell it. By that time, Tom had left and I said, "Fine, if you decide to sell it, let us know because we're very interested in that area." A deep Frio well had not been drilled in there as yet on these leases. So the next thing I know, Union Texas had sold that field to Houston Oil & Minerals. I got Jack Yovanovich to write a letter to Joe Walter, who owned Houston Oil & Minerals, which said, "We have always had an idea down there; if you're looking for a partner in developing that field we would like to be your partner." We got a nice letter back from Joe saying, "No, we've decided to drill it, we're going to drill a deep test in there and we're going to drill it heads up, we don't want partners, thank you very much." They drilled a big Frio discovery and that made Houston Oil & Minerals. It turned out to be about one-third of a trillion cubic feet of gas, which is about three times bigger than LaFitte's Gold in reserves. That made the company. We were always a day late and a dollar short on that field and it was very frustrating because we had an idea from back when those leases were not owned by anybody.

When was this?

This is during the time that we were developing LaFitte's Gold. I was in charge of the whole area and we were just a day late and a dollar short on that prospect. I've got to admit that our primary interest in that field was in the Miocene. They had quite a bit of production in the Miocene. We knew there was Frio potential down there, but our primary interest was in the Miocene, so to that extent we were wrong because most of the reserves turned out to be in the Frio. The Miocene had already been proven productive, and LaFitte's Gold was productive in the Miocene so we were right for some of the wrong reasons. Anyway, that was always kind of frustrating. Do you want to hear about some of the other areas?

Yes.

Another field that we found in Galveston Island was right under Pirates' Cove. This was a prospect that was developed by a geologist at Cabot Corporation, Floyd Wilcox. Humble (Exxon) had drilled a couple of wells right in the Pirates' Beach/Pirates' Cove area. One of them was a 19,000-foot well, but it encountered production up in the Miocene around 8,000 feet. That well produced for a long time. Then they drilled a couple of offsets and those wells continued to produce for a long time. It was halfway between the Bay and the Gulf, right under the Pirates' Beach area. This Cabot geologist came to us with the idea that the Humble field was bigger than anybody thought and extended northerly into the Bay and underneath the property we owned and were developing as Pirates' Beach/Pirates' Cove development. Cabot had bought the submerged lease, which was owned by the state, at a state lease sale, but it was a non-drillable lease because of the water depth and the lack of dredging permit. There wasn't any way they could drill it without getting a drillsite onshore, on the land, and we owned all the land. So they brought it to us. I knew Floyd; he was a friend of mine and he brought me the idea. I looked at it, and I thought he'd done some excellent work. I agreed with him and I took it to George and said, "George, we need to get in this deal because they need us and we need them and he's got a good idea." They offered us one-third of it and we liked it so well, we played hard-ball with them and got one-half. They didn't want to give us one-half, but we negotiated one-half, and basically we were 50-50 partners. Cabot was designated as the operator. One of the reasons is that they wanted to operate, and the second reason, I think, was that George didn't want to have Mitchell Energy operations, oil and gas operations, that could be in potential conflict with the real estate development. He didn't want to wear two hats down there at the same time, so we let Cabot operate and we had a silent 50 percent interest. We drilled the well underneath Lake Como, which was the state lease that Cabot owned and now we owned half of, and it turned out to be a nice discovery, but again we were right for the wrong reasons. We turned out to be in a different reservoir than Humble's production, but we had more reserves than Humble did. On our side there was a fault that we didn't know about. We ended up in a different fault block and had more reserves than Exxon did. That field has produced over 90 billion cubic feet of gas.

What's the name of that field?

Galveston Island Field. We called it Pirates' Cove area/Pirates' Beach area, but actually, the official Railroad Commission designation is Galveston Island Field. That field, last time I checked, was bigger than LaFitte's Gold in total reserves, not Mitchell's share, but total reserves. Exxon owned part, Cabot owned part and we owned part. That turned out to be a fine, fine field. We developed it very quietly, just like we did in Galveston, didn't make a big deal of it. A lot of people didn't realize what a wonderful discovery it was, but that provided a lot of cash flow for the company.

Give me a time period for that one.

Early '70s–'74, '75 and '76. Seven Oaks was another fine field that we discovered, and full credit for the discovery of this one can go to Tom Purcell, who incidentally is a fine geologist and now an independent. Seven Oaks is in Polk County and produces from the Woodbine formation. Woodbine is a Cretaceous age formation that is most famous for its production at the East Texas Field. This is way south of the East Texas Field, in Polk County, north of Livingston. He'd worked the area when he was with Tenneco and mapped a large stratigraphic trap down there in the Woodbine sand and sold the idea to George after he'd come to work for Mitchell. It was a very expensive, huge acreage position-type play. We didn't have the money to develop it, so we brought in some partners. We were just starting to try to develop that idea and buy leases when I came to work with the company. Then when Tom went off to Harvard, he put me in charge of it. We drilled a discovery well in Polk County, the Southland Paper Mills Number 1, and it blew out. It became Seven Oaks Field. But the first well blew out and we believe it blew out in the Austin Chalk, which, as you know, is getting a lot of play today through that very part of the state. Sid Walker was our chief engineer at that time and I remember Sid fighting that thing. Anyway, we had to plug that well and back off and drill another one. That happened while Tom was off at Harvard and I was in charge of the area, so I located the second well a few hundred feet to the south of the first well and that was the Number 2 Southland Paper Mills. We logged some Woodbine sand and it looked real tight, it looked like the sidewalk down there. We encountered a gas show and logged it and we didn't know whether it was going to be commercially productive or not because it was so tight and so hard. Even though it did have gas in it, we didn't know if it was an economic reservoir quality rock or not. In the Number 2 Southland Paper Mills, one of our partners in this project was Natural Gas Pipeline, who, of course, Mitchell has a long history with up in North Texas. They were one of our working interest partners in this project, along with Inexco and North Central. Mitchell was the operator and we set pipe on that well and completed it. One of the reasons we did was because Natural Gas Pipeline had a gas pipeline that ran just a few hundred feet from our location. If it had been farther away, we might not have gone to the expense of laying a pipeline because we really didn't know if this well was going to produce very much gas or not. We tested it, did some completion testing on it, some reservoir limits testing on it, and got an opinion from an old-time prominent consulting engineer. He looked at that thing and he didn't think it was worth anything. Last time I checked, the well had made about 10 billion cubic feet of gas. It was the discovery well for the Seven Oaks, Woodbine Field, Hortense, the whole Seven Oaks complex which includes probably the best well in the company, the Ike T. Smith Number 1, which has made a tremendous amount of gas and oil. That turned out to be a huge producer.

You say that's the best well in the company?

I think it is, but I've been away from the company for 19 years or so. If they've got a better well, I'd like to see it because that's one hell of a well. The whole complex in there is, and I developed that field and that was fun. That was one of the most satisfying things I've ever done because while it wasn't my prospect, it wasn't my discovery, I did successfully develop it and came up with the geological concepts necessary for us to develop the fields. Along with Mike Rush, our landman, and Ronnie Moore, our draftsman, we designed units in there, Railroad Commission productive units, that enabled us to hold our entire lease block together with the fewest number of wells and protect our tremendous lease holdings, which at that time were about 16,000 acres. I don't know what it is now. We developed it very quietly because we didn't want the competition. Eventually, some people came in there and got in our lease block and were able to pick around the edges of the field, but we had most of it under lease and we had, by far, the sweetest part of the field and the best production. We did such a good a job in developing that field that a few years ago as an independent, I looked at the area and tried to figure out a way to get back in there and maybe come up with some prospects, but we had done such a good job of tying up the acreage for Mitchell back then that I shut myself out. I couldn't get back in there because Mitchell controlled the whole thing. That was a wonderful experience. Also, moving on to a different area, the offshore Gulf of Mexico exploration program that we put together is something I'm pretty proud of. When we found LaFitte's Gold, it was obvious that these productive Miocene sands extended offshore along a trend that was perhaps 10 to 15 miles wide and parallel to the coast line in the state waters.

This is Galveston offshore?
Offshore Galveston, offshore Jefferson County, from Port Arthur all the way down to Matagorda Bay is where we started. Mainly offshore Jefferson and Galveston counties. The State of Texas is very fortunate because the state owns three leagues offshore of the Gulf of Mexico, which is 10.3 miles, 10 miles, roughly. Most states just own three miles. George decided that we should make a trend play and then continue the exploration for these Miocene sands into the Gulf of Mexico. Mitchell had never been in an offshore program except for a small interest in a program that Chambers and Kennedy had. This was quite an ambitious project and I headed that program up, along with Sam Meadows, our geophysicist. We participated in every state lease sale. The state had two sales a year and we participated in every state lease sale from about 1970 on until I left the company in 1976. We were quite active in the state waters of offshore Texas. We had some nice discoveries. I've got a few regrets. We missed out on some of the technology that we should have been paying more attention to. I think that we could even have been more successful had we been a little more technologically advanced in the area of seismic bright spots and things like that.

That was just coming out?
Seismic bright spot technology was just coming to the forefront at that time in our industry and it was something that was being developed in the major companies and not really being pursued by many independents. I regret that because there were a few independents who did embrace it and were quite successful in pursuing this bright spot technology. Our geophysical people were not that excited about it and we kind of ignored it and that was a mistake. I think we could have been even more successful offshore had we paid a little more attention to the bright spot technology, but still, we were quite successful. We found some nice fields out there and that was a wonderful experience for me, developing that offshore program.

It was in the late '70s, I think, when an offshore rig went over and a number of lives were lost. Were you with the company then?
That was after I left the company. Actually, that was out in federal waters at Block 189L. That was an old field that Mitchell acquired after I left the company. We had a very small interest in it through our Chambers and Kennedy program and George was always infatuated with the area, but I think they

actually acquired that field after I left. They were doing some remedial work on the platform, I think, when they had a fire, or explosion. While it was related to our state waters project, it was kind of an extension of it. I'm really not familiar with what happened, but I do remember that. We also were pretty aggressive in Galveston Bay and had some success in Galveston Bay along trend with some of the success that Houston Oil & Minerals was having. We did end up with a good well or two in the Texas City Dike Field because of some work that we did out in the bay. We had a good exploration program. I had the latitude to hire some people. I had an excellent staff. Also, when I came along, Mitchell had been isolated from the geological community and I've always liked to be involved in professional organizations, the Houston Geological Society, the AAPG.

Was that isolation by design, do you think?
No, they were just too busy doing other things, and they were small and they didn't have an image. As we grew, I was able to get involved, I wrote some papers and gave them orally and became involved in the Geological Society and it gave us more credibility, a better professional image. We were going public, and I think that helped to establish some credibility within the exploration and production community. We grew, I hired some people. I had at one time about a half-dozen geologists and support staff.

That was just Gulf Coast?
Gulf Coast and later, maybe for a year-and-a-half, two years before I left the company, I was actually in charge of West Texas, too. They redesigned the division boundaries and put that in my division, so at one time when I was vice president of exploration, before I left the company, I was in charge of the Gulf of Mexico, all the onshore Texas and Louisiana Gulf Coast as far north as Polk County, and then West Texas, also.

All that's left mainly would be North Texas . . .
North Texas and the Rocky Mountains, and Limestone County. I never worked on Limestone County or Madison County.

You said when you came there were six or seven geologists and when you left there must have been 20 to 30.
There were probably three in Midland, five or six in my division in Houston and four or five in the North Texas division, then there were some in the Rockies and Dallas and at one time we even had an office in New Orleans,

but that was right after I left. So I guess the company when I left had 20 geologists or so, plus geophysicists, probably 25 or 30 geoscientists, maybe, when I left. Continuing on just a little bit of history, in the late '70s, about 1975, we had grown to the extent where George and some of the other people associated with the company, primarily the banks, decided that George needed to relinquish some of the day-to-day control over everything and he brought in Leland Carter (Leland had been with the company for many, many years in Dallas) as president of the energy company. Up until that time, I had been working directly under George for several years. So Leland came in as president of the energy company and put another layer of management between me and George, although in practice, I still worked directly with George every day. Then after a year or so of that, Leland decided that he needed to bring in a senior vice president of exploration from the outside, somebody with major company management experience. And he hired a guy by the name of Lachlan Vass from Amerada Hess. Lachlan came in as senior vice president of exploration over me and the other two exploration vice presidents. This was a shock, although we dreaded this might happen, this was a shock to our system because now we reported to Lach Vass and not George anymore, and even though George didn't recognize that layer of responsibility when he came in to us, we had to recognize that layer of responsibility when we were going back through the various layers of bureaucracy to get back to him. It was kind of a mess, and that signaled to me that I had gone as far as I could with the company, as far as advancing and responsibilities and so forth. Basically, the handwriting was on the wall that they wanted somebody from the outside to run the exploration program and it became obvious to me that I had gone as far as I could go with the company. I had loved every minute of it and didn't want to work for anybody else in town, so I felt like it was time for me to make a change and the only obvious change was for me to go out on my own. So in December of 1976, or late November, I went in and talked to George and told him I decided to resign and he was disappointed. I told him I was going to go on my own and he wished me well. This letter is one of my treasured things, because I don't think he wrote too many of these but he wrote me a nice letter. I went on my own in December of 1976. Nevertheless, I wouldn't take anything for the 10 years I spent working for Mitchell. As far as the highlights, the number one highlight was working with George day to day, every day. He'd call me at the house several nights a week, we worked together very closely, we had a tremendous mutual respect for each other, and it enabled me to grow professionally. I learned so much about the oil business, in particular the business end of the business.

I learned never to give up. The persistent, tenacious spirit that he instilled in all of us has enabled me to get a lot of things done as an independent. I'll tackle things now that a lot of people won't tackle just because I learned from George that very few things are impossible if you've got enough persistence and creativity to make them happen. I worked on some great geological experiences, LaFitte's Gold, Galveston Island, Seven Oaks, Polk County area, starting the offshore program, taking the company public. I feel very fortunate at being able to do all those things with Mitchell. The people that I worked with there were tremendous people. I still have some good friends from those days. Some of them passed on, but I left Mitchell with a great deal of mutual respect, and I wasn't mad when I left. We had excellent relationships when I left, I just had different goals. So I've been able to maintain those relationships through the years. I've done a few things with the company since I've been on my own. In fact, the first deal I turned on my own was an interest in a well that was drilled, a discovery that was drilled on the tip of Pelican Island by Mitchell. It was a prospect of mine and when I left we weren't sure there was budget enough to drill it. So I asked them if I could have a farmout on that prospect which I had originated. Mitchell had decided to go ahead and drill it, but they didn't have the budget to drill it all and they were going to sell half the deal. So they just said to me, "Here, you take half the deal and go out and find us a partner and you can get whatever you can." So I turned the deal to Wainoco and we drilled a discovery well, an oil well, on the north end of Pelican Island and it turned out to be a discovery of maybe 150,000-160,000 barrels of oil. That was the first deal that I turned and I was able to get an override on that field. Mitchell Energy has helped me through the years in doing some things. I've worked with people over there, none of whom were there when I was in the exploration department. Also, several years after I went out on my own, George personally became one of my investors in some exploration programs that I put together. I can't remember exactly when he first became an investor, but it was at least six or seven years after I had left the company. We had some success and I put him into some production. He was a loyal investor and it was great to have him in my program because not only was he a great guy to work with, it was very flattering to me that he had confidence in me to put his own money in my deals. And it also helped me to bring other people in when they found out George Mitchell was in my program. That was one of the best things that ever happened to me is when George became an investor. A few years after he became an investor, we went to lunch one day and he said, "I've got a proposition for you. I have a son, my next to youngest son, Todd. He's getting ready to finish at the University of Texas with a master's degree in geology. Would you consider letting him come to work for you?" I was very flattered and told him so, but I said, "George, Todd can learn a lot more if he goes to work for Shell, Chevron, Exxon or somebody like that than he can learn working for me." He said, "No, I want him to work for you because I want him to learn how the independent operates. I want him to learn the business end of the business as well as the geology end of the business and Shell can't teach him that." I said, "Well, OK, we'll do it but let's send him to a lot of schools, a lot of professional schools." So Todd and I met a couple of times and I made him an offer to come to work for me and we sent him to a lot of schools. He worked for me, actually, as a salaried employee in my corporation for a year, and then he left to finish his master's thesis and go down and work on a couple of Johnny's mines down in South America and do some other things. A year later, he came back and shared an office with me and, at that time, they set up this family corporation named Dolomite Resources, which is owned by the 10 Mitchell children. Todd runs it. We still share office space together. Dolomite has been in some of my deals and I have invested in some of their deals. We have no formal relationship other than we share office space. That company has gone on to grow, they have several employees, we do some things apart and we do some things together, but we have an excellent relationship. This entire Mitchell experience has been a wonderful thing for me and I'll be eternally grateful for them.

When you were talking about geology, you said that MEDC had been kind of slow in picking up the bright spot technology. I have the idea that you felt the company was not among the leaders in utilizing advanced technology. It kind of runs counter to the company's perception of itself.

That's right, and I think that's true. I'm not trying to point fingers or cast blame. Sam Meadows was our chief geophysicist. Sam is a good guy, Sam and I worked together for so many years, we had a lot of success together and we have a great deal of mutual respect. But, for awhile, Sam was a one-man geophysical department. He was spread so thin he couldn't possibly carry out the entire responsibilities that were cast upon him and do that in addition to trying to keep up with the latest technology by getting out, going to schools, talking to people. We had so much to do that we were kind of operating in a vacuum. Looking back, we should have been more receptive and more inquisitive and more aggressive in seeking out what industry was doing in the latest technology. We just kept our heads down, doing it the old-fashioned way, taking care of the day-to-day brush fires and trying to do that offshore exploration program, keeping up with two lease sales a year. We underwrote and participated in a huge offshore seismic grid that we designed. Western Geophysical shot it. It covered several hundred miles in the state waters of offshore Galveston County and offshore Jefferson County and that was the basis of our exploration program. Sam designed that program and it was state-of-the-art at that time. He did all the geophysical interpretation, I did all the geological interpretations and we were so busy trying to do that that we neglected some of the advanced bright spot interpretation technology, and I think we suffered from it. I think by the time I left, we realized that we had perhaps lagged and I think that's the lesson that projected forward and the company soon corrected that, although I can't really make a call on what they're doing day-to-day now. I think we were slow to pick up the bright spot technology and I think we realized that and then we did embrace the bright spot technology before I left. Obviously, it was something that we realized later, that it was a very valid technology and we should have been using it from the beginning, but we did begin to use it before I left. Had we realized it earlier, we probably would have been more competitive and, perhaps, more successful in the early part of our offshore program. I do think you can put your finger on that one area and say that we lagged a little bit there, in the offshore, because in the offshore is where you get the best seismic data and that's where you can really apply the bright spot technology. Another comment: Most of the regrets I have stem from not being aggressive enough. There were some good plays that were developing at the time where we were being very aggressive and developing our own plays in the Upper Texas Coast and North Texas, and East Texas. The Lobo trend down in Webb County was developing at that time and we should have been down there trying to get involved in that. But we were spread so thin and we were just trying to tend to our day-to-day rat killing and perhaps didn't look beyond our office walls enough in trying to seek out other opportunities. There are only so many hours in the day and you try to set your own priorities and go on down the road, but looking back, I wish we would have been a little more aggressive in some of the other areas. Still, overall, I think we did a real good job. George and I always had a running argument about looking at outside deals. I liked to get a good staff of good geologists, geophysicists and let them develop internally their own plays. George felt that we should look at more deals from the outside. In retrospect, I think he was probably right. We didn't look at enough outside deals; we were, again, too busy with our limited staff trying to develop our own prospects. I think you need to do both and, perhaps, we didn't do enough outside looking. These are just my comments in retrospect on how we might have been more successful, but still, I think we were damned successful and I am proud of what we did.

It sounds like you worked with George a whole lot and it sounds like you did not have much traffic with Budd.

I had day-to-day contact with George and night-to-night contact with George and limited contact with Budd. We were in a lot of meetings together. He'd come by and ask questions for clarification on what we were doing, but George was my boss for the 10 years that I was there, essentially.

Clearly, in the very early days, George spent a lot of time with the geologists, going over all the details. Now, I think his love is more centered on developing The Woodlands and a few other projects that are very dear to his heart.

I think that back at the time that I worked for George, and I think this may be true today, it may not, but he was in charge, hands on. As a Mitchell employee, the best person you could be was a geologist. We all got our asses chewed out, but he was very sympathetic towards geologists, he never second-guessed me for a dry hole. He just said, "Well, we'll get it next time." I remember one time, I went out to his house out there in Willowick, in the Memorial area, one night. We'd logged a well in Montgomery county, north of Lake Creek Field, and it was a dry hole. I looked at the log, reworked the geology, came up with this idea of sidetracking the well south to test a kind of a harebrained prospect idea that I had, that I felt the data from the log we had run on the dry hole supported. The rig was still on location. I ran out to his house, went over this thing with him. It was like on a Sunday night, and I said, "George, I've been down to the office redoing all my maps and I really think if we sidetrack this well to the south, we've got a good chance of encountering the Wilcox formation that produces in Lake Creek Field high to the field and it's a novel idea, but I think it's got the potential to be a hell of a discovery." He was very skeptical, but I did a heck of a selling job on him and he said, "Well, if you feel that strongly about it, OK, let's do it." So we sidetracked the well and I couldn't have been more wrong. It was awful. The well, instead of being structurally high, the bottom fell out of the structure and it was real low by the time we got to Wilcox. I was totally wrong. He never said a word other than, "That's all right, we'll get it next time, it was a good idea, it didn't work. We'll just get on down the road and get one next time." That was typical.

Howard, that didn't apply only to geology, it applied to PR as well.
He would get most upset with us for not being aggressive enough. That's what upset him the most. If we weren't aggressive enough, if we didn't think of things we should have thought of, didn't have enough vision or foresight to anticipate problems, but mainly if we were not aggressive enough, that's what frustrated him more than anything. So working for Mitchell as a geologist was the best thing you could be as far as I'm concerned. He was always sympathetic to dry holes, never second-guessed us.

And he's a hell of a good geologist himself, too.
Very good geologist, quick to pick up things. It didn't take you long to show him a deal, he was way out ahead. He could grasp the concept instantly and be asking you questions about the business end of the deal and what happens way down the road. As the old saying goes, "It didn't take him all day to look at a horse shoe." Another important thing: He was not afraid to pursue exploration ideas in environmentally sensitive areas. We cut our teeth at Galveston in LaFitte's Gold, but this went on down to Pirates' Beach and Mesquite Bay down there in Aransas County. He was sensitive to oil and gas development in environmentally sensitive areas and I think he kind of led the way in these areas, and that was fun. Again, patience and persistence. Working with people, I'd get frustrated and just want to tell them to go to hell and go on to something else, but he wasn't that way. He was very patient, "Let's do the politicking, let's work with them, sit down and talk, and maybe not this year, we'll get it next year." While he doesn't appear to be one who has a lot of patience, on the overall long term, he does on projects like that.

Was Johnny working with the company when you moved from the Houston Club Building to One Shell Plaza?
When I came to work for the company, it was called George Mitchell & Associates. That was in 1967 when I came to work for them. A few years before that, George had acquired the interest of Merlyn Christie and Johnny, and the company was changed from Christie, Mitchell & Mitchell to George Mitchell & Associates. I'm not sure exactly when that occurred. It was before I came to work. Johnny was officing in the Houston Club Building and Merlyn Christie was officing up there. They both had offices up

there, but George had acquired their interests in all the production and renamed the company. Johnny was never very far away. I never got to know Mr. Christie that well. After a while, he closed his office and moved out and later passed away. Johnny always maintained his office nearby and was frequently down talking to us. I loved working with Johnny, too. He'd come in and ask me to look at some geology on a deal he was looking at or something out of Blue Ridge Field down in Fort Bend County where he had some production, and I used to enjoy working with Johnny. I remember one day over in One Shell Plaza when he was in there asking my opinion on some deal he was looking at. He was in there talking in my office and George came in and, of course, very impatiently wanted to know about some deal the company was working on, or something else, and he had a lot of questions. He just walked in, interrupted us and then started firing these questions at me. George didn't have a suit coat on, but he had on pants, shirt and tie like he always did, but the pocket on his suit was torn. Johnny looked over at him and said, "George, you look like crap, why don't you buy a new suit and some new clothes?" George looked at Johnny and said, "Johnny, if you'd pay me some of the money you owe me, I might be able to afford some new clothes." Never missed a beat, just went right on back to asking me questions, found out what he wanted to know and then he left, and Johnny and I continued the meeting. Johnny was always pretty dapper and George, although he always looked good, he never was concerned about being as dapper as Johnny was. Johnny was always around and I enjoyed my contact with Johnny a great deal, even though I didn't work for him. As a friendly favor to him I would look at his deals and help him on Blue Ridge stuff, just because I enjoyed working with him, just because he was kind of family. Johnny was a legend. I remember I loaned him my secretary when he was working on one of his books, to do a lot of the typing, proofreading, so we had a lot of contact with Johnny. That was fun. Changing the subject a little bit, when we moved from the Houston Club Building over to One Shell Plaza, that was about 1970, we went from individual private offices to the open office concept, which I detested and hated the whole time I was there. I just couldn't stand the noise and the distractions and lack of professional feeling that came with working in a cubbyhole where we didn't have any walls, but I had a good office location. All of the senior officers had the corner offices. Of course, George was in the southeast corner of One Shell Plaza on 39, but I had the office next to him, and the geological department just went on down that wall all the way around. It was kind of strung out along that wall next to George's office, but I had the first office next to

George and that was great because I had a tremendous amount of contact with him day to day. I was always bitching about the office situation and the noise. One day he came in and he said, "OK, we've got some space down on 37, or something. If you want to, you can move down there, you'll have a private office." I thought a minute and said, "No, I don't think so, I don't think I want to move from where I am right now." So, I didn't. I didn't like the office arrangement, but I sure did like being next to George. I wouldn't have traded it for anything. So it didn't take me long to think, "No, it wasn't such a good deal." One thing, and I don't know if you can ever find it or if you'd even want it, but when I was in charge of the geology department in my division, I kept a set of active files and a set of inactive files because our files were so big and we had a lot of inactive things that we didn't work with from time to time, but it was a handy reference file. It had records of the deals we had looked at, or been involved in, that were no longer active for one reason or another. I had inherited a lot of them, covering way back years ago in the exploration archives of the company. In one of those inactive files, in Louisiana, I believe it was Jeff Davis Parish, Louisiana, in the Elton Field area, there was a report written by George when he worked for Pan American Production, or Stanolind, whoever it was he worked for a short time right after he got out of school. There was a red bound reservoir report that he had prepared on the, I believe it was the Elton Field in Jeff Davis Parish in Louisiana. If they still have inactive files, it may still be there.That was the only time I've ever seen his signature where you could read it. It was signed, George P. Mitchell, as a young engineer right out of school and it was a good report. I would always save that and occasionally open it up to look at it.

Thank you, Howard.

Today is Wednesday, February 14, 1996. This is Joe Kutchin. I'm in my office in The Woodlands with Charles Lively, who until his retirement was a long-time employee of Mitchell Energy & Development and some of its predecessor companies. Charles, why don't you go back and just give me some of your personal background, even before you came with the company—where you are from, your education, what you did before you got here?

I'm originally from Palestine, Texas, and went to high school at Palestine High School, graduated from there in 1947. I am from a large family, 10 children, and I was the youngest of all of them. I was accused by some of being my mother's special because I was the caboose. We only had two boys and I was the baby boy. I enjoyed the time that I was in high school. I played football for Palestine High School and had some buddies there that I played football with and we are still in contact with each other. I graduated midterm, which would have been January of 1947, which means that I attend the reunion classes for both 1946 and 1947. We'll be having our 50th anniversary get-together this coming May. When I graduated from high school, a friend of mine who I played football with decided that we should go to Galveston and see if we could find a job down there during the summer time. So we went to Galveston right after school was out, when the season was right and Galveston was opening up the beaches. I happened to be the brother of Maureen Pine, who married Alleyne Mitchell's brother, Ralph Pine. Alleyne is Johnny Mitchell's wife. When they heard that my friend and I were in Galveston, they said, "Go by and visit Johnny, he might have something for you to do." So we got to know Johnny right away. He had just gotten out of World War II and was driving a maroon–maroon probably because he was a Texas Aggie at heart–1946 Ford. He wore short khaki pants and was very tan and was a very outgoing-type person. He was just getting started in the oil business and was good friends with the Maceo family in Galveston.

You had moved down there?

We were just down there for the summer time. Alleyne at that time was Alleyne Browder. She and Johnny weren't married yet. She had a place there that had several extra bedrooms, so she rented my friend and me a room. We stayed there in the same house where she lived. Johnny was a friend to

Sam Maceo. They used to have wide open gambling in Galveston, and Post Office Street was full of houses of ill repute. It was just a party place to go to, and anything went. You could gamble there at the Balinese Room or go down Post Office Street and walk down the street and have the girls whistle at you. Johnny had these two little places out on the beach where he sold cold drinks, beer, hot dogs and hamburgers and so forth. So my friend and I went to work for him out there on the beach and enjoyed it so much that summer that we came back the following summer and did the same thing. We just enjoyed meeting all the pretty girls in bathing suits. So that's how I got to know Johnny and how I got to know anything about the company at all. This was probably the summer of 1947.

By that time he was working with George and Merlyn Christie, I think.
He probably was. I just wasn't aware of that. Johnny made Galveston his home and he'd come up to Houston pretty often. But I didn't meet George then, it was just Johnny. The summer of 1949 would have been the last summer that we worked down in Galveston. After that, the Korean War broke out and I went into the Air Force.

What year did you go into the Air Force?
1950. And in 1954, I got out. Johnny sent word that when I had a chance to come to Houston, to come by, he'd like to visit with me.

Did you go back to Palestine after the Air Force?
After I was discharged from the Air Force, I went back to Palestine. You were asking if things had happened prior to my going into the Air Force. I went to Henderson Junior College in Athens on a football scholarship. I went there for two years, played football, and that was what happened from when I graduated in 1947 until when I went in the Air Force. After I got out of the Air Force in 1954, I went back to Palestine and went to Nixon Business College for a few months, taking some courses with them. That's when I got word that Johnny would like to talk with me. So I went to Houston and went to Johnny's office and his secretary said, "He wants to talk with you, but he has someone in there and will be a while." So I waited about 30 minutes and I said, "Look, I can come back and see Johnny some other time." She said, "Well, wait just a minute." So she went back and talked with Johnny and she came back to see me and she had an application-for-employment form and said, "Johnny wants you to fill that out." I wasn't expecting it, but I filled it out and as a result of that I went to work for Christie, Mitchell & Mitchell.

That would have been in 1956. I got out of the Air Force in 1954, and then I did other things and I didn't go to work for the company until May of 1956.

So you came in right alongside Budd Clark?
That's right. Pretty close. Rather than be employed by the company, I worked for Raymond Loomis, who was a brother-in-law of George's. He was working as an independent broker.

What kind of broker? Not a stockbroker?
No, land deals. What it really amounted to was that he was George's brother-in-law and he needed to have something to do, so he was hired and was working there. They didn't really have a place for me, so they put me with Raymond Loomis. What I was first doing was posting information on maps and doing some very uninteresting things. That went on for four or five months. They came to me one day and said, "Charles, we've got an opening for a landman's job up in Bridgeport, North Texas." I said, "Gosh, I don't know anything about land work." They said, "Well, if you want to try it, go up there and try it for a while and if you don't like it, come on back down here." So I said OK, and I think it was in November of 1956 that I went up to Bridgeport where most of the activity in oil and gas at that time was happening in Wise County and in and around that area. So I went up and after a couple of months up there I learned to like it and became a pretty good landman. I rode around with a fellow named Jack Brackett.

Jack was your boss?
He wasn't my boss, but he was a landman and had some experience, having worked for Texaco beforehand.

Who was your boss?
Dan Peters was in charge of that office. I rode around with Jack for probably a couple of months. Everywhere he'd go, I'd go with him and I'd listen to him negotiate with people and learn how to check records and abstracts at the courthouse, and they all complimented me on the fact that I learned so quickly, but I gave thanks also to Jack Brackett for having the patience to let me go along with him.

Very nice person.
Shortly after I got to know Jack, my girlfriend from Houston came up to the

Fort Worth area and got a job. Shortly thereafter she and I were married. Jack and his wife and Grace, my wife, and I were good friends and did a lot of things together. This was back at the time we were first involved up in Wise County. We were drilling a lot of gas wells, they were all shut in, and we didn't have a market for the gas at that time. Lone Star Gas was our adversary. They were dominant up there and they spoke not too well of Christie, Mitchell & Mitchell. But we were able to go out because the landmen who were working for Christie, Mitchell & Mitchell lived there, got to know the people and really associated with them. Went to church with them, socialized with them. We became their friends and as a result of that, we were very successful in being able to lease land and get some pretty good deals, such as get extensions and amendments to leases so we could unitize the land and shut in the gas wells 'til a market was available. During that period of time, Christie, Mitchell & Mitchell held an open house up there, had a big picnic, barbeque. Johnny Mitchell was the PR man for the company and he met with many of the landowners, who were invited to come to this barbeque and get-together. And they just really did fall for Johnny.

Do you remember when this was, what year?

That would have been somewhere around 1956 or 1957, as I recall. The landowners just liked the way Johnny talked, down to earth, and told things the way he saw it. And he said that Lone Star was trying to take advantage of the local people. He said, "I would rather be my own person and see if we can't get a market for this gas and get a decent price, and while we're doing that I'll eat cornbread with you, but hopefully, we'll have something better coming down the line." And they really did. We got a market for the gas, signed a good contract, the company did, with Natural Gas Pipeline Company of America.

You were not involved with those negotiations, were you?

No, I was strictly involved with going out and negotiating leases and working with other oil companies on farmout agreements and amendments to leasing and things of that nature. So I was up there probably for 11 years, working as a landman and getting exposed and experienced in more things all the time.

Tell again what years those were.

I went up there in 1956. I transferred to Houston in 1967. So it was about 11 years that I was up there. In the first part of 1967, there was a landman up there by the name of John Nolen who had graduated from the Petroleum Landman School at the University of Texas. George Mitchell was getting involved in acquisition of land. They'd made a deal for buying out Grogan-Cochran Lumber Company—that is, the stock of Grogan-Cochran Lumber Company.

It sounds, Charles, like you really hadn't had much dealings with George Mitchell up to this time.

I'd seen him, but I didn't know him. I didn't really get to know George Mitchell until 1967. The fellow who had the good educational background was a petroleum landman, worked in North Texas for about a year or two. They said they needed somebody real good to come down to Houston and work as a landman in the Houston office to do real estate work. So John Nolen went to Houston and stayed about six months but for some reason it didn't work out. So they said, "Well, we need someone else, do you have anybody you can spare?" They asked me if I would consider going back to Houston and I was reluctant because I had gotten to know a lot of people in the Bridgeport area and was active in the church and had a lot of friends. I talked with my wife about it and we decided yes, we might as well go ahead and try it. I transferred to Houston in October of 1967 and I still didn't know what to expect when I came down here, except that I was going to be working as a landman and didn't know exactly what my duties would be. I rented a place in Conroe and lived there for six months. I'd drive in to the Houston Club Building, which is where Christie, Mitchell & Mitchell had their main offices. I'd go in about once a week and get assignments. There was a fellow by the name of Joe Ramsey who was the head landman in Houston then.

You were doing petroleum land work at that time?

I was doing some of that, but mainly they were getting involved in real estate then.

Was this an easy transition? Sounds like it was a different kind of animal.

Whether you're working oil and gas or real estate, you're still dealing with land. Learning who owned the title to the minerals under real estate would be about the same as determining who the owner of the land itself is. You also learn to negotiate with people, get the best deals you can get for leases or amendments to leases, same for both petroleum and real estate. In dealing with oil and gas, you learn to deal with people, different personalities, you deal with corporations and you learn to kind of go with the flow sometimes. So they are closely related in that way. It's not like someone coming in and breaking new ground completely. Joe Ramsey was a helter-skelter type of guy. He had so many things going. He was also kind of a politician. He probably went to work for Mitchell during the early '60s some time. He had several people working under him. Some were employees and some were just brokers who did work as a broker and not as an employee, on a day basis, so much per day. After I worked with Joe for a little while, it was decided that we needed to build a little office for real estate up in Montgomery County.

To what purpose were you acquiring real estate in Montgomery County? What was the company going to use it for?
At that time, I didn't know. This was 1967 and early 1968 and I still wasn't aware of what George had planned, that is, The Woodlands. George, through one of the better deals that I guess had ever been negotiated anytime, had acquired Grogan-Cochran Lumber Company in 1964.

I didn't realize it was that early.
That's when George got interested in real estate and his appetite kept getting whetted, so when I came down in '67, about three years after the 1964 Grogan-Cochran acquisition, they started doing some little rural subdivisions—West Magnolia Forest, Clear Creek Forest and things like that. They were using land primarily that had been acquired through the Grogan-Cochran purchase, some 60,000 acres of land. George was using the fringe area real estate over in Waller and Grimes County to cut up into acreage tracts and sell off. He had a real estate broker by the name of Jimmy Dinkens (of Conroe) who was handling the sales of those subdivisions. There'd be some tracts of land adjoining acreage that had been acquired through Grogan-Cochran that were needed to fill out a block and make it contiguous in order to continue with the development. These were some of the first things that I was doing, going out and determining who the owners of those out-parcels were and making contacts and attempting to buy. We were getting active enough that they decided to build a little A-frame office building over at Pinehurst. That's where I officed for a while, between there and the Houston Club Building.

Do you believe that George had a major new town in mind by 1967?
I think that he did. I wasn't aware of it. I think that kind of grew on George as time went on. No one ever thought that we would be as successful in

acquiring the tracts that we did on the budget that we had to do it with. I had a little success getting some of those, not acreage that was within The Woodlands area, but outlying parcels around Magnolia and over in Grimes County, things like that.

You were kind of learning your trade.
I was getting my feet wet in this area and learning the trade. But I still wasn't aware of George's thoughts about The Woodlands project. I started having success getting some things that outside brokers had been trying to get for George, and I just lucked into it. I'd go and talk with them and was able to get a deal on it. Word got back to George that "Hey, we need to get approval of an AFE (Authorization for Expenditure) because we've got a deal made to buy this tract of land." And George would say, "What do you mean we have? We've been trying to get that for a long time. Who made that deal?" And they said that Lively went out and talked with them. He approved that AFE real quick and he told them, "When Lively comes back in the office, I want to talk with him." So, I was in Joe Ramsey's office one day and George got word that I was there. He came in and introduced himself. He was very friendly and I was impressed with him.

Was this still '67, '68?
This would probably have been '68, first part of '68.

How long had you been with the company now?
I went to work for the company in 1956.

So it was 13 years before you really got to know George Mitchell?
That's correct. I knew him, but I didn't really get to know him until 1968. I got to where I was just having a lot of luck, an old country boy and down home, and I wasn't trying to impress anyone, I was just being myself. But the people who I went out and dealt with, I was having some luck bringing in the contracts that a lot of these real estate agents just couldn't do. Word kind of spread around through the realtors around Conroe that George had hired this guy in real estate, buying real estate and was cutting the real estate agents out, that he was going through me. One of them laughed and told me, "There's a name we have for people like you, Charlie." I said, "What's that?" And he said, "It's called a scab." I said, "Oh well, so be it." I had some deals that came to me where a real estate agent would say, "Now, if we get this, there's going to be so much in it for me. How much of it do you want?"

I said, "I don't want anything. I'm working for the company on salary and if there's anything coming, let's just cut it off the price of the land because I'm representing Mitchell." That happened several times and I have wondered—I've never asked George—but I've wondered if maybe he was wise enough to say, "Hey, we're trying to find somebody who is going to be out here on this big project. We've got to find out what you're really like." But the agents found out that I wasn't the type who was going to take money under the table. It was offered to me a number of times and I'd say, "No, we'll just cut that off the price of the land, rather than pay a 6 percent commission to you guys, we'll pay you 3 percent and then take the other off of the price of the land." I think word got back to George about that, so I think he felt real comfortable working through me and trying to make a number of acquisitions for The Woodlands later on. Probably the first that I'd learned about the possibility of trying to block up land out here would have been mid-1968. I don't know if anyone has mentioned to you or not, but within The Woodlands complex here itself, there was a big survey called the John Taylor Survey. It's one of the major surveys here within The Woodlands. Within the John Taylor Survey, a large part was acquired by a man by the name of MacDonald. MacDonald made what was referred to as a mineral subdivision. He had it literally cut up into hundreds of 10-acre tracts. He'd sell the minerals—the idea was that it was a mineral subdivision but they were getting title to the land along with it. It was sold to people who lived all over the United States. I don't know how he sold it, if he advertised in papers all over the country or what. Right here where some of the office buildings are today would have been the MacDonald subdivision. They were all out-parcels.

10-acre out-parcels?

Yes. Just all over, and it seemed that no one could ever assemble all that land. It would be foolish to even try. So I met a man by the name of Frank A. Karnaky. I don't know if you've met him or not, he was an independent broker who worked for the land department. Jack Yovanovich knows him real well and they've been very close through the years. But Frank had done a lot of work for the company in Galveston, Montgomery County, a number of other places, oil-and-gas-wise. It so happened that years before 1968, when I really got involved here, Frank Karnaky had done a lot of work for companies, oil and gas, and he had made acreage take-offs which showed who the owners of those tracts were. I met Frank and hired Frank to work with me as a consultant. He's in his 90s now. He goes a long way back with George. He knows this area. He dealt oil and gas leases and did a lot of work with me, for me, after I got involved down here. It was through Frank's previous research in the area, and knowing about the minerals, that I got a lot of the information needed to go forward with getting the block put together.

You say a block. Did you know what the end target was?
None of us knew in our minds, Joe, what the end target would be, what the final configuration of The Woodlands would be. Is that what you mean?

Well, if you're assembling blocks it would seem that you were looking for land in a certain area and a certain amount of it.
We were. We didn't have anything going all the way out to Interstate 45 at that time. What we were trying to do was to fill in the holes in land that we had back over here in the interior. We had access via Sawdust Road to come into the southern parts of the acreage that we had then. As we had more success getting those little out-parcels out here, I got with Paul Wommack before he was a company attorney. He was working as a consulting attorney for Smith & Fulton. I'd go to Paul and we'd draw up a contract for an option to buy the land. We didn't want to commit to spend all this money for these little tracts unless we could get enough of them to make it fit in and make sense. So I said, "Paul, let's see if we can't get us some little option agreement contract that I can go out and we can get these people to sign on it and we maybe could put up $100 or $50 or something like that, and we could tie it up for 90 days and see if we have any success." Budd Clark said, "Yes, you're going to go out there and tie up all this land with green stamps." He was kidding George. So, I'd get little AFEs approved to go out and spend, get this option money to put on the land and I think it surprised me, it surprised everyone that we were as successful getting those option agreements put together as we were.

Let me ask you again. Did you know that you were building acreage for a new community then?
I did at that time.

There were a couple of land plans that George had done early. Were those in existence, either of those?
Along about that time, they came into existence. First of all came these little option contracts where we put up a few dollars and would have the land tied up for 90 days. Then all of a sudden we were successful and it was growing and growing and that 90-day period would be up on some of those original

ones that I got signed. So I had to go back and try to get those extended because we were still almost operating out of our hip pocket, more or less. Then the land plans that you were talking about came into being, and then we talked about having a 1,000-foot commercial strip of land coming from Interstate 45. We called it the commercial strip. It was owned by people named Sutton-Mann. You've probably heard their names mentioned before. I made contact with them and we had meetings in Houston trying to work up an agreement where we could buy this 1,000-foot strip of land, commercial strip, coming from Interstate 45 back to the big block of land that we had assembled back here to the west.

Let me get the picture. This was a strip leading in or facing on 45?
Yes, facing on I-45, and it would go from Interstate 45 for probably a mile-and-a-half back into where we had this other block of land that we were in the process of assembling. So, I contacted the Suttons and Manns and got them together down in Houston and we negotiated that. They were very particular. Robert Mann was a very difficult man to deal with. We finally got that 1,000-foot strip of land, that we called the commercial strip, under contract and we were to have built a road along that strip that would have served Sutton-Mann acreage on either side and made their land a lot more valuable. It grew from that into our negotiating for the acquisition of all of Sutton-Mann's acreage. It got very involved and complicated but we were able to finally come to an agreement for buying all of their land, which went all the way to Interstate 45.

Do you remember how many acres were involved?
Something like 2,000 acres.

Tell me the year if you remember that.
I'm going to guess that would have been early, probably '71, '72, somewhere in there. We had gone a long way in putting our block together back in the interior before we decided to go forward with this. We worked with Sutton and Mann for probably a couple of years before we ever got the land acquired. First, we got the 1,000-foot commercial strip and then after that we decided to go for the whole thing, so we did that. Probably didn't get that accomplished until maybe as late as '73 or '74.

The HUD loan came in in '72.
Maybe 1972 would have been the time that we finally got it put together. We

were negotiating on it for at least a couple of years.

Were there other people involved doing the same thing you were, or were you doing it all alone?
I had people working for me as consultants, and I was overseeing a lot of it. I would take certain ones. On Sutton-Mann, it got very complicated, so Paul Wommack got involved, and George personally got involved, as well as some of our accountants.

But the way things worked, it was essentially your responsibility, using your "consultants"—would they have been brokers?
Not all of them were. We had one that I would like to mention other than Frank Karnaky. I think his name should be in there somewhere, Gerson Haesly. He was from Tyler. He had done some oil and gas landman work for Christie, Mitchell & Mitchell and that's how I got to know him. He turned out to be a prince of a fellow, honest as the day is long, one that I could really trust and depend on. He worked with me a lot on getting the options in, going up after we bought the surface and then having to go and deal on minerals, getting the minerals in so we'd have control of the surface. He was one that I felt completely comfortable with, and he loved George and Johnny. He thought very highly of them. Gerson and I became good friends because I had so much respect for him. He was able to do a lot of work for me, real good work, very helpful. I had some others who had worked for various people in the oil and gas department. One in particular kind of turned out to be a bad egg and I had to threaten to take him to court because he'd come out and bought a tract of land through someone he knew and thought he was putting one over on us and I found out about it. But we were able to get it without having to take him to court. Frank Karnaky and Gerson Haesly were very good consultants who worked with me. I had numbers of others, but I would like to give some credit to those two particularly for helping us.

The major transactions—the first was the Grogan-Cochran deal. That was how many acres?
About 60,000. And then Sutton-Mann was around 2,000. Those were two major transactions.

Were there any other really sizable blocks?
Champion Paper Company had a lot of acreage, probably more than a thousand acres, I guess, that is located within what is now The Woodlands.

We had land in what is now the 1960 area. Is that the one where we did a switch?
There was an acreage exchange, a switch of land in eastern Montgomery County where Champion has a lot of acreage. It was appraised, and a certain value was put on acreage that we had acquired from Grogan-Cochran that adjoins some Champion land. We had it appraised and we had an exchange ratio where Champion would get maybe 2.7 acres of land out there for every acre of land we would get here.

Champion land is now a part of The Woodlands?
That's correct. That was another major transaction and acquisition. We had other tracts that would all be 100 acres and maybe 150 acres, maybe as much as 200 acres. There were a lot of those and many, many smaller ones.

The number of transactions was about 300 or so?
Something that I don't think was really factored into that is where we had to go back and deal with someone else on a tract of land to buy the minerals, or to work out a surface waiver or do something like that. It was a transaction unto itself that was separate and apart from the acquisition of the land. It was very important to us that we have control of the mineral ownership because in Texas the mineral is the prevalent right. I'd say that we're looking probably at as many as 500 transactions.

OK, but to acquire the land itself, would that be in the 300 neighborhood that I keep hearing?
I'd say that could be.

Plus another couple of hundred or so to clear up mineral rights and things like that.
Right.

And you were involved in all or at least most of those?
I was not involved in the acquisition of Grogan-Cochran. That was done in 1964, prior to my coming down here. Most all the rest of them I was involved in.

One of the other things that I hear is that no attempt was made to mask what the purpose of this was, that there was an attitude that we needed to be up-front and honest with all of the landowners.

That's not 100 percent accurate. After we were having a lot of success in blocking our acreage, word got out that Mitchell was blocking a lot of land down here. The first land that I was able to buy under option agreements I was able to get for as little as $400 per acre. Then as we had more success putting more of it together, the price was going higher because the word got out that Mitchell was creating a block. As a result of that, I suggested that we use someone we could trust to go out and put the land in their name as though they were buying it and Mitchell wouldn't be involved. We bought a lot of the out-parcels through brokers in that way. When I said a couple of minutes ago that we had a problem, I thought we were going to have possibly a lawsuit. The broker should have been out buying in his name for our account, but it got complicated because we were trying to put the land under options to buy. We didn't have the money to go out there and pay cash. Well, one of these consultants, or brokers, who would be working for me would go out and try to buy the land and the owner would say, "No, you'll have to buy it right now, I don't want a working option, you'll have to buy it and I only want cash for it." Well that's what happened. This guy said, "Well, I can only take an option or buy it under terms." And the seller would say, "We'll only sell for cash." So the broker felt, well, I can't deal for Mitchell, so I'll just buy it for my own account and pay cash for it. So he felt like maybe he wasn't doing anything wrong, but I said, "No, you don't buy anything for your own account, under any circumstance. You're working out there as our agent and you must understand that." So we went for a long time buying in other people's names. We bought a lot in Frank Karnaky's name. We had problems with Frank Karnaky's brother, Dr. Karl Karnaky, he had acquired some land down in the south part of The Woodlands that he didn't want to turn loose at a reasonable price. So, we had some problems there for a while that we had to work out. We were finally able to make a deal with Karl Karnaky by offering him, at our cost, a lot on lakefront property that we might develop in The Woodlands. But that wasn't assignable to anyone else and, unfortunately, Dr. Karl Karnaky has passed away, so that option to purchase was never exercised by him. Finally, Joe, we had the HUD agreement and many people knew about the plan for The Woodlands. At that time, it was decided, well, everybody knows that The Woodlands is going to be there so we might as well go ahead and deal heads up in our own name. That didn't make sense in all cases. I had to bring in some other third parties. One that comes to mind right now is Dean Couch, of Couch Mortgage. We took some contracts and banked them with him and got some money through Couch Mortgage to operate and make some of our acquisitions of land. It so happened that Couch bought what is Oak Ten Shopping Center on the east side of I-45 across from The Woodlands and started developing it. He also owned, in fee, the land where the approaches to Lake Woodlands Drive come up over the freeway. So we had to work through third parties, not let the owners of land know that The Woodlands was getting the land, or we would have never gotten it. So, I worked through a fellow by the name of Robert C. Watson, who was a real estate broker and an accountant, a CPA. He had someone from Mexico who was in Houston on business, and a meeting was arranged with Dean Couch's group. Robert Watson and this Mexican national came up—he was supposedly buying Couch's land for the account of someone in Mexico City—the land where the jughandle from the freeway is located, coming over I-45, leading into The Woodlands (Lake Woodlands Drive). Couch thought he was selling that to someone from Mexico. We had to devise something to throw him off in order to get that land.

No, I hadn't heard that. I'm surprised that your representatives would keep the confidentiality.
The representatives that I had working for me? It was with the understanding that if they bought it, they would be getting a 6 percent or 5 percent commission.

Did they also negotiate the deal?
They negotiated the deal through us. We told them what to do and they had to go think about it for a while and get back to the seller. Finally, we'd bless the deal or say, "Change it in this way" and they'd go back and say, "Well, we'll do this." But all the time, Couch thought he was dealing with this Mexican national, so we were able to get that accomplished. Did you have a chance to interview Dick Browne?

I'm trying to do that.
Dick and I worked together on that.

I didn't realize that Dick was involved in negotiations for land.
He thought it was so important to the overall success of the shopping center. He was involved in the planning process and frequently expressed how important he thought the "jughandle" acquisition was.

Where the big sculpture is now on the road into The Woodlands?

No, that one is on the Sutton-Mann acreage. The Couch land is down a little bit further, where we have the Mall. It's called the "jughandle." When we were successful in getting that, Dick Browne threw a party for me and called it the jughandle party and I got an award. It was real neat. He gave me this plaque for acquisition of the year or something like that.

How active was George in your acquisitions?
George got started calling me at home at night to get an update on how some of those things were going. Then I'd go to his office and we'd meet and he'd want an update on every one. I had a map that had all these parcels of land and he'd say, "What's the deal on that, on that and that and that?" I knew it well enough that I could tell him exactly what had happened, the status of negotiations and what we were waiting on, and I expected to have something back and on this one I've talked to the wife and the husband is out of the country, be back in two weeks, and it made George feel comfortable knowing that I was on top of all of it. He would call almost every night to get an update on things. I'm sure he was calling other people about other aspects of his enterprise. He's an amazing person. Has anyone else that you've interviewed said anything about prospects for building a lake on Spring Creek?

No.
I don't know if they've completely junked it or not, but there was talk about a lake there probably 15 years ago. We had some work done to determine whether it would have been feasible to have a lake on Spring Creek. Dick Browne worked on that quite a bit. We thought that perhaps we could build a lake there of maybe 800 to 1,000 acres and have the dam of the lake where Kuykendahl Road is. As a result of that, we wound up buying a lot of land south of Spring Creek, which no one would know was ever to be a part of The Woodlands, but we have several thousand acres over there now that we acquired through numerous acquisitions after the main block of land.

Tell me again, is that considered part of The Woodlands now or is it just kind of contiguous, sitting out there?
It's contiguous and I think it will be a part of The Woodlands.

Let's go back to the assembling of the land. You went through this process of getting options and, obviously, ultimately those options were exercised.
On all the options that we had, we had dates set where you had to get notice

within a certain period of time. Normally, where you have so many of them you would think that you're going to goof up on one of them and fail to timely exercise the option or give notice. Not one did we ever miss.

Before computers?
Before. It was me having a map, and a little calendar on my desk. We also had backup through the central records system. I'd send a copy of the option to central records and ask them to set up a file on it, asking to please flag that for notifying me that we had an option date of a certain time. If I had strictly relied on that we would have missed a few of them. I maintained my own records and used them as a backup in case something happened to me. I had my own calendar that I kept all those things on.

You, on behalf of the company, made sizable financial commitments?
Oh boy, we sure did and I have no idea how much.

So, it was not your thing to worry about where the money was coming from, it was your thing to make sure you got these options at a fair price.
At the best deal I could get under terms that I felt like we could live with.

So it was really the responsibility of the chief financial officer or Budd or somebody like that to come up with the cash that you needed?
George would run interference for me there. I never will forget, one day I went into One Shell Plaza, even after the real estate division moved out here and George still officed downtown. Budd saw me with my briefcase and he said, "Damn it, Lively, what are you going to do today? What building are you going to buy downtown today?" I laughed and said, "Budd, we're not involved in downtown at all." And he said, "Why don't you tell me that you're going on a trip around the world. We'd be a lot better off to send you on a trip around the world and hope you don't come back for two years." He wasn't happy with the fact that George was spending all the money, that he was buying the real estate. Budd knew that I was an integral part of getting all that accomplished. And Bill Houck made a statement to me one time, he said, "Lively, you're a very personable young man and it's all right to have success getting something every now and then, but do you have to be so damned efficient?" The financial people were having problems with the fact that we were so successful.

The HUD deal kind of cleared that up for a while?
It did for a while. After we got the Sutton-Mann, there were still some tracts and I suggested to George that it would make sense for us to get everything that we could from Sawdust Road north to FM 1488, to fill in everything that we could. Finally, we were successful in getting from Woodlands Parkway a good part of that way, anyway. Not quite all the way to 1488 but at least to State Road 242 that goes by the Montgomery County College. So we have a lot of frontage to play with there.

So you must feel that The Woodlands is one of your children?
Definitely. It's something that I'm proud of. When I look at it today, I'm proud to have been associated with it. It's something that I think anyone would be proud to say, "Hey, I had a part in making that happen."

Is it fair to say, Charles, at least on what we've been talking about, that you had a major role in assembling the land package?
Did you ever see that statement that George gave? I read in an airline magazine one time where George was talking about the success of The Woodlands. He said, "Well, Number 1, we have a good location and we've had excellent land planning. We've had outstanding acquisitions." Then George told who did the original land planning, talking about where it's located and so on. Then they asked, "Well, George, who did your land acquisitions?" George replied, "I did it. Me and one other man." I took a copy of that and circled it and sent it back to George and I said, "Who was this one other man? He must have been one hell of a person." I enjoyed it. It was a compliment, but when they asked who did the acquisition, he said he did it. George said, "I did it. Me and one other man." Well, George certainly had more to do with it than anyone else. Without George it would have never been done.

Understood. Everybody acknowledges George's creative and leadership role, but the fact is that, just for purposes of clarification, you were out there on the front lines implementing this whole thing and you were the point man for everything?
And it was my idea to go with the options. I'd come in with some ideas and mention them to George and tell him I thought we ought to try it that way. He would agree and then when you would talk with him tomorrow, it was his idea, which is fine because I'm his man, you know. At least he listened to me and we implemented it and as a result, it got done and it would have not

been done without him. I never did have a problem with that like some people might have.

Charles, you've been very helpful. Thank you.

WALTER A. LUBANKO

Today is Tuesday, February 20, 1996. This is Joe Kutchin. I'm in the downtown offices of Mitchell Energy & Development in Texas Commerce Tower with Walter Lubanko, a long-time member of the Board of Directors and a man who was considerably involved in the initial offering of the company's stock to the public in 1972. Walter, tell me your present business affiliation, please.
I'm chairman of the Board and president of my own firm called W. A. Lubanko and Co., Inc. It's an investment banking firm. I started this firm after I resigned as a partner of Lehman Brothers in 1978. Ever since then, I've been operating in that position.

Just give me a little bit of your personal history, where you were born, educated, things like that.
I was born in April of 1926, which makes this year my 70th on this earth. I was born in New York and I had elementary school education in New York City and went away to prep school in Virginia. After prep school, I entered Princeton six months after Pearl Harbor. During my college days I served in the U. S. Army for three years, from 1944 until February 1947, and was discharged as a first lieutenant. Thereafter, I graduated from Princeton in 1948 and entered Harvard Law School.

And your Princeton degree was in what?
B.A. I attended the Woodrow Wilson School of Public and International Affairs. Then from there I went on to Harvard and graduated in 1951 and I was invited to practice at a very large downtown firm called Simpson, Thacker and Bartlett. I worked there for about four-and-a-half years.

That was a law firm?
A law firm, a very large firm. However, now it's much larger. I think they have 400 lawyers now or something like that. I was fascinated by the investment banking business and fortunately, one of the clients at Simpson, Thacker was Lehman Brothers. I joined Lehman Brothers in 1956 and worked there for eight years. Thereafter, I changed to join F. Eberstadt & Co., an investment banking firm whose senior partner, Ferdinand Eberstadt, fascinated me. I stayed there for 12 years.

In what capacity?

As a partner in charge of the investment banking department. It was during that time that I was introduced to George Mitchell and Mitchell Energy in 1971. It wasn't called Mitchell Energy then—the name was devised in preparation for the public offering. I met George and dealt with Budd Clark, who was then very much the Number 2 man in the company.

How did you happen to meet George?

I met George through Shaker Khayatt. Shaker's firm was the other principal underwriter in the public offering. He was formerly at Eberstadt and then left, I believe, to join the firm of Coggeshall & Hicks.

I didn't realize that.

Shaker informed us that he had this very interesting company down in Houston and he thought he might like Eberstadt to be involved. I came down to visit and I liked what I saw.

What did you see?

In those days they were in the old Houston Club Building. They had a relatively small office and a small staff, although the company was growing nicely, and The Woodlands was just a gleam in the eye. They had acquired all of the land and did a lot of the planning. Before we decided to become underwriters, of course, we had to become better informed and perform our "due diligence." Shaker and I had a tour of everything. The Woodlands—there was nothing on it then, but we took a helicopter tour around the area and saw The Woodlands. We met with the people who were designing it. George was hiring the people who were to build it, most of whom are now gone, in one fashion or another. After a number of months, we proposed a public offering to George. George liked it.

Why would he do a public offering?

He needed some equity. He could have gotten it from other sources, but he thought there would be a number of advantages to the public having an ownership. Liquidity was one thing, plus it would be a source of capital, which George continuously needed. The public offering facilitated that task. I don't know whether you know this, but the first registration statement that was filed was significantly different than what the final offering was. It was the offering of units and the reason for that is, George had a firm idea as to the price he wanted for his shares. I don't remember the exact price but let us assume he wanted $17 or $18 a share for stock. When designing a public offering, you have to recognize that you have to price it within the confines of what comparable companies are selling at. The $17 or $18 wouldn't fly in terms of earnings per share and the then current market for an independent, relatively small oil company. So one way I thought of doing it was to offer the public a unit that would consist of different types of securities, such as a convertible debenture, some warrants and some common stock. The debenture, let's say, would be convertible to common stock at $20, the warrants would have an exercise price of, let's say, $17, and the common stock would have a designated value of $15 per share. The units would be offered at a price to raise the total amount of money that George wanted. If all of the debentures had been converted and all of the warrants would have been exercised and the common stock was then outstanding, all priced as a unit, the total amount of the offering divided by the total number of shares outstanding would equal approximately $17 a share. And that's the price which George wanted.

What percentage of the company's ownership?
The same percentage that we sold in the first place.

George still would have had his two-thirds?
He would have still had the same two-thirds. The difficulty with that was that the offering was not very large. The market was not very good and it was a rather complex offering. We knew that at the time, but that was the only way we could get George what he wanted. After discussing it with the people trying to sell the unit, including myself, we decided the best way to do it was to just simplify it with a straight common stock offer. So we did that. It was an offering that was not easily accomplished, primarily because it was a new company and the oil and gas business was not necessarily the greatest at that time. Still, we had a good story to tell and the offering came out. It was a successful offering. For the first five or six years after the offering, I think we had many splits and everything else.

One of the shares in the original offering became equivalent to 10 shares after splits and stock dividends.
Yes, a little more than that. But in any event, it was very successful and we were very proud to get it off. I can't say that the history of George Mitchell's company has been without its trials and tribulations, but on the other hand, it's one of proud things in my career. The fact that I'm still on the Board and

I built up a good relationship with the company and with the principals is something that I still am proud of after almost 25 years.

Tell me something about those trials and tribulations.

Mitchell Energy has been a company where the demand for capital was great. That largely accounts for its growth because it was able to take advantage of the various opportunities that were presented to it. Of course, the path of a company doesn't follow a straight line up with continued profitability and other things like that. There were times when the capital was short and we had difficulty meeting our needs. The original bank that Mitchell was involved with was the Chase Manhattan, the Chase Bank. The Chase became very upset because of the always-growing need for capital, and it questioned the ability of Mitchell to meet those needs.

That would have been, I would think, '74, early '75.

Yes, three or four years after the public offering. As a result of certain changes made in the method of operations, and I won't become too specific on it, we convinced the Chase to advance us the sum of an additional $5 million. But that left a scar. George and Budd and I and the others involved felt that they were drawing too tight a rein on us. As a result of that, we did change banks and we dealt with mainly, since that time, Manufacturers, and that relationship has been very satisfactory. David Rockefeller personally came to see George and woo his business back and George thanked him for the opportunity, but we continued the relationship with Manufacturers, which is now Chemical Bank.

Was First of Chicago involved?

Yes, as I recall. I remember personally Bob Halliday and myself going over to visit the Chase and suggesting that they should extend the line of credit. I'm not saying that both of us were solely responsible for that particular turnaround in the relationships with Chase, but I think it was helpful. Ever since then, I think, the relationships with the banks have been very, very friendly and very cooperative, and the problem with Mitchell from then has never been the inability to raise more money. We've always gotten a hell of a lot of cooperation from the banks. The offering in 1972 was the first and only public offering up until the last one which took place two or three years ago, when they raised a considerable amount of money.

Has finance been your main area in working on the Board?

No, although I think I contributed in suggesting the right firms to deal with and other things like that. One of the things I pride myself on is that the moment I came on the Board, although I represented a very fine firm, I recognized that there are conflicts in representing your own firm. For one reason or another, although I remained at Eberstadt and went back to Lehman Brothers again where I started out, until I started my own firm, I always felt that Mitchell Energy should have a great number of firms to consider. And they've never been wed to any one firm. They choose the best firms to do the job, and I assisted them in doing this. But I would like to think that my contribution to the Board was certainly more than finance itself. At the present time, and of course Shaker is included in this, I know more about the background and the relationships within the company itself and the people involved, and what they're involved in, perhaps than any other member of the Board with the exception of Budd Clark and George Mitchell. Now Bill Stevens is on the Board as the president of the company, and, of course, he's very well informed.

Have you and Shaker worked closely throughout your careers?
Not in parallel lines. We've remained friendly, but I don't think we've ever collaborated on anything after the Mitchell offering. Well, yes, we did, on one or two deals. I did some private placements when I was at Lehman Brothers for him and his firm. Yes, there were some other collaborative efforts. But not as close or as long lasting as our relationship with George.

What if the 1972 offering had not worked? It was no foregone conclusion that it was going to work.
No, not by any means. It was a tough sell. We recognized that because George came from Houston and Texas, it was very important to have a good demand in Texas for the common stock. And there was not that good demand forthcoming for one reason or another, and I'm sure there were various reasons. I personally went around to the Texas firms like Rotan-Mosle, First Southwest and Rauscher-Pierce and spoke to the principals about their attitude towards George and to explain to them why it was important and so on. As a result of that effort, and of many others, including Shaker's ability to interest institutions in the stock, we got the stock off, and it was a successful offering. Perhaps for a time it got below the offering price, but not for long.

Wasn't there a problem on the offering day?

Yes, it was very dramatic. We had to stabilize the market. Stabilize the market means that when you sell a common stock, you ordinarily sell more than you've got to sell. The reason for that is that many times people buy the stock and if they don't see it go up instantly, they dump it on the market in order to get their money out because they figure they can't make a quick dollar. So in order to take care of those temporary imbalances in the market, you sell more than you have to offer with the understanding, and the expectation, that you will buy stock in the open market to cover that technically short position. As a result of that, we did considerable stabilizing in that particular offering. Sometimes you don't have to. If you do have to, sometimes you lose money—because if the offering is very successful, then you have to buy the stock back at a higher price than what you sold it for. That difference is a loss to the underwriting group. In this instance we did not have a loss to the underwriting group, but we were able to maintain the offering price until the closing which occurred about a week after the commencement of the offering itself. We also had a very nice group of underwriters. Eberstadt and Coggeshall sold the bulk of the stock.

Were you co-leads on that?

Yes we were, and we ran the books. That means we were head of the syndicate, but the remuneration between the firms was equal, as I recall. Most importantly, the Texas underwriters did support the offering—not all of them equally, but a number of them sold a good deal of stock which helped solidify the feeling in the public that there was good support from the hometown, which is important.

Even then, I recall, it was a somewhat complicated story to tell, with The Woodlands and real estate.

That's exactly right. George had to convince them that there was some synergism between the two sides, and that's still a difficult story. On the other hand, some of the landmark achievements of the company have been in the real estate field, especially The Woodlands, which is a noble and very fine endeavor. Hopefully it will pay off in dollars. It's an outstanding example of what can be done by private enterprise, not only in the United States but possibly the world.

That's kind of interesting, using the company's assets that way when there's a quicker way to make money.

Yes, it is. Its ultimate goal is to increase the profitability of the company. But there is also a good deal of public benefit. You can argue that they might have been better without the real estate, but that was the situation we were in and we're trying to maximize the benefit of that particular investment as best we can. That story is still not finished.

It's getting to be more worth it, obviously.
Yes, but it's an unfinished story. It may take another 50 years to finish it.

In your experience in underwriting, would you say that the Mitchell offering was about average, more difficult, more challenging, what?
It was a difficult offering; certainly a great number are a lot easier done than this one was. On the other hand, it was finished and done and it was a success. It was not a hot issue in the sense that the stock immediately went to a premium of 20 or 30 percent, which many of them nowadays do. But that was a different era, too. A hot issue in a so-called IPO, Initial Public Offering, was not as frequent then as it is right now. Also, the oil and gas business never had the dynamics behind it that technology issues and other more esoteric issues sometimes do. This company really took off, of course, during the times when oil and gas prices seemed destined to go to $100 a barrel without question. That never quite happened, although people traded on that basis. As a result of that, we reached some astronomically high prices for both our stock and the products involved.

What are the challenges that you see now as a Board member?
I think that the challenges of the company are quite clear. We had the advantage of having negotiated a very large gas purchase contract with a very large and solid company, Natural Gas Pipeline Company, which supplies most of the natural gas to the Chicago area. That contract was negotiated at a time when they truly believed that natural gas was going to go to $10 per Mcf and oil was to go to $100 a barrel. So they were willing to pay us on an escalated basis, which would now be approaching $4 an Mcf while the market was in the $1 to $2 range. Now that contract has been ended and that foundation for cash flow of the company has been eliminated, although we received a significant amount of cash as a result of the settlement. But to replace that profitability is a very significant task. In addition to that, the involvement in real estate was more than only The Woodlands. A considerable amount of money was involved in a great number of less important fringe areas and as a result of the fact that we were not ready to develop these properties, it became apparent that the cost of carrying them was greater than what we'd

ever hoped to achieve in their sale. We've taken enormous write-downs in the last couple of years, to the magnitude of $200-$300 million dollars, which is a considerable amount of money. It was enormous and to fill that void we need to be meaner and leaner than we were before. That we can handle ourselves, but what we can't handle is the market price of oil and gas, which is still the most significant mainstay of our profitability. And to bring us to the levels where we were when OPEC controlled prices, where one year we made over $100 million, is going to require a considerable achievement. In addition to that, clearly our company has too much debt. We're taking steps to alleviate that burden, but it's going to take a considerable amount of time to do it. The challenges are there, but the solutions are there also. It's not going to be easy and it's going to take patience, there's not going to be, as we talk about it in Board meetings, a "silver bullet" that's going to correct the whole thing. We think that by chopping away at these problems and solving them as we are doing now, together with capitalizing on the opportunities within the company and also those which will present themselves on the outside, the company can again become a very significant company. It's still a significant company and one that has considerable respect in the marketplace. I think all of the things which are essential are there.

You mentioned how different the company is from what it was during the offering, and you've talked about some of those differences. Does anything else come to mind?

The real estate has changed considerably—we now are focusing almost entirely on The Woodlands and the sooner we can get there, the better off we're going to be and we realize that. The scope of our energy operations is broadened continuously, particularly in the field of natural gas liquids. And our ability to market: We're one of the significant marketers in the United States. And of course, the joint ventures, like the MTBE plant and the fractionation plants. We've had a big plant in Bridgeport, which helped us to expand the liquids business. We're much more of a basic supplier of hydrocarbon chemicals, raw materials, than we were in the past. We still have some transportation/pipeline facilities and, of course, our oil and gas production is still a pretty potent force in the company. But the magnitude in volumes hasn't changed quite as much as the prices have changed. At one time, our return on equity was one of the envies of the financial world. We were earning 35 or 40 percent on our equity. We're a long way from that now.

One of the things, obviously, that's happened as the years go by has been a change in management. Or has there been a change?
No, we still have our one guiding light in George. And George is still very much the chief executive officer. He and Budd obviously are getting to be a little older, as all of us are. I think we've had a good number of talented presidents of each of the divisions, and I think the ascendancy of Bill Stevens is a very welcome thing for the company. He brings a vast amount of experience, having been president of one of the more important divisions of Exxon, probably the largest corporation in the world and publicly owned. So he brings a vast deal of experience, plus the ability to lead men to execute what he thinks is the right direction. He's had the cooperation of both the Board and management under him, as well as George, in accomplishing these things. I think it's a very important step and a necessary step because, clearly, George being 76 years old now, of necessity will have to delegate his responsibilities even further. But he always likes to know what's going on and he will continue to know what's going on and he, of course, will continue to have the final say on what will ultimately go on. But we have, I think, an excellent Board of Directors, too—not because I'm on it—I think we have an independent Board and much to George Mitchell's credit, in spite of the fact that he has over 60 percent and clearly controls the voting power of the corporation, George could, with just a snap of his fingers, dismiss the entire Board of Directors. But he is a very intelligent man and a man who is capable of listening and he does listen to the independent Board, and we assert ourselves. In many instances, we do things that don't necessarily get a fully enthusiastic response from George. Many people on the Board would not sit there and be a rubber stamp.

Are there other things that we should talk about?
I don't think it's necessary to get into details any greater than we have, Joe. There are clearly things that have to be considered. Right now there is a considerable amount of talk on the outside about the desire or the willingness to split the company into two parts, one real estate part and the other energy part. Of course, we're very much aware of that and probably it would be in the best interest of the stockholders if this could be accomplished. But it has to be accomplished in a way that we're assured that each side of the equation, if you will, has the ability to stand and survive and prosper on its own two feet. It's certainly not being dismissed by the Board. We've got this as a possible solution, but we're nowhere near being in a position where we can say yes or no and have it happen in the appropriate fashion. This is very

much on our minds. We'll have two experts discussing this with the Board at the next day's meeting. Clearly, what we need is infusion of a significant amount of capital to reduce our indebtedness so that we will not be burdened with an extraordinary amount of interest payments, which of necessity consumes a great deal of our cash flow.

It means equity, doesn't it?

Yes, equity in some form or another. Equity can take many forms, as you know. We can sell more stock to the public under appropriate circumstances, or we can use our stock in further acquisitions. George is still reluctant, and understandably so, to pass control into public hands moreso than he has. The sole reason why we divided the equity into two classes of stock was for the purpose of using our stock to finance and to make acquisitions without passing control to other parties. I think we have to aggressively pursue this and I'm sure we will at the appropriate time. We've had a very great metamorphosis, don't forget, as to what happened in the last two years. We've sold an enormous amount of assets, we're still continuing to liquidate the non-core assets, and we're almost done with that job, but we still have some to go. And, of course, we'd like to do everything all at once, but we realize we can't do that and still do a proper job, but I'm sure we will become more active in improving our balance sheet by relieving some of the burden of debt.

With perfect hindsight, would you have changed the original offering in any way?

The original offering? No, no, I think that turned out to be a spectacular success. If I said yes, I think you might question my wisdom for being around for 25 years because I don't think there were very many public offerings that ultimately realized for their stockholders as much money as Mitchell investors. If they held on to their stock from the original offering, they would have 10 shares for every one they bought in 1972, which isn't bad. No, I wouldn't change the original public offering at all.

Walter, thank you very much.

(Walter Lubanko died in September 1999 while traveling abroad.)

This is Joe Kutchin. Today is Thursday, October 5, 1995. I'm in the South Shepherd Street offices of Jim McAlister and with Jim and his son, Jim, Jr. Let's get some of the basic data out of the way first, Jim. You came to the company when, and in what capacity?
1963, as assistant to the vice president, doing oil and gas evaluation and banking.

Which vice president?
Leon Lefkowsky. I'm a petroleum engineer, master's in petroleum, master's in finance. The company had no formal budget at that time, so I started a budget, which was picked up then by Bill Houck, who completed it. The assets of the company totaled $250 million dollars or less, with heavy debt. George Mitchell was in the process of acquiring acreage in Montgomery County at the time that I came in.

What was the name of the company then?
George Mitchell & Associates.

You left the company in what year?
I left the company in 1972.

And since that time, you've been in the real estate business?
I've been in the real estate business with my own company.

And prior to joining Mitchell?
I had been a petroleum engineer with several other companies.

And your degrees are from where?
Oklahoma University and the University of Houston.

Tell more about what you found when you joined the company.
Small company, high debt, absolute necessity for borrowing. We warehoused our loan at the Bank of the Southwest and did our master loans in either Chicago or at the Chase Bank in New York. It was a time when George was putting together the company and the company had to have debt. He was able to walk that line and to create the cash flow. During the time I was there, we

bought out a number of the older investors, the Smiths, the Hirsches, the Pulaskis, the Weingartens and Ann Alexander. George had already bought out Merlyn Christie, who was still officed there–in fact, one office down from me. Merlyn was part of the old Christie, Mitchell & Mitchell. George and Johnny Mitchell, in the early days, had a tag team. Johnny was able to go out and bring in the investors with a flair and George had the capacity to find successful drilling locations and come in with production. He was very successful for his investors.

Did your initial responsibilities involve real estate?

Well, there wasn't a dominance of real estate at that time. There was some red flag development happening at Cape Royale or West Magnolia Forest, some development at Pirates' Beach, but real estate was small. It was something that was in the corral, but it was certainly not an elephant. That part of the business got started in a big way with the acquisition of the Grogan-Cochran Lumber Company and its 50,000 acres. This acquisition was brought to Mitchell by Don Brooks, who was with the company, and a fellow by the name of Max Newland, who was an outside broker.

What year are you in now?

This is right at the beginning of my start with the company, early 1963. It was brought into the company under terms where Mitchell bought out the dissident factions within the Grogan and Cochran families, which owned the corporation.

This is the Cochran family?

This is the Grogan-Cochran family. The Grogan-Cochran Lumber Company was a timbering operation on 50,000 acres of land in southern Montgomery County, part of which is now The Woodlands. Don Brooks, Max Newland, Charles Lively became involved in helping with some additional acquisitions, all very capable people who did a wonderful job for George. The purchase of the Grogan-Cochran company stock was done by giving them a down payment and a purchase mortgage. Money was then borrowed at Great Southern Insurance Company which funded the amount of money required to buy the Grogan-Cochran stock. So there was no cash money involved from George.

Why did a reasonably successful oil and gas company want to get into 50,000 acres of real estate?

That started before I got there. It was a heck of a buy. You're talking about $125 an acre. So George had one terrific acquisition. George is a man of vision, and he just saw it. He's always been perceptive. I think it was George's vision of the coming growth of Houston to the north. He saw it as an opportunity, it happened to be in real estate, and he seized the moment. That is the long and short of it.

Is it your belief that at that time George had some idea of a major residential development?

No, my belief is he did not. In fact, I'm certain that he did not. He saw it as a great investment property. It also had timber on it which could be harvested. As it turned out, Morris Thompson, a Mitchell vice president, returned from a meeting with Georgia-Pacific with a deal where Georgia-Pacific would purchase timber from us. The contract called for a 12-12 timber cut (12 inches above the ground, 12 inches in diameter), which is a healthy cut for the forest. This is called a real estate cut. We were able to get $5 million for that cut, so now George is sitting with only $1 million invested in 50,000 acres of land, which makes quite an astute investment.

What you're saying is that he immediately had income of $1 million?

No. Five million dollars from the cut of the timber. Great credit goes to Morris Thompson for bringing in these people. Morris and I visited and then we talked about how to do it. The result was that the sale was consummated. At this time, I became more and more involved in company investments. I had been put into the position of director of planning and economics for the company. Part of the expertise I had to contribute was in investment decision-making. I set up profit standards for investment guidelines to begin a more sophisticated way to allocate capital for investments. That was an approach that I was trying to bring into the company.

That's solely in real estate?

Anywhere. There just weren't any defined standards up to this time. Everybody was flying by the seat of their pants. You have to remember that George has a lot of unusual qualities. Perception is one of them. But another one is a terrific capacity to recall. Therefore, George is comfortable flying by the seat of his pants. He carries files in his head that most people don't remember are in their file cabinets. The company was operating as a projection of George's personality. Well, we had to come out of that, we were growing, we were getting bigger, we had to have a budget and better defined investment criteria. How would we pay back the loans that I was having to do at the Chase? Where was the money going to come from?

You're no longer in 1963, are you?

We're in 1963-64. We're still in very early times. One of the things that Mitchell had started to do with some of the land purchased from Grogan-Cochran was "red flag" land development at West Magnolia Forest and Clear Creek Forest. The company took acreage tracts, divided them into smaller parcels, red flagged them and sold them on terms. One of my jobs then was to market the notes received from the sales. I would go bank the paper so that we would have money to do more developing. These were red flag developments for moderate income housing. Mitchell was selling one to ten acres at a time. This was not by any means a Woodlands concept. The question then came, "What else can we do?" I was directly involved and one answer was that we could possibly develop a "satellite city." We had some 3,000 to 5,000 acres, I don't remember the exact number, in one block close to Interstate 45 just north of the Harris County line. That tract was the core of what later became The Woodlands. Since the growth of Houston was going that way, the thought was to try to develop something up there to capture the growth. But we had not yet considered a concept of a totally integrated new community, and had no concept of what is now The Woodlands.

Let me catch up. The 3,000 or 5,000 acres on I-45, was that part of the 50,000?

Yes. The 50,000 acres was in several blocks, including some off of Highway 59, which I'll get to in a moment.

Didn't Mitchell own some acreage along 1960?

He owned some acreage south of 1960 at Imperial Valley, he owned some right at the blimp base on IH-45. At any rate, Mitchell had this core acreage off of IH-45, I can't remember whether it was 3,000 or 5,000 acres, certainly not 20,000. Then as things began to progress, it became apparent that this acreage was going to be a place of major development potential, and that presented all sorts of options. We had not conceived yet of what you now know as The Woodlands. In fact, out of a conversation one evening later on came the idea to do a satellite city.

Do you remember who was in that conversation?

I think just George and I were there, but this is something that George had in his mind, it was one of the options, and so we sort of focused on the satellite city idea. At first, throughout the company we called this project Satellite City. Few people in the company thought it would work. George and I thought it was possible to do something large on the property.

This was even before Columbia, Maryland, was done?
No, Columbia was already there, and that was one of the resources we used later on. Reston was in place, and that was another resource. In this initial process, however, we called this new development Satellite City. It was going to be a bedroom satellite of Houston. So, I started working on that. Few resources were allocated to the project. Initially, I worked on it in my spare time, period.

Can you explain why at that time the company would devote resources to something that's 30 miles from downtown Houston?
It was the next best place that was going to be developed as Houston grew to the north and that's one of the things that really sold the deal to HUD. I was deeply involved in the negotiation process with HUD and found they easily agreed to the quality of our location. Mitchell still had a shaky balance sheet at the time we went for the HUD loan guarantee. The thing that sold the loan guarantee was the fact that the location was so good. Even if Mitchell had a problem, someone would pick it up and go with it because the location spoke for itself.

You're talking about the loan guarantee with HUD?
Yes. In the ultimate analysis, location is what allowed us to get the loan guarantee, because Mitchell's statement was still heavily debt burdened.

It still is.
George is able to handle the pressure of heavy debt. That's his style. George's thought was that we needed to put together a larger acreage development. George and I were working on this by ourselves because nobody in the company thought a new community development was a possibility for us. And I will tell you, people made fun of us, and of George, during this process because no one thought this thing would fly. It was kind of a far-fetched idea.

People outside the company?
No, people inside the company. It was an interesting time because we didn't know it would fly. What did we know? Here's just an idea we're going to try. Of

course, from hindsight you see it happening, but from where we were then, it's a shock. And who knew how to build new cities? There wasn't anybody. There was one in Columbia, there was Reston, there was Irvine Ranch, there was West Lake, there were some large acreage developments by Avco, some developments by McCullough, but who knew about it? I was 30 years old and we knew as much as anybody, because nobody knew a great deal about developing a new city.

And your background was energy, finance?
Finance, energy and investments. There's not a broad gap between oil and gas logic and real estate logic in terms of the investment side of the equation. The result was that we just kind of trucked along. Dan Daniels, who was a friend of mine with Champion Paper, and I began to visit at the request of George. George's idea was that we should think about trading some really great timberland we had near Cleveland for some land that Champion had surrounding his acreage near IH-45. Champion had a sizable acreage block next to the Mitchell property adjacent to our core block. Champion is not in the real estate development business but in the timber business. After some period of time, and I don't know exactly what it was, we traded acreage we owned off Highway 59 around Cleveland, which was less valuable, for the more valuable acreage adjacent to the Mitchell core tract. The trade was 2.8 acres of Mitchell land for each acre of Champion land. But for Champion's purposes, they received the better timberland and gave us the worse timberland. So we traded away part of the original Grogan-Cochran purchase for the Champion land off IH-45.

Do you recall how many acres it was that you picked up?
I don't recall, but it was all of that Champion block off IH-45. You can go back and look at your map and you would know exactly. Then there were several other transactions that occurred that I think really brought the block together. A fellow by the name of Charles Lively has to be absolutely complimented with regard to one of these acquisitions. If anybody deserves a hero badge, Charles Lively deserves a hero badge. He was assigned the Taylor Survey, which had been subdivided into small tracts. He had the task to go out there and acquire all of the small tracts. There were over 1,000 acres. Hundreds of one-acre tracts had to be purchased one at a time, and Charles dealt with these people. He has such a personality and warmth about him, he was able to get it all done.

The story is, as I remember, there were 300 transactions required to put The Woodlands together. Were those mostly what Charles Lively was doing?
Yes

Is 300 a good number, would you guess?
I would guess at least that. I would think there would be that many in the Taylor Survey alone. Then there were a couple of other interesting transactions. George Strake had a piece of property, and a fellow by the name of Don Woucash had another. At that time, Woucash was a physician on DeBakey's staff, and Strake was from a well known oil producing family in Houston. At any rate, he felt he had to deal personally with George Strake and with Woucash. These were tough negotiations and George had to pay a premium.

You say he, is that Lively or Mitchell?
It's Mitchell. Mitchell pretty much had to deal with these people directly, it was a situation he had to handle. Lively was involved in part of it, I was involved in part of it, but with Strake and with Woucash, they focused in on George and that's where the deal was cut. So he got those two tracts. The purchase that took the longest was a tract towards the front. I have documentation in this file of how it happened and the numbers. It had frontage of 2,545 feet on IH-45 and was critical. That was the Sutton-Mann transaction, and it took a long time, tough negotiations. The front man for Mitchell was Lively, but Mitchell was heavily involved. That had to be dealt with from the top. That was a deal where George had to be personally involved. Sutton and Mann were from Woodville, Texas. They were bankers, and they were tough dealers on the property, and they knew George had to have it.

All of these transactions were with a view towards creating Satellite City?
At this point, we have not moved into a mode of a Satellite City. We're still not yet at The Woodlands concept.

The idea of a major development was pretty well established. What period of time are we in now?
Mid '60s. I have some dates of the Sutton-Mann deal. This Sutton-Mann deal went on for a long period of time. I have some exact documentation. I have reports and files with the chronology of The Woodlands here, I can

document each step of it. So now we have bought the initial core tract from Grogan-Cochran, the Taylor tract is being worked on, we brought in Woucash, we brought in Strake, we brought in Champion and are working on the Sutton-Mann. There was still another tract right where Needham Road intersected IH-45, on the west side of IH-45. It was an "L"-shaped tract fronting I-45 with the long end of the "L" along I-45.That property came in by a fellow named Homer Jackson, who was a broker. The agreement George and I had allowed me to buy investment land with my family and others. My agreement with George from Day 1, when we were doing this, was that I could buy some properties but I had to offer them to him first, which I did with no exception. In this case, he looked at the "L"-shaped Needham tract and said, "It's too expensive." I said, "Well, then I think I'll buy it." It wasn't that large of a tract, 30 or 40 acres. He said, "Well, all right." The next morning he called me in and said, "No, I don't think so." So, he bought it. Well, that's the tract that's now the Needham Road exit. Subsequently, I did buy a 50-acre tract just below that tract, where they are now building the SCI Funeral Home. I bought that tract, held it for a number of years, then George bought that tract back from my partnership. After I left the company, there were some additional acquisitions, I think, down towards Tamina but that's beyond my scope of knowledge. But that's where I left the scene. I had also personally bought 1,500 acres of land south of Spring Creek, west of Kuykendahl, fronting on Spring Creek. George subsequently bought this land from me. This acquisition extended The Woodlands down into Harris County.

How many acres had been accumulated at this point?
As I remember, at the time I left, there were about 17,000 acres. I'm giving round figures. I have in my personal files some exact numbers. I have exhibits which show our first economic models done at 7,000 acres, and I can show you the progression of our economic models. When we started to do the development, I was "it" for this project. There wasn't anybody else. We had real estate recreational properties, which was run by somebody else, and I was running this project for George.

There's nothing yet in Washington about federal support?
No. I will tell you how the federal support came in. You've heard it said that consistent luck is indistinguishable from talent, and we had a lot of luck. In the early days, we attempted to find financial sources through joint-venture partners for the development. I remember one meeting—and I think it was with the

Pick Realty people. We came in with absolutely the sorriest presentation that you ever saw. We just put together some information on the land with economic projections that weren't really thought out. We were trying to get them to come in to our office and, of course, recognize the obvious value of the land and the potential of the site and then step into it with us financially. Then we would undertake the detailed planning. We were trying to do just the minimum amount required because there was no money allocated. I am also doing a number of other projects. There's not even one person full-time on the project at this point. So, we tried unsuccessfully to get interest from possible joint-venture partners and finance sources. We simply had nothing to show them, we had nothing to give them. After one particularly embarrassing meeting, it just became obvious we had to do something. In this case, we called in a fellow by the name of Cerf Stanford Ross. Cerf Stanford Ross was an architect without a degree. We made a $50,000 commitment to utilize his services on the project at this time.

This was mid-late '60s, right?

Late '60s. We called in Cerf Ross and he was a person who bridged the gap. He was a Houston person, and a well-traveled individual. He had recently been involved in New Mexico, I think, or Arizona doing houses for Indians. They were bullet-proof, beat-up proof and bang-up proof. That's what he did and he did it with some government grants. Cerf developed the first land plan for the development of the project so we could go to Pick Realty and others who could finance the project. I have that plan here in my file. We were also working hard with First Mortgage Company of Houston to develop other financing sources.

Tell more about that, please.

In developing an interest with these potential financing or joint-venture sources, we did get some indication from a couple of them that they would be willing to provide financing. However, George would have been personally liable on the debt. The main thing that bothered me was that they had the right to quit funding at any time in the development process. If they can quit funding, then all of a sudden they can sink the ship. Then they have the land and they have George. It didn't make any sense when they had the right to pull the plug and end up with a reward, so we couldn't do those ventures. Therefore, we moved on developing a plan, really not knowing where we were going, just moving forward. We have a concept, we're moving forward. There's no road map for what we're doing. I'm 30 years old, George at this point is in his 40s, and we've never done a project like this before. There aren't many experts available to phone and say, "Come do this for us." We're out there by ourselves. One day, Cerf came into my office. Cerf, by the way, would be an interesting person to interview. He had just found the Title VII Program created by the New Community Act. It was initially Title VII, then they later changed it to Title IV. He found this New Community Act would allow the federal government to lend up to $50 million to a project like ours. We discussed it with the president of First Mortgage, and Travis Traylor. Everybody we talked to said, "Don't deal with the government." So, we were reluctant to go to Washington. Finally there were not any alternatives. Cerf and I caught an airplane, my first trip to Washington, D. C. We got the basic information.

Was this the Nixon administration?

Yes, Romney was in office as the head of HUD. I can tell you it's Romney because that's who my first contact was with. In fact, we had George write Romney a letter with regard to our interest in the New Community funding program. They were prepared to fund a new community in Texas at Flower Mound and two in Georgia. The people in Washington at HUD were, however, more interested in social results than economic viability. During the process, which covered several years, HUD began to close the gap on what they needed in terms of environmental information and financial considerations. In the early stages an extensive environmental study was not required. But our work with The Woodlands turned out to establish environmental standards for all the HUD new towns.

The organization at this point still was you?

I have one other person with me at this point, a fellow by the name of Autra Ayers. We were just going to computers. Autra had an education that did not reach high school, but he was a very bright, committed individual and a hardworking person. He had been with Mitchell for many years. He was absolutely a very bright person. Well, he was assigned to me to run numbers for our economic model. We were developing a computer program to run the economic model. There were no software packages. So, I designed a program to be able to do an evaluation of a new community. As we were programming, Autra did the calculations and spreadsheets by hand so we would know that the program was right. We had massive data. I have a copy of what we did here. I mean, it was hours and hours of calculations by Autra.

Was all this with a view to working a deal with HUD?

No, we started this before HUD. If we were going to get anybody interested in the community, we had to have a sense of plan, a sense of direction and some financial concept of where we were going.

OK, so you had Autra Ayers and on the outside you had Cerf Ross. No other consultants, no other staff?
Correct. Now, I'll show you this document. I regularly wrote a memorandum on the status of real estate to Mitchell in which you'll see the flow of how The Woodlands developed by date order. Anyway, I was in Washington meeting with HUD, when a fellow named Brace, one of Romney's assistants, called me aside to talk about planning. My memorandum relating to this conversation to George changed our direction. We then committed more funds and manpower and hired strong consultants.

1968, 1969, thereabouts?
It's after 1970, because it's in my files. I can't find my files pre-1970. At any rate, Brace said, "You need something stronger, the new community program at HUD has moved to greater sophistication in Washington." I came back and I called a fellow by the name of Glen House, who is an architect friend of mine, a Houstonian. Glen gave me a list of architects and planners to call, one of whom was Bill Pereira. The Glen House list is what we went from to find the new consultant team to plan the development. We had absolutely no idea of who would be a strong consultant. We initially tried screening them from Houston. Now we're beginning to form a team. We now know we were going to have to go to some serious planning. A lot of things I am, an architect I'm not. We had to have an architect on board, so I found Bob Hartsfield. Bob was with CRS at the time. He came up and met George. George and I were leaving the next morning for California to interview a number of architectural firms and consultants from the Glen House list. We needed somebody to help us grade these architects. I couldn't do it, George couldn't do it. George said, "You're hired, but you need to be on the plane with us in the morning." Bob Hartsfield was a wonderful asset in selecting the team to plan the project.

Through CRS, had he done some work for you, or some consulting?
If he did, it was minor. I had the contact with Bob. I cannot tell you how that originated.

He talks, as I recall, about some meeting, probably at Reston, where he was not an employee of the company but a consultant.

No, his employment happened here in this office. It happened in George's office when he was hired and he committed.

But he'd had some business with the company prior to that?
I don't know. He came through a contact that I had with him and I don't know what that was. I can't remember. I introduced him then to George and said we ought to hire him. George said, "Go for it," and off we went. There were two other people who became involved here. I may lose my time sequence a bit. I was given some additional help, as we began to run the financial numbers, through the assignment of Bob Wyrick to me. Bob Wyrick was a draftsman with Mitchell. He had gone 13 years to night school at U of H and would graduate at the end of the semester. Bob is an absolutely bright, bright man. In drafting he worked for Charlie McGann. His work was superb, his work ethic was impeccable. In fact, you'll see some notes in there, I wrote George notes every time his grades would come out—he'd have straight A's. This is a quality guy we have in Bob Wyrick. In a period of one year's time, Bob bridged the gap and he was now a considerable expert on doing financial evaluation. Bob came in, I think, before Hartsfield. The first person I employed from outside, however, was a fellow named David Hendricks. David Hendricks was a brother of my good friend, Clint Hendricks. I had told Clint I was looking for someone to come in to be my assistant.

Do you remember when you hired David?
I think I've got it in my notes. It must have been 1970. David came in as my assistant because of the recommendation by his brother, Clint. I had never met David before we met for an interview. He was a very likable fellow, and a very personable individual. As it turned out in The Woodlands process, we gave him the assignment of the infrastructure and dealing with some of the social things. That's where he was focused. He had to work very closely with Dick Browne's group at Dick Browne and Associates in Columbia, Maryland.

Dick Browne's group had already been retained?
I've kind of skipped, so let me go back. I'm trying to fill in two gaps. First was Autra Ayers, then there was Hartsfield, Bob Wyrick and David Hendricks. Plato Pappas, an engineer for the company, began to be involved as we began to need cost data for construction. That pretty well is the complement of people we have going into our HUD negotiations. Prior to this staffing, we

had used Cerf Ross, who had sent us up to Washington. Now his time was over. You're going to have to say something about Cerf Ross in the history of The Woodlands. He put us in Washington, D. C., and the new community funding. Without that, there would be no Woodlands, because there would be no other way that we would have ever been financed. So, Cerf Ross deserves a place of note in here, and he did it in a way that allowed us to continue. You see that plan? What does it say on there? What's the name of the community?

Woodland Village.

Woodland Village. You know how that was named? We had to go to Washington in 1970 after an all-night session where everybody was working. Cerf was trying to get all of the exhibits prepared. We had a 7 a.m. plane to catch in the morning. He called me about 2 a.m. in the morning. He was now ready to print. HUD had been very specific. They did not want a satellite community, so the name Satellite City had to go away, because that was exactly the wrong word. Cerf said, "We've got to have a new name." So at 2 in the morning, we just bounced some names and we came up with Woodlands Village. He did the thunderbird logo, printed it, and off we went to Washington. No talking to anybody, it didn't make any difference because all we needed was a name for the presentation. We could change it any time we wanted to. So, the Woodlands name really came from a 2 o'clock in the morning conversation back in 1970. We went to Washington and submitted the plan you see here with that logo back in 1970. OK, about the planning team. The planning team came together, partly out of Glen House. We got Ian McHarg. McHarg, Roberts and Todd, at that point, was the leading firm with regard to the environment. McHarg had written several books that woke people up to ecology. The word ecology was not well known. Ian McHarg made it known.

Hartsfield was a student of McHarg's, wasn't he?

That's right. Hartsfield brought in Ian McHarg, who is an Englishman. We made a very good friendship with Ian. Very personable man. We liked each other very much. Ian brought in a guy by the name of Jim Veltman to be his project coordinator and director. Jim Veltman is probably the man most responsible for the layout of The Woodlands. Bill Pereira, who was supposed to be the lead land planner, in my view, had minimal impact.

Tell more about what you mean when you say layout .

Jim Veltman laid out all the ecology and land characteristics, and then the final plan laid out on top of that. Bill Pereira, who was an architect and planner from California, was also part of our consultant team. He had been involved in the Houston Center. Now this was very important. I suggested that the team design to the fullest extent of their imagination. Put in infrastructure to the fullest extent of their imagination, conserve the forest, and then we're going to put the options in the economic model and work backward. The idea was a simple one. That's why I did it this way. If you stifle somebody before they start, they won't think all the way through. So what you want them to do is to think all the way through and then throw away the less important and unworkable ideas and end up with their best thoughts. McHarg called me "Cash Flow" because everything had to go through the model. Pereira was at one meeting at George's place in Montgomery County, his ranch. George was not present there. We had a whole team meeting. Pereira said he was going to quit unless he could take over the economic model. There was no way. The team must design to the economic criteria and that's going to be the goal. The project is going to have to be economically viable, and it has to be able to exist independently from other Mitchell activities. The planning team is going to be graded at the end on economics. If it doesn't make sense, it doesn't work.

Pereira was brought in through your trip to California?

George and I hired him in California, yes. Then came McHarg, because of Hartsfield. I forget how Browne came in.

He had done work for Columbia.

Browne was the infrastructure person there. That's specifically what he was there for. Veltman and McHarg were ecologists, the infrastructure was the niche that Dick Browne was going to fill. Dick Browne had a very bright young man there in his firm named Bob Tannenbaum. Tannenbaum was working for Dick, and Tannenbaum designed an intricate, well thought-out infrastructure, both people-to-place, people-to-people, the whole thing. David Hendricks relayed it back to us. Bob Tannenbaum did the job with Dick Browne. David Hendricks took what Tannenbaum had done and integrated it into the HUD presentation.

Browne had not yet moved to Texas?

No, Browne was still living in Columbia. He, I think, was very important to The Woodlands in helping us establish some relationships.

I'm sorry, "he" is who now?

Dick Browne. We didn't design the infrastructure at Mitchell. Browne invented it, we used it and David Hendricks was in charge of using the work that they had generated. The long and short of it is that Dick was very important in bringing in a lot of the pieces because of his experience at Columbia. We met with Jim Rouse. Jim Rouse was very giving. We sat for hours with Rouse and with his staff. The long and the short of it is that Rouse was very helpful, along with everybody else. The principal point that we found from talking with all of these new community people is that they had all had financial problems. Every one of the new communities had financial problems. Columbia's financial problem was that they built a little village and then they went way out somewhere else and built another little village, and then way out somewhere else and built another little village, etc. They had this huge infrastructure cost. You couldn't sell enough lots to pay the interest, let alone start paying principal, and therefore they had serious problems. There was just too much front-end cost. So we had to make The Woodlands financially viable. For sales purposes, you had to radiate out of a core in concentric circles to maintain expenses and develop a sense of community for all those people who are there.

Ivins was not involved at this point?

No. Ivins did not get involved until toward the very end. I was going to leave Mitchell. I was not into construction. That's not what I liked to do. My time was just coming to an end.

At this point are the majority of your efforts spent in preparation of the application to HUD, or are you still doing economic feasibility?

We're doing economic feasibility and dealing with HUD, not knowing whether HUD's our answer or not. We're keeping everything open. We're staying in touch with HUD. We're moving on all fronts now because the effort for HUD is consistent with what we had to do anywhere else. But the team is set up now, specifically for a HUD answer. That's why Pereira is in place, that's why McHarg is in place, our focus is HUD but not exclusively. We didn't know the financial answers ourselves. We didn't know whether it was an economic go or not, so we may not have needed HUD if it wasn't an economic venture. We still didn't have the answer that it was a go for us. But we are moving forward. In other words, we don't know exactly where we're going to end up, but we're going forward which, knowing George, is not hard to understand because that's what he does. He's a very courageous man and he moves forward. At any rate, we began assembling the package from the groups that we had. As it turned out, Senator Bentsen was a good friend of my wife's family, and my brother-in-law was in Washington as the Senator's aide. I went to my brother-in-law and to the Senator, and we started moving forward politically. Then my brother-in-law, Denman Moody, Jr., helped us get Senator John Tower involved. So then we had Tower and Bentsen involved. We were getting closer and closer. HUD had no economic models. They were doing Soul City, they were doing Peachtree City, they were doing Flower Mound, but they had no economic control.

Are you saying they had already committed guarantees?
No, they were looking at a few projects closely. So we took our economic model, the one that Autra Ayers and I had developed, to HUD so they could have a tool to be able to evaluate the projects they were considering. You will be able to see in the data that you have a paper that I wrote which showed them how to use the economic model and evaluate a new community or any community from a financial standpoint. That also came at a time when HUD and VPI had asked me, Jim Rouse and Robert Simon to speak to various people in the country to try to get other corporations to put money into new communities. My part of it was to talk about how do you put a new community together financially and make it viable. Robert Simon dealt with a lot of the social structures that were involved and Rouse dealt with how do you work with all of the authorities, such as zoning commissions.

Did your model contemplate 20,000 to 25,000 acres?
The last one was about 18,000. The first one was 7,000 acres. I have the exact submittal to HUD right there in those three books. We had the consultants but our Mitchell staff in here was very small. It was Bob Wyrick, Bob Hartsfield, Charles Kelly, Louis Brasher, David Hendricks, Plato Pappas and Larry Mahost. There were three secretaries. Again, it was Bob Wyrick who carried the big work task. Hartsfield did a lot of the drawings, architectural kind of things, but he also did one thing that was extremely important. He set up a critical path flow chart. He had control of where we set priorities. His chart told us, here's what we need to do and here's the time line. Because of his talent, he was able to put our work tasks in a prioritized time schedule and framework which allowed us to accomplish some things that we would not have otherwise. That was an extremely important contribution of Hartsfield. Hartsfield, as we began in the later stages to go to HUD, also dealt with a lot of the architectural things with Tony DeVito, who was

with HUD. He took Brace's place. You're talking about only a few Mitchell people on the project, but some very capable people. The hours we worked were humongous. In fact, my family throat doctor was at the building next door and we would work, we were tired, we would get sick. In the winter time, we all would have colds and congestion and his cure for your sore throat was gentian violet. Have you ever had that? They paint your throat with purple stuff and it hurts so bad, tears come down your eyes. We would go over there and keep ourselves going with Dr. Phillip Spence. When we were sick, everybody would go over there and get his shot and get his throat painted. We spent a whole lot of time. It was not a one-person deal. You're talking now about a team that's really working together, functioning together, one that cares. You're talking about a commitment of people who care for George, from A to Z. In my department, one of the ways we kept people stimulated was through personal exposure to George. When Wyrick or any of the other guys completed his phase of a study and it needed to be reviewed, he would present it to me and I would challenge him to the extent that I could. George is a very challenging man and I was preparing them to personally take their work to George and present it to him. I didn't present it, I sat on the couch and watched them present it. Sometimes I had to bail them out, but the point was, these people knew as they were working that they were going to get credit for what they did because they went directly to George to present the work that they completed. It is tremendously inspiring for them to be in front of the man. George does that to people, and with that inspiration he really did a whole lot of things with the very few people that were on staff. We came down towards the very end of the negotiations and George had never been to Washington for this project. They wanted to see George so I said, "George, you've got to go." George got on the plane, and we introduced George to HUD and it was nothing more than "Hello, how are you?" The second time that George went up there was for the final submittal. We are now ready to submit. We've done all of our work. By the way, in that interim we brought in a fellow who was supposed to be a big Washington lawyer. He, as far as I know, did very little compared to what Vinson, Elkins and all of our other people down here had done. He was supposed to come in and help out. I don't see where he made an impact.

We're in about 1971 now?

Yes, we're in 1971, 1972, and again, I can take you back in my files as to exactly what happened and when. At any rate, at that time the largest loan guarantee that had been given was $33 million to Flower Mound. If you worked out the comparative numbers, we deserved more than $50 million on the basis of what we had. So, we did a spread sheet for a submittal to hand out so people were able to see how HUD had funded the other new communities. It was very obvious that we qualified for the higher number, that $33 million was correct for Flower Mound, but on the same basis we would deserve something in the range of $50 million. George did not feel comfortable with going to $50 million. He did not feel comfortable, and we had many meetings about this. I was very positive, the rest of the staff was very positive and we finally prevailed, "Let's go for $50 million." So now comes the hour. We go into HUD. A fellow named Nicholson, who is Romney's assistant, sat in at the meeting. My brother-in-law, Denman Moody, was there representing Senator Bentsen. Senator Tower's representative was also present. Denman was very helpful to us in Washington, behind the scenes, and all the time he was doing things for us. At any rate, the time for the submittal arrived. The way that I had it set up was first, to have Jim Veltman get up and go through the ecology, then second, we went through the social part, which was Hendricks's, so he made his presentation. This was followed by the presentation of the development plan with all exhibits. At this point, we had never told them how much money we were going to request.

George was present?

George was present. This was his second time at HUD. George was nervous all the way on the airplane going up there about the $50 million he wanted. At any rate, my economic presentation was the last part. So I went through the economics and was the last speaker. I had to close and state our request. I said, "And, therefore, for the reasons you've just heard, we respectfully request $50 million." Total, total silence for the longest period. Denman tells the story over and over. Mitchell begins to look around and looks at me like, Oh! Oh! He gave me the Oh! Oh! look and I was standing at the podium and the silence just went on. I didn't say one thing. If you know about sales, you know that the first person to speak, loses, and I just stood there and thought, "God, I'm going to get fired, my life's over, it's done." Then Nicholson speaks up and says, "That was a very fine presentation and we will take it under advisement." That's all he said. Then they had some questions as they went through the process and, ultimately, we got the loan guarantee. At this point, Len Ivins had already come to Mitchell and I was phasing out of the company.

Can you put a time on that?
That was 1971 or 1972. So, at any rate, Len came to Mitchell. We were now preparing to go into construction. I stayed with the company to assist in the transition.

I'm trying to get straight. Do you know when Len came and HUD said yes?
We knew HUD was going to say yes, and I think Len was instrumental in bringing in Jones, the Washington attorney.

Just for what it's worth, I first started coming down from Chicago in the first quarter of 1972, approximately. The HUD grant had not been approved yet. At that point, Len, I think, had not moved yet to Houston.
That's right.

Jim, was one set of economics submitted to HUD and then the Columbia people came in and the operations were based on another set of economics?
My understanding is yes. Again, I don't know what's right. The numbers, as we saw them, were much more conservative than the numbers as Len saw them. They based their decisions on the numbers as they saw them, which is all you can do, and it turned out that those things didn't happen.

Before I turned the tape recorder on you said something about a gathering of ex-employees.
Yes. I will tell you an interesting thing about George. Gladys Halbouty, who is an ex-Mitchell employee, called me earlier this year. She wanted to put together a group of ex-Mitchell employees. You had to be "old" and you had to be "ex." A large number of us met for dinner about a month ago. I'm talking about people who were with George in the early days, where they were sharing offices, having to swap seats. I'm talking about the early, early, early guys and girls. What started out as a conversation, like "Let me tell you a funny story about working with Mitchell," people would add, "Let me tell you the benefits I received." So we went through about two-and-a-half hours of an absolutely marvelous testimonial of what Mitchell had meant in each of these people's lives. The shame of it all was that it was not recorded for George Mitchell to hear. If I ran my company in such a manner that the people who worked for me would get up and say that for me, I'd say I've really

done something right. There wasn't anything negative. George has impacted my life. Because of George, I made millions. I was a very young man when the project began to grow and began to be recognized as a significant project with great potential. Others in the company would have liked to take over the project. George and Budd Clark stood between me and them during the entire time and never once pulled away their support. Not one single time. And because of that, I would fight a bear for Mitchell in the parking lot, and these guys in this "Mitchell Ex" meeting would do the same thing. That was the kind of environment we had. We liked working for George, he was inspiring, he gave us our challenges. When you did something today, that was forgotten, you had to do something tomorrow. He wasn't long on giving personal credits, but he was long on giving inspiration and it was awe-inspiring. But we came in and pulled The Woodlands off with a young group of people. One guy had no high school education, one guy 13 years in night school to get a degree, nobody very special, just dedicated. They were very bright, so they did have that common denominator. There's another guy who needs to be mentioned–Louie Brasher. Louie Brasher who was a friend of Maria Ballantyne's son, she's George's sister, came to work for me in the summer. He did his graduate work, master's degree at Stanford, and then came to work permanently after completing his graduate work. I kept him informed about the project while he was at graduate school, so he stayed up to date. This guy was brilliant, there's no other way to describe him, he was just a brilliant individual. I should mention Charles Kelly as part of the team we had assembled. Ivins came in and cleaned house with the exception of a few people he just clutched around him. Everybody else went away. Louie Brasher went to Gerald Hines. He went to work with Gerald Hines as a project manager, moving up in the ranks very quickly. He was in an automobile accident in San Antonio and died.

George continues to speak very highly of Ivins. He regards him as a visionary. Anyway, I remember that we went public in February of 1972, and the HUD approval came about two or three months later.
The Woodlands has had a great impact on Mitchell and I will tell you one other thing. When we went into the final meeting to decide on the commitment to go forward with HUD, George did something that was very interesting. We had three alternatives to develop the city.

Which final meeting are you talking about?
The final meeting to decide, are we going to commit to developing The

Woodlands. It's now committing time. It's simultaneous with the HUD approval. We either had to go left or right or up or down, we had to make a decision. The new city option was there, and there were other options: One was to make an FM 1960 kind of development area. In this case, Mitchell would only need to put in a major road, sell off 300 acres here and there with few restrictions, very little risk and a very high profit margin. We had another one where we simply sold off land, same analysis, without requiring front-end expense. The third option was The Woodlands. The internal rate of return on The Woodlands was less than 18 percent per year. It had high risk, and risk has two factors. One is the probability of risk, the other is the severity of risk. So you have two components to the risk. The Woodlands had the greatest probability of a negative result than any of the other options. The consequence of that risk's happening could be the total loss of the company. George chose to do The Woodlands and his answer was a simple one. He'd been so fortunate, and he was putting something back into the pot. This was 1972. That's the way he is. Look into his Galveston experience, what is he doing with Galveston? What is he doing for the people in Galveston?

There are different views.
My own view of Mitchell, my picture of him, is you can have great success and you can be a Donald Trump and I don't think that's very much of a compliment to anybody. That's just simply being rich and self-serving. George's monument, if he is building one, is reflected in something better for other people. He's going back home to Galveston where he had a very hard life and where he grew up with modest means. He is making a statement in Galveston by investing his money and improving the city. But isn't it good that the statement he chose to make is the improvement of life for many Galvestonians? All of the downtown, the renovation, the Tremont.

George's efforts insofar as the Strand is concerned and reviving the business district are certainly laudable.
I didn't speak economics. I only speak to benefits for people, from him.

But if you don't stay afloat, then all your good intentions go to hell.
But I'm looking at the man's heart. Anyway, I guess, generally, that's the overview. I have my entire reading file for 1970, '71, '72, which is the chronology of going into the HUD venture and the chronology of the investment analysis, and it says who did what.

Before talking much about that, let me try to pin down some things. This oversimplifies, but your major role during your years at the company: Were you the guy in charge, responsible for the preparation of the HUD submission?

Well, yes and no. The way we set it up, I was the person dealing with the project from the early '60s until 1970 when the requirements by HUD necessitated more personnel. The answer there is unequivocally yes, because there weren't many alternatives but me and then later David Hendricks. Of course, Mitchell is always part of the equation. I was just doing what he wanted done. Then we brought in Hartsfield, and this is when, in a conversation with George, I said, "George, this needs to be set up where those people can think their thoughts and I can challenge them. He can't work for me and do this because I'm going to grade him." So, we kind of had a separation in here. Hartsfield went into and coordinated all of the architectural efforts, worked very closely with Ian McHarg, all the environmental studies. I was on the other side saying, "Here's what we can and can't do from a financial standpoint," and testing the model and incurring their wrath when things didn't work. I did the initial work with HUD, then later Bob and I jointly coordinated the presentation. I still focused and controlled the negotiation strategy.

OK, with help from the people that you've named, Autra and Bob Wyrick, the economic model is your baby?

Absolutely. Autra and I sat there and ran numbers till we looked cross-eyed.

The part about Cerf Ross being the guy who led the HUD submission is new to me, or at least if it's not, I had forgotten it. George credits Kamrath with the early plan.

I don't understand. Kamrath was in and out of George's office and was his tennis friend. I have all the documents. I don't see a footprint of Kamrath.

In the U of H professors' book about The Woodlands, George was very specific about wanting the Kamrath plan included, and George doesn't forget a hell of a lot.

Kamrath was his very good friend, and neither do I forget a lot. Not only that, I have the files. I don't remember Kamrath. I think you will see from the files pretty much everything that happened.

You've kept extensive files?

Yes. At a time when there is a setup to keep these records properly and at a time when the complete story is available, then I'll be glad to give them up.

That makes sense.

(From this point forward, Mr. McAlister largely refers to and quotes from various documents developed in the early years of The Woodlands.) Those three books that are the HUD submittal, I have two of the three volumes of the Gladstone & Associates study, "Market Opportunity and the Base of the City." I have the first environmental studies, I have the first submittal for the University of Houston north campus, I have the first spendable analysis for the $50 million loan guarantee, I have the loan amortization, Woodland Village base case, based upon 7,312 acres, I have the appraisal of '71 for The Woodlands. Here is the first presentation of Cerf Ross, and if you'll notice the configuration, it's not too different from what you see now, including putting the lake in place. This is the first economic model and this is the paper that I had to write to get it done, in which we went through the whole process, and here are the assumptions for the Woodland Village. That's our first economic model. This is what Autra Ayers did by hand and this is where we did the model, here's the Sutton-Mann tract being brought in. Let me just show you what we have. This is my reading file, 1970 through 1972. In this I have pulled out important things: April 7, 1970, departmental structure . . . this is where we were looking at trying to do 15,000 acres . . . personnel requirements—this is where we need a staff man and where I write a memo to George saying, "Here are the requirements for the people that we should hire." This is to the University of Houston, where we told the university we were trying to get them involved . . . this one is to Mike Spear of the Rouse Company, on what we were trying to do . . . this is a memorandum to George P. Mitchell, June 5, 1970. This date with Paul Brace, Director of Planning for HUD: "I spoke with Mr. Brace in Washington this date with regard to procedure for updating the design presentation to be presented to HUD." Brace and I got this thing through up there because we had good personal relationships. Brace became my friend. This is about some sales of industrial sites. Employment of David Hendricks, June 22, 1970. That's when I hired David Hendricks as my assistant. Choice of planning team, Woodland Village. "It appears to me that we will be interviewing two basic types of companies, research and marketing companies such as Stanford Research, or the architects and graphically-oriented, such as CRS." That's where I went and interviewed CRS and that's where I met Bob Hartsfield. Here is an interesting occurrence. I went to a seminar in Austin. At the

seminar, an individual talked about land he was developing on the west side of Austin. He had finished the subdivision and he was now ready to do his high-density phase in the front where he was going to make his profit. The community he had established had enough people to create their own city. They did so, put in zoning, and he was never able to develop the remainder of that property. I came back from Austin and started meeting with Paul Wommack, a Mitchell attorney, and said, "Paul, this is a problem." We then began to work through an attorney, Will Sears, who was very political. I think at one time he was attorney general for Texas. He was a very bright man, very well connected in Austin. We found that if any part of a property touches the extraterritorial jurisdiction of another city, you can ask that the city incorporate all of your property. There were three cities that touched The Woodlands–Conroe, Shenandoah and Houston. Obviously, you wanted to go with Houston because of no-zoning. Then Paul Wommack and I went to work on getting The Woodlands land in the ETJ of Houston.

Why would you want to be annexed by anybody?

Because if you put over "X" number of people in a square mile, they can form their own city and that city has extraterritorial jurisdiction and can control land development and building codes. Exxon previously made a different mistake in Friendswood. They had one Municipal Utility District for the whole city. Members on your MUD Board have three-year terms. So after a time, the developer has no say on the board. Exxon's Friendswood Development lost control, because now that board was dictating what would be built where. We, therefore, set up one master district that produced water and processed sewage. It sold service to other districts. Then we created a number of smaller districts. We would be through developing in a district by the time we lost control. That was the plan and you'll see in this file, I worked with Don Howell, with Vinson & Elkins, on that matter. Here's information on all the things that happened when we tried to do an industrial park with MoPac. I sent Mitchell from time to time the status of real estate projects, of which The Woodlands is one. Here's the extraterritorial jurisdiction part, here is some of the master planning. So I've got the first and second stage, even in the reviews. These are master planning proposals in connection with the University of Houston. I was the speaker who presented the University of Houston proposal. It got accepted by the U of H trustees but we botched it in Austin by failing to do our follow-up work. As a result, it was turned down by the State. Now you're seeing The Woodlands getting more and more of my time in this file. Tommy Robertson: George and I met with him because we thought that First Mortgage could bring in the capital to do the deal. They liked the deal. We kept trying to get them to finance it privately and he was saying, Don't, for goodness sakes, go to public money. He just hated public money. OK, Richard Browne. Here is where he first comes in, August 6, 1970. This is about coordination of real estate investment activities. We're now getting more complicated. It talks about how marketing must be set up and a recommendation to George about how to do that. These files go on and on. Marketing, organizational implications of new community development, this is what I submitted to George, organizing the financial analysis, the treasury budget, the accounting, the master plan, the engineering development. I suggested we set up these particular things, which we did do. This is a news release, when I became director of investment real estate right after I became vice president. Federal programs which were available. And so this just went on and on into preliminary offerings, U. S. Government purchase, Sealy-Smith purchase. Here's the whole chronology of The Woodlands, 1970, 1971 and 1972.

All right, very good. Thank you, Jim.

George Mitchell

This is Wednesday, August 23, 1995. This is Joe Kutchin, and I'm in the office of George P. Mitchell, who is chairman and chief executive officer of Mitchell Energy & Development Corp. George, when were you born?
I was born May 21, 1919, in Galveston, Texas.

Tell me about your early life.
My early childhood was in Galveston. I lived with my mother and father, two older brothers and a younger sister. We lived in a very modest area above my father's old pressing shop. He had a little shoeshine parlor there, too. We went to grammar school and high school in Galveston.

All four of you?
All four of us went to school there. My parents were Greek immigrants and their main motivation in coming to this country was having opportunity for their children, although the children were born after they were married here. My father came over in 1901, when he was 20 years old. He was a Greek immigrant who worked on the railroad for three or four years.

Savvas Paraskevopoulos?
Yes, his name was Savvas Paraskevopoulos. The first week after he left Ellis Island, he began work on the railroad, first in Arkansas. Then he went to Utah. When the Irish paymaster started to give him his pay, he said, "What's your name?" When he said, "Savvas Paraskevopoulos," the paymaster said, "I can't say that. You're Mike Mitchell from now on, and that's it." That's where the name came from. So I think that he worked on the railroad three or four years, then went down to Houston and met his cousin, who came from the little town of Nestani, Greece. They both came from there.

His cousin was here?
The cousin was in Houston. They had a little pressing shop. Actually, they had two of them. Then he saw in the weekly Greek newspaper about this Greek beauty who had just come to Florida from Greece. But he couldn't read very well. In fact, he could read hardly at all. He not only couldn't speak English, or read or write English (he learned English gradually), but

he couldn't read or write Greek, either, because he had never been to school. But he did have a foreman on the railroad who taught him.

He had been a sheep herder or goat herder in Greece?

In Nestani. It's mountainous country, a beautiful little village. Cynthia and I went there years later. He worked in the mountains with the sheep and they'd pick up all the firewood in the summer months. They had a little place in the valley there, in the little mountain village. He knew that no opportunity was there. He had three older brothers and a sister, he was the youngest. So he left because there was nothing but a half-acre of land for him to make a living on.

How old was he?

He left when he was 20. He got on the ship at Kalamata after walking 70 miles from the little mountain village of Nestani, which is 40 miles north of Triplis in the Peloponnesus. So, when he got to this country, he was able to get a job on the railroad because in those days, the railroad needed able-bodied people as workers. Gangs were putting tracks in and such.

The brothers had not come here before? Was he the first one?

His brothers had never been here before.

So he was the first one?

The first one. In fact, he was really the rebel in the family. He's the one who struck out for America when he saw that he had no opportunity in Greece. He just had to figure out what to do with his life. He had no education, but he had a good mind. When he came here, the foreman taught him, crudely, how to read and write Greek in one year of effort when he was on the railroad. The foreman saw him crying one day during the first couple of weeks. He said that he wanted to write his mother, but he didn't know how. That's when the foreman helped him. The foreman was Greek, too. So, it was not the Irish paymaster but the foreman. Then, in 1905, he moved to Houston with his cousin. They opened a shoeshine parlor near the Rice Hotel. Then they opened one in Galveston. So finally in 1908 or 1909, he saw my mother's picture in a Greek newspaper. She was in Florida, a very attractive person. He saw a picture of this Greek immigrant beauty. My mother's sister had invited her over because she had a potential husband for her in Florida. My father had enough gumption about him to go to Florida. She already had an engagement lined up for my mother.

What was your mother's name?

My mother's name was Katina Eleftheriou. My mother's oldest sister's name was Eleftheria. What my father was able to do, he convinced my mother to marry him and come to Texas. And that's what happened. She came to Texas, and they moved to Galveston and took care of the pressing shop-shoeshine parlor in Galveston. He split up with his partner, and that's when my father and mother started the family. Mother had a sixth grade education. She came from the town of Argos, in the Peloponnesus of Greece, too. She was a very pretty woman, and she was very warm. My parents later on were able to bring her younger sister and her younger brother over here, too, to help them out. And my mother's mother, too—I remember my grandmother when I was young. And she came to live with us, too. She died when I was about five years old. That's the way it worked in those days. My mother and father and the Caravageli family—they were the pioneer Greek families of Galveston—were always helping Greeks to get located in Galveston. So that's the background of the family.

What was the language at home when you were growing up?

My mother couldn't speak English at all. I spoke Greek until I was 13.

Do you still speak Greek?

In fact, when I've gone back to Greece, after two or three days I can understand and converse a little bit with them. Now Maria remembers it better. My mother died in 1932, when I was 13 years old. I went to live with my aunt and uncle. My uncle was a younger brother of my mother. We spoke English in that household, because his wife was of Irish and English descent. But Maria went to live with my mother's sister in San Antonio, so she still spoke Greek for a long time. She remembers Greek a little better. I've lost most of my Greek. My mother couldn't speak English, very little English. My father spoke broken English. From the standpoint of schooling, that's all you could hear them say was, you've got to go through school, you've got to go through school. That was the attitude they had about what they wanted their children to do.

So that applied to all four of you, Christie, Johnny, Maria and you?

That's right, even though we didn't have any money. I'd make a little money in the summer fishing.

How old were you?

Well, about 14 or 15. I'd go fishing, catch fish and sell them to Houstonians. I sold to Gaido's daddy and things like that. So I made my summer money fishing. We'd go out to this place, Bettison's Fishing Pier. The owner wanted us out there because we caught fish and would sell them to Houston people who couldn't catch anything. We'd go down to West Beach and go fishing or hunting and things like that. It was a wonderful place to be raised as a kid, and we had a closeknit family.

By this time your mother had died? Wasn't she quite young?
She died young. She died when she was 44. And I was 13. I stayed in Galveston with my aunt. Two years later, I graduated from high school. I graduated in 1935, but I was too young to go off to the university. I was barely 16, by only a few weeks. So I did postgraduate work for a year.

High school postgraduate?
I had been accepted at Rice and my mother wanted me to become a doctor. I was going to take pre-med at Rice. I was too young to go to college no matter where I went—I didn't know where I was going to go, but I thought it would be Rice—so I returned to high school and took third year Latin, because I wanted to be in pre-med, and I took trigonometry and solid geometry and advanced algebra. That's post-graduate. I was good in mathematics. I graduated with the first class, but I stayed another year at the school because I was too young to go to college. I'd skipped in grammar school a couple of times. And then I went to work one summer in the oil field with Johnny.

At your mother's death, the kids were split up, weren't they?
Shortly after my mother's death, my father was hit by a car and his leg was badly shattered. As a result, it was very difficult for him to take care of Maria and me, so I went to live with my aunt and uncle. Since their home was only a block away, I was still able to spend a lot of time with Pop. Maria stayed with my other uncle and aunt who owned a cafe called the College Inn, right next to the Medical Center. About a year-and-a-half later, they moved to San Antonio. And then that's when we split up, but then Pop was still right there within a block from where I lived, and I was with him all the time. At that time, Christie and Johnny were in college. Christie had gone to A&M first, and he took engineering. But he said life's too short, so he transferred to UT and took journalism. Then Johnny went to A&M about a year-and-a-half-later and he took chemical engineering. I went on to A&M

when I was about 16.

By the time you got out of your postgraduate high school, were you and your sister living together with your father?

No, we were separated. We didn't live with my father. I lived with my aunt and uncle. And Maria lived with an aunt and uncle who lived on the other side of town, but then she went on to San Antonio right after that, and then a year-and-a-half after I went to the university, she went to Mary Hardin-Baylor College. We never lived together after my mother died. The thing is, that's all we could do. Anyway, the summer after I got through the postgraduate work, Johnny was working for Exxon. He had been out of the university for a couple of years, maybe a year-and-a-half.

Did he have his degree then?

Yes. He had a degree in chemical engineering but they were making a petroleum engineer out of him. All the oil companies used to take chemical engineers and make petroleum engineers out of them. That's before there were petroleum engineering courses available in school. Then he left Exxon to start working on workover wells in Louisiana. He raised a little money here and there to help finance the workovers. And so I went to work with him one summer as a roustabout. I was just a 17-year-old kid.

Johnny's five years older than you?

Johnny is about six years older. So, I got involved in the oil fields and got intrigued. I thought, no, I didn't want to be a doctor, I wanted to take petroleum engineering and geology.

That's awfully hard work, isn't it?

Oh, yes, being a roustabout was very hard work. Johnny was trying to drill these wells and I was roustabouting all around them. He wouldn't let me on the drilling platform because it was too dangerous, but I worked around there for six straight hours helping to lay the pipeline or whatever had to be done around these leases, cleaning up, and he was actually working the well over, making a little gas out of them, making a little oil.

Do you remember what it was that turned your interest from medicine? What was so fascinating about oil?

When I went to work for Johnny, I started reading a lot about the oil and gas industry and petroleum engineering. I started looking at all the magazines I could get my hands on. *Petroleum Engineer* and *Oil Weekly* used to be two of them. And then I tried to understand more about geology. The field operation intrigued me and the fact that I was good at mathematics had an effect. I changed my mind about what I wanted to do, and Rice wasn't the place to learn about petroleum engineering and geology.

Do you remember the price of oil at that time?
Yes. The price of oil was about $1.75.

Not that much better now when you adjust for inflation.
Gas was 2 cents. I switched that summer. I decided that I wanted to become an engineer and geologist. That's what I wanted to be.

So let me get it straight. You did your postgraduate year and then was there a lapse between that time?
No, just one summer. I went to work with Johnny and I got more intrigued about the oil business because Johnny was involved. I worked with Johnny for about two to three months, and I decided that Rice wasn't the place to take petroleum engineering and geology. I had gone up to Rice with my father and Rice had accepted me late in that postgraduate year for the next year. And then I decided to apply to A&M because I knew from Johnny's discussion of being there that it was a good school. And then when I checked it out, it probably had one of the best reputations for petroleum engineering and geology in the country, so I went up there.

You were 17?
Really 16.

Did you have a scholarship?
No, no, I didn't have a scholarship. I did have good grades in high school. I was in the top 10 percent. But anyway, at that time you didn't hear about scholarships much at all. At least I didn't. That summer I went on up there on the train. My aunt took me to the train in Houston, by myself. I got off the train in College Station and I remember walking across that campus thinking, "OK, what am I headed for? Where am I going? What is my future going to be?"

Had you been to the campus before?
No. I hadn't been to the campus before. But I got off the train, heading

across the campus to go to wherever you're supposed to go to register and then I remember thinking about the future. I got registered there. A&M was known as a tough place. I got a job waiting on tables.

Where did you wait on tables?
In Sbisa Hall, in the mess hall.

That's a men's residence?
Yes. I got 26 cents an hour. Right around there. I also did two other things. In my dormitory, you see, you worked in a battery. You actually were in an Army unit. I was a private and I lived in the dormitory as part of an Army unit. I was in B Battery, Field Artillery. That was the same battery Johnny was in six years earlier. He graduated two or two-and-a-half years before I got there. So I got in his battery, because I heard from Johnny about Field Artillery and B Battery. Anyway, I had to figure out where the money would come from because my father didn't have any money to speak of and Johnny had hardly any money and Christie had a poor job.

You were getting almost no support from home. Is that right?
Very little support from home. I made a little money that summer working for Johnny with these people he had been working for, not much. But it only cost me $29 a month up there plus a few dollars for each course you took. The $29 was for room, board and laundry. So I was in the dormitory with my battery, 150 people. You had a captain, you had a top sergeant. It was just like you were in the Army. Then I built some book cases that I'd sell to people, and I got permission to sell candy in the dormitory. I had the tailor concession, and then I was working the mess hall to try to make enough money to make it work.

This was your first year?
This was my first year. Very tough. Very often I wouldn't have any money if I didn't make it in candy, or they'd steal my candy and wouldn't pay me for the damned clothing. They were about to kick me out two or three times. I would have to go borrow from Buddy.

Buddy Bornefeld?
Buddy Bornefeld. You see, he came from a fairly rich family in Galveston. He had the only car. I used to kid Buddy: We'd go to Galveston once in a while and he'd charge a dollar for a round-trip, but as we'd go back, if the

wind blew from the north he would charge a dollar-and-a-half. But anyway, several times during the year I had to borrow money from Buddy to be able to make it, because you needed books, you needed this, you needed that. Buddy is a fine friend of mine. I knew that I had to get out in four years because I just had to get out and get started. But I wanted to do geology and engineering. So I took petroleum engineering and then I took all the geology that was available. So that meant that I had to go to school, every year for four years, believe it or not, 23 hours a semester. I went to school every day until 5, every Saturday until 1, and tried to work out for tennis to be on the tennis team.

You were doing things to make money besides?
Yes. I was trying to work in the mess hall, make book cases, haul those big book cases and try to sell them from dormitory to dormitory and couldn't get anyone to buy the damned things, and I would get so mad. But, I sold a few, and, of course, the clothing.

And candy?
The candy thing, you'd put a stand up and you'd put up a sign saying, drop your money in, and you'd have 20 or 30 different candy bars and sometimes they dropped the money and sometimes they didn't. My senior year I was in the same dormitory and I still did that my senior year because it was kind of fun. But all these big football players, the national champions, Kimbrough and all that bunch, big bruisers, they'd eat that candy and never put any money in.

And you didn't argue with them.
But by that time I was making good money.

So when did you start your stationery business?
That started about my junior year. The first year was very hard. The second year was a little less hard, because I worked every summer with Johnny to make a little money. And then I had these other things going a little bit. My father never did have any money, but he knew all the people in Galveston and I'd send him a wire that I believed that I was about to get kicked out of the school because I owed Buddy and I couldn't pay him and I couldn't pay the school their $29 and they were pretty severe about it—if you didn't pay them in about 45 days you were out. So he'd take my wire and he'd go over to Sam Maceo and say, "OK, my son is in the top of his class and he needs

a hundred dollars." And so Maceo would give him a hundred dollars and he would send me $50 and keep $50.

So that's where you learned it.

That's how he made his money. He played by his wits. He had a very sharp mind, completely untrained. He really had tremendous perseverance, and he just had a great mind. He was friends with everybody in town. All the doctors and all the lawyers. They all did things for him for nothing. He would always need some help of some sort. He was just that sort of person. Anyway, I was taking these enormously tough courses. Engineering, petroleum engineering, is very difficult. I did well in that but I took so many hours, and then I tried to play tennis. I practiced tennis when I was a kid, hitting the ball up against the wall where we lived.

In Galveston?

Yes. And I also went to the tennis courts, which were close to me in Galveston, not too far from where I lived, but I've never had a tennis lesson in my life. I'd watch good tennis players and try to emulate their style. I became a pretty good tennis player. In fact, I became Number 1 on the tennis team at A&M, the captain. I was never that good a player–they never had a very strong team then like they do now. Much stronger now because they give scholarships and they didn't give scholarships when I was there.

You still play today?

Yes, I still play today. Even though my knees are giving me trouble, I still play. I played tough tennis for 55 or 60 years, since I was 13, so you begin to wear your knees down. Cartilage gets worn down. But, anyway, I still play. I don't play golf, I play tennis. Everybody wonders why I have seven golf courses but don't play golf–they think I'm crazy. But anyway, going through school was very interesting. You can imagine having to work all these courses I had, all the labs I had, all the work I had, so I was pretty busy. And then I wanted to get out in four years. It was very important to me to get out in four years, to get started. I knew what I wanted to do.

There were some professors that you've recalled very fondly.

Yes, in my junior and senior years I had Harold Vance, professor of petroleum engineering. Very unusual person and he had some good philosophy that was very interesting to me. He had a little private practice and still was head of petroleum engineering. He was able to do consulting work on the side. And we'd do a lot of his logs for him. He'd work us to death and collect money as a consultant. He had a real homespun philosophy. He said, "If you want to go to work for Exxon (or Humble at that time), fine, then you can drive around in a pretty good Chevrolet, but if you really want to drive around in a Cadillac you'd better go out on your own some day." And he had done some independent work himself, too. He was a very interesting person, a good professor, very talented. But getting back to schooling, I made good grades and I was top of the class with my grades all the time. I was still really working on engineering and geology because those were the two fields that I wanted to follow. And then by my senior year I was doing engineering and some geology for Johnny because he was a chemical engineer. He knew the oil business well, but he had concentrated on the mechanical part of it and he was struggling. He went bankrupt one time, got a little piece of property my father had all tied up in a mortgage mess. My father didn't have any money but he had assigned the property to the four kids and Johnny got it all tied up in his bankruptcy deal—about $8,000. I ended up having to pay off the damned thing after about five years. I helped Maria go through school, I had to help her out. The money I borrowed for her for Mary Hardin-Baylor, I had to pay off. I ended up being a major on the staff of the battalion that I was in. This was '38, '39 and the war was getting close. So the Army was recruiting some of the top students to become permanent officers at that time. But I didn't want that. I knew the war was imminent, maybe in a year or two. You could see the rumbling in '39, but I didn't want an Army career. I wanted to get out in the oil business and get going so I didn't accept the Army offer. Buddy did. Buddy got in it and Ed Dreiss did.

I know that Buddy and Ed are your friends since college. Is there anybody else who goes back that far?

I met Ed Dreiss at school and we roomed together the last year because he was captain of the track team and I was captain of the tennis team. I had my battalion right next to me, but I stayed in the athletic dormitory with that tremendous football team. They were national champions that year, 1939. Ed Dreiss and I became close friends. He was in petroleum engineering and I was in petroleum engineering and so was Leon Lefkofsky, a very bright man. So Dreiss and Lefkofsky and I competed for the scholarship award that I won in petroleum engineering. I got a watch. I give a scholarship every year, with the same type of watch, for the top student in petroleum engineering. Ed Dreiss was from San Antonio and we'd go there sometimes, to his father's place. Anyway, I didn't want to get in the Army. I thought that there was a good chance that the war was imminent, but we didn't know for sure what was going to happen. So then I went to work with Amoco. Here I was at the top of the class, won a number of awards. I had interviewed for two or three jobs because it was a tough time. That was when things were really tough with the Depression. Amoco said, "OK, we will take you on. You have to go to south Louisiana." Johnny had to drive me because he had a car and I didn't have one. "But you have to report tomorrow." They had had a blowout down there. We called my boss "Wooly Red" Bedford, that son of a gun. He didn't need me down there. But he wanted to show me that you had to be right on the ball. So I had to be down there right before the final review. I got my diploma, but with a battalion of 800 students under me, I had to miss the final review. I was battalion commander and my family came and my aunt and Pop and all of them came there and then I had to leave the day before the final review. It was very disappointing. It made me so damned mad, but I had to do it because jobs were very scarce and it was a good opportunity. So that's why I went to work with Amoco.

What was your job at Amoco?

Exploitation engineer with the geology and engineering together. The way they worked the fields, you'd do the geology and engineering in the field. I went down to Jennings, Louisiana, and we worked out of Lafayette and we had 22 drilling rigs running at that time. We had just bought the Yount Lee Oil Company and had all that south Louisiana acreage. Yount was an individual out of Beaumont. In a year-and-a-half I got tremendous experience. Because they had 22 rigs running, I was involved in geology all over the place, I was doing a lot of work. I had a good senior officer there, an engineering geologist. I learned a lot working with them.

We're talking 1939 now?

I'd say really '40, probably May '40. Then the war came in December 1941. And then I went into the Army. I was an Army reserve officer. So they notified me in September '41 that I was going in the service and I reported, I guess, in October.

For Amoco where did you live?

I lived in Jennings, Louisiana, but we operated—I had to roughneck for three months—out of Hackberry. What a miserable place. I can remember going at night, it was raining and sleeting and I had to work on that rig. It was dangerous as hell. I thought, my God, if I thought I'd have to do this all my life I'd go jump off this damned rig right now. But I had these wonderful Cajun workers who were with me and they protected me. They were tough. They would always give me trouble. You'd have to drink out of that dirty cup and they would spit out tobacco juice. But I tell you, they were wonderful people. And the people I lived with were a Cajun family, wonderful, in south Louisiana. The Creoles were friendly people. I worked all over that area, about two or three months roughnecking. Then I was doing engineering and geology there. And then, of course, the war. The Army called me in. With my background they put me in engineering, the Corps of Engineers, and they assigned me to the San Jacinto Ordnance Depot in Houston. We had about a thousand civilians working for us, we had big engineering firms, Lockwood and Andrews and Turner and Collie, working for me. There were about four other Army officers. My boss's name was Jack Malloy. He was a captain and I was a second lieutenant. We were responsible for building this enormous, complex 5,000-acre San Jacinto Ordnance Depot and big ship channel. A very large project.

How old were you then, George?

I guess that was in late '41, early '42, so I was 21 years old, something like that. After we finished that job they put me on the Dixon gun plant. We were still on the ship channel, with the same team. The Corps of Engineers out of Galveston handled all the construction in the Southwest, all the air fields all over Louisiana all the way to West Texas. All these projects we had were with the Army Corps of Engineers. We were maybe a half a dozen Army officers and maybe 2,000 civilians. So I got tremendous experience managing people because here I was, a 22-year-old having to worry about a bunch of people under me, working on the engineering firms, working with the contractors. So I was able to get a lot of experience that way. The Dixon gun plant was the next project and after that they put me in military production. I had two of my ex-colonels ask for me, one for the Burma Road and one for Europe. But my commanding officer wouldn't let me go because I was doing work that he thought was very good.

You were married by this time?

I was married a little later. I met Cynthia when I was in the Army in '41 and we were married in 1943. I didn't get called overseas and Cynthia kids me, she says, well, they sent you down to Galveston, so you went overseas. What we did was all military production. I helped work on the building at Bergstrom Airfield, the New Orleans airfield. I was in charge of all production around this area and we worked on getting the pipelines to Alaska to get the oil from the Imperial Field up there. We were concerned about a threat there from the Japanese. I was in on the preparation of a prefab military hospital for overseas. I was in charge of about 100 plants that were doing operations for the Army in this area and I had a staff of about 40 people in Galveston. We worked on the contracts, the military production and the supervision, even though the contracts came out of Washington. So that was great experience too.

Pretty far distance from the oil patch?

Yes. It was a great distance from the oil patch. But we did build the pipeline system to go to the upper part of Canada to take that oil to that little refinery up there. Those people now are the Smith Company. I got to know him so well that he wanted me to come up and manage his company after I got out. He offered me a position to manage his company. I said, "No, I want to go back into engineering and geology." I said I can always go back to Amoco. Amoco wanted me to come back, because they thought I had talent. What bothered me was that I was in the Army for four years, from late '41 until I got out in January '46. All my compatriots at Amoco who didn't go into the service for some reason, for whatever reason, had gone up, and here I was down there. I said to hell with them. With Johnny just getting started, I said, "All right, Johnny, I'll work with you and we'll try to get this thing together. We'll get a little money together and we'll create a little team. I'm going to Houston and then we will work together and see if we can't create something that way." And he had convinced a few people to put a little money behind him for these little wells he was drilling. But he wasn't making any money.

Johnny got out of the service about the same time as you?

He got out before. You see, he went to Europe and he was in Patton's Fifth Infantry Division. They had a tough time. They had 21,000 casualties in his division. Lucky he didn't have any problem. He got out about six months earlier. The war in Europe was about over in May and I was still in until that following January. I had to close out all those contracts. Even after the Japanese surrendered in August, I had to close out all of these contracts, so they wouldn't let me go until January of '46. So in the meantime I started doing some geology when the war was over and Johnny would try to sell. We did the geology of the Caplan Field. I did that one when I was still working to close out these contracts. Then I decided that I was going to be a consultant in geology and engineering. No experience to speak of, but I was determined that I was going to do it. If it didn't work, I could always go back to Amoco. I was confident I had enough talent to go back to engineering or geology with some major company and do it that way. I felt that with the experience and background I had, that someone would hire me at a fairly reasonable sum. I didn't have any money, and I had Cynthia and two kids and one on the way and I was worried about that.

The company was founded in February 1946?
Yes, that's right.

So, that happened right after you got out.
When I got out Merlyn Christie and Johnny were working on some little deals. They asked me to come in.

We'll cover that in our next session. Thank you.

* * * * * * *

Today is Tuesday, August 29, 1995. This is Joe Kutchin and I'm in the Texas Commerce Tower office of George Mitchell. Last week, George, we brought you just about up to 1946. You got out of the Army and you were about to start getting into the business. Please tell me what the circumstances were then.
As I said, during the last few months I was in the Army I was closing down military contracts on the Gulf Coast. I started doing some geology. I was working with Johnny and Merlyn Christie.

You were still in the Army?
I was still Army, yes. The war was over in August and then I had to close down these contracts, settle them all. I was scheduled to be in until about January of 1946. While I was settling the contracts, I began to get data on the Caplan Field, started doing some geology in that area and I worked up a prospect. I turned to Johnny and Merlyn to promote it.

What was Johnny doing?
Johnny got out of the Army before I did. He had been in Europe in Patton's Army, they let him out earlier. He came back here early and started working

in the oil fields again—a little production at Pierce Junction, production in Vinton. Anyway, we were going to try to put some deals together, so I did the geology, and Merlyn Christie and Johnny got Zinn and Pulaski and all of those people. When I came to Houston, Johnny was in Galveston. I came to Houston because I knew that the best opportunities in oil and gas were going to be in Houston. I was going to do consulting work and also work on organizing with Merlyn Christie and Johnny this little company called Oil Drilling.

Merlyn Christie kind of comes out of the blue. Did you know him, or did Johnny?

Yes, we knew Merlyn Christie even before the war. He was a broker in town and he'd come to Galveston and he'd see Johnny on the beach and Christie. He was a friend of Bill Moran and Glenn McCarthy and had friends such as that. He had been a broker who had traded some deals.

A stock broker?

No. An oil and gas broker. And he had a little royalty, he had a little income. And then he wanted to kind of work with me, thinking I could do the geology and engineering and we could work out some things together. In the meantime, I was bent on becoming a consulting geologist and I got my friends, six people who would put up $50 a month for me, and a small override, to do geology and engineering for them. And that included Morris Rauch, Louis Pulaski, Harry Pulaski. Merlyn Christie was involved in that and I had Jimmy Gray and one other, I think. I had a guarantee from them for one year to try that out to see if it would work, to do consulting work for them. Also, at the same time we were going to start seeing if we couldn't participate in some of the things on some basis by working deals and letting Johnny and Merlyn sell the deals. I would do the geology and they would sell the deals. And that's how we began Oil Drilling in '46.

Did Oil Drilling exist prior to that?

No, Oil Drilling was formed by Merlyn and Roxie Wright, who had the drilling rig, and then Jimmy Gray. Those are the three that formed Oil Drilling. I was doing their consulting work and Jimmy Gray after the first three or four months decided that he didn't want to be in it because he didn't think that it was going anywhere. He had gone to Oklahoma and I bought his interest out for $9,000. He had a one-sixth interest. I borrowed the money and I had Pulaski and Rauch endorse my note for me at Commerce Bank. And I bought Jimmy Gray's working interest out. This little company had a rig, one little rig, and they were going to do deals. I would do the geology and engineering and Merlyn and Johnny would help sell the deals.

Let me catch up with something. I always thought that Roxoil was the first company.

Well, it may have been Roxoil first and then it became Oil Drilling. Yes, Roxoil was named for Roxie Wright. So it was Roxoil and then Roxie put his rig in the company. That's how it began. He was a drilling contractor and he drilled deals for Bob Smith. He was a deal maker but mainly he was a drilling contractor.

But did you then get an interest in Roxoil?

No, I came in when it was merged into Oil Drilling. My interest was in Oil Drilling. We'd put a deal together and if we made a well, we'd go to the bank and borrow money on the well. If we made a dry hole, we didn't have much money in it anyway. So we were putting deals together. That was one phase of what I was doing. The other phase was, I was consulting geologist for Floyd Karston and Pulaski, Rauch and Glenn McCarthy, and Eddie Scurlock. I was getting a little overriding royalty myself. So I had two things going: the company that Johnny and Merlyn and I were working to create called Oil Drilling and then I was doing consulting work to try to get enough money ahead so that I could make it myself.

The consulting was solely a George Mitchell enterprise?

It was my enterprise, but mainly I would work with Stewart Boyle. We'd work on deals with other people. I would work the geology and they would go get deals from the major companies usually and put the package together and then Johnny and Merlyn would help sell the deal. The people who did the consulting work, they would try to sell the deal directly without Oil Drilling having anything to do with it. I did a lot of work for Floyd Karston, who was a big operator early on. He had three Jewish partners and they wanted me to be sure that I looked at the engineering and geology for them because he was the operator and they didn't exactly trust his expertise.

Who were they?

That was Morris Rauch—that's Leonard and Gerald Rauch's father—and Louis Pulaski and Harry Pulaski. Two Pulaskis and Rauch. They had a fairly good-size independent operation going on and we drilled a lot of wells and made a lot of discoveries. I did the geology, engineering and the seismic supervision for them. Karston was a pretty good operator and they were just silent partners, so to speak. They worried about what was going on. I made some nice little overrides out of that. That's when I knew that I was going to be able to survive as an individual, and that was especially important because I had children.

How many by then, George?
When I came to Houston I had one and one on the way and then by the first year or so we had two children then and probably another on the way. Oil Drilling, Inc., didn't have any income to speak of. Then we would take deals. What Johnny would do with Oil Drilling and others, he created the Big Nine, the Jewish group. He had Irving Alexander, the Weingartens. We had nine Jewish people, damned good partners. It was a hell of a deal. Man, they were tough. Old man Zinn in Galveston was tough. I'd come in and say, "Mr. Zinn, sorry, but that well we drilled got salt water," and Mr. Zinn would say, "Hell, I've got all the salt water I need in Galveston and I don't need to spend my money to get salt water." And Roxie Wright, we'd drill a well and I would say, "Roxie, we got a dry hole, we cut a fault," and he'd look at me and say that the fault's behind that desk–you. It was quite a wild group. Johnny and Merlyn would go down to the Esperson Drug Store and sell the deals.

You were in the Esperson Building?
The Esperson Building, yes. We had one little office. Merlyn and I and Johnny, we had one secretary. That was it. We were dealing in Oil Drilling and Johnny was working with the Big Nine and I was also doing consulting work for another group. I did work for McCarthy, I did work for Eddy Scurlock and they, all of them, took our deals. Roxie was an old friend of Bob Smith's and Merlyn was, too. I didn't know Bob Smith, but I'd heard of him. Bob was a very smart cookie. He was tough, tough as nails, not a mean person, a good person. Roxie would want to have a deal, and he'd come flying in and say, damn, give me some geology, we are at 3,000 feet, hurry up. Bob Smith would take a lot of deals from Roxie. He knew Roxie from the old drilling days. We got acquainted with Smith pretty well. Merlyn Christie knew him some, but he had no confidence in Merlyn. I started doing geology and engineering for Bob Smith. I did a lot of work in geology and engineering. I didn't get any overrides, he just paid me for it.

By this time Smith was wealthy.
Yes, he was already doing well. He had bought a bunch of production in East Texas. He came out of the drilling business, had drilling rigs. His main interest was oil, he had no real estate at that time. He bought some wells in East Texas on the good side of the field. He drilled some wells in Carthage when he first started, and Carthage was a good field then. I promoted a lot of Carthage royalty that I sold to a lot of my good friends around town and they still get the royalty, still doing well—we bought it early at $80-$100 an acre. Smith was a well-to-do operator at that time and he would take deals. He would end up taking a few of our deals and finally he would take more, more and more. Finally we ended up working with Smith to take 25 percent of all our deals on a net profit basis. For example, we would put a package together and drill a 6,000-foot well for $80,000 and they would carry us for interest on the 1/32nd oil drilling. Later on as we got better, we'd get him to carry us for 1/8 interest and then we started taking a little bit of interest ourselves. If we made a well we'd go to the banks and get some money; if we didn't make a well, we didn't get anything. We bought the Vienna property early on.

Now this was Oil Drilling, not George Mitchell?
Oil Drilling, yes.

Were you still doing consulting on the side?
Yes, and at the same time, we were building up Oil Drilling gradually. Then we bought out Roxie Wright—we worked it out with Merlyn that Johnny and I had half and he had half. I had 40 percent of the half and Johnny had 10 percent of the half. We built it up and then that's when we started really moving ahead—we found the Buffalo Field, Palacios Field, and we had Scurlock as a partner. Mr. Riddell was one of our partners, we had Barbara Hutton as one of our partners. You go through the history, you'll see all sorts of partners that we had.

Put this in time.
I would say that it was even before Wise County, which was 1952. We had discovered a number of fields including the Palacios Field, the Buffalo Field, a number of fields. We drilled some wells in Jackson County which were some of our first wells with Oil Drilling that made wells. The King Ranch Field that we just plugged out last year or so. We made fields because we were able to get gas prospects, farm-outs from companies that didn't care

about gas. All we'd get was 3 cents for the gas. We knew they were good prospects and that we could make wells, and by doing it real cheap we could make some money.

This was the period between '46 and '55?

I would say between '46 and '51, '52 or '53 we made hits on maybe two or three dozen wells. There were dry holes, too . . . always a scattering of dry holes, of course. We would always try to drill four nearby wells for every wildcat. It's just a matter of principle. If you drilled 12 straight dry holes you were out of business. You didn't have any more customers, clients. We started building Oil Drilling. About that time we had already made a number of fields and had pretty good production going. We had a lot of partners. In Palacios we had the Riddells, we had the Pulaskis in La Sal Vieja. The Big Nine, they were still partners in a lot of these deals. They still have production down in the Valley area, La Sal Vieja. Early on, some of them were in the Wise County wells. The big farmout on the Hughes Ranch in Wise County came to us through "General" Louis Pulaski. He sent a Chicago bookie to me who told me about this deal in Wise County and he asked me to go check it out. They called him General. He wasn't a general, but he was named after the famous Polish general, I guess. When I checked it out it looked very promising.

Did you go to Wise County?

No. The geology was sent to me here. I looked at it and it looked promising and I found out that the guy who was hawking it was Ellison Miles who I had known in school, but I didn't know him very well. He was an old A&M ex. He and John Jackson had been hawking this deal for two years all over the country. What the hell, why would a bookie from Chicago call you on a deal? I think it had been shopped everywhere in the world.

What was the deal like, George?

The deal was that Jackson and Miles said they had 3,000 acres on the Hughes Ranch available. Whether they had the lease in hand at that time I don't know. If everything worked, the bookie would get a small overriding royalty. The geology he showed me looked pretty good, the stratigraphic looked good. Some wells had been made to the north, and they looked pretty good, and it looked like some 6,000-foot wells on the 3,000 acres would be a pretty good shot. So we made the deal. We had this Big Nine group, the Jewish group, and then we had Smith in at that time and we had several others—I think maybe Barbara Hutton was in it, but I'm not sure. We had a whole list of partners in the first Wise County-Hughes Ranch well. We made the well, and I could tell immediately by looking at the geology that there was probably a major stratigraphic trap.

Nobody had figured that out before? There had been a fair amount of drilling in Wise County, hadn't there?

Yes. There had been some dry holes, but what we found was that they really were not dry holes. Within 90 days of the Hughes Ranch discovery, we leased 300,000 acres and during the first year we went back into 10 of the wells drilled by major companies and made wells with each one.

How much did you pay for the leases?

About $3 an acre. We had all these brokers working. We turned around and we got farm-ins all around the area from 10 major companies. The dry holes they drilled, we went in and made wells in every one of them that they passed up. But it was a combined engineering answer as well as a geological answer. Hydraulic fracturing had just come in about two or three years before. Without hydraulic fracturing you couldn't make decent wells. So this is where we combined the engineering with the geology to make it feasible. And then, of course, we had this big play. We burned everybody out and we ran out of money right quick, except for Smith. He stayed in.

Where did the money come from?

When we got all the acreage in and got all these people in we started seeing we had trouble. We went to the banks. We were up to about 40-50 percent interest ourselves which we couldn't digest because we couldn't get a gas contract. We tried to get Natural to get the contract in from Chicago through the Panhandle and that took a two-year fight to get that through. In the meantime, we didn't have the money, so Natural advanced us, the whole group, a $7 million production payment to drill the wells. We had 21 shut-in wells. It was a tough time. Lone Star was fighting us because they wanted to dominate the area. They wanted to take the gas at 11 cents and they wanted all the processing and that's when we balked. We had to fight them and the Railroad Commission and we had to fight up in Washington, too. It took us two years to get Natural to come through.

Let me step back. You saw the stratigraphic trap when you drilled the first well?

After the first well. And within three months, we had 300,000 acres at $3 an acre. We had brokers working all over the place. Maybe it took as long as six months. Then we saw all this potential, these old holes that had been drilled, and we were getting partners, but we were getting in pretty deep and a lot of them just couldn't go that deep. Some of them dropped out. We finally had to work a deal with a pipeline company to advance $7 million to drill wells to keep this thing going. I couldn't get everybody. Smith might have come up with the money, but the rest of them couldn't and we couldn't come up with the money. In the meantime we had to fight Lone Star and the Federal Power Commission to get our permit through.

Why Lone Star?

Because Lone Star monopolized the area. They didn't want us in the area. They were the ones that had all the gas. They wanted all the gas for their Dallas market. They were going to fight us in Washington and fight us in the state, which they did for two years. We finally got Natural to go through. Furthermore, at that time a Supreme Court ruling came out that a producer was a regulated utility and Mobil had made a deal with Natural to help get that line from the Panhandle through Oklahoma into North Texas. We were the only company to stick with them, and Natural appreciated that many years later, because no one else stuck with them. All the others bailed out except us. We stuck with them and they advanced the money to drill wells and we built up more and more reserves. We finally got it through Washington—13 cents, I think it was, the initial price for gas. We worked a deal with Warren Petroleum to build a plant where our people had 80 percent and they had 20 percent. We had Smith, Miles and other people in the deal. We ended up buying all those out of the processing plant as the years went by.

Was there some problem with the wetness of the gas causing a problem with transmission?

Yes. The gas was rich, 1300 BTU, which is about 80 or 100 barrels a million of liquids and you can't haul that much liquids in a pipeline. It drops out and then the pipelines don't work right. You have to really knock that material out of it, which is profitable. Processing reduces the BTU content and makes the gas qualify for the pipelines to take the gas. Otherwise, they would have too much liquids in the line.

So there was both an economic reason and a practical operating

reason.

Yes. From an engineering standpoint Natural had to clean it up, but they would not take the gas unless the liquids were extracted. Warren Petroleum would have done the whole plant, but we didn't want it that way. We let Warren in because of their expertise and they were able to go to the bank in Chicago and borrow $7 or $10 million dollars to build the initial plant, because of Warren's experience. Gulf bought them out later on. Warren was our partner for 20 percent and then we had a bunch of partners for the rest, including our company, Oil Drilling. No, by that time it may have been Christie, Mitchell & Mitchell. I don't remember what year we changed the name. I bought Roxie Wright out, I made a deal with Merlyn and we called it Christie, Mitchell & Mitchell. Merlyn had half and we had half, Johnny and I. And then I bought Merlyn's half out later on.

Was Merlyn an easy guy to live with?

He was difficult. He was very smart, a good trader, but then he could be very upsetting to the whole company.

But, to his death, did you folks stay relatively friendly?

Yes, we were friendly. Merlyn said that we were going too fast and I think we bought him out. I've forgotten what year. I think we paid $7, $8 or $12 million at that time and then we bought Bob Smith out for whatever price he wanted. You'd have to go back to the records for what we paid Merlyn and, later, Smith. Smith had about 25 percent of Wise county and about that much of the plant. We had production all around. Smith had been taking deals for a number of years. Some of the Big Nine people sold out, some didn't. Some of them still have their interest in Wise County and La Sal Vieja and other places. When I bought Merlyn out, I went and made a deal with Johnny that I had 80 percent and he had 20. Later on, when we became Mitchell Energy & Development, Johnny had Jade Oil and he had a 20 percent interest in Mitchell Energy. I had no interest in Jade. What we did was build Mitchell Energy from that time on. And then, of course, we did a lot more drilling in other places.

You really knew you had a major operation going, though, when you were successful in Wise County?

We knew Wise County was a major operation, but I will say this: We had found a lot of other fields that were good fields, a lot of expansion, a lot of deeper zones, a lot of other areas.

For instance, give me one or two.

Well, Buffalo, Alba, Palacios, La Sal Vieja. We have a whole list of them, but Wise County was by far the major one. I believe we brought Vienna in in '46 or '47. They had drilled one well, this guy who had this, from East Texas, I forgot his name. I had all the Shell data and seismic for that area that had been bootlegged by someone, but it was common knowledge all over town so it wasn't anything that they could do anything about. I could see the structures and . . . McMurray was his name. In '46 or '47, we bought the Vienna Field for not too much money, about $600 or $800 an acre. They had one well and about 2,000 acres and we bought the field and we drilled about 10 wells and made good wells, oil wells, and we had the gas behind the pipe, and finally made the gas wells, too. What we would do was make the deals with this guy and then we'd go to the bank and borrow the money. We wouldn't have the money to pay him off. Mr. Zinn still has part of the field. He stayed in with us to buy it.

Say that name again.

Mr. Zinn. His son, Robert Zinn, now here in town, still has a little bit in the Vienna Field. Zinn was Johnny's original partner. Some of the established wells he did included Pierce Junction and Vinton in Louisiana. Vienna was a good field that we acquired for a little or almost nothing, and then we drilled it up. And we've been drilling deep wells there in the last five years.

So we come to what, late '50s early '60s now?

I would say that what I am talking about was probably the late '60s. Oil prices went down in the '60s. We were on eight-day production. Tough again. Imports were very high. Gas was just beginning to go up a little bit. You had controls on interstate commerce, gas was 13 cents, 14 cents. Ridiculous price. That was about the early '70s.

I think you need to talk more about the deal you made with Natural Gas Pipeline. It was a 20-year deal and then another 20-year deal.

The Natural Gas Pipeline deal was made in '55. Production started in '57.

It would have to be something like that.

We had drilled a number of wells. The acreage and wells we drilled, we could identify a potential of several hundred billion feet of gas. We planned to drill a lot more wells.

You say "we." Was there somebody else?
No, just us plus all the partners we had.

But you were the geologist?
John Jackson did a lot of the geology with me. In fact, John Jackson worked with Ellison Miles on the original prospect. But, it was our company that made the deal with Natural. Phillips had production from Oklahoma that comes through the Panhandle from Oklahoma into here, and then it goes on to Chicago. Anyway, I heard that Natural was looking to expand and needed gas. They just had that one line that goes to Chicago from the Panhandle. The Panhandle reserves had gone down. So we talked to them and apparently Phillips had talked to them, too. I think maybe it was Mobil as well as Phillips. I'm not sure. That was in '54. We had negotiated, the three of us, our company and the other two companies, a deal with Natural to expand that line all the way down from the Panhandle to Wise County, going through Oklahoma, picking up that gas en route through Phillips and Mobil. The Supreme Court decision came in and said that producers were regulated companies, so Phillips and Mobil pulled out of the deal. They didn't want to be regulated. That just left us and Natural. We didn't have enough power to keep proving up our reserves and that's why they helped us by loaning us money. In the meantime, we made a commitment to them to fight it through the Federal Power Commission. A 13-cent price was the price at that time, about 11 cents all over, everywhere, and we had the right to process the gas. That was the key to Natural's deal.

So you were getting 13 cents for pipeline quality gas, and you had a lot of BTUs left over.
Yes. Normally pipeline quality is 1,000 BTUs per cubic foot. With 1,300 you have the equivalent of 80 barrels a million BTUs of liquids—ethane, propane, so on—which was very profitable even in those days. When we first made the deal with Gulf-Warren, the price of liquids, propane, was 2 cents in the summer time and 3 cents in the winter time. I remember September 1, just sitting around waiting, for God's sake, hurry up till the winter season comes on to get the price up to 3 cents for liquids. Gulf-Warren was the biggest in the business at that time. So we struck a deal with them—they had nothing to do with that area—but we struck a deal, because they knew plants and so on, to help us supervise it and help us get financing, which was the main thing. Then we struck the deal with Natural and we stuck with them and they appreciated it and they helped us by giving us the production pay-

ment to drill wells. We paid that out years later. In fact, even before it was paid out completely they waived the rest of it. We still owed them about a million-and-a-half. They tried to figure it out, but it was too complicated and they said just forget it. We gave them something, too. The deal with Natural was a very good deal. That was before the gas price went up, but then what happened was that intrastate gas ballooned to $1.10 and there I was with 13 cents.

When was this?
Before '73. I've forgotten when intrastate gas went up above 13 cents.

I didn't realize that intrastate was that much.
Intrastate gas skyrocketed. Here we were, sitting there with 13 cents and everybody selling intrastate was getting $1.10. We were mad as hell. We were trying to figure out how we could get it. Finally the Federal Power Commission woke up and then they gave a gas price increase, but mainly because of the crisis in '73 and '74, and then the Power Commission finally allowed prices to shoot up to a fairly decent position.

You followed the price of oil?
We followed the price of oil, yes. When oil went up in '73 from $2.50 a barrel to $12 a barrel in '74, that's when it all started going up. The Commission was always lagging behind intrastate prices. In the meantime, we kept a good drilling program going on and we could make money even at 13 cents by watching our costs. We would use second-hand pipe and do anything we could to keep the cost down. We drilled those wells for $80,000, $70,000, and now the cost is $400,000. All costs have gone up. In the meantime, we did a lot of drilling in other fields.

Do you have 100 percent or 80 percent of the company at this time? Does Johnny still have the 20?
When we bought Merlyn Christie out, that's when we did the 80-20.

Did Johnny have 20 when the company went public in '72?
Yes. And then Johnny sold off some. Johnny, when we worked the 80-20, I let him have a net profit out of some production up there, different than what I had. So he had some stock and he had production net profits on some of the wells, which he still has. Now we had done a lot of drilling and the plant was a very good money maker. We created those plant packages

because we didn't have the money to drill ourselves. The plant had an outside operator, Warren, with 20 percent. We worked out a deal with the plant where they would drill the wells and help furnish the plant with gas and when they got their money back, 80 percent went back to all of us, the owners, so we could hold the acreage. That was a very big program.

Those were smaller plants?
No, that was the big plant. I don't know what year it was, but we bought Southwestern Gas from Walter Davis. There was a gathering system that we wanted to get to bring gas into the plant, besides our own gathering system that Natural had. The big deal we made with Natural was most important. They would gather the gas and let us build a plant after it was gathered. So that saved us all that gathering cost and we could process. That is one reason that the plant was very important to us. The Natural deal was very good because we owned most of the plant, plus they brought their gas to us and we processed it, so we didn't have all that cost.

We're talking now late '60s, I would guess. Did you have 100, 200 employees by then?
Oh, I think we had probably at least that many. We had operations everywhere in a lot of other areas, drilling wells in Alba, Buffalo, La Sal Vieja, Palacios, we had a lot of things going on. And many other little fields here and there. Then I think we finally started acquiring the Barbara Hutton interests. We bought them out of Fallon. We bought other people out of the Fallon area. The wells didn't do so well until we got this massive hydraulic fracturing there, too. We held the blocks together, just like in North Texas we held the acreage together. We'd buy wells that might not have been very good wells, but we held the acreage and later on, with better technology, we did better on them, drilled deeper and things like that. The Barnett play we have now is a deeper zone below where we have the Atoka zone. Those are the things where you benefit by the fact that you hung on to your blocks and didn't allow them to expire on you. Getting back to Southwestern Gas: Permian had its pipeline in that region and they sold. So we made a deal, I think a $5 or $6 million buyout. A pretty good gathering system that could get production from other wells to our plant or bring it to another area where we had processing. We built that big plant with Gulf-Warren's help and talent, too. And then I got to thinking that there must be a lot of other gas streams you could process. We needed to figure out how to use these little turbo-expander plants that were just coming

in. That's when I ran into Bruce Withers over at Tenneco. He had tried to get Tenneco to do just exactly what I was talking about and they wouldn't do it. Bruce said, "Well, I'll come over with you and we'll try to do these smaller plants and design turbo-expanders for our use."

Morris Thompson previously had been the person in charge of processing for the company, hadn't he?

Yes, working with Bill Hudson over at the consulting firm of Butler, Miller and Lentz. They were our consultants looking over Gulf-Warren's operation. We bought out some small pipeline companies before we acquired Southwestern Gas. We began getting a little pipeline network going. We had gas streams a lot of other places and then we figured out that if you could build a little turbo-expander plant for 10 or 20 million cubic feet per day it would be profitable and you could haul it in to the main area and extract the propane before you did ethane. That's when we started that technology with Bruce here. We built plants in a number of places.

I know Bruce started in 1974 with Mitchell Energy, so that puts it in some time frame.

We were going to take advantage of the small areas of gas that the big companies didn't want. They didn't care about having the small plants, so we did these plants for 20 million feet and found places to put them. So we ended up with 54 plants, very large–45,000 barrels a day–and we've been the 13th or 14th largest processor in the country. That is a big part of the company now, processing. If prices get back up again, if oil prices go back up some, processing should do a lot better and we are in a good position on that . . . and we bought this deal with Conoco in the last few years. Processing is a big part of the business. Oil and gas exploration is a big part of the business. We haven't been big in acquisitions. We've bid on a lot of them, but have lost a lot of them. Sometimes we're glad we lost them. We have not diluted our stock equity by merging with stock companies. We could have and some people said we should have, but you lose control when things get very tough, and we would have been gone, someone would have taken us over. The advantage of keeping controlling interest was that we didn't have the threat of someone taking over.

Why did you go public?

We got to the point in '70 that I owed the bank so much personally that I had to endorse all the paper, all the North Texas operation, because I probably owed at that time maybe 30, 40, 50 or 60 million dollars. I either had to sell out or merge with someone or go public. The banks would say, well, you've got to do one or the other. Either you become a public company, merge with someone or sell.

Thank you, George. We'll pick up again at a later time.

* * * * * * *

Today is Wednesday, October 11, 1995. This is Joe Kutchin. I'm with George Mitchell in his office in The Woodlands. I've just reviewed the two earlier tapes, George, and the first one was essentially about your early life in Galveston, going through A&M and so on. The second tape covers the time from when you left the Army in January 1946 and deals largely with your work in the energy field and the early days of the company.

I was doing the geology and engineering for Roxoil after I left the Army in 1946 even though I didn't have an interest in the company at that time. Then I bought Jimmy Gray out. I was getting an override and a little interest for whatever deals I put together for Roxoil. But then I bought a working interest when Jimmy Gray decided to go back to Oklahoma, and eventually we bought Roxie Wright out.

Was Johnny in then?

Johnny was in then but Johnny only had 20 percent of what I had, or something like that. I may have brought him in later, I don't remember for sure. Merlyn had one-third interest, Roxie had one-third, and Gray had one-third. I was doing the engineering and geology for them. Roxie just had a drilling rig. He'd weave these tales about the Burkburnett boom and the Sour Lake boom and the High Island boom and things like that. He was a fascinating person, knew the history of those boom towns. But anyway, we bought Jimmy Gray out and then later on I bought Roxie out. That's when Merlyn had 50 percent and Johnny and I had 50 percent—I had 40 and Johnny had 10. We started Christie, Mitchell & Mitchell at that time.

Didn't Oil Drilling come before Christie, Mitchell & Mitchell?

Oil Drilling came in right off the bat; Roxoil converted to Oil Drilling. Right after I bought Jimmy Gray out, we changed from Roxoil. So Roxie was gone and we named it Oil Drilling in late '46. Then when we bought Roxie out in 1953, Merlyn had 50 and we had 50, Johnny and I—I had 40 and he had 10. Then later on when I bought out Merlyn, then I had 80 percent and Johnny had 20. And, of course, we were drilling wells, finding fields, and ended up buying a few. We bought Barbara Hutton out, we bought some of the Big Nine out, we bought people out, we found all this oil and gas for everybody and we eventually bought Bob Smith out.

When you say "we," do you really mean you or you and Johnny?
Johnny and I and Merlyn Christie. Then when Merlyn wanted to sell out, I think that was before I bought out Bob Smith, Merlyn sold out for about $6 million. He didn't want to have any more debt structure. He was worried about debt structure. We bought him out in 1962 on a payment schedule, $5 or $6 million. That's how it eventually ended up, Johnny had 20 percent, I had 80 percent. Then Johnny sold off stock a lot earlier than I did—I didn't sell any stock to speak of—and then when we went public is when we all went down somewhat on ownership.

And that's about the time that our last session ended, when you went public. And you said you went public because you needed the cash because the banks were coming in on you.
At that time we had fairly large debt, all endorsed by me, and Johnny, too. Johnny would sign anything, I was reluctant to sign. You knew you either had to merge with someone or sell out, because you couldn't keep going to the bank. We had difficulty with the bank in 1975 when The Woodlands first opened. That was the Chase. The other banks weren't as difficult, but a lot of times the bank would say no. The banks are no lead pipe cinch. When we bought the first piece of Vienna in 1946 or 1947, we really had to do a selling job to the bank because we didn't have any money to speak of. And then when we'd make a well we'd go back, but if you made a dry hole you couldn't bank anything. That's the way we finally built it up. I would say this, that the reason we went public was that you can't grow too easily as a private company. Some people have done it but it's very difficult.

I remember it was 770,000 shares.
I ended up with 62 percent, where I had had 80 percent. We had trouble getting good underwriters. Shaker Khayatt was one of them. Walter Lubanko came on the Board. There wasn't that much money. We were lucky to get it public. Between Walter and Shaker, they said, "Sell down to 35 percent, get some money out of it." I said, "No, I'm not going to listen to these people." Had we sold down to 35 percent, when the first downturn

came, it would have been gone. The only reason we survived is that we had 50.1 percent. Great Southern Life Insurance in town, owned by the Greenwalls, or some name like that, is a good example of what happens when you sell control. They were owned 35 percent by three or four family members. Hell, there came a takeover and they were wiped out. So that was one of the smart moves I made, that I refused to sell down below 50.1. And I kept a little leverage of 62 percent in case I did have to sell down.

And even now you have about 62 percent of the voting stock.
Sixty percent. I still have 60 percent of the vote, and I've kept some of the nonvoting. My average is 47.40 and my voting is 60 or 62.

What we've been doing is trying to follow a reasonably straight line story of the energy operations and then probably come back and talk about real estate and even another time I'd like to talk about your extracurricular activities like HARC and the Pavilion. But let's talk about energy since 1972. The company was doing OK, but there was a substantial drain starting from real estate operations at that time. Why don't you talk about what the situation was?
Starting construction of The Woodlands in '72 and opening it in '74 took a lot of energy, plus a lot of financing besides the HUD-guaranteed money. The HUD money was mainly to buy the land and a few other things. The energy company was doing quite well. Then '73 came and you had a spike in two things. One was the spike in oil, with the price going almost overnight from a few dollars to $12. And later on, in 1977, it went to $25 and $30. But the thing is that interest rates spiked, too. When we opened The Woodlands, interest rates had gone through the ceiling. Some of the money came from energy and some from other assets that we had mortgaged—for instance, the Grogan's land block was bought at a very favorable price and we were able to finance the land because we were still selling timber. Land takes a lot of money and when you don't have income, it gets you in the hole deeper and deeper. So we had to furnish money from energy. But in the meantime, energy became very profitable because gas prices broke between '72 and '73 and broke more in '77, higher.

These were intrastate prices, right?
Intrastate. It's very peculiar how this thing worked. For a while, for intrastate prices in Wise County we were getting 13 cents and Lone Star was paying 11. Later on what happened, the intrastate price broke and went up

to $1 or $1.25. The Power Commission price was still at 13, 15, 18 cents and then finally went up higher than the intrastate prices, kind of a leapfrog effect. And of course, all the differentials on gas that helped us were very important. A good example of what we were able to do has to do with Galveston Island. We had acquired Pelican Island and we had Pirates' Beach in '65 and had Pelican Island about '63, '64.

Those were with a view to real estate investment?

They were real estate investments. To show you how real estate investments led us into very profitable oil and gas, we looked at Pelican Island and we saw that Gulf had drilled a dry hole on the very east side of Pelican Island that had a very good looking sand section that was wet. And then we looked at some of the seismic work and we could see a lead onto Galveston Island. So, we took the lead and did the seismic work which showed some evidence of having maybe just a little structure building up on Galveston Island. So I said to Sam Meadows, "Sam, see if you can go down and get me permission to run a Vibroseis," which was a new technology that I'd just heard of, where you didn't use dynamite (it couldn't have been done with dynamite), to run a line down Broadway and then run one line on 37th Street, north-south. He came back and said, "We can't shoot down there, too many houses, too many people, too many this, too many that." I said, "The hell you can't. Go down those streets and get Vibroseis and do it at 2 o'clock in the morning." They went back and they finally got a permit and they got it done. They shot the two lines.

That's basically a truck pounding with some device?

With Vibroseis, a pumper truck backs up and creates a shock wave. You can only read maybe 7,000 or 8,000 feet, you can't read 15,000. All we needed was to get down through the miocene. So they did it and then we could see the structure evolving on the north-south line, east-west line. And we could see the structure looked very promising on Galveston. Then we got some of the drillsite leases around the first well, only about four or five. They were all town lots except for city property. A few pieces were 50 acres, 100 acres of city property. Most of them were nine lots per acre. And we had to trade on every acre, every lot. So we drilled the first well very tight. In fact, I ran the logs myself. I could see the two zones looked good. I even took the kids there when I ran the logs and told them, "Don't you tell anybody." The kids were young then. "This looks like a good discovery, so don't tell anybody." I didn't tell anybody in the company. Nobody in the company knew what we had for sure, but they knew we were interested. Then I had Jack Yovanovich go down and he had 10 brokers and we made 9,000 leases in a year to block up the whole area we thought was promising. We never completed the well.

This was the late '60s and you were still the guy doing the geology?
This was '68, '69. Then we got the block together and then we made the well, then we drilled two or three other wells. On the other hand, we noticed from the geology of the block that the other half of the structure was offshore. So I said to Sam Meadows, "Let's go offshore and shoot the other half of the structure," which we did and it showed a beautiful anticline falling offshore. So we went and got the leases from the State. Now I've got to drill a well a mile in front of a beach that handles 4 million people. Shell had just had a big blowout down at Louisiana, and Santa Barbara had been about three years before. Everybody was going crazy over environmental issues. So, how am I going to do this? I've got all these beachfront people. If I got oil on that beach they'd hang me. The first thing I'm going to do is go see Mike Gaido. I'd known Mike all my life. I said, "Mike, I want to drill a well right in front of you. I think I can do it without hurting you. If we make any production, it'll be gas mainly. It'll come around the end of the island and the platform won't be ugly. We'll fix it where all you'll see is what looks like a little buoy." He said, "George, I've got confidence in you. I've known you all my life. OK, it's all right with me." Then I had the 10 brokers go to all the beachfront people, all the owners of property, commercial business, and all except one agreed to let me have a lease and put them in a unit. One guy gave me a bad time. So we finished leasing up the townsite–we'd already drilled two or three wells within the townsite, but we hadn't done any offshore at that time. I was trying to get permission to drill offshore. So I asked the City Council, "I want to drill a well offshore right here." They looked at it and said, "God almighty, a mile in front of all these businesses? Come on, George, wake up. Go find a piece on land that you can drill." So I found out that Bert Wheeler owned land at Fort Crockett where the San Luis Hotel is now. He had just made a trade with Tenneco, they took him out, so I made a trade with him. Tenneco was somewhat involved, but I got it worked out. So I bought that six or eight or 10 acres. I had a beautiful design of a hotel, condominiums all around it, five wells in the middle, sidetracks around it. So I had a hearing. The church next door came unglued. I mean they tore me up in the hearing. So there were three or four months gone. The City said, "Go back offshore." So then I went off-

shore and asked the Corps of Engineers for a permit for the first well. We had a hearing, we had a thousand people protesting. And Bob Moore, the environmentalist, tearing me up every way he can. They had people on the witness stand, and I'll never forget, they had two hearings and this woman had the same story every time. She'd get on the stand and say, "It's going to catch fire and burn my children" and tears would come down her eyes. A thousand people. What could I do? I loaded up three bus loads of our people to cheer for me and we finally, after two years, got the permit to drill that first well offshore.

Was that before or after you drilled from the cotton warehouse?
It was after. We had already drilled two or three wells onshore, and we knew they were pretty damn good wells. We produced about 70 billion feet out of the field. Then I knew part of the structure was offshore, probably a third of it. So I wanted to finish up the other part of the structure and get permission, so I quit drilling onshore for a while and tried like hell to get the offshore permit, before it got too many oil people nosing around too much. So we finally, after two years of hearings and screaming and working on environment, working with the news media . . . they were tearing me up in 15 different directions. You can imagine, after Santa Barbara and that big Shell blowout in Louisiana, the resentment they had. Anyway, it was very important to get Gaido on my side to start with, then we got everybody else signed up except one guy that had a service station who never did give in, but we had enough that it didn't kill us too much. So then we drilled the discovery well offshore. And then we filled in the other wells onshore mainly. We drilled three. They only allowed us three wells offshore, had to be a mile-and-a-half out, and I could slant-hole at a half mile, which I did. We could drill only between October and April, only drill three wells; if one goes out I can drill one more. And that's the way it worked out, and it's worked real good.

Was there something special about the Christmas trees, too?
The Christmas trees were low profile. They came up and there was a little separating tool. We laid an extra pipeline running under the water all the way down Eight Mile Road and that's where we separated the condensate from the gas. So all you can see, if you look right now, it looks like a little buoy out there. You've seen them, and that's it, rather than a big, ugly tank. What screws up the West Coast, I've told the majors this, it isn't the environmental concern, that's history, Santa Barbara won't happen again, it's

the ugliness of the damned tanks and the platforms. Now if they would do it right and not have those ugly platforms and have a streamlined appearance offshore and have nothing but a little buoy and do the separation onshore, you wouldn't have that opposition, but they won't listen to anybody. That's what we did, but it cost a lot of extra money to do it that way. That's the first environmentally sensitive deal that I'd say we did, and we did a great job. The City now picks up $200,000 a year in taxes. They tax my offshore oil and gas and it's not even in the City limits. We paid taxes for 10-15 years, a couple of hundred thousand a year. So the City picked up all the taxes onshore and all the royalty owners picked up probably $10 or $15 million of royalty out of the 1/8 and 1/6 royalties we had out, probably more than that. So, anyway, it was an economic benefit to the City. We didn't screw up anything to speak of. If something blew out, you could cause some waste problem, but it worked very well. So that's a good example of what happened.

Before you leave Fort Crockett, I just want to clarify one point. Somebody I've talked to, I don't remember who, said that as a boy you worked on Fort Crockett. You poured concrete or something like that?

When World War II grabbed me, I became an engineering officer, located here in Houston, at the Dixon gun plant. I had a thousand people working for me. I worked for two of the Army officers and with two big contracting engineering firms here. That was a very exciting thing because we were running a thousand people. I got a lot of experience. I guess I was 22 then. Then the next assignment with Dixon gun plant was on the ship channel where we were building a big gun plant for centrifugal casting of anti-aircraft guns and that was a big operation, too. So for 18 months or two years of the war I was in those locations. Then the Corps shifted me to Galveston. Cynthia says that was my overseas duty. The U.S. Army Corps of Engineers in Texas and Louisiana handled all the airfields . . . all the construction at Marshant Airport, the field we built. I did a lot of engineering on those air fields, all in west Texas. It was all handled by the Corps out of Galveston. So I got a lot of experience there. In came an AFE from Washington saying, "We want you to take Fort Crockett (they were seeing some submarines off of Louisiana) and we want to add two more feet of concrete on top, with another four feet of sand, to protect these big guns we have." How stupid can those people be? Doing that would cost 3 or 4 million dollars. To do that is ridiculous. I said, "I can't believe the stupidity."

But anyway, we did it. So that's the story about Fort Crockett. Now, I knew Fort Crockett as a boy because they would fire those guns. When I was 10 or 12 years old, they would fire those guns about once every two years. The 12-inch guns that used to come out of that big snout we got there, you know. And man, they'd break windows all over town. I could see that boat out there about 20 miles, that poor soul towing a target about a thousand feet behind. I said, "These guys got guts, towing that target." Anyway, you could see the shells land and all that stuff, kind of interesting really, history. But anyway, I knew that as a boy. We did design of the condominium and hotel but we didn't drill there. And someone said, "All right, let's put the package together and sell it as a hotel site to someone." We had the conceptual design done and we tried for four years to get someone to build a hotel and they wouldn't do it, so I said, "Let's try to do it ourselves." That's when we started the San Luis, we had this limited partnership. We built a beautiful hotel, great design. Getting back to oil and gas, it was right after '72 that Galveston was a big operation. We went back into the Palacios Field which McCarthy had given up on, we made some damned good wells there, and Buffalo and Alba. Vienna we had, we had a lot of other fields over there in that county, the one right below Colorado County. We did Polk County later on. But we did all of La Sal Vieja, way down in the Valley. And we got maybe half a dozen fields in the Vienna area, little fields here and there. And we were drilling Holloway in west Texas and then, of course, we had what I started in Galveston in '68, '69. And then Wise County started in '55, '56, and we were expanding Wise County all the time, and drilling, and working with the difficult situation up there on the gas price and giving gas interconnection and Lone Star stopping us and blocking us and fighting it, and then working with the banks to stay alive. So that's what went on, but there was a lot of activity, geologically. The first well we drilled at Fallon–it was a good-looking reef, I looked at it and I thought we had a good chance at making a play out of it. And we made two or three wells, but they didn't hold up.

That's very tight sand there.

Very tight and rock, it's limestone. And so everybody got out, we bought all our partners out with a little bit of money.

What period of time are we in, George?

Mid '70s. It was discouraging at first. We didn't make any money on the first well. Then we decided to use massive hydraulic fracturing, which has made it successful. That was in 1978. We used to acidize some wells, and that didn't work. So that's been important.

OK, the spike in prices brought about by the Arabs in '73 or '74 really opened things up.

Two things made it really very, very good. Right after the war there was such a pent up demand because there had been no wells drilled for four-and-a-half years. The pent up demand created a situation where, even though the price of oil was only $1.50-$2 a barrel, you could get a lot of good prospects and a lot of people wanted to put money in it. Then it was really tough in the '60s, when imports began to kill us. It got pretty tough, everywhere. We went to eight-day allowables in Texas. Gas takes were curtailed some and prices were bad. The first gas well we made, we sold the gas for 3 cents a thousand. We made a little money out of them because we drilled them cheap. And there were prospects the majors didn't want and we would take them on and make some wells. Right after the war, everybody knew there was a pent up demand and then in '73 the first spike came and in '77 the second spike came and those were the heydays. Those are cycles you go through, but the thing is that we found a lot of discoveries, new reservoirs, new blocks, new fields.

I believe this company has found and produced 2.5 trillion cubic feet of gas. That's an impressive number.

We've made a lot of discoveries and found a lot of new reservoirs. We've built up a pretty good talented team and I think technology-wise, I'd say we've been good.

Wasn't there a period not too long ago when we were really kind of right with the leaders in the use of technology, five years ago, seven years ago?

I would say we were on hydraulic fracturing, I think we did some bright spot. On 3-D we haven't been as good as most of the majors but we've been earlier than most all independents. Now they're all catching on. Todd Mitchell, he talked to his counterparts and it's a whole new world. And Todd is right in the middle of it, and really it's very interesting to see what these young people are doing. I think it's more difficult because the prospects are smaller, the chances are tougher, the raising of money is very tough. I'll tell you this, they are really a bright bunch.

I said to somebody the other day that you've had people in charge of geology who couldn't find oil in an Exxon station.
That's true. You know, I've had some Ph.D.s who worked for me who were pessimists, they couldn't find oil in a 50,000-barrel tank. That's true, you have to have a nose to find oil and gas, you have to have a little knack, and your judgment's got to be pretty good. You can't be a wildcatter. You know what a wildcatter is, I told you, didn't I? A wildcatter is a guy who's drilling at 10,000 feet, he's out of money, things are tough, he thinks he's three feet from a million dollars but he's really a million feet from three dollars. You've got to have judgment, you can't be a wild wildcatter, you can't make it. I did a lot of consulting work with Glenn McCarthy, Eddie Scurlock, Bob Smith, besides what they took with me. I did a lot of good work for them. I had overrides on quite a bit of that stuff. In other words, the thing is this, that you just can't drill all wildcats, it's too dangerous. You've got to have a mix, you've got to have judgment and a mix. Maybe one wildcat for three outstep wells or three deepenings because you'll burn out your people pretty quick if you get all dry holes. You get six dry holes in a row and you're out of business.

There was one year, I think in the '70s, when we had a capital budget of about $400 million, most of it for drilling. Do I remember correctly?
Yes, more than that–it was $500 million.

Did that work?
No, because you throw money too fast and become inefficient, and the cost went out of sight. I think that should have been constrained more. We were able to do one thing before it blew up–it collapsed in '84, '85–we cut the budget down to $119 million. We had to lay off 1,000 people. That shows that if we had to do it, we could do it. We laid off 300 now, but that's less than what we had to do in '84.

Would it be accurate to say that with the advent of the early '70s and the opening of The Woodlands, you started to pay a lot of attention to The Woodlands and real estate and a little less to energy?
Yes, I'd say that probably took some of my attention. The Woodlands itself could be nearly as important as the oil and gas over a period of time. You have to look at different maturities. In energy, I've got to find it every 10 years because the energy depletes. In real estate, if I get it going good, it's

good for 30-40 years. If you've got a building that's doing well, it could be good for 30-40 years. In energy, you run out of oil and gas, it's dead, it's run out of production. So you look at different investment philosophies. On energy you'd better make a 25 percent rate of return because you've only got 10 years to make that in. In real estate, you make a 15 percent rate of return, you've got 30 years. Of course, real estate can be very risky, too. But you have to have a different concept and most people can't see the difference. Even bankers don't understand the difference. We've got some of our buildings that are five or six years old and they're probably going to be good here for 30 years, at The Woodlands. And if I had oil and gas fall down, then in 10 years they're out, gone. And I may have made a 25 percent rate of return those years on what I invested in oil, but it's gone. I've got to go out and do it again. And I've got to have the talent to do it again, and again and again.

What do you see as the future for energy?

I think the future looks good because I think gas prices are going to get somewhat better. I can't believe they're going to stay at $1.65. You can't explore at $1.65. Yet, while I used to say it would take $2.50-$3, you may not get there, but we've become more efficient with 3-D. I think if you get $2.25, which I think we'll get in a couple of years, although everybody says you're not, I think it still looks good. I think certainly on our infill wells, which we've got a lot to go, we can really make money at $2. And now we're looking at Wise County to make more than $1.75. That's where we've got 600 wells to go, maybe more than that. So I think that you've got to be very particular now in exploration, much more than used to be true. You've got to make it work at $2 or $2.25, $2.50, and that's hard to do. Petroleum reserves have been picked over but our technology is better. Drilling costs are down some but not much, drilling time is down to half of what it was, but our cost hasn't come down half. Costs may have come down maybe 10 or 15 percent in Wise County for a deep well.

Give me some quick highlights.

I would say that one of the big deals we made that helped make the company, in my opinion, is when I refused to let Lone Star have a processing plant in Wise County. And I fought like hell and we got a part of that new business with Warren Petroleum. They knew the business. They came in as a 20 percent partner and they helped to get the financing from First Chicago, which was very important because we didn't know anything about a liquids plant. First Chicago wouldn't give us $8 or $10 million to build that plant, whatever it was. So that was a very important move because that's when I got the idea of these little turbo-expander plants that are around now, and not refrigeration. And that's when I convinced Bruce Withers to come with us to do something he wanted to do at Tenneco and they wouldn't do it. That's when we came out with these 20 million-cubic-foot plants. So that was a big move away from being strictly in exploration and production. We began to take advantage of the pipelines we had by extracting the liquids and making the gas more valuable. And of course we have the MTBE now and the fractionator, those have become important parts of the company. Those two facilities will make $12 million a year combined, no matter what the price of oil and gas is. It's a service, so that's good. But then there's no doubt that with gas prices at $1.50, $1.60, that hurts you, especially when you need to get $4.

Over time, is energy going to be two-thirds, real estate one-third?
Yes, I'd say energy two-thirds, real estate one-third. You're going to have exploration and development and the pipelining and the processing, and then you're going to have real estate. What we're trying to do is to get rid of some odds and ends in energy and buy some of the better pieces. I'd like to find a $100 million good acquisition. We've bid on some, but we've never been able to get one. And some of them I'm glad we didn't get.

OK, my last question for today is, is the liquids business a fundamental part of the company and is it going to grow?
A very big, fundamental part. I think it will grow, particularly when we have things like the MTBE plant and the fractionator that add to the growth. Another important deal is the one we just did with Union Pacific. When we get finished with that in another two or three months, we'll be producing probably 300 or 400 million feet of gas a day and they're drilling these wells with 10 or 20 million cubic foot open flow. They're taking the risk and they're doing pretty well. Our problem is that you're hurt with the price of gas at $1.50 or $1.60 like we get now for the new exploratory wells or infill wells. We'll be more cautious on exploration, using the latest technology and we'll need to look at prospects more carefully. You won't do as much exploration, we'll need to be more sure of ourselves and do more infill wells for a while to see what happens to the price. That's what I think we should do.

All right, thank you, George. Next time we'll go back to the late '60s, early '70s and start talking about real estate.

* * * * * * *

Today is Wednesday, May 15, 1996. This is Joe Kutchin in the Texas Commerce Tower office of George Mitchell. Today we are going to concentrate on real estate. Let's talk about the Cynthia Woods Mitchell Pavilion.

Dick Browne and I went to Columbia when we were designing The Woodlands. We saw the Merriweather Post Pavilion there.

What time are you talking about? What year?

That was back in probably 1970-1971, before I hired all the Columbia people. We said, "Some day we'll get something like that in The Woodlands." Then about 10 or 12 years ago, we had some people come in here to talk, including Dr. Phil Hoffman and John Cater and maybe Ben Woodson, too. They said, "Would you try to figure out how to build a pavilion in The Woodlands as a venue for some of the cultural arts?" I said, "Well, we might think about that." That was probably '81 or '82, when things were hot in energy. Then '84 and '85 got here, and we forgot about it. So, right after that we had someone make a study for us. They looked at all the pavilions around the country–Ravinia near Chicago, Wolf Trap around Washington, Tanglewood near Boston, the one in Concord in California. So they came in and gave us information on how those things worked. We had some information on Saratoga in New York, too.

Are you saying Mitchell people made trips there?

No, we had someone make an analysis for us. Somewhere in our files you've got a report.

I didn't know that.

So we had the idea of maybe doing a 5,000- or 6,000-seat pavilion. By that time, the oil crunch hit and we gave up on the project. In 1984, 1985, we just decided we couldn't afford it at that time. Then about 1988-1989, when things got a little better, Roger Galatas said, "I'd like to revive this pavilion idea–a small one, maybe 1,000 or 2,000 seats." So we looked at it and said, well, that won't work. Then Dick Browne and I thought, well, maybe 6,000 seats would be fine. We did some work on it. Then we heard of Allen Becker and the Pace group. So we asked Allen Becker to come in and consult with us. We asked him if he would have an interest in putting on the events if we built a pavilion. He was very enthused because he had four or five pavilions around the country that he managed around that time, and he was very successful at that. He said he would work with us.

Brian Becker was not part of the original discussion?
No, Allen was. Brian came in right after that. Brian's father, Allen, is the head man. And he said, if you don't have 10,000 seats, you can't get the attractive entertainment groups, you can't afford to pay enough money to them to get what you need here. And that was very important–that was probably the most important thing that we found out. So then we started putting the package together and we designed it for 10,000 seats. Pace made a deal with us and then they got the beer distributor with us, then they got the food concession with us. They knew the business because they had five or six outdoor entertainment centers around the country, and they had clout. We knew they could book the shows, rock groups, country-western groups and the other groups, the middle-of-the-road groups, a lot better than we could because that was their business. We kept responsibility for the cultural presentations. We started construction, I guess, in '88, '89, and we opened in '90.

It opened in '90 but it was announced publicly in '87 or '88 and then it got put on the back burner because of the energy market.
Well, it was opened in '90. It took us a year-and-a-half to build it. A good example of how important the arrangements are with the various suppliers and why it took so long to get things going: I don't remember who the original beer concession was with, but under the newest three-year beer contract, the Pavilion received $1 million for just the right to sell beer. On the expansion we just did, from 10,000 seats to 13,000, we got a new food concessionaire who gave us $1.8 million, plus the $1 million for the beer concession, and Pace gave us $1.5 million. We were able to do the expansion at no cost to us. Apparently, it's been successful so far and we think it's a very good project for The Woodlands.

Let's add to that the fact that you and Cynthia have given a major endowment to the Houston Symphony which has resulted in the Pavilion's becoming the summer home of the Symphony.
That all started at the time we began negotiating to get the Symphony out there for some events. Dancie Ware was involved originally in that, in getting the two parties together, and we worked out a deal that we think was very good for both parties. We also gave a three-year contract to the Houston Ballet to perform out there. We gave them, I think, a quarter-million

dollars. And we've been working with the Houston Grand Opera, too, to some extent.

The keystone event was changing it from a relatively small capacity of 6,000 to a more commercial size of 10,000. And even that turned out to be somewhat small, but we didn't know that at the time.
No, Pace was very important in telling us how these things work around the country and what it takes to command the best entertainers.

All right, let's go back to the beginning. In the past we talked about the corporation, about your personal life, about the energy company, but we haven't talked much about real estate. In the '60s, you had a prosperous energy company going and then you went into real estate. What did you do and why did you do it?
There was a downturn in domestic energy, with the big imports of oil, in the '60s. Our proration was down to producing oil eight days per month. So we thought, well, maybe we could do something else, real estate, too. We did a couple of small projects with Pacesetter Homes and we started Galveston, then we bought Pelican Island at that time, around '62 or '63. We had done Pacesetter a couple of years before that. That's when we had someone bring in the Grogan-Cochran land. That was a 50,000 acre purchase we made in about '63, a very good buy.

You said somebody brought that in?
Yes, what happened was we heard about it through Max Newland, our forester up there.

If you had a forester, you already owned a good chunk of land.
We had a couple of pieces of land, but he was busy trying to sell deals. We had one little piece of timberland where the Trade Center is now. We bought it from someone in Galveston, I've forgotten her name, she was in bankruptcy. She got us to buy this piece to keep the bank away, to settle with them. They got another piece. We bought that very cheap, where the Trade Center is now. And then Max Newland cut the timber on it, that's how he became acquainted with us. We bought that at $60, $70 an acre. Unbelievable story. Her husband in Galveston left all his estate to the dogs. I'm serious. Sophie's her name and she got mad as a hornet. With one thing and another between the bank and the district judge, I showed up and was able to buy this piece of land for about $60–$70 an acre. Raymond

Loomis was involved at that time. I let him have a piece of it, then bought him out later. I bought it personally, I didn't buy it in the company. I sold it to the company later, when we started The Woodlands, about five years later. That's the first piece of real estate we had. We had done a couple of developments with Pacesetter Homes in Houston, a little housing development, and hadn't made any money. We started Pirates' Beach about '65 or '67 and it was coming along, but it was a tough, long-term deal. Just a resort development around the golf course. We didn't have much acreage at that time, maybe 300 or 400 acres. The big play came when we bought Pelican Island, we bought 5,000 acres in '62 and '63.

Talk more about how the Grogan-Cochran land came to you.

The Grogan-Cochran families had all that land since 1903. They had a couple of lumber mills, including one near Magnolia that burned after we got it. There were three family groups and about 90 heirs and they were fighting each other. We worked with Max. A lot of people had tried to buy this tract for four or five years but couldn't do it. One of the heirs, Poindexter Grogan, lived up in East Texas and I worked with him. Paul Grogan was another and there was the third Grogan, I've forgotten his first name, maybe Horace. Two of them lived in Conroe, one up in East Texas. And they had all the heirs under them. So what I did was go around to all of them and work out a settlement on their lawsuits between them and said, "OK, I will give you $125 an acre for the 50,000 acres." That's $6 million. I calculated that based on what timber sales I could make and what gravel sales I could make to pay the interest. After a lot of effort I got them all to sign up and settle the lawsuits.

How long did that process last?

It took about a year. We didn't have any money, very little money, but the Bank of Southwest advanced me the money for the down payment, which was 20 percent, $1.2 million, and for that I got a release on some land. So I took the release and banked it at the Bank of Southwest to get the money. Then right after that, I worked out a deal with Louisiana Pacific to sell the timber for 20 years, or 10 years, whatever it was, and the bank helped on that. Then I got Great Southern Insurance Company to pay off the bank within a year because I had made this timber contract and I had the land and the timber contract. So that's how we ended up with the 50,000 acres.

Why did you buy 50,000 acres?

Because it was a good buy. I could pay for it out of the timber production, out of gravel. I hadn't even thought about The Woodlands at that time. I knew the land was close to Houston and it was beautiful land. The Grogans had bought it in 1903, 1904, 1905 and never paid over $3 per acre and they didn't buy any wetlands—it was all high land. They weren't paying $3 an acre for a piece of swamp. It was all good land. And then we nearly bought the Kingwood tract. I went after that, about a year later. And I had 21 percent of it under option. They lived in St. Louis, I went up there two or three times.

You're talking about the Kingwood tract?
Yes, the Kingwood tract—50,000 acres again. The Foster Lumber Company owned it. I had already bought the Grogan-Cochran. I had it lined up for $150 an acre. So I had 21 percent of the stock under option and Boise Cascade was going to hold my hand. Bob Halliday, I think his position then was chairman of Boise, was on the Mitchell Board then. They agreed to hold my hand, but they said you've got to get 51 percent before we agree to come up with any money, and I made a joint venture deal with them. But we couldn't get the 51 percent because a guy here in town found out and blocked us and went to King Ranch who went to Exxon and they bought it for $300 an acre. It was a doggy piece of land. I was going to red-flag it because it had bad drainage, it's a bad piece of land compared to The Woodlands. But anyway, we made $1.7 million because I sold my options to them at a $1.7 million profit, and we needed the money then. Boise got half of that and half the minerals. They did nothing.

Let's go back to the Grogan-Cochran. Talk about selling the timber.
We sold the timber on a long-term contract. The terms were that they would take the timber at so much a thousand for 10 years, or 20 years. I didn't have the money until they cut it every year, but I had the timber contract guaranteed, so Great Southern Insurance Company put up $6 million to pay off the bank.

But you owed the money to Great Southern?
I owed it to Great Southern through the bank. The bank held my hand, Mr. Vance held my hand, with no equity from us. We had to put the package together and we leveraged the 20 percent release of the land we got with the first bank loan. Then within a year we sold the timber contract and we refinanced the land purchase out of the bank. The land mass in The Wood-

lands includes only 2,500 acres of Grogans-Cochran land. Then another 300 transactions were needed to put the full block of 25,000 acres together . . . really it's 20,000 acres. I bought some from the Sealy-Smith Foundation, some from the Catholic church, a lot of people. Where the commercial center is now, that was 640 acres, 64 tracts of 10 acres each in an old subdivision. I had to figure out how to get it put together; it was right where the commercial area is. So we worked on that for about a year-and-a-half and we bought all but two or three of them at a fair price.

But you didn't know there was going to be The Woodlands yet, did you?
Oh yes, we had already made that decision.

So we are in the late '60s now?
We are in the late '60s, that's right, after I decided in 1966 how many acres would it take to build a town of 100,000, after I had gone to Bedford-Stuyvesant and gone to Watts and decided from my YPO experience that all our cities were being destroyed by what's happening. The flight to the suburbs by the brains and the middle-class whites, and the loss of tax base. How do you turn it around? I can't turn it around, but I can set an example. Let The Woodlands be designed to show the country you can do it better.

You physically went to Bedford-Stuyvesant?
I went there and then I went to Watts with a YPO meeting. I was impressed with the fact that all our cities are in trouble—New York, Detroit, Washington, Cleveland, they are all being destroyed. The concentration of the disadvantaged and the flight to the suburbs of the middle-class whites—that's destroying all our cities. And then we have political subdivisions all the way around these cities—you have 100 political subdivisions that won't work together. But The Woodlands' being a part of Houston provides a chance that it can work with Houston. That was the concept. We sold HUD on that idea. So then we put the block together and that took a lot of effort.

You said how many acres?
About 2,500 of the original Grogan-Cochran purchase is part of The Woodlands.

So what happened to the other 47,500?
That's what we're selling off now. And then for my ranch near Magnolia, I

picked up 2,500 acres. I bought my ranch at the same time the company bought its acreage. The remainder–that's what we're selling off now, you see. We paid $125 an acre for it and we've cut probably $800 an acre timber off of it already. So that was a good deal. Then we had timber, we had grazing, we had hunting, we had all kinds of things that went with it. So those were good buys. Now when we bought some of the later land in The Woodlands itself, for some places we had to go up to $3,000-$4,000 an acre, $5,000 an acre, but not big tracts. But the thing is that building the block was very hard. We couldn't get the Couch piece which is where Oak Ridge is, which disappointed me because it was sold to George Butler, my banker friend, and he sold it to Mickey Couch before I could get it.

That's the other side of I-45?

Yes, but on this side, too. I tried to get Oak Ridge to protect against second rate development and keep the subdivision rules for another 25 years, but they wouldn't do it. But anyway, we got most everything. I couldn't get the corner at Grogan's Mill. I had a chance to get that, but it was $1.25 per foot and I couldn't afford it at that time, for 150 acres.

Let me come back a minute. Here you are, a successful oil man, dabbling a little in real estate, and it sounds like your attitude kind of undergoes a sea change where rather than red-flagging and junky development, you begin to devote a great part of your life to dealing with one of society's most difficult problems.

I was very concerned about what's happened to all of our cities.

Did the YPO—Young Presidents' Organization—lead you to that?

The YPO led me to that. And then when we got into planning for The Woodlands, HUD came up with this Title VII Program, which would provide a loan guarantee of up to $50 million. Then we went after that and we were successful, one of the largest grants they gave–it was a loan guarantee, not a grant–because they felt our concept was the best concept they'd heard of, and it's near Houston which was important. You see, I knew where the airport was going to be. It hadn't been built yet. I knew where I-45 was going to be. At one time, I optioned 2,500 acres on 1960 and I-45 for $2,500 an acre and I couldn't get enough land, so I gave up there and went back to where I had 2,500 acres already and I built a block around it.

There was a trade there, wasn't there, that 2,500 on 1960?

No, I gave up my options, that's all. I wanted to build at least a 10,000-acre block at 1960, and when I got into it, it was impossible. So then I moved from there up north. But I knew where the airport was going to be and I knew we had to be outside of the sonic boom area, probably. I knew where I-45 was going to be, it hadn't been built yet. I knew where Houston was. There was no toll road at that time. But I figured we had to be part of Houston some day. So that's the concept we put together. Then we had McHarg come and do our planning. We put a hell of a team together.

McHarg was the first one?

What I did was I hired Dick Browne from Columbia and Len Ivins and all that bunch. McAlister and Lively helped me put the block together. Hartsfield did the original work with me, and then he's the one that recommended McHarg to me. I read the book *Design with Nature*, which intrigued me, so I got McHarg to come down and we worked out a deal with him and then worked out a deal with Pereira. McHarg had about 10 scientists work on the concept plan, environmentally first, which impressed the hell out of HUD. We prepared the first environmental impact statement that had been written. We wrote it and HUD used it as an example. We had Pereira do the master plan, recognizing the constraints imposed by the environmental plan, which is what everybody tries to do now.

So you had a bunch of all-stars planning this whole thing, plus your existing team, which was comprised of strong people.

We had Plato, Hartsfield was with me already, and then we hired some of the Columbia people later. But the original planning started out with McHarg and Hartsfield and ourselves. I got Dick Browne to come down.

Dick came at the same time as the others from Columbia?

Dick came in the early '70s. And Mike Richmond. I hired about six or seven Columbia people. And the people at Columbia sent a man down to tell me that if I hired any more of their people, he was going to sue me. We already had the HUD deal and we were ready to go. The last piece to get was where Lamar School is, 320 acres. The seller had me by the neck. It was a banker in Woodville, very tough. I bought about 2,000 acres from him, but he kept the heart. With the contract I had, I had to build him a gutter road, six-lane road, right through the middle of his 320 acres, right in Woodlands Parkway. He was going to keep that as commercial land. I was trying to close the HUD deal, Paul Wommack will tell you. I had to meet him at the airport before I closed the HUD deal to try to resolve the situation, and I had to give him about $10,000 per acre—it was a terrible deal. But I had to do it because it enclosed the whole front end of the block, and getting it made the HUD deal work easier. (I've forgotten his name but it's in the records somewhere. He was a banker in Woodville and went to Waco later on. He got in trouble with the Bank of Southwest on some deals that he had at the Waco bank.) He had about 3,000 acres and we bought all of it at a pretty reasonable price, about $1,000 an acre, or $2,000. But the last 320 acres he really nailed me for about $10,000 an acre, because I had to get rid of that commitment I had to build a road on the 320 acres. When I bought the other acres, he had kept 320 with that commitment in the middle. The HUD deal made it feasible for us to get the financing, although we didn't get enough financing. By the time we opened in '74, the crunch had hit on interest rates as a result of what was going on in the Mideast and gasoline prices had jumped. On the other hand, we had energy income which was inflated. But The Woodlands was squeezed because interest rates went to 12 and 14 percent, and we couldn't sell anything. And we had a strain with the banks—Chase Bank—and they gave us trouble, threatened to take control of the company. So I got, at that time, Manufacturers Hanover to take our loan over from the Chase Bank.

Manufacturers Hanover?

Yes, to take the Chase out. We had to fire some people and had to agree to let the bank help control the development. They originally wanted to take our stock control, which I wouldn't do. You were here then, you know the problems.

Do you feel it's turned around now?

Yes, it's turned around, The Woodlands itself.

Are you getting positive cash out of it?

Yes, we're getting positive cash out of the development part, but we put it back in buildings. Just like the real estate we bought right outside The Woodlands. We made good money out of that, but we put the money in buildings and other things. It will take a while, but the cash flow of The Woodlands is moving up nicely. All of our real estate together is positive now. That includes everything—Pirates' Beach, Cape Royale, Point Venture, everything. Now we're slimming back and selling stuff on the edges and concentrating on The Woodlands and maybe Pirates' Beach and Wood

Trace, a couple like that. It will be 90 percent Woodlands and 10 percent other real estate.

You had some very high goals when you started The Woodlands and began working with HUD—everything from innovation in housing to, I remember, making the population a mirror image of the Houston population. After 22 years how well have you done?
I would say that we're behind schedule, probably 8 or 10 years, because we had two real serious spikes in interest rates. But we've still done quite well. I would say 10 or 12 years ago we had about 11 percent rent-assist people in The Woodlands. That ratio has gone down, which is disturbing, and we're working very hard now, because when the crunch came in and all of these foreclosures around The Woodlands came in, all these $70,000-$80,000 houses were selling for $30,000-$40,000, and we couldn't compete with them and we had to trim back on our efforts. The value of The Woodlands held pretty well, the values outside The Woodlands deteriorated severely. But we had to cut back on what we were doing. The socioeconomic thing we did quite well, but less well the last seven or eight years and we're trying to get back on track. The minority thing–there's still a long way to go. It'll be 50 years before we have 10 or 15 percent minorities in there, I think. We had hoped to get more and have worked at it, but still it's a matter of, you don't have any more government programs to speak of. We have some pretty good rent-assist apartments, as you know, right now. We have fine programs, but I'd say out of the total population of 45,000, I don't know, we may still have 10 percent on rent-assist programs. We had housing as low as $29,000 to start with. I remember the teachers in the Conroe school district couldn't afford a house, and I had to try to build a house to make them fit.

Can they now?
I think they can now. They probably make $25,000-$30,000 a year. They can probably buy an $80,000-$90,000 house.

Do we have $90,000 houses?
We have some $90,000, yes. We just came back within the last couple of years and we're getting more into it. And we've got apartments, of course, we've got a couple of thousand apartments which would help them. That bothered me, not having school teachers able to buy a house. We really had the prices down pretty low to start with, and then when the crunch hit, we

couldn't afford to build them because there were too many cheap ones around here.

Beginning in the late, mid-'60s, you devoted a great part of your energy, your talent, your time to The Woodlands. If you had it to do all over again, would you do that?
Yes, I think The Woodlands is a very important part of what I've tried to do. Naturally, energy is very important but you don't see energy, it's under the ground, and we've done a lot of good on energy on the national level. And I think that Galveston has been very interesting to me, too. But I think The Woodlands has a chance of becoming a standout example of how you urbanize better, how you become a part of Houston. We opted for that in 1972, to make Houston extend their boundaries around the whole 25,000 acres so that some day they will annex it. Not too soon though, I hope.

The residents wouldn't like that.
Of course they won't. A majority of the residents think that annexation would be unfair, but failure to annex would be unfair, too. All outlying communities parasite off the airport, where there was a billion dollars spent and they're going to spend another billion. They parasite off very costly toll roads, they parasite off the port, which makes Houston go. And they're not willing to be part of the whole? Come on, what's the matter with them? That's the philosophy I've had all the way from the beginning, since I saw Bedford-Stuyvesant, since I saw Watts, since all these cities are being destroyed all over the country, you name it, they're all being destroyed. Unless we figure out some way to relate to that, we've got problems. We'd better make Houston work. Those who live in The Woodlands had better make Houston work, in my opinion. They may not like it, and like you say, 90 percent would probably throw us out if they could, but it's shortsighted on their part.

I'm in the 10 percent. I believed what you said in the beginning.
Why should you parasite off those wonderful things? You couldn't do The Woodlands without the airport. You couldn't do The Woodlands without the toll road. You couldn't do without the City of Houston, its cultural base. But they're not seeing things clearly. Anyway, we'll keep hammering away and we'll see what luck we have. It's a microcosm of what could be done all over the country. Were you here when we had all the bishops here?

I think not.

When we first went to Columbia, we saw their ecumenical church. I talked to them and I found out there was like a war up there among the denominations, and I said that won't work. We had all the bishops come here at two or three meetings. And we got them all to agree they would do an interfaith arrangement. Every church denomination would do a certain thing. One would do early learning, another would do senior care, another would do this and another would do that, and they would coordinate everything through Interfaith and that would help them fund Interfaith. And now it's part of the Houston Open, they get part of the money from there. Interfaith has been remarkable, what we've done in The Woodlands. We now have 29 churches; we had a Jewish congregation starting in a Baptist church, believe it or not. I'll have to give Dick Browne and Hartsfield credit. How do you develop an area of beauty like this and be able to have people who have $30,000-$50,000 houses? Well, we had to create a pollyanna effect. The pollyanna was that you'd have $50,000 to $80,000 houses, and they all have trees, and then houses $80,000 to $120,000, and so on up. And the kids all go to one school, one neighborhood shopping center, one everything. That's the concept we put together. You couldn't have a $50,000 house next to a $500,000 house, that won't work, you know that. But the way we did it, thanks to the trees, we were able to isolate the areas into 50- or 100-family cul-de-sacs. So the concept that Dick Browne and our planners put together, they did a great job.

Roger has said that to build another new town, not necessarily here but anywhere, going over 5,000 acres is just about impossible now.

Let me say this. We're doing Wood Trace now, you know, that's a good example, 3,000 acres. When we first started working with McHarg, I had him working on what we called the string of pearls. We would do five projects of 5,000 acres in various parts of southern Montgomery county where the beauty is, and interweave them into Houston on rail mass transit. They'd all have their identity as part of the urbanization of Houston. If you listen to Barton Smith at the University of Houston, and I believe him, in 30 years we'll be 6 million to 6.5 million people. How do you make it work? That's the problem.

If you don't solve the transportation problem, which is acute, nothing is going to work.

The argument that I have with Lanier is that if you think it's going to get better with just HOV lanes, you're wrong. You've got to have rail and HOV lanes.

That's a change in your view, isn't it? You used to say everything could be done on buses.

Oh no, I've never said that. I've been hammering rail with the downtown people since I can't remember when. The rail has to be from The Woodlands. You know why I'm fighting for this people mover and the Union Station? Because the toll road is going to be plugged up in seven years or so. And if you don't get rail to the airport, you won't be able to get to the airport from downtown or the west side of town. So, therefore, we must go rail to get to the airport and to The Woodlands, and then it has to go to Montgomery, Fort Bend and Galveston, or Clear Lake, anyway. Rail is extremely important over a 20-30 year plan. But then Lanier doesn't believe in heavy rail, he wants concrete everywhere. But, rail is very important, I think. The only way you're going to urbanize like Chicago, you've got to have rail available in all your outlying areas. With 6 million people you must have bus ways, HOV lanes, highways and rail. And you have to have a lot of areas where you work, live and play in some kind of mall area, like we have in The Woodlands. The Woodlands mall area will be the downtown of a million people in 15 years. That's the way you ought to do it, same with First Colony, same with Clear Lake.

Those you just named, are they in effect the string of pearls now?

No, they're not. They're not what I'm talking about. They just came along and are doing well. But my concern is how do you take care of the next million people? You'd have, say, a string of five pearls of 5,000 acres each and each with 50,000 people. That's 250,000 people in southern Montgomery County besides what we're going to have in The Woodlands, 150,000, all related to Houston.

150,000 in The Woodlands?

Yes, that's what it's designed for. You see, we've only got 16,000 dwelling units and we expect to have 54,000.

Say again about the string of pearls. That's not going to happen in our lifetime, is it?

No, because I've given up on it except for Wood Trace. That's the one up to the north, 7,000 acres, that'll be good when the Aggie expressway gets through. The place where the Sealy-Smith Foundation was, next to my ranch, would have been a great place, but we're having to sell it off because of financial pressure. And we never did do the design that McHarg was willing to do. We just couldn't afford to get into it, but that's what ought to be done. Each one should have its own character and all of them should be related to Houston. The beauty is in southern Montgomery County. See, when I was in Memorial, I looked at it and said, "How would you do Memorial over again if you had 10,000 acres out there of wooded area, not the way it's been done, all upper class?" And the beauty of what we did at my ranch, how we developed that—the lakes and the open land and everything—made you realize the beauty of the woods. It was interesting when we first started Pacesetter Homes. We built some subdivisions on the west side of town, 100-lot subdivisions, small ones, never made much money. To do the subdivision, you put in lots, roads and streets, you have one lot here that's open land and one lot there with three trees, you get $1,000-$1,500 more for the lot with the three trees. I knew that damned well. I saw them destroying FM 1960 early. I said, "I don't want to do that." And 1960 is just Westheimer warmed over. That isn't the way to do it.

Let's wrap it up. We've been talking about The Woodlands, which is, of course, your signature achievement. Name some aspects where you think it's if not unique, most unusual, anyway.

Cynthia and I go out and eat a lot, and never a week passes by that some people don't come up and say, "I've been here 10 years, 15, and I want you to know it's one of the most wonderful places I've ever lived in." And they're sincere, I can tell. And I think it's because of the quality of life for the people, particularly those who have families. Now we're getting a very big senior citizen group coming in. Their families may be somewhere in The Woodlands. These seniors aren't in a big senior citizen compound, but they're scattered around. The quality of life and the lifestyle is what's doing it. And now we've got the mall, we've got this and we've got that. I'd say that the Research Forest is extremely important to The Woodlands. That's true of the conference center, too.

You talk about the grouping of the neighborhoods, too, so that the $500,000 house can be next to the less expensive homes.

It won't be next to, but it will be within the same area of shopping and same area of schools and so on. And I think that the research center, HARC and the Cynthia Woods Mitchell Pavlion could be very important. It's still embryonic, we'll see. I think that the conference center is working well, especially with the new Shell commitment. There's the beauty of golf, tennis, everything else, great school system, and the Cooper School has been a big asset. One of the biggest things we did was the Montgomery Community College. We tried to get U of H there for 20 years, you know that, never could get them to do it. The community college really took the ball and ran with it and we got them and gave them the concept of the University Center. Pickelman really did a job on that. Ashworth has always been trying to help me get something going with the U of H, but the U of H dragged their heels and the Coordinating Board wouldn't do it. But the University Center is the way they should educate all the people in the State of Texas. You can't cram people into A&M and The UT anymore, they're too big already. You've got to allow people to go to school near their job or near where they live, even through the master's level, in order to serve the people. So this can be a tremendous asset. Leonard Rauch, who's head of the Coordinating Board, said this is the most unique plan we've seen in Texas, that we've put together, north of The Woodlands.

Hardy Toll Road has certainly been a major factor.

It has been major and what happened on that is an interesting story. When Kathy Whitmire came into office, during her first month, I called her and said, "I'd like to come talk to you, Kathy, there are two things I want to talk about." So I went and had lunch with her. The first thing was, she was going to de-annex Clear Lake. The politicians were on her side. I said, "Kathy, don't do that. Houston cannot grow if you de-annex Clear Lake." NASA was trying to get started down there. Well, she lost on that issue, luckily. Clear Lake is part of Houston now. Then I said, "Kathy, there needs to be a toll road running north of the city." She refused to do the toll road. I said, "Kathy, you're going to spend a billion dollars on the airport and you can't get there. I can't understand what is wrong with you." She wouldn't do it, so I got Lindsay to do it. I said, "Lindsay, really, there's no way to get to the airport. You remember how tough it was 15-20 years ago when traffic was so jammed?" So Lindsay had enough guts to do it. He had to put that and the Sam Houston tollway together and they're working beautifully now. But anyway, the tollroad is extremely important for the airport as much as anything and for The Woodlands, secondarily.

I think I've got to come back to you one more time.

* * * * * * *

Today is July 11, 1996. This is Joe Kutchin and I'm with George Mitchell in his office in Texas Commerce Tower. Today I'd like to get into your extra-curricular activities. I don't think we can cover everything you've ever done but I've got a list here of what I think are your most important accomplishments, and certainly what you and Cynthia have done in Galveston would be right near the top of the list. Why do you put so much effort in it?

It's my hometown and I have a great interest in my hometown, if I can help it get on its feet, because it's had a terrible time. Second, historic preservation is interesting to me. I knew all those structures when I was young. I never paid much attention to them. Then after World War II, Cynthia and I would go down there and then we built a house in about 1951 there. We'd drive around and we'd see these beautiful homes on Broadway being torn down for convenience stores and service stations and it kind of made us sick to see that happen. I couldn't do much about it at that time, but it did bother me, the sadness. Then Barnstone wrote this book, *The Galveston That Was*. It's a wonderful book. In 1962 or 1963, I read it and it impressed me. He had Henri Cartier-Bresson and Ezra Stoller, both of them great photographers, come into town to take pictures of historic Galveston buildings. Barnstone was an architect here, died about 15 years ago. His book had all these great photographs of existing Galveston structures and many of those that had been torn down. It was a very sad book about Galveston's losing its entire heritage. It was a beautiful heritage because of the wealth of Galveston back in 1880 and the beauty of the structures and the homes, and now it was being destroyed. That impressed me. Then in 1971, I became acquainted with the Chief Executives Organization, went to Hilton Head– this was the organization's first year and this fellow Mills Lane, who was head of the C&S Bank, the biggest bank in the Southeast, said, "I want you all to go to Savannah with me, let me show you what's going on, what we're doing." So a group of three or four of us went over there and it was very impressive. He had created what he called the Savannah Plan. What they did was they got the Federal Government to put in a couple of million dollars and the citizens to put in a couple of million dollars and they were buying and restoring houses in Savannah. They had restored 2,000 homes. That impressed me and I said, "Well, damn it, why can't we do something like that?" Galveston had a ladies' organization which was wonderful but very small. They've restored the Brown Ashton Villa home that you've probably seen. They did create the East End Historical District, which was important. And they did the Samuel May Williams house. Those are the main things they did, which is impressive considering their size. So I said to them, "I'll tell you what. I'll pay your way if you'll go to Savannah and look at the Savannah Plan to see if it's something that we could do in Galveston." It took them seven months to take me up on my offer, but they did go. Anyway, they came back all enthused and they talked to Mrs. Northen, the Moody heir, to see if they couldn't get someone to make a study of Galveston to see if we could do the same thing as the Savannah Plan. So Mrs. Northen gave them some money to go look around the country to see if they could hire someone. They hired Peter Brink from New York, who was a lawyer and had done restorations. He came to Galveston and became enamored with Galveston. He made a study and was very impressed with the town and its potential. I can remember during my time in Galveston in the late '30s, '40s, even the rats wouldn't go down the Strand, it was so bad . . . decrepit and the winos, just a very run-down area. So he came back and was very impressed with what was possible. Then he got the Moody Foundation to put up $200,000. Mrs. Northen was the Moody family heir, great dame of the Moody family. She was on the Moody Foundation Board at that time. She died just a few years ago. So she had the Moody Foundation put up the $200,000 to hire Peter Brink who decided to come to Galveston and start organizing, in a more effective way, the restoration. The way it had been done for 50 or 80 or 100 years was by a very small group of ladies, really wonderful people, but they didn't have the dynamics that Peter Brink had, the knowledge that he had. So they introduced the Savannah Plan in 1973 or 1974, and started acquiring structures that were old and dilapidated. A structure that used to cost $400,000 you could buy for $20,000, $30,000. They'd buy them and resell them. Then we got involved with Peter Brink, a very dynamic person, and then we worked on the Elissa Project, we worked on the restoration of the Opera House. Also, I bought the Wentletrap Building. The first one I bought, in 1975 or 1977, was the Thomas Jefferson League Building, where we built the Wentletrap Restaurant. We had Boone Powell work with us. He had been working on Galveston with Peter Brink for three or four years before I hired him.

You were buying these old buildings at Brink's recommendation?

What I would do, I'd look at a building that I thought made sense and try to figure what to do with it. If it was available I'd buy it from the Foundation if they owned it. They owned most of those old buildings. Then I would agree to keep the facade and do the restoration according to what the rules were.

But the general concept was Peter Brink's?

The general concept was the Savannah Plan. Peter Brink knew what should be done. He knew which buildings were worthwhile. You won't believe it, but I had to help Peter fight hard on saving Old Red, that beautiful first medical school west of the Mississippi. They were going to tear it down, The UT was going to tear it down. I couldn't believe it, Joe. Then we saved the Santa Fe Building at the last minute. They had a contract already awarded to tear it down for $200,000 and I worked with Peter to stop Santa Fe from doing it. Mrs. Northen bought it and her Foundation put up the money for the restoration. Just by the skin of our teeth. Anyway, Cynthia and I got involved with the next big project which was the Tremont House Hotel. We acquired that and we started that construction, I think, in 1979 or 1980-81. It went through Hurricane Alicia. That's when it nearly fell down. I have a slide show of before and after. You can't believe the slide show that I have on about 15 buildings before and after we took them over. You could tear these buildings down and build something at half the cost, but they wouldn't have the beauty, they wouldn't do well over the long term. I haven't seen the whole restoration program turn around yet but I'll say this, I still think that if we can ride out the shortfalls that occur over the years, the restored buildings will do better than if you'd torn them down.

There's no personal gain in this for you anyway, is there?

Right now there's no gain at all. Oh, there's a tax credit, but that's not really a big help. In the end, the Bacon-Davis Act costs more than you get back in tax credits. If we get the tax credit, which we've done with several of them, we have to pay $17 an hour under the Davis-Bacon Act for labor, versus $9 an hour otherwise, and therefore you're a net loser. But then we made progress on the restorations and we worked with the Texas State Historical group and the Federal Government. The Tremont House is a beautiful hotel. Condé Nast says it's one of the 500 best in the world. They're the largest travel magazine, and they made a survey of all the people who had been anywhere in the world, and the Tremont House was voted as one of the 500 best in the world. Anyway, I had a big fight with the government. They require that you restore the building both inside and outside, and that got tougher as time went on. In that hotel we had a beautiful brick wall in the atrium and we had three or four rooms with brick walls. I wanted to leave the brick walls there because they were there in 1900. They argued no, they weren't there when they built those hotels, everything had stucco. They threatened to withhold my $3 million tax deduction, and I went to Washington twice to see Senator Gramm and others to say, what the hell is the matter with these people? We'd done about 98 percent of what they wanted and I didn't want to cave in on that because lots of the beauty of the place depended on preserving the brick. We finally agreed I'd put sheetrock over the walls in the restaurant and they'd leave the atrium with the brick exposed, so we settled it. But that's how difficult it is to work with restoration. But we've had great architects—Boone Powell did some great work for us.

Are other parts of the community involved in restoring the Strand?
Let's say this. I've restored about 16 structures. We did the Washington Hotel that burned, we restored that and that's why we call it the Phoenix. It came out of the ashes. Then we did a lot of the other buildings and I think that others are beginning to do more now. Some of the houses were done even before I got involved. I didn't do houses. A lot more have been done lately. There are some more downtown things that have been done the last 10 years, and the whole effort is beginning to pick up steam. We got the trolley downtown, through federal grants, mainly. That's really made the downtown. Then we did the trolley to Fisherman's Wharf the last year. We tied it all together and now we have the trolley running on weekends free, and they're continually serving more passengers. So it's beginning to come alive, but it'll still take another three or four years to see if it's going to be economic or not. In the meantime, we'll just have to ride out the storm.

The Savannah Plan is kind of the overall concept, right?
Yes, under the Savannah Plan you have a pot of maybe up to $1 to $2 million dollars. They had $2 million. We had about $400,000-$500,000 to start out with. They'd buy houses and sell them, buy them and sell them, buy them and sell them. When they sell them, they require you to do the architecture like the exterior is supposed to be. And they'd help get financing on some of them.

Is there an organization in Galveston that shepherds this whole thing along?
Yes, Peter Brink did for years. He went to Washington to be second in command of the National Trust, which is very important. See, what many people don't realize is that the Galveston Trust has one of the greatest aggregations

of Victorian structures in the country and has the greatest number of iron-front buildings in the county because there was such wealth here in 1880-1890. There are 2,200 Galveston structures on the National Registry, and there have only been 50 commercial buildings restored out of 300. Only about 400 houses out of 1,700 have been restored. We have a long way to go, but the momentum is increasing.

Does the Galveston Historical Association have a role?
Yes, they are very important. They do the paint partnership for the residential area, they try to get people to know what to do, they try to help control buying buildings and selling buildings. They've always been deeply involved. Peter Brink was very strong and he was the one who helped set it up and it continued even beyond him. Cynthia and I received the National Award about two years ago up in Boston, we received the State Award two years ago for best restoration done in the state. We've received a lot of recognition.

It sounds like you provided the original impetus . . .
On the commercial part of it, yes, and then the non-commercial took hold.

And that job's not done yet?
I think that some of those buildings are beginning to show evidence of turnaround. It will be three or four or five years before you see the cash flow look like you're making a pretty fair deal on what you've got in it.

Is the creation of Mardi Gras part of the restoration package?
Yes, that was part of it. When I was young, Mardi Gras was a very famous social event in Texas, in Galveston. It was one of biggest, other than Mardi Gras in New Orleans and Mobile. In about 1940, when World War II came on, it died for four or five years. And then a small group, like O'Connell School, would have a Mardi Gras, but nothing of any consequence. So when we were getting ready to open the Tremont House Hotel in 1985, I said, well, why don't we start Mardi Gras again and see if we can't get it to catch fire. So we underwrote the first year or two of Mardi Gras, Cynthia and I and Dancie Ware and others who worked on it. We used it as the opening of the Tremont House Hotel in 1985.

Did Mardi Gras and opening of the Tremont coincide?
Yes, within a short time of each other. Each year since it's grown. The first

year there were about 150,000 people. It's gone up to about 500,000 people over two weekends. We've now got two weekends. It's been very successful in livening up the town. I noticed that the town in February was absolutely dead and the Mardi Gras gave a big lift for the people, because people have a terrible time making a living. So it's been a shot in the arm in the worst part of the season. It's been very successful. Galveston has had Mardi Gras since 1867, so there's a great history of Mardi Gras. It was very popular when I was a young boy. Christie and Johnny used to go to the balls, but I was too young for that, about 15 or 16 years old. I saw the parades and everything. It was a big social event for the whole State of Texas and we just brought it back again in 1985.

You provided the original impetus to bring it back? You put a lot of money into it?
Yes, we put it together, Cynthia and I, Dancie Ware. We had to spend about $150,000 to $200,000 the first year but we were opening the hotel, too, and Mardi Gras was good promotionally. Also, we wanted to see if it would take hold and help Galveston in the off-season. It has done a lot.

And your personal support has not been as essential as it was?
No, we've got many other people involved. It's been mostly self sufficient recently.

Is it a similar story for Dickens on the Strand?
That came before I got involved. The original impetus came from GHF, the Galveston Historical Foundation. They were in existence a little bit before Peter Brink was there but it was just a group getting started, primarily to work on housing in the east end. Then they began to get together on a little Dickens party at Christmas because of the historic Victorian structures. I think Dickens is in about its 20th year now. It began to evolve and it's come to be a very major event, too, and it draws about 100,000 people a year. That's the first weekend in December.

And again, you and Cynthia are major supporters.
We support it, but they had started it even before we were major supporters. It was just a very small thing before but when all these things started going on down on the Strand, why it got more and more life to it.

Talk about what you're doing with the Galvez. Or maybe you should

talk about the San Luis first.

The history of the San Luis is interesting because it was an outgrowth of the company's drilling wells offshore. We picked up the property originally to drill directional wells after we made the discovery back on the railroad tracks in Galveston. First, I tried to get permission to drill offshore, right in front of Gaido's, right after the Shell blowout. Fat chance. The City said, "Go find someplace onshore." So we went and bought this piece of land and we designed an elaborate hotel with condominiums around it, all in order to drill seven slant wells. The church on the corner raised hell with us and the Council said, "Go back offshore." So we went back offshore and tried to get a permit. We went through two years of hearings, 2,000 people opposing it. Any problem at the beach, it would have been a disaster. It worried people and I don't blame them. So we had these hearings and we had a lot of opposition. You can't believe the arguments they put on. Anyway, we won out and finally got permits to drill. We had discovered the field onshore. We could see the structure. Then we looked offshore and then that's when we had to get the hotel site. After we got that, we couldn't drill onshore so we drilled offshore and made the discovery. We tried to get someone to take on building the hotel, but we weren't successful. We didn't want to do it, but we couldn't get anybody else to do it, so we said, "All right, let's see if we can do it." So we got a syndicate together and built the hotel and condominiums. It's a beautiful design, but it hasn't been successful. It probably will be in the future. You know, the hotel business went to hell anyway.

But building the San Luis was another major event for Galveston. I remember the fireworks display when it opened.

The most beautiful fireworks display I've ever seen, really it was a piece of art. The Galvez Hotel is different. We built the Tremont House, we built Harbor House, we did these 15 buildings down on the Strand, we got 10 factory outlet stores, a Discovery store and other stores.

Are those Mitchell properties?

No, we just lease those structures. We bought the Discovery Company and put the store in and it's done very well. Then the Discovery Channel wanted it and we sold it to them. We sold it because we bought it for about $600,000-$700,000 and we sold it two years later for $3.5 million. I did pretty good and that's the only time I made money out of Galveston so far. Some people keep saying, "Why do you keep fooling with Galveston?" Well, the company has made $150 million out of oil and gas on the island. So I've still got a big stake in Galveston. Everybody just thinks it's another sinkhole, but we've made good money there. We have all our land at Pelican Island and Pirates' Beach and Galveston and we have oil fields there. Anyway, the Galvez was one of the last ones we've done. Cynthia heard that it was about to be acquired. Aetna was trying to sell it. Dr. Cooley and Archie Bennett had bought it 12 years ago and had redone it and it wasn't successful. Our economy here went down during that period of time, you see. So it went back to Aetna who had put up the money for the loan. Cooley and Bennett sold it back to Aetna and got out of the obligation. So, Aetna was trying to sell it. They had tried to sell me the Galvez Mall and I said no but I might be interested in the hotel. Cynthia had heard that Aetna was about to sell it to a group of foreign investors. She said, "If you care anything about Galveston you ought to buy that hotel." She was afraid it would be destroyed, ruined, and she was probably right. So anyway, we acquired it and at a good price.

Did the company still have the San Luis at that time?
Yes, it had the San Luis. I had the Tremont House and then I bought the Galvez. Cooley and Archie Bennett had really changed the design of the original hotel by moving the main entrance to the back. And we had to go back and restore the entrance and get rid of the pool and redo the whole front. There's a beautiful entrance now and beautiful pool. We haven't finished all the interior yet, but it's still in very good shape. It does have the beauty that it once had and it looks like one of the big resort hotels, like at Boca Raton. It's just beautiful, with the landscaping in front, the great big palms and the pool. It's one of the few great structures on the beachfront.

Is this another case where profit has been a secondary motive?
Profit is secondary but still I was able to buy it from Aetna cheap, very reasonable, $3 million, and it's worth $35 million if you'd try to replace it. Of course, it needed a lot of work. So we spent $6 million more just doing the pool and the front entrance. Then there's another $2-$3 million to finish the interior. Our main thought is that it has such beauty that we didn't want it destroyed. I think that one will really make some money. Give it a couple, three years and it will do well. It's doing pretty well now. Some make money, some lose money. It takes awhile.

I know you're involved also in the theater down there.

The Strand Theater. We worked on the little Strand Theater which was really so run down, just like the Strand was. It's across from the Tremont House, so we wanted to fix it up. And the little theater interested us, so we had Boone Powell do a beautiful design with a fellow named Lloyd Ball, who worked for Boone. But you'll have to go down and see it. Have you seen it lately?

No, not lately.

It's gorgeous. It's probably one of the most beautiful little 200-seat theaters in this part of the country. And now we're trying to revive it. They're trying to get a new board, we'll see how it works out. We spent considerable money and had some UDAG money we used, Urban Development and Action Grant money, but mainly our money, to restore it and it is really gorgeous. And we worked on helping restore the Grand Opera House. You've seen that. A lot of people did that. Jim Ware was in charge of that. And one of the biggest things we did was to help restore the Elissa. For the Opera House, we were able to get a little federal money, and we've got about $800,000 in it. It ended up costing about $7.5 million to get finished. The Elissa–we were trying to find a ship that had been in Galveston in the 1800s. They found the ship and it was over in Greece in a junkyard, about ready to be junked. So Peter Brink sent about five people over there to make it sailable so we could get it hauled back. And I warned him, I said, "Boy, you go with that Greek ship, you're going to get eaten up." And they got eaten up. They came back, and they said those Greeks ate them up. Anyway, I helped them get a tug to tow it back. So we got it back to Galveston. Everybody pitched in, helped out. The Houston Endowment did a lot for that, and the Moody Foundation helped and so did the foundation in Dallas, I can't think of the name right off. Anyway, the Elissa was restored and the final cost was $6.5 million. It's a great ship for all of Texas. One of the most beautiful trips we ever made was when the Elissa sailed from Galveston all the way up to New York and paraded in front of Mitterand and Reagan when they rededicated the Statue of Liberty. About 100 of us from Galveston got on in New Jersey and saw 22 tall ships in review before President Reagan and President Mitterand. What a thrill that was. And Elissa was the oldest ship among the 22 tall ships. Anyway, it's a great ship for Texas, a great ship for the country. Galveston's got a lot of good things going for it, the Seaport Museum that we helped put together with Peter Brink, the waterfront has been restored, we did Pier 21, which was beautifully done. That's our own private interests, not a restoration, a new one but we designed it to be of the 1900 era. It was beautifully done. So all those little things are working together.

Is it fair to say without bragging on yourself that you've been the instigator and motivator and a financial contributor to a lot of this?
I think a lot of the downtown restoration was probably started by me, financially and also concept-wise. And I'm not saying it's fully operational, but I'll let you know in three or four years. Also, we're involved in this beach nourishment and I was involved in getting beaches alcohol-free. A lot of people are mad at me about that. I helped the concept to get the beach nourishment done. If you go down there now and see all those umbrellas on the beach, a 300-foot wide beach, it's beautiful down there. Now I've got to figure out how to keep the beach from washing away.

I figure from what I hear that half the people there would like to hang you and the other half would like to elect you as president.
There are some detractors, but I think most of the people feel we've helped a lot.

Let's talk a bit about your interest in growth studies. You got started on that in the early '70s, didn't you?
I read the book *Limits to Growth* by Dennis Meadows and it just impressed the hell out of me. I recognized that the book could have all sorts of things wrong with it but the concept was important.

Do you recall what caused you to read it? Somehow there was a YPO involvement, wasn't there?
There may have been, I'm not sure. Somewhere at a YPO meeting, *Limits to Growth* was a topic. But it impressed me, especially his thoughts about exponential growth. But the first thing that really stirred my imagination about this came at a YPO meeting possibly 35 years ago, at a seminar in Aspen, at the Aspen Institute. It was the first year I was in YPO, I think. Probably in 1961. Buckminster Fuller was there. He spent four days with us, the most fascinating person I have ever known. He began to make it very clear that the Spaceship Earth couldn't tolerate what was going on, just couldn't. It took two or three days to understand what he was trying to bring to you, and that stirred my imagination. Maybe my interest in growth issues and sustainable societies came out of that. There's no question in my mind that we are in trouble. It all comes down to this simple question: If

you can't make the world work with 5.6 billion people, how are you going to make it work with 11 billion? It's not going to work. World population is headed that way and you've got to cope with it, Joe. How do you cope with it? There are big population pressure problems, especially with the developing nations. The whole Southwest will be overrun in 50 years and what bothers me is that nobody in the business world gives a damn. That's what bothers me. I've got to go back and find 100 business leaders who are willing to look beyond the next quarter's earnings statement, Joe.

That's a steep uphill climb.

Well, we're making progress. When I started the growth issues conferences, "Limits to Growth" was our first topic. The next one concerned "Alternatives to Growth." Cynthia and I gave a $20,000 prize for the first conference, $50,000 for the second and $100,000 for each of the conferences after that. I knew our focus was not quite right. I know you can't have limits to growth. Alternatives to growth didn't make sense because some growth is important, some growth is not. Finally I came up with what I think truly describes the right direction: the "Nature of Sustainable Societies." You hear about sustainable societies everywhere now.

Gerald Barney ran the conference based on that theme.

He did the Global 2000 report for the Carter administration. Let me say this: The job he did for Carter was quite good, but it was a first-cut. He said it's full of errors, but he tried his best. You know he's head of a big church organization now that does similar work?

I've pretty much become a believer in your approach, but not too much. You know, Julian Simon has been right more often than your team has been, George.

No. I think you mean Herman Kahn. Meadows out-debated him all over the place. The thing that impresses me now is that when we started working on Alternatives to Growth, Limits to Growth, there may have been two or three organizations in Washington that were involved, and now there are 150 different organizations. We've gone a long way, Joe, but we're still not there yet. We've got to get the private sector involved; academia and environmentalists are not going to solve it. I tried for 10 years to get the National Academy of Sciences to take on this project. We were working with Frank Press, who was head of the Academy. And do you know what woke them up? What really turned the subject back on was that the Academy of Sciences went to the New Delhi conference four years ago. The 59 national academies said, "We're desperate. We've got to do something about this whole problem: population growth, resource depletion, everything else. Would you lead that for us?" I've been trying to get them for 10 years to do something. They came back all enthused before the Cairo conference and said, "Gosh, let's get to work on this." Our Academy of Sciences. That's the best vehicle I know that is impartial, though it has its share of politics, too. So the 59 national academies said, "If you'll lead the parade, we'll join and work with you."

That's a major accomplishment, to get everyone headed in the same direction.

It's so big, Joe, it's very difficult. I've always said it would take 20 years just to get a handle on it, but we've got to begin somewhere. That's why the private sector has to get involved, the people who have money, because it's going to affect them and their families. I believe that in 50 years, every major company in the Southwest will be affected. Not long ago, Mexico had 44 million people. Now it's 92 million. That's a true population explosion. There's no doubt about how vast changes have been. At another YPO meeting, Marshall McLuhan talked to us about the media's being the message. He convinced me that we in America may think that the people of Africa and South America, for instance, aren't aware of what's going on in this country. The fact is that everybody knows about it now, and they're all going to try to get here. And that's going to change the whole complexion of everything.

Your interest in growth issues predates 1974, but the first hard evidence of that interest was the first conference. Since then you've personally put in a lot of money through the Mitchell Prize, you've invested your own efforts and the company has been a supporter in various ways. Are you satisfied?

I would say this, it's frustrating not to get more of the private sector to realize, "What the hell's the matter with us, there's something going to happen. It's going to happen to our families, to our future." We've got to cope with it, you're not going to have much chance of changing it, you've got to cope with it. Everybody thinks that technology is going save us. Technology is important, it'll keep us going longer, but it won't save us unless we figure out how to get the political issues, the religious and social and economic issues worked out. It's a very difficult situation.

Here again is another case where you have put forth substantial personal effort and resources and none of that is going to put a dollar in your pocket. In fact, it's taken a lot of dollars out.
I just think it's important and we should do as much as we can to help out.

You're responsible for $1 million for this current project concerning the global commons and the National Academy?
Plus part of the money that goes to HARC is also being used for that.

And another $100,000 for the Mitchell Prize.
Yes, I'll have to fund that next year again. But it's an extremely important project. Rice University's DeLange Conference has joined us. That's fine, that helps. The Academy of Sciences has agreed to be very vocal, very out front, it's very important to work with them. The Academy of Sciences has the best reputation. We can't do it ourselves, Carter can't do it, it needs the stature of the Academy.

Is Jimmy Carter playing a role?
I wrote a letter to him the other day asking if he'd join me to go see our friend Ted Turner. His CNN did a lot of reporting on one of our earlier conferences, the one we had here 10 years ago. He's involved in environmental and growth issues, but he must be doing a big merger now. And he said when the merger is complete he's going to put $4 billion of his money into a foundation for these issues. I'm trying to get him to help us now because he would be great for publicity on CNN. He has already given us $300,000 in financial assistance. I've asked Carter to help, but Carter's got his own axe to grind. He's got his own Carter Center in Atlanta, so I don't know if I'll be able to get him to do anything.

I remember you made a sizable contribution to the Carter Center, and he was a speaker at the Mexico conference a number of years ago.
He's a great person, he really is. Out of all our presidents, I think he's done the best job of all. That is, after his presidency.

Thank you, George. I'll have to come back one more time.

* * * * * * *

Today is Tuesday, September 24, 1996. This is Joe Kutchin and I'm in

the Texas Commerce Tower office of George Mitchell. George, I believe HARC is at the top of the list of what we still need to talk about. You've said in the past that of all the things you've done, you felt HARC—the Houston Advanced Research Center—was at least potentially the most important.

Yes, I still think HARC could be the most important. But what's frustrating is that I can't seem to get, and Skip Porter can't seem to get, the Houston foundations and major companies to buy in on the concept. We have 15,000 jobs in The Woodlands now. We have 3,000 that are related to high-tech, biomedical and other technical companies. Together there are 300 Ph.D.s, as I mentioned before. But we have the chance to achieve critical mass, and HARC is at the center of that, if the large universities will work with us. But that is very difficult, and I can't get the Houston group to buy in on the idea that if you're going to really try to have a high-tech research area, the best place to put it in this region is in The Woodlands. Austin has that critical mass, the Research Triangle has it, we don't have it. I'm disappointed that we haven't been able to get enough support. But it's not over yet. We're pushing ahead.

You said the Houston group. What do you mean by that?

I meant the private sector and the foundations, except the Houston Endowment, which has given us half a million dollars a year for the last several years for environmental issues. And then the Fondren group just gave us $300,000 for telemedicine. But we haven't been able to get the type of funding that's needed. I can't do it all myself, nor can the company. We're working on it, and we keep them funded and we still have in mind to help them a lot more in the future. But I'd like to see HARC start doing more and earn their own way. We never planned to be, for instance, like the Southwest Research Institute, which is strictly for profit. They don't get foundation money, they just have a research institute that does work for the private sector for profit. We have academia involved with us and that's hardly ever for profit. We're trying to work it now to where they would try to do some things for profit, like we do in geophysical now. We've got 10 companies helping us there. And then you still would do some of the start-up ideas, like the data compression concept. That hasn't worked out yet, but that's been on the right track, with university people working on the concepts and ideas that might be useful in the future. I still think HARC has a lot of potential. HARC is the anchor of the 32 companies around here that employ the 3,000 scientists and assistants who are here now. We have to build that critical mass to at least 5,000 or 7,000 high tech jobs.

Isn't part of your point, too, George, that as is true everywhere, some of those jobs you're talking about in The Woodlands are at the very low end?

Yes, part of the 3,000 are low-paying jobs.

So you're looking for the higher-paying jobs?

Yes, I'm looking for more of those for the Houston region and I do think that The Woodlands Research Forest is the best place for that to happen. We've got plenty of other kinds of jobs—we've got restaurants all over the place, offices, we've got a shopping mall. You name it, we've got it. But HARC and the Research Forest are all about higher end, professional jobs. They've done an enormous job at the Research Triangle in North Carolina. I'm trying to convince Houstonians that the same thing can be done here. It's not an easy message to get across.

As I understand it, HARC primarily seeks grants rather than sponsored research. Is that right?

We have some sponsored research and some grants, as well, but we need more sponsored research. In sponsored research, the sponsors pay direct costs plus overhead. We're doing other research with academicians and that's focused on developing new ideas. You have to have seed money to do that. And there's a problem if you have too much of that. Right now, organizations like the National Science Foundation, the Navy and others are cutting way back. I've talked to the people around the country, and I've found that all the universities are having trouble finding funding. I think that pressure exists because of government cutbacks. What we're doing now at HARC is, we're saying to major companies, oil companies, consolidate some of your research at HARC. Texaco, for instance, built a $2 million lab, a geochemistry lab—it's the latest, state-of-the-art lab of its kind in the world. And we're trying to get other major companies to join them in the work being done there. No one seems to be concerned any more about proprietary research. So I think that it's feasible to get that done at HARC and we're trying to get other companies to do the same kind of thing as Texaco. We have a rock physics lab started by Unocal, and we've helped to get other people join in the rock physics lab. We think we can get three or four major corporate involvements like Texaco and Unocal and get other major companies involved. Our geotechnology center is growing, our global studies component is growing. Jurgen Schmandt has been doing a good job there, and the Houston Endowment has provided some important funding. But they've had some problems connected with building a magnet for the University of Illinois and that wound up costing HARC $2.5 million. That's hard to overcome. Anyway, we settled without big lawsuits. And then this data compression controversy has cost them $1.5 million, which I'm not sure we're going to recover. It's great science. Shell just made a deal with us to offer that to the geophysics market. It starts as a $75,000 deal, which is not a lot. We get a 40 percent royalty if they use it somehow. But over all, I'm disappointed in the reaction to HARC in Houston. We've tried to get every foundation to help us and we've tried to get the private sector. We got some help, but it takes more than what we've gotten so far to really do a good job.

Go back to the beginning for a couple of minutes. You founded HARC. Give a little of that early history.

I thought that the region needed something like HARC. MCC made Austin come alive. They got a lot of the chip manufacturers, a lot of the computer companies working together there in a big research effort. I think that we could have done the same kind of thing here with more community support. Anyway, it was clear that the concept was still good, that it's working in North Carolina. But I think that Houston has such outstanding dynamics that over time we'll be successful. Remember, we've only been in it for 10 or 11 years. The Research Triangle is 50 years old now. Stanford Research started before World War II. MIT and Route 128 started before World War II. So, we're very young and still we've done quite well. We've done about $120-$150 million worth of research in the last 11 years.

Harvey McMains, who was originally in charge of HARC, goes back farther than 11 years, doesn't he?

McMains started with a little something, but it wasn't much. I feel that HARC really got started with Skip Porter's arrival about 11 years ago when we started on the SSC. I think the potential is still there and what is interesting is that other things in the Research Forest are working pretty well. We've got a lot of biotechnology over there, a lot of medical technology, we've got some energy people in there now other than HARC. HARC is kind of the nucleus of what should be around it.

Do you want to say a few words about the original thinking?

We had Arthur D. Little do some studies about the medical complex, probably 15 years ago. We did a study on how the medical profession in Houston might respond to some of the research going out to The Woodlands. Then we had them do a study on whether it was feasible to have a research center here, working with three universities. The report cost $300,000 and was very complete. Basically, the report said there was a place for a research center if we could get all the universities to cooperate.

At that time it was three universities?
Yes, Rice, U of H and A&M, and after they came in I asked A&M if it was all right if I invited UT and they said, "OK, go ahead." The UT lent some help, particularly in the global issues.

Another of your major undertakings has been the Cynthia Woods Mitchell Pavilion, and now you're considering a theater. What was your thinking in founding the Pavilion?
It's interesting. When I was recruiting Richard Browne from Columbia, in 1969 or 1970, I went to Columbia, Maryland, and saw what was going on. That's when I recruited Len Ivins and all that bunch. Browne and I went to Merriweather Post Pavilion in Columbia, very small, that Rouse had built. It would hold about 1,000 people. I saw what was there and felt it was very interesting. After we opened The Woodlands in 1974 and after we had the 1975 cash crisis with the banks, then Dick and I started thinking about an outdoor entertainment center again. And after a while, Roger Galatas became interested, too. The oil crunch was about that time, maybe '84, '85. So we said, "We'll put it on hold." After the oil situation began looking a little better, in the mid-to-late '80s, we started thinking about it again and started to gather data. I think maybe about that time, Phil Hoffman, John Cater and Ben Woodson came to me to talk about a summer home for cultural arts events in The Woodlands. Dick Browne and I had already done preliminary thinking and we had information about other places–Tanglewood in Massachusetts, where the Boston Symphony plays in the summer; Ravinia near Chicago; Concord in California. Hoffman and Cater and Woodson came to see me and said, "We'd like to see if you would do a performing arts center in The Woodlands and create a summer home for the Houston Symphony." And I said, "Well, we've given some thought to that, and we'll give it some more thought." But that was still during the crisis. We finally started to move ahead with the idea of building something with possibly 6,000 seats, spending maybe a couple of million dollars total. So, we

talked to the Pace organization and Pace made us smart. They said, "It won't work, you won't have enough size to get any decent acts." And they were right, they gave us some good information. So, we had to change the design to 10,000 seats and it ended up costing $10 million or $12 million. The Pavilion has done very well. After a while, we got Pace and the food people to give us $3 million to expand it to 13,000 capacity, and even with that, it's frequently filled.

That was the Ogden group and Pace group.

And Cynthia and I gave the Houston Symphony a $5 million, 10-year endowment so the Symphony could perform at the Pavilion and make it their summer home. And we've worked with the Ballet and we've worked with Houston Grand Opera to expand their presence at the Pavilion. Next year they're planning to do a whole opera, "Carmen," out here. They're going to tour it around the country and the Houston Endowment is giving them a half-million dollars to build the sets. We'll get a sponsorship to pay the costs and we'll actually have a full-blown "Carmen," with sets and all at the Pavilion. The Ballet has done a great job with the few performances they've had. We're trying to build that up even more. I think next year they're going to try to do "Swan Lake." That would be great. So, we're getting there; it's coming along a little better. We give them all some endowments, the Ballet a $250,000 endowment. Attendance is going up about 20 or 25 percent a year and, hopefully, the performing arts someday will be even more successful. We've got about 25 non-Pace-type programs there. We have middle-of-the-road programs like Linda Ronstadt and Bette Midler, we've got Alan Jackson coming next week, and Hootie and the Blowfish. The noise is terrible. But let me say this, they pay for the Symphony. It's been a very fine venue for The Woodlands and for the whole north side of Houston; in fact, for all of Houston. Beautiful design—it really has a great deal of beauty to it.

It has just been a bang-up success from the moment the place opened.

It's one of the few really "art" things in town that break even, and we hope to keep it that way.

There are a lot of different reasons why you might want to have something like the Pavilion. What was on your mind?

For one thing, I wanted to help build the classical performing arts on the north side. And we wanted it to be an entertainment center for all kinds of performances for the people in The Woodlands, plus people around here.

Although The Woodlands Corporation put money in, I know the Mitchell family was very generous in helping to finance the Pavilion.
Well, it's called the Cynthia Woods Mitchell Pavilion, named after Cynthia. We've put in several million and still give them about $70,000 a year to help them out. The Woodlands Corporation gave them the land and provided quite a bit of the construction costs to get the Pavilion off to a good start.

What about the theater?
Until the lawsuit against Mitchell Energy & Development is out of the way, we're doing the design on the little theater, but I don't want to start on construction until I see how we can raise half of the $5.5 million cost, or if we have to put up the full amount. But no matter what, I want to wait and see how this lawsuit comes around.

There are going to be 500 seats?
Yes, I think so. One thing is that both the little theater and the Pavilion are on the waterway. And the waterway will be started in four or five months because we just received a $4.9 million grant from the government to build the transit way to the Town Center, and the waterway is just part of it. Really, it's the transit way that they give the money for. And we got another $1.3 million just last week. So we've got about $5 million or $6 million, which will do most of the waterway. This whole waterway over a period of years will be, I think, a tremendous success for the whole area. The downtown of The Woodlands, where the mall is, will eventually be the downtown of a million people within a 20-mile radius. It's beginning to look that way already.

There were some rocky times in the early years, but The Woodlands just seems to get more and more momentum each year. I think you're a super geologist, George, but you're going to be remembered for The Woodlands.
That could be.

You've mentioned the endowment that you gave to the Houston Symphony. I know that you and your family have been generous to other institutions.

Well, we've done something with the Ballet, something for HGO, but the biggest thing we've done was with the Houston Symphony. And we've done a lot of work in Galveston, too.

I know, too, that you've been involved in things like the Texas A&M Target 2000 Project, you've been on the state higher education board, you've been chairman of TIPRO.

On October 7, I'm to receive the Lamar Award for my involvement in public education. It's given by the Association of Texas Colleges and Universities. My award is mainly for my work with John Pickelman, chancellor of the community college, to do the University Center at The Woodlands. The University Center is very important at The Woodlands and the community college is very important out there.

Do you want to say a couple of things about Target 2000? That was a major undertaking.

That was a very big two-year project. I was chairman of a committee that called on 250 people from throughout the state to help plan A&M's future. That was about 15 years ago. It's amazing that they used a lot of the report and actually implemented it. Many of the chancellors told me later that the report was very important because when the faculty said, "Well, you can't do that to us," they'd say, "Well, that's what Target 2000 says." I will say this: It has been a very strong report for A&M and I think it's helped the university a lot. It had a lot of interesting things about the funding needed for A&M. I remember they had a development program of $64 million at that time. I made the analysis and said, "In 10 years you're going to need $600 million to be a competitive university," and they nearly fell over dead. Today they're up to $590 million.

You've been involved in education at the state level.

I was on the Select Committee for Higher Education for two years in Austin, I was chairman and president of TIPRO for four years, on the National Petroleum Council for 20 years, or more. One of the most important things is that I've been a member of the All-American Wildcatters. That award is given by geologists and petroleum engineers and land people. They pick three All-American Wildcatters each year, and to me that was very important because energy has been my main profession. But probably the most important award was the Lorax Award that President Carter gave me. The Lorax was given by the Global Tomorrow Coalition, Don Lesh's group, and Carter was there to present it to me. It was a surprise. The Lorax is from a Dr. Seuss story. The Lorax said, I speak for the trees because they have no tongues. That's a great statement.

You have felt that YPO—the Young Presidents' Organization—was very important in your life. And its offshoot, the Chief Executives' Organization.

YPO and CEO are pretty much the same. You come into YPO before you're 40 if your company's business volume is at a certain level and you have at least 100 people working for you. And you become part of CEO after you reach age 50. Probably the most important broadening that I've received was from the YPO university where you would spend five days every year. The courses were half business and half humanities. And through that involvement I went to Bedford-Stuyvesant to see the disaster in New York. And I went to Watts to see the disaster there. You and I have talked about de-annexation and the flight of the middle class to the suburbs, and that's what was destroying all of our cities. I could see in '66 what had happened to Bedford-Stuyvesant and Watts and that's where the concept of The Woodlands came from. I hadn't been in the real estate business on any large scale at that time. But the idea for The Woodlands came through the connection I had at YPO. They had seminars that were fascinating and had nothing to do with business.

Was it at a YPO seminar that you met Buckminster Fuller?

No, that came at an Aspen Institute seminar 35 years ago. Buckminister Fuller was one of the most fascinating men I've ever met. He had a three-day lecture. It took three days to understand what he was talking about. He had a tremendous mind. He's the one who first gave me the concept that the the globe is in really deep trouble. He described it as Spaceship Earth, hurtling through space, chewing up resources. He said we were on an overpopulated planet and that disaster is heading our way. That was very fascinating and I think the significance of growth issues first hit me at that time. And later on, of course, I read the *Limits to Growth* back in the early '70s and then we got Don Lesh aboard. But the first thought came out of the Aspen Institute, I guess, about 10 years before that.

You say the idea of The Woodlands came out of some YPO seminars and that your interest in growth issues came from your exposure to Buckminster Fuller at an Aspen Institute event. Are they related in

any way?

Not really. Not to minimize the significance of The Woodlands development, growth problems are fundamental and universal. They affect everyone on earth today and will for generations to come. We're having trouble making the world work with 5.5 billion people. How are you going to make it work with 11 billion people, Joe? One day we're going to have that population. How will we cope with it? Science is important, population control is important, you have to look at resource depletion. All of those things are important. How do you get a handle on this big thing? It's an enormous thing, an enormous concept. In fact, we're working with the National Academy of Sciences now on a three-year project. I keep telling them it's going to take 12 years before we have any idea, really, of how we're going to even tackle this thing.

About the basic idea for The Woodlands: That came after you saw Bedford-Stuyvesant and Watts. And it was YPO that took you there?

That's right. I got to thinking, what the hell's the matter, and I got to thinking about Detroit, Cleveland, New York. All the big cities are in deep trouble. And why? Because the political subdivisions surrounding the city and the flight to the suburbs by middle-class whites were destroying the central core. And that's what we're arguing about today, Houston's annexation policy. See, I'm the one who says, "You have to have annexation, although you ought to try to leave as much freedom as possible to the people being annexed." But that's a tough issue.

George, in 22 years you've made a convert of me but I don't know if anybody else believes it.

Well, I agree. Still, you have to understand that Houston will not survive properly when we get to six million people unless we begin to figure out how to make things work. Houston and Kingwood are in a controversy now about the city's plan to annex the suburb. They're working on a compromise now which may have some merit. The idea would be for the suburb to be part of the regional concept but you still give it some autonomy so that it is responsible for garbage collection and things like that. Maybe something can be struck that would be acceptable. Otherwise, there's going to be a tough fight.

What about Universities Research Association? Is that a big thing?

It's very important, yes. It's very intriguing for me to see how 72 universities try to work together on big scientific projects. It came out of the Superconducting Super Collider program.

And you're a member of the Board of that group?
I'm a member of the Board. It's about a 15-man Board. They have about four members from the private sector, five out of universities and five others. They're in charge of the Fermi Laboratory. They were in charge of the SSC Project, which unfortunately was scuttled, and I think that was a mistake. It's fascinating to see how they try to work together and have cohesion to go after major projects that benefit the nation. The SSC was the main interest I had at the time I got involved. I'll probably continue for awhile. I only go there twice a year, but they still have the Fermi Lab and they've still got a lot of things going on.

Gas Research Institute—are you still active in that?
Yes, I'm on the board of the Gas Research Institute. They have the best research programs for natural gas in the country. Even a major company can do only so much research on its own; GRI is able to do much more. It is funded by a tax on consumers that amounts to about $200 million a year. They're in Chicago. About one-third of the money is spent on the supply side, which is oil and gas exploration; one-third on hauling it across the country–pipelines; and one-third on end use. They've done a lot on coal seam gas, and Mitchell Energy has benefited from that; they've done a lot on hydraulic fracturing; they do a lot for the end user. Now they're working on compressed natural gas vehicles, they're doing a lot of work now with gas air conditioning.

Is the main function of that organization to direct the flow of research money?
Yes, to projects that affect the gas industry.

I just want to say that you've got an honorary doctorate from the University of Houston. I keep meaning to be more respectful.
Yes, it's in the field of humanities because of my work on The Woodlands and things like that.

And you founded or were among the founders of the Houston Racquet Club?
Yes, that's an interesting story. Just as I turned 45, I played in the Senior

Grass Tournament, national seniors, in Brookline, Massachusetts, on grass. And we did pretty well, we got to round 16, and that's pretty tough competition. I had a pretty good player to help me. I wasn't that good, but he was good. But what I noticed startled me. Here's this very prestigious tournament but in a facility that you could tell was 100 years old. No air conditioning. It was pathetic. I thought it could have been restored and done better. So I said, "Hell, we can do better than that in Houston." Country clubs like River Oaks in Houston hated tennis members and I didn't play golf. I played tennis with A&M, I was captain of the tennis team. So I said, "Let's see if we can't start something here." I was the head of the Houston Tennis Patrons here in town about a year or two before. Rennie Baker, who was the head of the Tennis Patrons at that time, and Ruddy Cravens were my tennis buddies. I said to them, "Let's see if we can't organize a tennis center. A beautiful center." The first thing I did, I got an option on 10 acres right where Voss hits I-10, that little triangle. I was going to build right there. The Houston Tennis Patrons were promoting tennis here in town, helping disadvantaged kids, and among us we agreed that we needed something bigger than that triangle of land. Besides that, the site that we finally got was near my home. So I went to see Bering about 22 acres he had, but there was no access on Memorial. I made a deal with him for about $7,000 an acre on 22 acres, plus an option. Then I went to my friend George Butler, who was with Butler, Botts, Binion and Cook here. He had seven acres on the front and I gave him $12,000 an acre. That still made the whole thing very reasonable: seven acres at $12,000 and others at $7,000 or $8,000. In the end we had bought options on 29 acres at an average of about $8,000 or $9,000. Then I went to the Village of Hunter's Creek and said, "We want to build a tennis center." They replied that that would violate an ordinance. So I said we wouldn't have lights to bother people, we won't do this, we won't do that. And after two or three tough hearings they gave us permission to build. Normally, that land would be worth $88,000 an acre if you wanted to build houses, but we got it for $8,000 to build a tennis center. Then with help from American General we organized. I had Karl Kamrath do the rendering. He did it for free. I was trying to coordinate it all and to organize it and start a sales campaign. We sold about 300 memberships to our Tennis Patron people. We needed a little bit more than that, we needed to sell at least 600 or 700. I went to the Sagewood Club—they were at Sage and San Felipe or somewhere near there, where a big tower is located now. They wanted to sell out because it was just a drinking club, and they were having trouble making it. I had lunch with some of their leadership one day and told them, "If you're willing to be Tennis Patrons, we'll take you in." And they said they'd go along. So we merged with them and they sold that property and we gave them a good deal. They provided us with 700 members. We borrowed $1.3 million from American General and built the most beautiful club in the state and maybe the country.

Do you still have your interest? Is that a for-profit?
It's owned by the members and I still have my membership. It's been a beautiful club. They have about 1,000 members now.

I'd like to move away from your extracurricular activities and talk about business again. You've always used debt in your business, is that correct? A lot of debt?
You have to when you start out with nothing. How else do you do anything?

But haven't you kept that true, even when you had something?
Yes, we've always had considerable debt, although the company is working now to reduce debt because our stock would be better accepted if we had a lower level of debt. We've had to be careful, to keep a close eye on our reserve situation, we've had to see that the debt was manageable and that can be tricky. When we first started The Woodlands, that gave us trouble. But we had the energy business to help us get through the trouble. Yes, we've had to use debt to grow and we still use debt but our debt is coming down now. Even my personal stuff is down, but I still use debt.

So it hasn't been a matter of preference, of leveraging, so that you could do more with less?
No, to do good projects, you have to use debt. You've got to be cautious, you can't go on a run of 10 consecutive dry holes. See, I'm not a wildcatter. I wildcat a lot, but I've tried to use judgment as to what the risk factor is. We would try to drill three or four extension wells or deeper reservoir wells for every wildcat we'd drill. If you drill a string of wildcats, you lose all your participants. Of course, we're doing a lot better nowadays with 3-D seismic and other technical improvements, but you've still got to have judgment. You have to have the ability to find oil and you have to understand the geology in order to find oil and gas.

From what I know during my years with Mitchell Energy & Development Corp., the bulk of the drilling has not been rank wildcats, it's

been development and redevelopment and extensions and such. We made a lot of field extensions, 250 extensions, and a lot of new field discoveries. We've also tried to exploit more in areas of better opportunity and keep our wildcatting risks within reason. You have to have judgment as to what your chances are. You have to understand the geology and the reservoirs. And even when you do find oil or gas, sometimes you don't make any money out of it. It's a tough business.

It's a good business. George, thank you very much.

INTERVIEW FOR TIPRO ARCHIVES

The following is a condensation of an interview of George P. Mitchell in his office at Mitchell Energy & Development Corp. in Houston on April 16, 1993, by Lawrence Goodwyn and Barbara Griffith. The interview is now part of the Texas Independent Producers & Royalty Owners Association archives at the Center for American History at The University of Texas, Austin.

The ship that brought your father to the United States—did it land at Galveston?
No, at Ellis Island.

How did he get to this part of the world?
He was the fourth child, the youngest son out of three boys and a girl. His father had died. They lived in the little town of Nastani, which is right near Tripolis, right in the middle of Peloponnesus, and he knew there were no opportunities. When he was a 21-year-old man, in 1901, he said he heard about America, and he went down to Kalamata and shipped out. I think you had to have about $30 or $40 for fare to get here. So he got to New York City. And the railroads were employing all the immigrants. He was strong. He went to work on the railroad, starting in Arkansas. And then he worked in Utah. He worked on railroads about four years, 1901-1905.

Learning the language?
He never went to school a day in his life. He not only didn't know English, but he couldn't read or write Greek, either. Very difficult to overcome.

You must've been proud of your father.
He was a very unusual person. He had great perseverance and a good mind. Somehow he made a living, in a little shoe-shine parlor, and he raised three boys and one girl in our family.

And you—did you get to A&M about '36 or '37?
I got to A&M in '36, the class of '40.

Did your brother Johnny have a petroleum engineering degree from A&M?

He had a chemical engineering degree, but by that time, he was trained in petroleum engineering. Exxon made petroleum engineers out of all types of engineers. He worked with Exxon a couple of years. He graduated from A&M in the class of '34, and then he worked with Exxon for a couple of years, and then he went on his own. In my senior year, I was doing some of his geology for him. Johnny was a wildcatter of the first order.

After the war you came to Houston. Why?

Because I knew that we weren't drilling any wells during the war. And Houston was the hot spot for energy. Therefore, I wanted to come to Houston to get started here.

Well, wouldn't some people say it'd be Dallas?

No, I knew Houston was the place because most of the major companies were here. Dallas had some. Houston was the area that I thought was going to be the centroid of the energy business. They'd just discovered the Hastings and the Friendswood Fields. I was still working with the Army to finish up all the contracts, and I started doing some geology for Johnny and four or five other people in town, even before I got out, while I was cleaning up those government contracts. I got out in January of '46 and I came to Houston. I had two or three offers from people who had known me during the war.

They'd watched you build these things.

I said no, I wanted to be in engineering and geology. So what I did is this: Since we had some friends here for whom I had done a little work during the tail end of the war, and Johnny knew some of them, I got six of them to give me a $50 a month retainer and a little override. So I became a consulting geologist with only a year-and-a-half experience, and that's pretty short experience.

What was the deal? That you were going to go out and find oil?

Oil and gas. I'd do the geology and the engineering for them, and for that each of the six would pay me $50 a month, plus a little overriding royalty. And then, since I apparently was talented in finding oil and gas, I found a lot of oil and gas for them. About six months later, Johnny and I and H. Merlyn Christie started a company called Oil Drilling--in 1946. That was a predecessor to Mitchell Energy. What we would do is very interesting. We didn't have any money. I had one child here and one on the way and was making $300 a month. Didn't know whether I was going to make it or not, but I knew I could get a job with Amoco. They wanted me back, Amoco did, but I didn't want to go back with them. I wanted to take a chance. I'd lost four years in the war. All my counterparts who didn't go to war were up the ladder. I figured I could get a job with any oil company, or at least I thought I could, as an engineer and geologist, exploitation geologist or engineer. So I was going to take a chance to see if I could make it as a consultant. We started our company, and what we would do—we didn't have any money—I would do the geology and engineering, get the deals together, and then we'd have some land man help us get the leases together and Johnny and Merlyn would go down to the Esperson Drugstore and sell the deals. That's how we started. They'd sell an eighth here and an eighth there, they'd sell over coffee, and the first thing you know, we had a deal that cost $30,000. We had one partner who had a little rig and they'd use the rig. And he would drill the wells. We'd get maybe at first a 32nd carried interest, and then a 16th carried interest. And then eventually we got Bob Smith. I did some consulting work for Glenn McCarthy and for Eddy Scurlock—I did consulting work for everybody in town. We found a lot of oil and gas for Bob Smith and many others.

Take me behind the scenes a minute. You see things and you turn them down, and you see other things and you go for it. What are you seeing?

Well, you just have to have, first, good technical skills and then understand the geology and the potential. You have to judge the risk factors. Because I was knowledgeable in both geology and engineering, I was able to see things that the geologists alone couldn't see or the engineers couldn't see.

Can you tell me about that?

At one point, 60 people worked for me in geology and engineering. Most didn't know how to bridge the gap between the geology and engineering. It takes both, in my opinion, to really understand. Engineers are too conservative. They have to have everything spreadsheeted out for them. A geologist may be too optimistic, although you have to be optimistic. You have to have optimism but you have to have good judgment. And most engineers don't really think that way. So you really have to do both. One of the big failures of our universities is that they don't train the people to be able to do both.

What you get out of school, that's just the beginning of your education in geology and engineering. Anyway, first thing you know, we started building, and then we got our quarter net profits, and soon we started taking more of the deals ourselves. I've been involved in the geology of drilling of 8,000 wells and a thousand wildcats. I don't know of anybody who's drilled that many. I knew certainly this region well, and I knew the country well. I had a lot of people in town who thought I had the talent to find oil and gas for them and worry about the engineering, too. And I did that. Most little companies at that time didn't have engineers or geologists. They used people like me. So that is the way it started. Johnny was my partner. He is older than I by seven years. He did a lot of the work, working on the wells, but Johnny was principally a wildcatter.

Can we go back? You have these six people giving you $50 and the override. And you're 25 years old, in Houston, Texas, and out there, somewhere, maybe, is oil. Now, what'd you do? I mean, you got in the car and where'd you go, what did you do?

The first thing I did when I came here, I went to see Mike Halbouty. Johnny and Mike had been fighting, they're both Aggies, because Johnny had made a deal with Mike about two years earlier. I got out in '46, he got out in late '45, right after VJ Day, actually before the Japanese surrendered. He went to Mike Halbouty and said, "Mike, I got a deal in Vinton." That's where I first started working in the oil field with Johnny, and he had some pumping wells there, a little bit of production–but he wasn't making any money to speak of, just hanging on. And he said, "I got a deal on the flank of Vinton. Could I see some map on Vinton?" Well, Mike said, "Yes, I'll show you a map here." So Mike Halbouty showed him a map and Johnny took a photostat of it–and sold the deal. Mike heard about it and Mike said, "I want my override." Johnny said, "Hell, your override for what?" He says, "For the map." Johnny said, "Hell, I could buy that map down the street for $2.50. Why should I give you an override?" So they had a big fight, and Johnny'd write a hot letter about Halbouty and send a thousand copies out to all his friends. He called him a Lebanese rug peddler and he said, "I'll tell all of my Greek restaurant friends here not to let you eat, you'll starve to death," and Mike would be incensed. In fact, one time, Mike called him on the telephone, screaming, and Johnny–over at the Esperson building–and Johnny told him, "Mike, you get off the telephone. Just open the window, I can hear you from right here." [Laughs.] And this went on–Johnny just needling Mike, and Mike can't take the needling. So here I'm going to see Mike Halbouty.

He was a known geologist, and here I am, trying to get started. I said, "Mike, I'm just starting, I'm an Aggie." He says, "God damn, if you're like your brother, boy, just get out the front door!"—but Mike and I've been friends ever since. It's rather interesting because of the personality differences between Johnny and me and Mike. Johnny's gregarious, he knew everybody in the business. He knew every TIPRO member who ever existed. I don't, I'm not that way. I know a lot of people, but I'm not as gregarious as Johnny was. Johnny and Mike made up about two or three years later. They're the closest of friends, after that.

Was seeing Mike Halbouty the first thing you did after you got the six people together?

I saw Mike Halbouty and then the other thing I did, I made some very important contacts, including John Todd, who had a log reproduction library here. I couldn't afford to buy the logs, so I made a deal with him. He'd loan me the logs, at five o'clock. I'd go home, work till four o'clock in the morning and bring them back the next morning by nine o'clock. I couldn't afford $500 or $600 to buy all these logs to do the geology on various areas I was working on. And then I made a lot of other contacts around here, brokers, Stewart Boyle and others who would be getting deals from major companies. I'd look at them and I would screen deals. Then I worked with a Jewish group, the Floyd Karston company, and we found a lot of oil and gas for them, I did a lot of geology and engineering for them. So my override began to build up. By the time I was 28 years old, I had about $3,000 a month income. I knew I was over the hump by that time, and our company had started picking up, too, the predecessor to Mitchell Energy. So Johnny and I and Merlyn kept building the company gradually. If we'd drill a well, make a well, we'd run to the bank to get some money. If you drill a dry hole you get nothing.

Tell me about Wise County, Boonsville.

First, let me say this. I've been involved in 244 discoveries. We had Palacios, which McCarthy had abandoned, and we found a big field there.

When was that?

I'd say about '49 or '50. And we made good wells in the Vienna Field, on a farmout back in '47 or '48. I finally had a lot of people backing me, including Bob Smith, who was in almost all my early deals. Bob Smith was the big operator in town at that time.

Big independent.

He was a big independent and a dramatic person. Very smart and difficult to work with, very difficult. But I was able to get along with him—you had to use finesse with Bob Smith. I'll give you an example. He'd get mad about some cost or something, and he'd call me up. "Come on, come on, dammit, I want to talk to you, come on" and he'd raise hell about the problem. I would say, "You know, this thing here—you may be wrong on this." He says, "I've never been wrong." And he was sincere about that. He said, "I've never been wrong." Well, he was just browbeating me, you see. But I was his partner for 14 years. No one else had ever been that long a partner with Bob Smith. You had to know how to work with Bob Smith. You had to know how to take a lot of criticism. He had complete confidence in my ability. And Scurlock and McCarthy and all those people did. As an independent, you really have to understand how to work with people, too. Over time, we did a lot of discoveries, fairly good discoveries, you know, in which we had a quarter interest, and we grew. Discoveries in Vienna, Palacios and Buffalo, and we had partners like Riddell from New York, and Barbara Hutton, the Woolworth heiress, was our partner at one time.

Who got that, Johnny?

Johnny knew her. Johnny was kind of a playboy. Anyway, Merlyn Christie and Johnny got her in. We had a lot of local people take deals, and some of my original partners were Jewish people—they were wonderful partners. Very difficult, intolerant of anything that you did wrong, but real gamblers. If you drill a dry hole, well, OK, where's the next one, where're we going to go next? They were very good partners, and they helped me get started. In fact, out of my first six partners, I'd say three of them were Jewish people. We had a group called the Big Nine, all Jewish people, and we started some of the Wise County effort with them. The big deal came through a person named Pulaski. I used to call him General Pulaski, but he wasn't a general. He called me one day. He was a gambler, and he had the Pulaski Iron Works here in town and was one of the six men who underwrote me early on. He said, "I've got a Chicago bookie who's got a hot deal for you." "Oh, come on, what do you mean a Chicago bookie?" "He's from Chicago, and I want you to listen to him." And I said, "OK, send him over." So he came over and he showed me the Wise County deal, and I saw that it had some merit. So I got interested and I looked at it. I said, "Well, this thing is worth taking on," so we took it on. We had about 10 partners.

After you talked to him, did you go up there and look at it on the ground?
No, I had all the geology.

Just the geology?
They brought the geology of the area. I found out from a person I knew at A&M, who had been peddling it for two years around the country and it finally ended up in Chicago, with the Chicago bookie. Someone showed him the deal and said, "Why don't you talk to your friend, the General, and see if someone down there might take it?" But I saw the merit, and then I found out he had come through two fellows named John Jackson and Ellison Miles, who was an Aggie. And that's where the deal originated. It had been going around the country for two years. John Jackson was the geologist. He and Ellison Miles had offered it through the Chicago bookie. They'd tried to sell that deal to everybody.

Were they petroleum engineers?
Jackson was a geologist, and a good one.

You knew them from school?
I knew Ellison Miles at A&M.

You knew they were sound?
That's right. And they had enough data that I could really analyze what was going on, and I saw it had merit. So we started with 3,000 acres.

This is fascinating.
We drilled the first well, and I could see it had pay. This was a combined engineering and geological success. What happened, there'd been 11 dry holes drilled by major companies that didn't understand hydraulic fracturing, and they didn't do it. We drilled this well, and I could see it was a big stratigraphic trap. So within 90 days we bought 300,000 acres at $3 an acre.

How could you do that?
I had 20 brokers working to get the acreage in.

Would you come back to Houston to get those reports?
We kept it very confidential, very confidential. Then we got farm-ins from 10 other companies, like Signal Oil and Gas and Cities Service—old wells

they had drilled and plugged. We went in and made gas in every one of them.

They didn't go deep enough or they didn't know how to . . . ?
No, they didn't understand the engineering technique we had to use.

Fracturing the sand?
Fracturing—that's right, fracturing the sand. So that was a combined geological-engineering success. Do you see what I mean? Had I been a pure geologist, I may not've understood that very well. People who drilled wells in this country tended to look for a big structure to trap the oil, like an inverted saucer. But very often you have a big monocline, and the traps are caused by the sand's pinching out. And it doesn't have any reversal. Therefore, the big fields in the nation, like East Texas, are stratigraphic traps. The big discoveries are mainly in big stratigraphic traps. So I could tell right quickly, after the first well. I looked at all those other wells and I said, why didn't they test those wells? They would've made wells, too. So I analyzed the situation quickly and moved quickly.

You mean, they didn't even test those wells?
No, they didn't even test them. They thought they were too tight, that they wouldn't make a well.

They could smell oil in the core.
Well, it was gas mainly. But let's put it this way—they were looking for oil, and not gas. You couldn't sell the gas for over three cents. No market for it, too. Another thing is that we had to fight with Lone Star. We had 21 wells shut in and were out of money. And they were fighting us to keep the price down. They wanted to take our processing plant away from us if we ever built one, and we refused to do that. I wore out the Big Nine—they stopped investing—but I finally got Bob Smith to take more.

They were trying to break you.
They were trying to break us. They held us up in the Railroad Commission for two years and with the Federal Power Commission for two years. In the meantime, I made a deal with the People's Gas Company and Natural Gas Pipe to come down from Amarillo and pick up our gas. And for that, they gave me some money to drill some wells to shut in until they could get there with the new pipeline and fight out the battle with Lone Star.

Now what year is this?
That was in about '56. We made the discovery in '55, and we finally got on production about '57. We had a two-year battle with Lone Star, it took time to get the pipeline down here–it went to Chicago but it came from Amarillo. In the meantime, the Supreme Court decision came out, which had the effect on us of saying we were in interstate commerce and therefore couldn't get a price increase. Mobil and everybody backed out of that contract with Natural Gas and they didn't know if they were going to build a line at all, after two years, because the rest of them backed out. Mobil and Amoco backed out. We stayed with 'em. That's why we've been such great partners, because they knew we stayed with them and they stayed with us. They gave us money to help develop some wells because we were out of money. By that time, our company took 35 to 50 percent interest in drilling those wells. So, as more partners dropped out, I would step up and take more and we'd get more farm-ins. We have about 2,000 wells up there and we've still got another thousand to drill. So that's our biggest discovery. But we've had a lot of discoveries in this part of the country.

When did you know that you were safe—that you, in effect, had defeated Lone Star?
I would say that we didn't know till about '56 when we finally beat them at the Federal Power Commission and beat them at the Railroad Commission. They tried to keep our price down to 11 cents instead of 14 cents. They tried to take our processing. I tried to make a deal with Lone Star. They said, "We'll give you 11 cents but we want all your liquids." I said, "To hell with you. We're going to build our own plant." I had Warren Petroleum come and say they would help me get the financing to build a plant. That's when we started our processing business.

(The following question comes after Mr. Mitchell describes how his company obtained permission to drill in Galveston.) We need to understand this as a society. You can work with the people in Galveston, and with the city fathers and the restaurant owners, the beach front people, and people who just like their city and don't want to see it spoiled. And you can make an arrangement that benefits the city treasury and the beach front people and yourself, and you do it with grace. And here we have a major company that hires public relations people just to throw dust in people's eyes. Why do they create a Santa Barbara?

They could do a much better job of public relations and also develop something right, even though it's more costly. Galveston probably cost us $3 million more to develop than it would have cost if we hadn't been concerned about the community and the environment. We didn't want to put a tank battery right in front of Gaido's and leave it looking like hell. In Santa Barbara, it's a scenic problem, not a spill problem.

Right.

It isn't the spill that bothers people. I've been to Santa Barbara. I think those tank batteries are ugly out there. You can do it in a better way. They don't do it because it costs a little more. Furthermore, they're not smart about PR. Even before the first crisis, I always said this. We have a good image because I work at it and we get our people to be PR conscious. What they do, they probably take some executive who isn't very good at other things and make him head of PR. They should make the president of the company the head of PR.

Yes.

They've got a terrible image. Now, the major companies are very smart, they're secretive, they're resource-rich, they've got a tremendous organization—and they're hated. They ought to know that. That's exactly their problem.

Let me try and get a couple of relationships clear in my mind about the history of wildcatting. Did Bob Smith get his start in the East Texas field?

No, he did Pierce Junction, before my time. He bought the wells in East Texas on the upswing. He built the production. Pierce Junction is where he started. He had some rigs with Claude Hammel, drilled a few wells. That was before I got acquainted with Bob Smith. And he bought some production in East Texas on the upside edge. He was smart enough to buy it where the water would be the last to reach those wells. He bought 100 wells and bought them right. Then he started working on Scurry County. I didn't have any interest in that, but he asked my advice and I looked into the geology. That was his big discovery, Scurry County. And then, of course, he had 25 percent of what I did in Wise County. I bought him out later on. And Bob Smith, he was a hell of a good wildcatter, and a smart guy. Tough as nails, but a wonderful person.

Scurlock. Where did he get his start?
Eddie Scurlock used me as a consulting geologist to drill a lot of wells. He knew how to buy low and sell high. He had very little education. Came from Tenaha in East Texas. Every major company, every pipeline company loved him. He was a wonderful person. He put together a big pipeline system and he'd get 10 cents a barrel or 25 cents a barrel, to haul the oil to the pipelines of major companies. They all appreciated that. He built his reputation and image that way and built his company that way. Then he started drilling wells, and I would do all the geology for him. In the Buffalo area and the Victoria area, I did geology for him. In New York, there was a friend of his called "Penny-a-Barrel" Riddell. He was the biggest oil broker in the business back in the '40s, and he got a penny a barrel for trading oil, foreign oil, and that's why he was called "Penny-a-Barrel" Riddell. He took some of our deals. That's the way you did it in those days. We didn't worry about SEC in those days. And we'd always leverage ourselves to where we'd get a carried interest, maybe a 16th interest, and then we finally got a quarter net profits and things like that. And we built our company. We now take most of the deals ourselves. We still might joint venture with someone, though.

Did you continue doing consulting work?
I had a few friends, like McCarthy. I did work for him on the New Ulm Field. I thought he was doing some things wrong. But then I got to where we were so busy doing our deals after about seven or eight or ten years, I was mostly doing things for our own account, to try to get people to come in on our account, to build our company.

So that would be perhaps, I say, 1960, when you more or less did not have time to do any outside work?
I was still doing some consulting work for some people, like Scurlock and others and friends, including Smith. Smith took a lot of our deals, but I'd help him on other things, he was looking at other deals, too, up to about 1960, and then we probably got too busy with our own stuff.

I'm going to assume that geology attracts certain kinds of people, and in the course of going to school and learning the discipline, it changes people. And, petroleum engineers become cautious, and geologists become optimistic.
That's true.

Is that because of the discipline, or because it attracts certain kinds of people?

I don't think that comes from the discipline because you have to have the courage to say you have high hopes, and that's very difficult for some people. And you have to figure out what are my chances and what are my odds. You have to have your own computer in your mind to understand your chances, what you're trying to hit, knowing that it's a risky business. Over a period of years, you'd better have your judgment right or you'll be out of business. Engineering is completely different. They look at all the facts. They're not very much risk-taking in engineering. They look at the facts and then they go on and make it work. But by having a lot of facts, they usually are right. Geologists? You know, four out of five wells are dry holes or maybe six out of seven are dry holes, if they're wildcatting. So I think that it does take really a different mind set, just like you say. And I think engineering pulls them to be more conservative. And geologists? If you're a good one, you've got to be reasonably optimistic, with good judgment.

I've got one question. Do we have a national energy policy? We never had one.

Yes, you got one.

What is it?

Go to war. That's our policy.

All right, thank you very much.

* * * * * * *

INTERVIEW ON THE *JOYCE GAY REPORT*

In 1983, Joyce Gay of the *Joyce Gay Report*, a television series originating in The Woodlands, interviewed George P. Mitchell. In the following highlights, some parts have been deleted and a number of questions and answers are paraphrased.

Welcome to the *Joyce Gay Report,* the report about people, places and events. Our guest today is George P. Mitchell, chairman, president and chief executive officer of Mitchell Energy & Development Corporation, which is based in The Woodlands, Texas. Mr. Mitchell, please tell

about some of the early thinking behind The Woodlands.

What interested me back in 1964 when we started thinking about The Woodlands was how do we evolve an area that is now 3.5 million people, which is what the Houston metro region is, how do we evolve to an area of 6 million people and do a better job than what's been done before.

I feel that you think new towns are one of the answers. Is that right?

Why do people think they have to go downtown every day to work and come back 30, 40, 50 minutes in terrible traffic to their homes? You're not going to get everyone to reside in a downtown high-rise. The social structure that goes with high density may work in some areas but it won't work in the Houston region too well. There will be more downtown apartments and so on, but not near as many as some people predict. What people really want is the rural lifestyle and to be part of the big city, and I think The Woodlands is an answer in that direction.

Let me ask you something about the planning process that went into The Woodlands. I understand you brought together people like social scientists, city planners, ecologists. Tell us about that process.

We began in 1964 when I had Karl Kamrath and Hugh Pickford do some planning. I asked them, "What would it take to build a town of 100,000?" They did some planning and said 20,000 acres. We asked Ian McHarg to study all the constraints that we would face if we were going to urbanize this beautiful forest with 180,000 people in the next 30-40 years, how do we do it right? So what he did was to get 10 scientists together. They spent a year studying the hydrology of the woods and the trees, studied the soils, studied the drainage problems and came up with a design that showed where the high-density should be, where the low-density should be, how do you recharge the rains, which is necessary for the future water supply of the City of Houston and others, how do we protect the wildlife. A&M made a study showing that if you keep corridors in flood plains—7,000 acres would be left in the flood plains—if you keep corridors of at least 600 feet of forest, the deer will remain if you control the dogs, and that's always a difficult problem. We had A&M do infrared photography to show us where the best areas of trees were. Then we laid the roads and the other systems out around that. We had Rice scientists work with McHarg. They said if you do drainage properly where you build lakes and you have the water run off flowing slowly, say 48 hours or 60 hours, by retaining the waters and 7,000 acres of flood plain that's left in its natural and wooded state, that we would not flood like Cypress Creek is being flooded now in the 1960 area, like the whole City of Houston is being flooded by upstream development with a lot of asphalt, a lot of rooftops. This lack of planning results in water being drawn immediately into those floodways and causing flooding downstream. When The Woodlands is fully developed, no more water would run off The Woodlands than would have run off naturally. You don't see that in any other project in this part of the country.

When will you have reached the completion point?
I would say some time in the next 30-35 years.

And where are you now?
We are just under 20,000 people now. We think there's no Utopia, there are all sorts of problems everywhere and people should not think they can escape whatever problems they may have. It's just, I think, a better quality of life. Transportation is better. In designing the road systems of The Woodlands, we did the reverse of what they did on 1960 or the Galleria, where every little developer had 200 acres, 100 acres. They're shelter builders, they're not community builders. They developed their land with no cohesion as to the whole road system and how it fits. In The Woodlands we designed the road system and then we designed how many square feet of office space and how many houses we could put into it, and dealt with the question of how do we save the 7,000 acres that's natural, how do we save the beauty of the woods as much as possible, how do we build the community by building areas that have maybe 40 houses and a cul-de-sac, where they relate with each other, where all children go to the same school system, everyone shops in the same area. I've always talked about The Woodlands this way: If you owned all of Memorial and could do it over again, what would you do? This is what we're trying to do, knowing it's very difficult to execute, a very complicated, $12 billion project.

It must be difficult. I understand only one out of three new towns succeeds. Why did you succeed?
Many of the other new-town projects around the country that HUD has guaranteed have had trouble because they're ill-conceived, poorly planned and didn't have the staying power that we were able to furnish The Woodlands. We've had some tough times when the money market got to 15-16 percent in '72 and '73 and then again in '78 and '79. But the parent company, Mitchell Energy, had enough resources to go through those tough times. But

the reason the others didn't succeed was due to bad planning, political awards.

I understand one of your more ambitious undertakings is something you call a Research Forest.

We had the concept some five years ago that energy would not provide the job formation in Houston in the next 25 years that it did during the last 25 years. We felt that a research center here would help Houston broaden its base of high technology. When I say high technology I don't mean just electronics. There are probably 100 varieties of high technology—medical high technology, many energy technologies, including those connected with offshore drilling or the service companies or the mud companies or geophysical data. There are so many things that Houston can do to broaden its base. It took us five years to put together the four universities which comprise the Houston Area Research Center. One of the role models we used following a very exhaustive study by Arthur D. Little was the Research Triangle in North Carolina, which was started by Governor Hodges 25 years ago; it now has 2,100 scientists working between Duke, which is private; North Carolina State, which is public; and the University of North Carolina, which is public. We think it's timely that Houston's private sector work with the universities to reach the Houston urban region. Houston is 60 percent of the entire economy in the State of Texas and people don't realize that. Therefore, Houston should work with these universities and the universities should reach into the central core of the cities. The next step we took, we talked the Texas Medical Center into joining this same sort of setup in our new Research Forest, providing them with a gift of land. I had the great thrill of seeing this enormous Medical Center develop in the prairie over the last 35 years. My wife and I and the company made a gift of 150 acres in The Woodlands and $5 million, and my challenge to the heads of the Medical Center institutions was to come up with a good answer to the question, What do you do in the next 35 years? They feel that maybe they can evolve as it gets crowded in the Medical Center. Everybody knows that future medical research can be done in The Woodlands and yet be part of the Texas Medical Center. What's happening in the medical field is that they're graduating too many doctors. Therefore, the bright students ought to go into medical research. With such enormous strides being made now in medical research, that's where the new M.D.s and Ph.D.s should be. For instance, The University of Texas now is working on positron imaging. They think they can diagnose heart attacks four years before they happen and do it non-invasively. They do it with imaging that is very sophisticated, very complicated. It takes a doctor and a physicist to design the imaging, which is very highly computerized. Now they're working with Genentech on a way to prevent the blockage that accompanies a heart attack, the blockage that destroys the heart muscle, and they think they can do it with an injection—they're doing it with catheters now at Hermann Hospital. They think a paramedic responding to an emergency call can probably do it by injection, and if that injection is given within two or three hours after a heart attack, they can relieve the blockage of the heart. In NMR, Baylor is now looking at images that are 10 times clearer than CAT scans with the new research tools they're working on. That type of research, I think, has a great future in this region and that is why we're trying to support it and make The Woodlands a part of it.

Thank you, Mr. Mitchell.

* * * * * * *

A&M MAGAZINE

The following is a condensation of an article in the Winter 1994 issue of *A&M Magazine,* which was published in College Station, Texas, under a license granted by Texas A&M University.

GEORGE MITCHELL
THE LEGEND

An Exclusive Interview by Michelle L. Brenckman

When he first started planning The Woodlands, George Mitchell was faced with uncertain times in the energy industry. He had created a significant company from 1946 to 1962, but tough times loomed ahead—prompting him to diversify his operation.

Mitchell had already started developing inexpensive housing with a partner, and he acquired the 3,000-acre Pelican Island waterfront industrial complex on Galveston Bay in a leveraged deal. He also began developing Pirates' Beach, a subdivision of second homes for upper income Houstonians. Five years later, he donated 100 acres to Texas A&M to establish the Mike Mitchell Campus for Marine Sciences. Mitchell convinced Pennzoil, Shell and other offshore operators and service companies to locate facilities in Galveston, persuading them that the island provided centralized access to the Gulf of Mexico.

The Woodlands is just one of several sub urban developments surrounding Houston, including Friendswood Development's 13,000-acre Kingwood in Humble, and Clear Lake City near Galveston Bay. Mitchell expects that The Woodlands will be annexed by Houston in the future. The Woodlands consists of five separate villages; another village is yet to come. Each village includes such retail businesses as grocery and drug stores, service stations, doctors and dentists, hair salons, hardware stores, dance and gymnastics studios, child care facilities, and nursing homes, among others. Stores and offices are separated from residential areas by green belts, which serve as wooded buffers between businesses and homes. Mitchell and his wife Cynthia live in a contemporary house on one of the golf courses.

Every village offers at least one elementary school, where Mitchell said children from families of different economic levels attend school together. There are approximately 24 churches in The Woodlands. Building covenants in the form of deed restrictions provide some zoning protection so homeowners will not suddenly find themselves next door to a McDonald's. HUD subsidized rental property and lower priced housing are available in two of the villages.

Mitchell said the poorest family in a federal housing project would be welcomed in The Woodlands, although anyone who applies for subsidized housing must follow strict guidelines and may face long waiting lists. One of Mitchell's main goals in developing The Woodlands has been to maintain the same ethnic and economic mix that exists in Houston–which is 35 percent white, 28 percent black, 30 percent Hispanic and 7 percent Asian or other minorities. In reality, The Woodlands populace is 90 percent white, 2 percent black, 5 percent Hispanic, and 4 percent others. Planners have said that the arrival of the new mall will create sales jobs that may be filled by minorities.

The median income for residents in The Woodlands was $50,000, according to the 1990 Census Bureau report, $19,000 higher than the Harris County average of $31,434. Kingwood's median income, by comparison, was $86,830 based on statistics from 1992 real estate transactions. Homes in The Woodlands range from $80,000 to $1 million and more. The larger homes are situated on choice lots, overlooking fairways from the development's scenic golf courses, or bordering numerous lakes. Smaller homesites are clustered on crowded cul-de-sacs and long, narrow roads. Of approximately 2,500 rental units in The Woodlands, some 45 percent are rent-assisted apartments ranging from $400 to $800 per month. Almost 1,000 of those dwellings are senior housing, nearly 70 percent of which are rent assisted.

Two 18-hole golf courses are offered as part of the Resort: the North Course, designed by Joe Lee; and the Tournament Players Course, designed by renowned architects Von Hagge and Devlin.

Two more 18-hole golf courses are reserved for members of The Woodlands Country Club. At The Woodlands Athletic Center, open to members only, there is a 25-yard indoor pool open year round; a 50-meter Olympic-sized, heated outdoor pool; a children's wading pool; and a diving tank with one- and three-meter springboards and three diving platforms. The WAC has fielded competitive swim teams since 1977 when it hosted the First International FINA Age Group Diving Championships with young divers from 14 different countries.

Roger Galatas, president of The Woodlands Corporation, said The Woodlands is a specific example of a planned community that has integrated federal air, water and land requirements into planning guidelines. He explained that The Woodlands Corporation established a policy requiring that 25 percent of community development be designated as green belt or open reserves. Roughly 50 different neighborhood parks are interspersed within The Woodlands.

The $2.5 billion Woodlands Town Center, a 1,200-acre urban center under construction, will feature a regional shopping mall, a 1.5-mile riverwalk and ultimately up to 18 million sq. ft. of commercial, retail, entertainment, recreational, civic and office space by the year 2020. Town Center already includes more than 200 businesses occupying 1.5 million sq. ft. of office space. The long-term project is expected to bring an estimated 10,000 jobs to the area, with an annual local economic impact of $400-$500 million.

The 20-acre Lake Robbins, under construction near the mall, and 200-acre Lake Woodlands will eventually connect with a canal running adjacent to Timberloch Place. Ornamental street lighting, sculptures, colorful awnings, banners, gardens, fountains, plazas, benches, and landscaping are planned for the canal. Water taxis will ferry shoppers and workers along the canal, which will reach a maximum depth of five feet. The bulkheaded canal is a key element of Town Center's drainage system. A tram will provide transportation to the mall and along the canal.

"The Woodlands is in the Conroe Independent School District and it is a good school district," Mitchell said. "Because a portion of The Woodlands fell in the Magnolia school district, we've worked out a deal to let them become part of our school district in Montgomery County. We also offer private educational facilities in the form of the John Cooper School in the tradition of St. John's in Houston."

The John Cooper School, named for the late headmaster of the renowned Kinkaid School in Houston, opened in September, 1988 on a 40-acre forested campus. Several preschools are also located in The Woodlands, including The Woodlands Christian Academy and the Interfaith Child Development Center–two private schools which have outstanding reputations and long waiting lists for enrollment.

Mitchell, whose vision for The Woodlands included an environment suitable for intellectual pursuits and scientific research–not unlike a traditional college campus–recognized the need for diversification during the late 1970s energy boom in the Houston region. Sustainable economic stability could not be achieved as long as the area depended so strongly on a single industry. He reasoned that the smartest move was to establish an academic and scientific environment which welcomed ideas for new technologies, but with the resources and ability to nurture them and transfer the resulting technology into the commercial marketplace sooner and more efficiently. Mitchell managed to convince Pennzoil, The Western Company and other energy-related corporations to establish research and technology facilities in The Woodlands, a victory unto itself, since Montgomery County was still considered relatively remote at the time. Mitchell also managed to lure light manufacturing and distribution companies to the 900-acre Trade Center, located across I-45 from The Woodlands. Serviced by the Union Pacific and Missouri Pacific Railroads, it is located 16 miles north of the Houston Intercontinental Airport and 38 miles north of the Houston Ship Channel.

Not satisfied with his progress, Mitchell incorporated a project into his original concept for a Trade Center that had helped North Carolina during the late 1950s. The state–which was successfully educating scientists at Duke, North Carolina State and the University of North Carolina–found that Ph.D.s were leaving North Carolina for greener pastures and better wages elsewhere. To turn the tide, the governor established a research park with several thousand acres of government, university, non-profit and corporate entities. Mitchell seized upon this notion, partly because he admired the ideal of university brain trusts working in harmony and partly because of the potential for technology transfer to industry and the commercialization of applied research.

Throughout his business career, Mitchell had lent his support to Texas A&M, Rice University, the University of Houston and The UT. In 1982, Mitchell made a lucrative offer to the universities–100 acres of land and $8.2 million in seed money to create the Houston Advanced Research Center (HARC). Besides providing the start-up capital, Mitchell's reputation and leadership attracted a core of notable scientists, engineers and philanthropists who also worked to support HARC's mission.

HARC, with an annual operating budget of $15 million, is a collaborative research consortium comprised of 11 institutions, including founding members Texas A&M University; The UT at Austin; the University of Houston, and Rice University.

* * * * * * *

Despite his personal longevity, there has been speculation in *The Wall Street Journal* and other media as to who will succeed the 74-year-old founder of Mitchell Energy and whether or not the successor will be a family member.

"Operations would continue successfully without me or members of my family because we have such a strong outside Board and strong presidents of each operation," explained Mitchell. "We've got some good people in the company, 3,500 talented team members. I work with all of them, but they handle operations. I work on concepts and strategy."

"We have a meeting once a year with all ten of my children and my wife Cynthia to discuss strategy on successions, foundations and everything. We just lay it right down the line as to what I would like to see them to do," Mitchell added.

For the first time since Houston's boom days of the early 1980s, the company has positioned itself for growth again. Mitchell Energy is aggressively pursuing new downstream ventures which would allow the company to make money from the gas as it travels from the wellhead through the pipeline into processing facilities–where gas liquids like propane and butane are extracted–then on to petrochemical manufacturing plants where the liquids are converted into gasoline additives or building blocks for plastics.

In partnership with Conoco and Trident NGL, Mitchell Energy expanded a large gas fractionation plant in Mont Belvieu by 50 percent. The facility separates mixed gas liquids, stripped off at field processing plants, into pure products like propane and butane. Mitchell Energy has also entered into a partnership with Sun Co. and Enterprise Products Co. to construct a gasoline additives plant at Mont Belvieu that will ultimately create a secure market for half of the company's butane at a guaranteed profit for five years.

* * * * * * *

In 1880, Galveston was the largest city in Texas with a population of 38,000 and the second wealthiest city in the United States. While most of Texas was still wild frontier, Galveston boasted the state's first gas, electricity, telephones, trolley lines, and medical school. The Strand, named for London's famous Strand, was known worldwide as the "Wall Street of the Southwest." Ornate cast concrete, polychrome brick designs, stately stone carvings, imposing ironwork and neoclassical columns fronted the buildings that flourished along The Strand.

Tourism flourished, as did casino gambling and the mafia, and Galveston became a destination city for affluent vacationers. But, on September 1, 1900, a hurricane ravaged the island, killing an estimated 6,000 people.

After natives rebuilt the island, they turned to the task of reconstructing the economy. Before long, cotton was again booming and Galvestonians ignored the imposing threat of competition from a proposed deep water port in nearby Houston. In 1914, President Woodrow Wilson opened the Houston Ship Channel and sealed Galveston's fate for decades to come. By the 1970s, Galveston had become an economically depressed, ramshackle island with more boarded-up buildings than open ones.

Of those buildings that survived the 1900 storm, more than 1,000 were still standing—saved from demolition by benign neglect. Spending an estimated $60 million, Mitchell restored 16 historical buildings in and around The Strand, Galveston's main street in the downtown district. His most ambitious restoration, the $20 million Tremont House, is located one block from The Strand and assumed its name from an 1839 Galveston hostelry that housed six presidents. The Mitchell structure is actually a former merchandise warehouse, owned in the 1880s by Leon J. Blum—once hailed as the "Merchant Prince of the Southwest."

To draw people to the island and protect his investments, Mitchell resurrected one annual event in 1986 and established another in 1973—Mardi Gras! Galveston and Dickens on The Strand, respectively.

Galveston is an asset to Houston and could be the top tourist destination in Texas if tourism is properly nurtured, said Mitchell. Initially, Mitchell decided that if he carefully promoted the island's history and capitalized on the public's interest, then they would not care about the less than spectacular Galveston beach. If tourists were not interested as much in the beach, he reasoned, then they might come to the island in November to see history instead.

Using $9.5 million in federal transportation grants, Mitchell convinced city managers to install more than four miles of trolley tracks across the island. Diesel-powered replicas of turn-of-the century electric trolley cars were purchased to transport tourists from beach hotels to The Strand's shops and restaurants, and to shuttle guests from the Tremont House to the beach. Mitchell and the Moody Foundation chipped in approximately $700,000 of the total $10.7 million cost.

Mitchell's newest development in Galveston is the 42-room Harbor House, a seaside inn overlooking the wharf at Pier 21, near The Strand. Pier 21 features the Texas Seaport Museum and the tall ship Elissa—an 1877, square-rigger cargo ship.

* * * * * * *

Less than two weeks before Pearl Harbor, traveling aboard a chartered train on Thanksgiving Day to Houston from the Texas A&M/UT football game in College Station, Mitchell met his future wife, Cynthia Woods. The stylish 19-year-old from New York was a student of art and psychology at the University of Houston. She had accompanied her twin sister, Pamela, and her boy friend on a blind date, and was trying desperately to rid herself of the drunken lout. An hour north of Houston, Cynthia leaned over to her sister's boyfriend–Raymond Loomis (whom Pamela later married)–and whispered, "Can't you find a poker game to lose your friend in?"

As the two men stumbled toward the gambling car, a 22-year-old, uniformed Mitchell politely introduced himself and took the empty seat next to Cynthia. Two years later, Cynthia and "Mitch" (as she now refers to her husband) were wed. Together, they became the parents of ten children, the grandparents of 18 grandchildren, and engineers of a family empire.

Mitchell served three terms as president and two terms as chairman of the Texas Independent Producers & Royalty Owners Association; and is a member of the All-American Wildcatters, American Association of Petroleum Geologists, the American Institute of Mining Engineers, the National Petroleum Council; as well as a member of numerous professional, educational and civic organizations. He served as a member of Texas' Select Committee on Higher Education and the Texas Governor's Science and Technology Council. In 1988, he was elected trustee of the 72-member Universities Research Association, named a director of the Gas Research Institute, and accepted for membership with the World Resources Institute Council. In 1989, he was selected as a member of the President's Circle of the National Academy of Sciences.

Honors accorded Mitchell include Texas A&M University's Distinguished Alumni Award in 1977; the medal for distinguished achievement from the Texas A&M Geosciences and Earth Resources Advisory Council in 1980; the Horatio Alger award in 1984; the American Society of Mechanical Engineers' first award for distinguished service to the petroleum industry in 1984; an honorary doctoral degree from the University of Houston in 1984; *Houton City* Magazine's Houstonian of the Year Award in 1985; the Ima Hogg Historical Achievement Award in 1988; the *Galveston Daily News* Citizen of the Year Award in 1988; Global Tomorrow Coalition's Lorax

Award for contributions to the environment in 1989; Inc. Magazine's Master Entrepreneur of the Year Award in 1992; the Hornaday Award for Environmental Achievement from the Boy Scouts of America in 1993; the Galveston Grand Opera's Leonora Kempner Thompson Community Enrichment Award in 1993; and, of course, the first *A&M Magazines's* Aggie of the Year Award in 1994.

* * * * * * *

VIDEO BIOGRAPHY, 1988

JG Studios, based in The Woodlands, prepared a video biography of George Mitchell which was shown to employees at a barbecue in support of the United Way in the spring of 1988. Subsequently it was broadcast on Houston's Public Broadcasting channel, among others. The following is a condensation of the audio portion of the video.

"A LIFE OF VISION"

Michel Halbouty: I must know thousands of geologists by their first names, all over the world, and there are just a few, a real few, that I would refer to as an oil finder.

Bobby Jackson: I remember the times when our credit was cut off at the ice house. You know, we owed them $20 or $30.

Ruddy Cravens: Ferocious. He's very much a competitor as I think everybody who's done business with him or played any game with him certainly knows.

Frank Vandiver: George, to me, is one of the most authentic futurists I have ever known. I would like to say of him that for him the present is solved. It may not be solved the way he likes it but it's solved and he really doesn't worry about it. He's worried about 30 years down the road.

NARRATOR: George Mitchell, by almost anyone's reckoning, is an important man. He's one of the world's most successful oil men. He owns about 62 percent of the stock in Houston-based Mitchell Energy & Development Corp., a company with more than $2 billion in

assets, both in energy and real estate. He has other considerable business holdings. George Mitchell is also known for his interest in the future of our world, for his interest in global environmental and social issues. He's known for his concern about the Texas economy and the Houston-Galveston region. Yet for all that, Mitchell is modest and soft spoken.

Jack Yovanovich: There's nothing pompous or demanding about him. He just blends in with wherever he's going and seems like the normal folk around him.

NARRATOR: George enrolled at Texas A&M. Financially, he had a tough time. His family would borrow money now and then to help pay the bills but George had to earn a lot of his college money himself.

George Mitchell: The first year or two were very tough at A&M, but the third year I sold stationery that had gold embossed, like field artillery, coast artillery, beautiful embossed stationery. So all the lovesick freshmen would come in–at A&M, there were no girls up there–and we would sell stationery all over the place. That was really a good venture. In fact, I was wondering about taking $126 a month salary from Amoco when I was making $300 a month my last year at A&M.

NARRATOR: The Mitchells had two children and a third on the way by the time George was mustered out of the Army in early 1946. With his family growing, George needed all his entrepreneurial drive, so he decided to take a chance. He managed to convince half a dozen oil investors to pay him $50 each per month in return for his geologic and engineering work on prospects. The new consulting geologist was in business, and within a short time George, his brother Johnny and friend Merlyn Christie were putting together drilling deals.

Jack Yovanovich: If you go back to the early days when George initially started, he had investors who invested with him and his main method of compensation was through net profits which meant that George had to find production in order to get compensated. And this he did.

NARRATOR: But despite some early success, the young company still

had to struggle hard. Mitchell Energy employee Bobby Jackson, who joined the firm when he was only 22, remembers how short of cash it often was.

Bobby Jackson: I remember the time when our credit was cut off at the ice house. You know, I think we owed them $20 or $30. We went down to get ice and they said, "Sorry, boys. You can't have ice because you owe us." It was like that all the time; it was a struggle. We worked on a shoestring and we didn't know when we went into a supply store if we were going to get supplies or not.

NARRATOR: The Wise County find and other fields found by Mitchell Energy added to George Mitchell's reputation. World renowned geologist and oil man Michel Halbouty calls him an oil finder.

Michel Halbouty: I must know thousands of geologists by their first names all over the world and there are just a few, a real few, that I would refer to as an oil finder. Someone that has the knack, someone that has the deduction, someone that has the inner relationship to the earth, that can be able to understand what he is looking for. And those few are oil finders.

NARRATOR: Yet for all his risk-taking, his associates, like long-time employee Jack Yovanovich, say he's not reckless.

Jack Yovanovich: I really think George is more of a conservative-type geologist than what I would say would be just a total out-and-out wildcatter, although we participate in a lot of wildcats and get out in the forefront from time to time, but we have done a lot of development drilling, which is expanding fields that you find initially with your exploratory wells.

NARRATOR: Just as George continued to expand his business, he and wife Cynthia continued to expand their family. Their children are a source of great pride and pleasure to the Mitchells and like most big families they develop their own ways to keep life functioning smoothly. When the family traveled together, even on short trips, the children would count off so that one wouldn't be left behind in the confusion. Scott Mitchell, George's oldest son, also remembers his father as one who built trust in the family by always believing his children.

Scott Mitchell: He believed us even when we told a lie, and I think that gave us all the strength to believe in other people. He built trust.

NARRATOR: George Mitchell, the successful Houston businessman, also continued with the pursuits of his Galveston childhood. He loved spending time on the weekends fishing the Galveston waters. His son Scott remembers, though, that dad brought more than just the average level of intensity to fishing.

Scott Mitchell: With sports or recreation, Dad is just as serious about that as he is with business. On a typical morning we would get in the boat and go out to the jetties, which would usually take about 30 minutes to get there, and within an hour, hour-and-a-half of fishing we might have pulled up anchor in five spots. I remember that because I was usually anchor boy. If we hadn't caught a fish in about five minutes, it was up anchor and we were off to another spot.

NARRATOR: Another childhood pastime, tennis, also remains important to Mitchell the businessman. He plays regularly and an old friend reports that as with everything else, Mitchell brings intensity to the game.

Ruddy Cravens: Ferocious. He's very much a competitor, as I think everybody who has done business with him or played any game with him certainly knows, that if you're out there one against one, why he'll fight you to the last minute. Now, if you go to him and you want to borrow a nickel or even a dime from him, no problem. He's a patsy.

Frank Vandiver: George to me is one of the most authentic futurists I have ever known. I would like to say of him that for him the present is solved. It may not be solved the way he likes it, but it's solved and he really doesn't worry about it. He's worried about 30 years down the road.

NARRATOR: George Mitchell, almost everyone agrees, is a visionary. He's also a worker. His early partners stepped back from the business, but George stayed on. He went to work to make his energy company grow and he started looking to the future again.

George Mitchell: In the early '60s, things were tough. They were importing oil from overseas, beginning in '55, on a very large scale and we were only producing 8 days out of 30-day months for oil, so I couldn't see how in the world we were going to survive as a company if we could only produce 8 days out of 30 days. You can't make it work. So, we started looking at real estate. I had these brokers who were starving to death around here and I said, "Well, go find us some real estate real cheap, on leverage, and maybe we will work out and acquire some.

NARRATOR: The Woodlands opened in 1974. Late that year, Mitchell Energy was in default on some of its borrowings for The Woodlands, and a severe cash bind threatened the company's future, but George Mitchell persevered, and today The Woodlands is not only surviving but prospering. In the early 1970s he and his wife Cynthia began to focus on how to handle the growth of population and the destruction of the environment of our planet.

George Mitchell: And I'm concerned that the growth issues are going to be very severe in the next 20, 30, 50 years. The population worldwide is five billion people now and the world can only handle so many people and do it well, environmentally and other reasons.

NARRATOR: Mitchell decided that he would help bring together the best minds to address these issues. Since the mid-1970s, he's been the force behind a series of international conferences with notables in politics, economics, history and natural resources. And he and Cynthia have offered the Mitchell prize, $100,000 in cash, to the authors of papers that make the most significant contribution to the study of growth-related problems. The president of Texas A&M has attended some of those conferences.

Frank Vandiver: If we don't concern ourselves with the energy problems, he feels, in the next 30 years, we're going to be in real trouble. If we don't concern ourselves with world problems like AIDS and general health and really get into it and do something about it, we're going to be, maybe, beyond help.

NARRATOR: Galveston's future has also been a concern to Mitchell. Many years ago he saw the need for the island city to become a tourist attraction year around if it were to prosper, and he was also

concerned that the beauty of Galveston, its Victorian buildings and its history were being destroyed. In 1970, Mitchell visited the City of Savannah, Georgia, which was restoring its old buildings and later putting them on the market. He took that idea and eventually convinced others in Galveston that it would work for that city. Peter Brink is the executive director of the Galveston Historical Foundation. He recently talked with Mitchell about the restorations along the old Strand area.

Peter Brink: George, this is one of the really special buildings on the Strand. It was built in the 1890s by Nicholas Clayton and really shows the sophistication and ornateness that his styled developed.

George Mitchell: It's one of our favorites, too, because, as I said, Cynthia really looked at the Rosenberg library and saw all this beautiful architecture and we think that this is the most difficult building to restore because it is so massive and the terra cotta is very difficult to handle.

NARRATOR: George Mitchell also initiated such events as Mardi Gras, which has begun attracting hundreds of thousands of people to Galveston.

Jan Coggeshall: George and Cynthia are not only interested in the economic activities in helping us restore these beautiful buildings and bringing new tourism back to Galveston, but they've taken a very personal interest. I was so intrigued when Cynthia got involved at the Tremont House with the kinds of bedspreads that were going to go on the beds and the wardrobes that she ordered and her personal interest in the decor of the hotel. And I think the two of them have really gone out of their way to be involved in everything in Galveston.

George Mitchell: I've always heard the expression, "The business of business is business." I don't buy that at all. I think that you can really run a business and have a social duty that requires you to really think about how you help improve society, knowing that the main efforts, they have to be economic efforts, have to be, because you don't have the resources to do every other thing.

NARRATOR: George Mitchell is a successful businessman. He's be-

come a leader of his industry and a spokesman for independent oil men on many issues. He's held many influential positions but he is more than that. Throughout his life this mild, modest man has looked at the world and seen what could be if we would only try.

Frank Vandiver: I think one of his notable achievements is George Mitchell, and he doesn't know that, and he would deny that. But he has made himself, in my eyes, a truly renaissance man and there are not many in this modern world. I mean that in the most complimentary sense. He's a man who thinks in so many different subjects on so many different levels, comprises in his head a tremendous amount of knowledge, thinks about things that most of us are too bothered with the present to worry about. My favorite word for him as I said before: He's a visionary, he works toward what he sees.

Plato Pappas

Today is Tuesday, August 22, 1995. This is Joe Kutchin. I'm in my office in The Woodlands with Plato Pappas, who this past spring retired after 28 years with the company. His position was senior vice president of engineering and construction. Tell me a little bit about your personal history.

I was born on August 28, 1928—in fact, next week I'm going to have a birthday—in a little town known as Kiaton by the shore on the island of Peloponnesus in Greece. My parents were Americans. My older brother was born here in the United States. When my parents went back to Greece and they proceeded to have me and two more boys after me, they made their permanent home in Greece until after World War II, when the family was brought back to the States as American citizens and set up housekeeping in Astoria, New York.

How old were you then?

I was 17 years old. I had just graduated from high school and I had intentions to go to college in Greece where I was attending college preparatory school for entrance examinations into the Polytechnic Institute of Greece. However, a month before that would take place, we got on a boat and we came to New York.

George Mitchell's family is from Peloponnesus, too, isn't it?

Not too far from where I was born. As the crow flies, his father's little town is not more than 15 miles, up in the mountains. Of course, I had no knowledge of English whatsoever and when we came to New York, I attended high school a couple of months, and then I went to Queens College where they had a program for foreign students. I attended that for a year and after that I applied at many universities and colleges on the eastern seaboard because my home was in New York. Although my high school records were good enough for acceptance, the schools were having a tremendous influx of World War II veterans and I would be told to try again in 1950, try again in 1951, try again in 1952. I had no desire to wait until then, so I applied at the Universities of Oklahoma, Missouri and Kansas, and all of them took me, and I chose the University of Missouri. In 1948, in February, the spring semester of 1948, I started attending the University of Missouri, from which I graduated in 1952.

What kind of degree?
I got a degree in civil engineering and I spent two years in ROTC. Then when I returned to New York, I had a job with the aviation department of the City of New York for two months, and I was interviewed by the Corps of Engineers to get a job in Newfoundland in Canada. When I went to pick up my papers, I was told that I could not leave the country because I was to be drafted in the Army.

This was what year?
1952. That was August 28, on my birthday. First, I got my papers to go to Canada, and by the time I got home, there were greetings from President Truman awaiting me. So, I spent two years in the Army where I went to Germany for a year-and-a-half. Then I returned to the States and started working as a civil engineer.

Where, what company?
I went with the New York City Transit Authority. I worked there for about two years, 1954 to 1956, then I got married in New York and I left my position there in 1956 and I accepted a position with the engineering/consulting firm of Brown and Blouvelt in New York City. They were doing a lot of domestic work and a lot of overseas work. I went overseas for them three times, twice to Liberia and once to Indonesia. Then I opened up an office for the company in Youngstown, Ohio. Then an opportunity came for me to come to Texas with a land developer, Coventry Development Corporation, a subsidiary of Triton Shipping; that was in 1961.

Where was that?
That was in New York. However, I came to Houston, Texas, where Coventry had a tract of land on Spring Creek and I-45, across Spring Creek from The Woodlands. It was the Kuhlman tract of about 1,500 acres. Coventry had acquired it in 1961 in the name of Chrimirene Corporation, which was a combination of names of the three ladies in the Lemos family, Christine, Mary and Irene. Also, the corporation had another 2,000 acres of land on I-45 by the NASA site down near Webster, plus another tract of land across the street from the Shamrock Hotel. They were in the mode of developing the properties and that's how I came to Texas. It was at that time that I met Mr. Mitchell for the first time.

Do you remember what year that was?

Oh, yes. It was in the spring of 1962, and I met Mr. Mitchell through his partner, Norman Dobbins, who was president of the Houston Homebuilders Association at that time, and Pacesetter Homes.

I didn't realize George was involved in real estate or homebuilding at that time.

Pacesetter Homes was a homebuilding company. Norman Dobbins introduced me to George. At that time, George did not have other real estate going at all, other than Pacesetter Homes.

Was that solely homebuilding?

Homebuilding, yes, only home building. And that's how I met Mr. Mitchell. In fact, the day I met him he was walking on air in the Houston Club Building offices of the corporation for he had just bought out his partner, Merlyn Christie, and he told me in the corridor as I met him that he had just bought him out for $16 million cash. I congratulated him and in parting he said to me, "You people are in the transportation business and land business, we are in the oil and gas business and housing business. Think of something that we can work together on." I thanked him and I went on my way. Coventry Development Corporation was in the mode of developing those tracts of land that we had already acquired. And for that purpose, I had contracted with a company from California, a market analyst by the name of Goodkin and Associates, to come to Texas and advise us as to what kind of projects we could initiate that would have a good possibility of success. They came down in the summer of 1962 and they conducted a market analysis for us with the following results: The property that we had in the north part of Harris County by The Woodlands, but remember, at that time, The Woodlands was non-existent, it was suggested that it would be maybe 30 years before development could take place. This was back in 1962. The property down in Webster, about 2,200 acres–that was recommended for development on a very guarded basis. Meanwhile the NASA site was announced. The property was purchased by the corporation in August of 1961, the NASA site was announced in November of 1961 and I came down here between Christmas and New Year's, same year. Mr. Goodkin suggested that some development could take place there because of the NASA effort. However, he said, we must be aware that defense-oriented industry was going to go in the proximity of the site and we ought to be prepared in the case that defense effort, in any way, were to be pulled back because of economics or because peace is around the corner, then any development in that region could suffer because of a pull-back by the federal government. That left only the tract of land that we had across the street from the Shamrock Hotel, a 15-acre tract of land. We paid a very high price for it and Mr. Goodkin suggested that we play a waiting game, to wait on the market a little bit more until positive signs indicated to us that development ought to take place. And that was about the end of the summer, beginning of fall, in 1962. The decision was made to do no development in this area and to prepare to pull back to the main office in New York. That took place on the day after President Kennedy died. My family was put on the plane the day that President Kennedy was assassinated and my wife and I drove back the day after.

Drove back to New York?

Yes. While in New York from 1962 to 1964, 1965 and 1966, I built two high-rise luxury apartment buildings for Coventry. At Christmas 1966, I got an invitation from a very good friend of mine here in Houston, an attorney by the name of Dempsey Prappas, to come to Houston for a New Year's Eve party. My wife and I came down. We renewed old friendships, people that we had met in 1962. I went up to the office to meet Mr. Mitchell again, visited with him a little bit, and I saw his partner, Norman Dobbins. A year or two after I returned to New York, I got a call from a guy by the name of Ernest Wiseberger, a realtor here in Houston, the person from whom Coventry bought the Chrimirene properties and the Webster Ranch in Webster. He came up to New York and told me that 65,000 acres of land, scattered all over Montgomery, Harris and Grimes County was for sale. So I packaged the whole deal together as he gave it to me and I sent it to Mr. Lemos, of Triton Shipping, who at that time was in London. I told Mr. Lemos that we have an opportunity to buy several tracts of land from one ownership and I don't know whether or not we have an interest to get involved in such a large purchase. He said, "Mr. Pappas, I'm coming back home for Thanksgiving and spending the holidays in the States and at that time we'll go down to Texas and take a look at it." And that land package was the purchase that Mr. Mitchell made, where he bought the Cochrans' holdings. At the time, in 1966 when I came to Houston for the party, the purchase had already been made, and Norman Dobbins suggested that I might see George now that he had a lot of land, to see if there might be an interest in creating some kind of real estate company for the purpose of developing it. So that was something that was discussed with Mr. Mitchell at that time and then I left. I went back to New York in the beginning of January 1967 but stayed in touch with Norman on the phone until I returned for a visit in

April of 1967 when I saw Norman again. We went upstairs to see Mr. Mitchell, Morris Thompson and Budd Clark, and we all discussed the possibility of me being hired as an engineer for the company. At that time, of course, Norman was developing some land for the Mitchell-Dobbins Corporation and one of the projects was where the Esplanade is now, Imperial Valley. He was doing all the residential parts of that property, and there was another development that he was doing in Missouri City which he called The Woodlands.

Are you talking about Mitchell, Dobbins or both?
Mitchell-Dobbins, the company. He was doing the subdivision in Missouri City called The Woodlands and there was not a tree on site. That was on South Main. He was really spread out.

Let me interrupt. At the time we're talking about, the late '60s, what was George's company called?
I believe it was George Mitchell & Associates

So that was one corporate entity and off to the side he was involved in Mitchell-Dobbins. OK, go ahead.
I don't think Dobbins had any interest in George Mitchell & Associates. The only interest that was common to both was Mitchell-Dobbins and Pacesetter Homes. However, the purchase was not made in the name of Mitchell-Dobbins. The purchase was made for George Mitchell & Associates or some other corporation. Norman was involved in those developments going for Mitchell-Dobbins. An offer was made to me in April of the same year by Morris Thompson to come and work for them for $14,000 a year. That's what I was making in New York and they matched it.

To do what? What was your responsibility?
I was to be the engineer for the land development projects that they had. I'm talking about Mitchell-Dobbins.

So Morris had a role in Mitchell-Dobbins?
Yes, as an officer of the corporation. In fact, I was working directly for Morris. I started over here on Memorial Day 1967. Memorial Day at that time was not a holiday here; it was a holiday in New York, but here they were working that day and I did not know, so I stayed in my motel room. This was 1967.

What kind of company did you find?
A company that had a lot of realty holdings, a company that had a lot of development. In fact, the homebuilding part of Mitchell-Dobbins Pacesetter Homes at that time was building 300-400 homes a year. It was a sizable undertaking.

You didn't go with GM&A, you went with Mitchell-Dobbins, or did it make any difference?
It did not make any difference at that time. In fact, I was upstairs in the office next to George and across the hall from Morris. But I was offering my services to the corporation whenever the real estate entities needed some help. Especially so with the land development portions because they were hiring engineers, including Bernard Johnson, at that time, and they had nobody really in charge. When I looked back to see what had been done previously, I saw that large sums of money had been spent in the engineering of those subdivisions that the company was developing and building homes on. I recall that fees were in excess of 20 percent of the development costs. I was accustomed to paying 3 to 4 to 5 percent of construction cost, not 20 percent. When I started inquiring why those high costs, I was told by the engineer—and it was verified by Mr. Dobbins and Mr. Mitchell—that a lot of changes had taken place. The engineer was instructed to design something and when that was finished, Mr. Dobbins or Mr. Mitchell were there to change it. Changes cost money. You put all those costs together and you relate that to the development costs and the percent is very, very high.

Before you go any further, let's go back. You had been with the company in New York for six years or something like that. I presume it was a fairly satisfactory job, so what caused you to move? You didn't get any more money.
When I came to Houston, I really liked Houston and that's what attracted me, and I wanted to return to Houston. I did not ever like New York and because of that I wanted to leave New York. In the nine months that we stayed here we really developed some good friendships. A friend of mine, Dempsey Prappas, compared New York to Houston. He said, "New York is an apple tree. All the apples have been picked, there might be one here and there, but there are too many hands reaching for those ripe apples. Houston is also an apple tree, but all the apples are still green. When the time for picking comes, there will be some good picking." And really that made an impression with me. Aside from that I saw that Houston was a very vibrant

place, a place for young people, people who wanted to get ahead in life, and that's what really attracted me.

And you saw the 20 percent vs. what should be 3 or 4 percent?

In the engineering part of it we started taking control of our destiny and tried to minimize the changes. We developed a better handle on what we were asking the engineers to do. At that time, Pirates' Beach and Pirates' Cove were under development. The engineers were working for the corporation, but the corporation, I found out, was not paying the engineers because they were short of cash. The engineers had to carry the corporation.

These were not employees, they were consultants?

Yes. They were carrying the company for nine months, a year. So in order to maintain some kind of profitability, they were charging the company without the charges being thoroughly investigated. One of the engineers who was the project manager with Bernard Johnson was a guy by the name of Wilson Wendell. He was a very fine engineer, a person who I really thought very highly of because of his engineering ability. He had a very good understanding, and he was doing most of the work for our company. I did not think that the corporation should be paying such high prices. I thought that the corporation could do much better with a smaller size engineer, where we would not have all that overhead. Slowly, I started holding back some of their work and looking for other engineers. I talked to Dannenbaum, another big engineering firm. He had the same outlook as Mr. Johnson did; Mr. Johnson and Mr. Mitchell went to school together. Mr. Dannenbaum wrote the book on how the engineers ought to be charging their clients—he was at that time president of the Texas section of the American Society of Professional Engineers. I didn't have very much success there, either. Then I started talking to a guy by the name of Connie Hinshaw, who was a one-engineer, one-man shop, except in his case, he had two or three other engineers in his fold. He was willing to work with us on the basis of not exceeding 3 percent of the construction costs. And then I found out that Mr. Mitchell was the best man at Mr. Hinshaw's wedding. Connie and I would work together. I took the work away from Bernard Johnson and I gave a lot of it to Hinshaw, especially the Cape Royale subdivision that was purchased in September of 1967. The Imperial Valley subdivision work was given to a guy by the name of Irvin Peabody, who also had a small shop. Between those two engineers, we were able to bring down the fees considerably. No work was given out to engineers without going through me and, of course, I was not about to give any work out without knowing that the principals already had decided what kind of subdivision they wanted, what kind of work they wanted done by the engineer. And that eliminated a lot of the duplication of work.

So you were successful in bringing the relative cost down from 20 percent to 3 percent?

Yes. In fact, Mr. Hinshaw said, "Plato, I like working for you; however, I have a contract at Cape Royale that I would like to give back to you. I cannot make any money on that." I said, "Connie, I'm sorry to hear that you want to give that up, because there is a lot more work there." He said, "Plato, that fee will not make me any profit." So I took that contract and gave it to Wilson Wendell, who wanted to leave Bernard Johnson. He opened up his own office and picked up that contract. So Wilson Wendell came to the fold, along with Irvin Peabody, and did most of the work.

At this time, you were not involved in The Woodlands?

No, that came in 1969 when we started putting the land together. Mr. Mitchell was the focal point. By that time, we only had about 13,000 acres accumulated and, of course, he was dealing with a banker from east of Livingston, a Mr. McCann. McCann had a bank there and he owned this tract of land on both sides of Woodlands Parkway, all the way out. We bought the land in pieces, Mr. Mitchell did, and then he assigned me the task of finding two sites in The Woodlands where we could create two lakes. Wilson Wendell and I looked at the maps, looked at the topography.

There was a land plan then?

No, just the land itself. We looked at USGS maps and other topographical maps.

Kamrath hadn't done his plan yet?

No. We identified two potential sites, one at Spring Creek where Spring Creek Lake is supposed to be and the other where Lake Woodlands is today. We identified them as early as 1969, maybe the end of 1968. We did some good work, inexpensively, to get a handle on it.

The Columbia group had not yet arrived?

No. As we were picking up properties, Mr. Mitchell authorized an architect, Mr. Kamrath, to do sketches. He gave his work to Mr. Mitchell, primarily. I think his guidance was primarily given to Mr. Mitchell to decide on acquisitions in order to put together a large block of land.

I know he did the land plan, I don't remember the year. It's about the time you're talking about. And George considers that the granddaddy of the present plan.
Mr. Mitchell and I differ on that, because the first land plan was put together by an architect whose name was Cerf Ross. He was given $5,000 at the time to put a plan together. He went and leased a big home in River Oaks while he was working on putting together a master plan for The Woodlands. At any rate, the first master plan that I recall was by this architect. We spoon fed him with material and money in order for him to put a plan together that could go to Washington and qualify the project under HUD's New Community Act. At that time, I believe that Governor Romney was secretary of HUD, Governor Romney from Michigan, who, I understand, just died about two or three months ago. At that time we only had 13,000 acres in the block that we called The Woodlands. In fact, if my memory serves me well, we did not have the access to I-45 that we have today from Woodlands Parkway. It was not in the original plan.

Why not?
We did not own that land. It was further north where we accessed I-45 in the first plan.

Who were members of the planning team?
It was the architect, Cerf Ross, Mr. Mitchell, Jim McAlister, myself and Morris Thompson. They were the only people involved at that time. We had a meeting or two with HUD, with Mr. Mitchell being there, Mr. McAlister and Jerry Coleman.

Coleman was not a company employee?
No, Jerry Coleman was the director of the Houston-Galveston Area Council at that time, the first director of the Houston-Galveston Council, and more or less under his guidance we made the presentations to HUD.

I'd never heard of Jerry Coleman's role in that.
In fact, after we made the first presentation to HUD, in order to get input on what the application ought to look like and be like, we had dinner together, the four of us, Mr. Mitchell, myself, Jerry and Jim McAlister, in the

basement of the Sheraton in a Polynesian restaurant, Trader Vic. In fact, I was introduced to mai-tais for the first time. We even have a restaurant like that here in Houston, it was a chain. As we were having dinner, I remember that the previous week I was in what is now The Woodlands, looking at the site that Wilson and I had designated for a lake. On my way out, I saw two turkeys. I used to carry a revolver under my car seat, the old car that Morris Thompson used to have, an old white Chevy. I took the revolver out and shot one of the damn turkeys and I related the story to George. I said, "George, you know, there are turkeys up in The Woodlands, and as a matter of fact I killed one last week." He asked, "What did you do?" I said, "I killed one, George, a turkey . . ." He said, "You killed my turkey?" I said, "What do you mean your turkey, George, it was in The Woodlands, it was wild." He said, "That was my turkey." He almost fired me right there on the spot and then he explained to me that he had the previous year let out a lot of small turkeys in order to see whether they would propagate. I never knew that, and when I saw the turkey I didn't even know there was a season on turkeys. I killed the damned thing. And George Mitchell, to this day, still relates that story.

You were talking about dinner at the Sheraton, in the basement.
From that point on, we started massaging the plan with Cerf Ross, spending no more than $1,000 at a time, while we were looking to see how we could put a team together to do this new town. Morris Thompson was the person who would provide us with the funding as needed, through the corporation. I was providing the engineering discipline and McAlister was doing the economics, with Ross doing the bubble diagrams while George Mitchell kept acquiring properties to complete the parcel. I believe it was at this time when Mr. Mitchell, myself and McAlister went to Columbia, Maryland, after we made the presentation to HUD, in order to get some ideas on how a new town was to be done.

Give me the year.
That was still 1969. Columbia was just starting up. It must have been either 1969 or 1970 when we went up to Washington with the first application. But first, we went to Columbia. Mr. Mitchell had made the arrangements for us to meet with Mr. Jim Rouse who was the developer of Columbia. He invited us to his company's outing at the pavilion that they had built. It was a company-wide affair and he invited us as his guests. It was an event where he explained to his staff and to his people where they were going with

Columbia from that point on. After the meeting, we went up to his office and he told his people to open up whatever material, whatever information we wanted and make it available to us. We stayed the whole afternoon in Columbia where we were escorted by his chief planner, Ted Baker, reviewing the development process. Subsequently, Ted left Rouse and joined HUD. He was one of the people who was reviewing our plans once we applied to HUD. Ted Baker came here several times and introduced us to the new town type of evolution, as far as planning and engineering was concerned. At that time, I asked Baker, who was the engineer involved with Columbia, and he told me about Richard Browne and Associates. When I came back, I put a call in to Richard Browne who was not there, but I talked to someone in his office by the name of Bob Tannenbaum. He was Browne's assistant. And I told him that I would like to have a call from Mr. Browne when he came back and the reason for it, that we were trying to find a team of experts to guide us through the planning and preliminary engineering for The Woodlands, a new town. Dick Browne called me back days afterwards, and he came to Houston with Bob Tannenbaum. They were introduced to Mr. Mitchell, myself, Morris Thompson, Jim McAlister and, most likely, Budd Clark might have sat in on that meeting. We proceeded interviewing other people at the same time.

What were you looking for?

We were trying to put a team of consultants together in order to develop the plan. We interviewed, Mitchell and I together without anybody else, engineering firms, local firms, Turner, Collie & Braden, Lockwood, Andrews & Newman and Bernard Johnson, three big outfits. I don't believe we interviewed anybody else. Both Mr. Mitchell and I agreed that the most prestigious and knowledgeable company to do this work with us would have been Turner, Collie & Braden. Mr. Turner used to be an officer with the Corps of Engineers under whom Mr. Mitchell spent the war years. Newman was the officer between Mr. Mitchell and Mr. Turner. After we reviewed all the capabilities primarily of the two companies, and having experience with Bernard Johnson, we decided on Turner, Collie & Braden. Dick Browne came to Houston to assist us in the preliminary engineering and putting the plan together for the application to go to HUD, as did Mr. William Pereira from the planning standpoint.

Were you part of the group that selected him?

Yes, both William Pereira and Ian McHarg. McHarg came in later. Then we interviewed CRS Sirrine, a local architectural planning type of company, and the presentation was made for that company by Bob Hartsfield. CRS Sirrine, of course, was primarily an architectural company with very little planning abilities other than what Mr. Hartsfield brought as the chief planner for them. Mr. Mitchell thought highly of Mr. Hartsfield and he proceeded to hire him as a member of our company. And then Mr. Hartsfield brought in McHarg and more or less at that time, Bob Hartsfield took the initiative and became the point person in putting the team together. Dick Browne already had been interviewed as a consultant. That team was put together and Bob Hartsfield, McHarg, a financial advisor from Washington, D.C., Bob Goodman, Richard Browne and William Pereira. That group constituted the team that would put the new plan together.

Do you remember when the team of consultants came—the Brownes, the Pereiras, the McHargs?
1970, 1971.

So, George's plan was taking shape. He was getting the best in the country to do the job.
That's correct. We had a lot of meetings in Washington and we would meet in Philadelphia in McHarg's office and we would meet in Pereira's office in Newport, California. We would have two- or three-day meetings there, and then we'd have the consultants revise the plan.

These were good times?
They were taxing times. You could see the money going out the window even though we were trying to do the best that we possibly could in order to make something that would stay afloat. Of course, those people did not come cheaply. Our in-house team was Bob Hartsfield, myself, Morris Thompson most of the time–I don't think Jim McAlister attended those technical meetings. George would come in and out. The team would take the ideas and massage them and then would have another meeting until the program was put together.

At this time, George had in mind the creation of a new town?
Certainly. In fact, the first approach to HUD had already been made, and now we were massaging the plan with all these primary consultants in order to come up with a plan that, first of all, would make economic sense; secondly, would make market sense, and thirdly, that HUD would buy into. We

made the presentation to HUD. Pereira wasn't in. McHarg was the main person making the presentation. I don't recall now whether or not the HUD commitment was at hand when George started hiring the people from Columbia, Maryland. It was in 1972.

I think the company went public in February 1972. The HUD commitment came in later, in the spring.

And of course, one of the persons George brought in first was Len Ivins. Len Ivins then proceeded to bring in Bob Grace, Bob Everhardt, Sam Calleri. They took the plan and there was a major effort after that to revise the plan. In fact, my memory might be a little fuzzy. Maybe we did not go in at first for $50 million to HUD. I think we were going in for $25-$30 million. However, when the boys came in from Columbia, they massaged the plan to the extent that we amended the application. I want to say that we amended the application and went in for a guarantee of $50 million in debentures. At that time, they had brought in Bob Hinton and attorneys like Wanamaker and some other attorneys who were involved with Columbia in revising the plan to the point that could justify the $50 million debentures.

There are mixed reports on the impact of the Columbia group. How did they affect what you were doing?

In the beginning, there were difficult times. There were difficult times to the extent that they didn't feel secure that their undertakings and decisions would stay within the confines of the walls that they were made in. They were reasonably certain that if the decisions were made in my presence, that those decisions would become known to George or to other members of high management. They wanted to have a freer hand in doing things. I had no intentions in any way of taking information from one group of people to another unless I was asked specifically why I did certain things. Then I would have to say that the decision was made and that's why I did it. However, they had visions that I could not be trusted.

Was Vern Robbins part of their group?

No. The group was administrative primarily and somewhat engineering, which is how to put an organization together to build a new town. They brought in managers, McGee, Lawrence Kash and David Lawrence. All that group came in for administration. And, another corporation was put together, Community Development and Construction Corporation. This new corporation had made a contract with George Mitchell whereby all the work of the new town, as well as other parts of the Mitchell family organizations, and all the construction would be handled by the Community Development and Construction Corporation. The new corporation was to enjoy something like a 10 percent override over and above the construction costs of the projects as they came along.

That corporation was independent of George's holdings?
I think George had a percentage of that. I don't know whether George was to receive part of that 10 percent. At any rate, I became one of two vice presidents of that corporation. Vern Robbins's background was land development, and I was working more or less for him in the land development part of the business, the Community Development and Construction Corporation.

How did Vern get involved with the company?
When Vern came, they created Community Development and Construction Corporation. Vern Robbins was the president, Jack Nutting was executive vice president and there were two vice presidents; I was one on the land development part of it, and Ray Gutosky was on the above-ground facilities. Office Building One was built under those conditions. CDCC was first at One Shell Plaza, then was moved to Two Shell Plaza. They wanted to keep it arm's length from other real estate operations. The first week that we moved to The Woodlands, I was moved here, outside from the school there, Lamar Elementary. We moved two or three trailers out of which we worked. That was the end of 1972, 1973. At the same time, the country club was started. Also, that's when Carlton Gipson came with us. In fact, the first piece of concrete that we put down in The Woodlands was on New Year's Eve of 1972, at the intersection of Robinson Road and Woodlands Parkway on that ramp. It was only a truckload; Vern Robbins wanted to have 1972 as the beginning.

Was your relationship with Vern good?
Excellent relationship. Vern had told me that I was given to him so that he could fire me. Vern never did fire me. We became very close. Vern had a lot of savvy, a lot of experience. I learned a lot from Vern. He was a very, very decent person. If there is a word to describe Vern, it would be decent.

What was his prior relationship with the people from Columbia?
He was in Columbia in charge of planning, engineering and construction.

All those guys, including Len Ivins, worked for Vern at one time or another. Of course, we worked hard. In fact, Vern, on his own, rented and paid for a room at the Holiday Inn just before we opened The Woodlands. For about a month, we were working day and night, around the clock. We had floodlights, and he would go and sleep there for about four or five hours, relieve me, and I would go back and sleep in the other bed in the same hotel room. That's how we started The Woodlands. At any rate, we started land development work in 1972, September 14, I believe, when the first bulldozers arrived. Vern told me to have them here at 9 o'clock, to start rolling in at that time of the morning on a Monday, because there would be the county judge and a lot of dignitaries here. So I had made arrangements for all the heavy equipment to be lined up at Highway FM 1488 in order to roll in at 9 a.m. on Monday morning. However, they forgot to get a license from the Highway Department to travel from FM 1488 to here along I-45, and they had to run all the way in to Houston to get a permit from the Highway Department before they could use I-45. By the time they rolled in here, it was about 10 or 10:30. I was in the downtown office, not part of the festivities. Vern came in about 12 or 1 and said to me, "What happened to the equipment?" Meanwhile, I knew what happened to the equipment. I told him. He said, "But I told you I want them rolling in at 9, I didn't tell you 10 or 10:30." That was the first and only time that he chewed my ass out. We had, from that point on until he died, the most respectful relationship that you could find among two people.

Your good relationship with Robbins did not necessarily apply to other members of the Columbia group?
No, not necessarily so at all.

* * * * * * *

This is Tuesday, August 29, 1995, Joe Kutchin in my office in The Woodlands. Plato Pappas and I are continuing a conversation that we started last week. We had pretty well brought things to the point where work was under way on The Woodlands. Toward the end of our conversation, you talked about the late arrival of the construction equipment to get started and talked about your relationship with and the role of Vern Robbins. So you got under way with infrastructure? This was about a year before the actual opening?
Two years before the actual opening. We started in 1972, in the fall of 1972, and we did not open The Woodlands until 1974, and I believe that was Octo-

ber 19, or thereabouts. At the opening, we had 12 townhouses built and people moved in on opening date. They were the only residents that we had at that time, 12 families all residing in what was appropriately named Settlers' Corner. At the same time the country club was also finished.

Describe what infrastructure there was.
A central water facility with one well, one storage tank and the associated facilities that would start sustaining the community were in place. The sewer plant was in place with sufficient capacity so as not in any way to impede a normal and ordinary development. And, of course, the streets associated with access to the property from the outside and from within the community were in place. All the infrastructure was designed and well under construction in order to accommodate the marketplace, so to speak. At that time, the design was severed from the construction. There was a group of engineers, in-house engineers who were administering, and the consulting engineers, who were designing the subsequent phases of the development. After the bids were taken, then those plans were handed over to us, to the construction company, to execute in the field.

And to step back a moment, were you, CDCC, responsible for the roads, for the MUD, for the Inn, as well?
All the public roads were constructed by CDCC. All the water and sewer were constructed by CDCC. The sewage treatment plant was constructed by CDCC. It was very advanced, and I will go into that a little more in detail, if I possibly can. In the beginning of The Woodlands, the idea was not only to meet the minimum criteria for good design but also to exceed wherever possible.

Was this part of the HUD agreement, or just an attitude?
It was an attitude, and indirectly, of course, it came all the way down from the top, that we were going to make a community here that would survive the century and go beyond it. We were fully aware of the deterioration that was happening in the existing communities, lack of adequate sewer capacities or treatment that prevailed all over the world, polluting the streams of the country, and of the world, and we wanted to do something better. We were trying to demonstrate that we could live with nature and leave nature in a better condition than we found it, and that was the motivating factor behind our undertakings. We went to HUD, and after examination of different processes for waste water treatment, we decided . . .

We—were you one of them? Who else beside you was involved?
Yes, I was there. Bob Grace was another one who was with me at that time and we were guided by a guy named Bill Cheracles. He was a professor at Rice University and his expertise was waste water treatment. He suggested the Univax system that was put on the market by DuPont. This was based on a pure oxygen type of process whereby you induce pure oxygen into the mass of effluent because in the biological treatment process, that's what you're doing, inducing oxygen. And, of course, the cheapest form of oxygen is to get it from the air. However, when you have a concentration like pure oxygen, it requires smaller basins and the treatment is much more certain to kill the bacteria. And we all bought into that. In fact, I made a trip to see a demonstration like that in Washington, D.C.; the only plant that they had at that time in actual use was in Tahoe. I went to Tahoe with Bob Grace, Bill Cheracles and a man from S&B Engineers to see the system in actual practice. No discharge was made into Lake Tahoe at all, it was totally prohibited. The effluent that was generated in that valley was all concentrated in one place, treated with pure oxygen and then pumped over the mountains 28 miles to lakes up in a very high plateau for further processing. They were growing Rainbow trout in that effluent. We saw all that process, from A to Z. We came back and introduced the system here. For further kill of the bacteria, we introduced an ozone method, which had a tremendous kill power, instead of chlorine. Chlorine, of course, is the agent that has been introduced universally. You have to maintain a certain amount of chlorine some 20-30 minutes after you introduce the treated effluent into the stream. However, you cannot hold ozone in that kind of molecular state because it dissipates right away, but the kill is universal and absolute. That's what we had planned. It was quite expensive and because of that, because we were trying to introduce something new, we had some help from HUD in deferring some of our costs. That was the first increment that we introduced. That was the first phase that we introduced, one-and-a-half-million gallon per day capacity in the plant. Then we went for an expansion.

One-and-a-half-million gallon capacity—is that what it was on opening day?
No, on opening day we had the demonstration plant over here on a semi-trailer that was given to us by HUD. It was a complete plant. However, it did not have the basins, because we only had 12 connections. No plant–unless you load it with 25 percent of its ultimate capacity–is going to work. On opening day we were using a demonstration plant on a low-boy. After a year-

and-a-half or two years, about 70 percent of the capacity of the first plant was exhausted. We wanted to increase the capacity of the plant and we went back to HUD and asked them to help us on several phases of it. We wanted to bring it up to 3 million gallons. HUD said to us, "That's all good and fine. However, the stream that you're discharging your effluent into is quite healthy. And there is no reason for you to discharge effluent into that stream of higher quality than the stream can stand. If you want our money, then you are going to have to reduce your effluent standards from two parts per million BOD–biological oxygen demand–and suspended solids, you've got to increase it to 15 parts per million and 20 parts per million." To the novice, probably that doesn't mean very much. However, in order to reduce from 15 parts per million down to 2 or 3 parts per million will cost you three times as much as it does to bring it from 400 parts per million down to 15 parts per million in the first place. There is a tremendous cost associated with the reductions beyond the 15 parts per million that the local criteria call for. So we packed up our papers and went back to HUD and asked them for money and they said, "No, if you want our money, then you've got to reduce your criteria," which would put more pollutants into the stream. We dismantled the ozonator, we dismantled the Univac System, and then we went to a standard biological process with basins, contact stabilization basins.

That's pretty standard stuff when you go to the basins?
Absolutely.

You had gone from something state of the art . . .?
We came back to whatever everybody else was doing across the country.

Was that discharge used for the golf courses?
We wanted to reuse all that effluent to water the golf courses and we had the system in place to bring back all the flow from the sewer plant instead of having it drain into Panther Creek and then into Spring Creek. The idea was to pump it up to Lake Harrison where it could be used for irrigation of the golf courses, but that part was never activated. The pipe is still underground. By now, it's deteriorated, never used in that fashion.

What is the total capacity now that we are a community of 43,000?
Six million gallons, and we are pushing another 1.2 million out of a temporary plant. A second permanent plant has been designed and is in the

process of being funded right now through the San Jacinto River Authority. The other system that we tried to introduce into the land development process was to make certain that we had adequate water. But let me back up just a little and talk about what happened after the opening in 1974-1975, when a lot of things came to a screeching halt because we found that we had over-expended our budgets.

That was true throughout the company.

I am speaking of that, yes. After the opening of The Woodlands, the realities began to set in. We had a reversal, the economy became very stagnant. We were out in The Woodlands with expenses in excess of $30 million, and that's a wild guess on my part. It may be more. New houses were not being built because the market dictated that the housing market would be very slow, so it was a time to regroup. The regrouping took place. Some of the ways that we were doing business had to be changed, and there were severe cuts in personnel.

Did that affect CDCC?

Very much so. At that time, CDCC disappeared, or more or less was dissolved. Mr. Nutting and his assistant started a new company and they were building shopping centers. They had a contract to build a shopping center in the southwest part of Houston and they went down there. Vern ended his association with CDCC and became an officer of The Woodlands Corporation, or the predecessor to The Woodlands Corporation, and I stayed with Vern and became his assistant. During that reorganization, a lot of people from the engineering and construction phase of our business over here were let go, a lot of marketing people were let go, and there was a reorganization whereby The Woodlands Corporation's engineering and construction business was handled through Vern Robbins's office. I had taken all the non-Woodlands engineering and construction from Vern's various technical responsibilities because Vern was not an engineer to begin with. He did not know the non-Woodlands properties as well as I did because that's how I broke in with the corporation. So, aside from The Woodlands, the non-Woodlands properties became my responsibility at that time. Cape Royale, Pirates' Beach and Pirates' Cove, Pelican Island, the timberlands, Clear Creek, West Magnolia Forest–all those areas that did not require the degree of engineering sophistication that The Woodlands required. They were under my direct responsibility, and administratively they were also under Vern Robbins. When we joined the engineering and construction department together under Vern, we found it necessary to do away with a lot of things in order to be able to administer them a little better. At the end of 1974, beginning of 1975, there were 15 different engineering firms that were doing work for us here in The Woodlands. Every one of them had very good engineers. However, when we engaged them for one job, then they would introduce into the design phase what they knew best from their own experience, and, of course, you would have 15 different ways of doing the same thing, in the same community. That created tremendous disarray, there was no unanimity in the product. Every street looked different than the streets in the other section, because they were designed by different engineers. So one of the first decisions was to start getting away from the multitude of engineers.

You weren't paying them anyway—cash flow had slowed to a trickle.
That's right. At that time, no. But we did bring some normalcy into the design phase of it. Of course, practice had it and still has it here in our area, where the engineer more or less is the alter ego of the developer. The developers by and large are not engineers, they are not attorneys, they are people who more or less have a way of getting the money to develop a piece of land. They depend on the engineer to bring in the lawyer, financial advisor, and to be everything that the developer is not. The developer is likely to have nothing except money. However, in The Woodlands we found that we had everything, including money, at least a little, and therefore, we could keep the engineer doing just engineering work, which is what we hired him for, and so we were able to bring his fees down considerably.

It is a practice to have engineers as consultants rather than on your own staff, is that correct?
That's correct.

Is that an industry-wide practice?
Developers don't have the volume to carry their own engineers. An exception, of course, is developers of master-planned communities, and even they tend to rely heavily on consultants. At that time, there were not many master-planned communities in place. The other thing that was more or less in practice at the time was that the design engineer usually also had the responsibility for supervising the construction work, and it was not uncommon at all, during the early parts of our business here in The Woodlands, to see an inspector or a contract administrator on every job, because different jobs

were done by different engineers, and, therefore, on two adjacent jobs you would find two construction administrators. No administrators would necessarily be there full time, at any one time, anyway. Therefore, you found inspectors and contract administrators loafing around and having parties of their own while the work was progressing. I asked Vern for his permission to go to Austin and meet with the Water Commission, which imposed upon us that all work had to be supervised by a professional engineer. I argued the point that this was overkill, and finally I was permitted to introduce to The Woodlands the practice where one company supervises all the jobs of the engineers. And under his name, the Water Commission would accept the facilities on behalf of MUDs. And to this day, we are operating under the same understanding and agreement. That saved us a lot of money and a lot of headaches.

Let's summarize the parts of The Woodlands that you had a sizable role in . . . certainly the road building, the storm and sanitary and water. What about the Inn?
No, the Inn and Country Club was a separate deal altogether. We did only the infrastructure that supported it. In other words, we brought the water, the sewer and the drains and the roads right to the tract of land in order for the Inn to receive all the utilities from us.

Is it accurate to say that those are the kinds of thing that you've been involved in through your career here?
Yes, absolutely. And that brings us up to opening day of The Woodlands. At that point, of course, there was very little commercial activity. In fact, there was none until we started getting residential base to support commercial activity. Still, it was the developer's responsibility to bring in the utilities necessary to support that tract of land. Within that tract of land, where the commercial activity was taking place, it was the responsibility of that developer whether the developer was ourselves or a third party.

And when you say utilities, you don't mean electric and gas?
No, I'm talking wet utilities. Electricity and gas are dry utilities and are the responsibility of Gulf States or Entex.

One of the things that was well publicized was the self-draining parking lot. Did that really work?
Yes, definitely. In fact, I was instrumental in making that happen. When we

made the application to HUD, I found out from the HUD engineer that there was a porous pavement design that HUD wanted to demonstrate. What a beautiful opportunity for The Woodlands it would be to have that kind of pavement demonstrated here in The Woodlands. So, I went to Elizabeth, New Jersey, where the pavement offices of EPA were at that time. I got hold of the design and we negotiated with HUD for the pavement for the parking lot. They asked us to do one-half acre of parking lot and they gave me the design. At that time, we were doing the Grogan's Mill shopping center and we designated Rice University as the entity that would monitor the system and write a report on their findings. Austin Bridge Company was the contractor that installed the parking lot.

This covered all of the Grogan's Mill shopping center parking lot?

No, only one-half acre of it, 100 x 200. It was well isolated. A big hole was dug into the ground and filled in with sand. We put down polyethylene to drape the whole bottom, and then a very thick layer of sand, about three feet of it, and then on top there was one foot of gravel, well graded gravel, and then that was coated with some black top, and then a three-inch layer of well graded black top, for surface. The gradation and the mix allowed the water to penetrate through that system and go into the gravel base, then into the sand base, which acted as a filtering system. The whole system was graded towards one corner where there was a device that would grab samples of water in order to test its quality, and from there it went to Lake Harrison. When we opened the parking lot we took a hose, a four-inch hose, from the fire truck and we put it on the pavement and opened it up full force and the puddle that was created was only 12 inches in diameter. All the water was sucked down. And we kept that open for either two or three years. A paper was written; I helped write it with Phil Bedient of Rice University, who is head now of the environmental engineering department of Rice. The people who worked on that system, their names are on the reports. Once we pulled out the monitor and equipment, we paved over it.

Did its use become widespread?

It was very costly. As I recall, the cost would have been twice as much as a concrete pavement.

It worked as design, but it was not economically feasible?

That's correct. Those things need to be carried down to the next step, but the federal government gets excited about something, does something and then lets it go. That's the very unfortunate part about it. But again, you need to have money in order to do all those good things. Research is good provided you translate that research and the findings of that research into something for the common good, an application. And we had an application. However, nobody has pursued it to make a profit on it. That's about the extent of it.

If I understand from Michael Richmond, you're the best source about the background of the Hardy Toll Road. I believe you played a very sizable role in bringing it out here.

I didn't do very much, just with the engineers who made the decisions. We demonstrated the need concerning where those roads ought to go. I had some bad experiences myself and I didn't want those experiences to be repeated. One time, I was sent to Indonesia to do a route survey in Sumatra, 1,000 miles of it, to find ways to access the island with a good highway system. We were three teams of two. Once we did our job there, I got a telegram from the New York main office to stay behind, that further orders would follow. I stayed behind with another engineer and we got orders to design a full cloverleaf interchange for the Indonesian government and to get further instructions from a certain person in the Department of Public Works. I went to meet this certain person who was the assistant secretary of public works, and he said, "We need a full cloverleaf interchange." I said, "That's fine, we know how to do that. Where do you want it? What streets, what main thoroughfare will it accommodate?" He said, "The Russians are designing a gymnasium to house the Pan-Asian Games. The United States government offered to make that site accessible from the harbor all the way to the gymnasium so people traveling could get there." I asked, "But what intersection might we design it for?" He said, "It doesn't make any difference, just design it to bring the roads in." So we designed a full cloverleaf interchange that was on wheels, so to speak. That's bad engineering, and I didn't want that to be repeated anywhere, so we tried to reason with the highway engineers and the toll road engineers about where the people are going to be and the needs that needed to be met. For those people who believed that The Woodlands would be a success, it was quite obvious that's where the toll road belonged, and that's where it is. But a lot of people didn't believe in that, if you'll recall. A lot of people thought that we were going to fall on our faces and our asses. We did not do that. That included some of the competition and even some of the people who work here today.

You didn't have any political role in dealing with the Harris County people in causing the toll road to happen? At the Pavilion, when somebody wanted something to happen with the highways, they would say, "Make Plato do it. Everybody does what he says should be done."

Only within reason. Over our many years in the field of engineering and construction here in Houston, yes, we got to know a lot of people. I got to know the county engineers, the Toll Road Authority people and what have you, and it was personal contact that we have had, so really, we did not have to start from scratch in developing those relationships. When a reasonable need was demonstrated, we had their ears because they knew that we would not ask for anything that was not safe and reasonable. We had demonstrated that to them. Of course, we had personal contacts and they trusted us and we trusted them. On mutual trust, a lot of things can happen that otherwise would not.

You talked about a couple of important innovative things that you were closely involved in, the sewage treatment, the self-draining parking lot. Does building the Pavilion hill fall into that category, or was that just a straight engineering job?

The Pavilion was nothing too complicated to require any degree of extraordinary sophistication or intelligence. It was a matter of how to utilize the extra dirt that was available to us and place it in such a way to meet what those people needed and yet get rid of the dirt. It was a matter of coordination, nothing more than that. From a technical standpoint, we knew the compaction that we had to have and make the fill on. The person that I relied very heavily on to make certain that all the excess dirt was going to a specific place was Virgil Yoakum. He worked for me for 13 years before I retired. He worked very closely with me and he was the organizer-coordinating person. Before that, we stored a lot of material that was excess material from the development process. Where we knew there would be golf courses, for example, and because we did not know how much dirt each hole would need, then we just stockpiled it and then we used dirt from the "big mound" that we had on the other side of Lake Woodlands, by the high school. We had a tremendous mound of dirt there, which we used eventually . . . for the golf courses, among other things.

In 28 years with the company, what are one or two things that you're proudest of? Your own accomplishments.

The Woodlands is at the top of the pyramid. One thing that I would recall very fondly in developing The Woodlands was our ability to organize the engineers and the contractors and the marketing people in preparing the product that The Woodlands wanted to market, to get it in the marketplace in the form that the marketplace would accept. That effort really was done in all strata of our organization, from engineering with guidance of the marketing people, what they wanted, and, of course, the important guidance that we had from George Mitchell in seeing that The Woodlands was what we all had hoped that The Woodlands would be. That effort, really to this day, provides the central direction of all the young people now that we have carrying on the theme of The Woodlands. It started from the very top and has gone all the way down to those people who are out in the trenches right now. Every one of them is so much aware of the kind of product that we want The Woodlands to offer in the marketplace because that feeling started at the very top.

Any major disappointments?
The only disappointment that I had is that I wanted to see The Woodlands finished during my productive years, while I was still in The Woodlands. *(Pappas participated in the voluntary early retirement program offered by the company in the spring of 1995.)* If you recall, our application to HUD was for a development period of 20 years.

That turned out to be beyond reach.
Of course. That dream did not materialize and to that extent there is a small disappointment. A large disappointment is not to see the riverwalk completed. I think that will be the jewel of The Woodlands.

Was that concept in the original plans?
No. If you recall, the original plan had us developing from the inside out. The original concept, Joe, was to try to isolate ourselves from experiences on the outside. We turned everything inside. In fact, if you see the earlier road networks, although we connected with the outside world via I-45, there were only two areas where we would interface with the outside world. Everything else would come around in circles, all the way to the centroid of The Woodlands, about where Lake Woodlands is. In fact, the shopping center, the mall, in the original sketches you will see it very close to Lake Woodlands. The fact that we were able to change, to make certain that we would not make those horrible mistakes of excluding the rest of the world, I thought that was a

demonstration of good planning. I would say that Dick Browne did a hell of a good job because the previous planner had concentrated everything within, while Dick Browne was the person who started opening up to the outside world. The disappointment is that I will not be around to see the project completed.

Have we failed to talk about something that's important?

No, I'm very happy to tell you that having observed the operational people for the last three or four months after my retirement, I'm very happy to say that we are leaving the operations of The Woodlands in very good and capable hands, in the hands of Alex Sutton, in the hands of Bob Heineman and all those people who have demonstrated a good way of doing very good business.

Plato, thank you very much.

W. ARTHUR 'SKIP' PORTER

Today is Tuesday, August 15, 1995. This is Joe Kutchin, and I'm in the office of W. Arthur Porter, "Skip" Porter, as he's widely known. He is president of Houston Advanced Research Center. Tell me a little bit about your background, what you did before you came to HARC.
After graduating from Irving High School, I did a bachelor's and master's in physics at North Texas in Denton, now the University of North Texas. That was in '63 and '64 for the bachelor's and master's. Then I went to Texas A&M in '64 and started my Ph.D. there. All of this was in physics. I worked until '66 at Texas A&M in the physics department, ran out of gas. I had finished my course work for my Ph.D. in physics and was doing my research. Just really got tired of the educational grind. I went to Texas Instruments, in their R&D labs.

What year was that?
6-6-66. Easy date to remember. My first working day at Texas Instruments.

Is that in Dallas?
Yes, in what was then called SRDL, the Semi-Conductor Research and Development Laboratories. I was assigned a project termed, internally, "continual diffusion." The idea was, somebody grab a bag of sand and pour it in a funnel, nobody touch anything and out would pop integrated circuits. We got pretty close. My first patent was on the continuous diffusion system. Exciting time to be at TI, exciting project. It was a period where they gave young whippersnappers like me just a tad more responsibility and authority than they were supposed to have. In my case, for me it worked and I think for TI it worked. That led to their major front-end program. That's how I got to know Fred Busey, who later came on HARC's Board. Fred was then vice president of Texas Instruments and was the heir-apparent to become president, and did. In '68, when I was at the end of having developed that system, Texas A&M asked me to come and set up their solid state program. This was a transition period from the vacuum tube to the solid state device, back in the late '60s. I took a leave of absence from TI after completing the development of six of the systems that I designed that were installed in the Sherman plant, building the IBM chips that TI had a big contract for. I went to Texas A&M and started building what became the Institute of Solid State Electronics and completed my Ph.D. in Interdisciplinary Engineering, which

was a combination of building on the Ph.D. work I had done in physics and adding material science and electrical engineering. Then in 1970, when I finished my Ph.D., A&M offered me a tenured track position, and so I stayed there. I stayed on leave of absence at TI for a number of years. Jack Kilby and I worked together there. Jack had been the director of the laboratories at Texas Instruments. Jack, of course, as you know, is the inventor of the integrated circuit and the hand-held calculator.

This was during your TI period?

Yes, that's when Jack and I met. Jack subsequently left in the mid '70s. He and I got together during the energy crisis, when we were all waiting in block-long lines to buy gasoline. We came out with a concept for a spherical solar cell. And Jack Kilby and Jay Lathrop, who had also been at TI—matter of fact, was my immediate boss at TI, had also left Texas Instruments and had gone to Clemson as professor of electrical engineering—Jack and Jay and I started working together on the idea. Then Jack came to Texas A&M in a transitional mode, bouncing between Dallas and College Station, and in my laboratories we developed the device. During that period, Jack became a professor of electrical engineering at A&M on an adjunct basis. He was named a distinguished professor. Jack loved to go to the ball games, he liked all that Aggie tradition.

I think he has a lot of Illinois tradition, too. He and I got off to the side one day and talked a lot about Champaign-Urbana.

Well, of course, he did his master's work there and is a distinguished alumnus of that place. He has a lot of fond memories there, as well. As a matter of fact, after we developed this spherical solar cell in my labs at A&M, Texas Instruments had a right of first refusal on whatever Jack did. So it was presented, they took their options and then we started transferring the technology from Texas A&M to Texas Instruments. We did that through the mid-'70s, late '70s. The code name for the project inside Texas Instruments was the Illinois project. There were about three guys who had graduated from Champaign-Urbana. We got the patents on that spherical solar cell project and contracted at Texas Instruments to support the development work. The government gave up its rights to the technology, which is now standard practice. We also got the patent in record time because Congress had passed a law requiring the Patent Office to act on any application dealing with energy within a year. In 1980, I was still professor of electrical engineering at A&M and was asked to direct the Texas Engineering Experiment

Station. I did that for five years until January of '85, when I took the presidency at HARC.

During that period, you were on the Board of HARC, weren't you?

Correct. HARC was first created on paper. As you'll recall, George had hired Arthur D. Little to do a study of what could be done in the Houston area. They interviewed me at A&M, along with a number of other people there. We've got a copy of the original proposal around here somewhere. Something along the lines of the Research Triangle Park in North Carolina with the three universities, North Carolina, North Carolina State and Duke. Incidentally, as you know, Duke is our newest collaborating institution at HARC, joined last year. The Little recommendation was to create HARC, then called the Houston Area Research Center, and I was the second person from A&M on the Board. John Calhoun, I believe, was the first representative from Texas A&M to serve on HARC's Board. That was probably in '82, and then in '83, Harvey McMains came on as the acting director. He had come from The University of Texas at Austin. He had worked with George Kozmetsky at UT. I think he was brought in on a half-time basis, or something like that. And at that time, I believe, David Gottlieb was the U of H representative, John Margrave was the Rice representative, Gerry Foncken was the UT representative and I was the Texas A&M representative. Every now and then, whenever George got an urge, I guess, he would call up a meeting of the executive committee of the Board.

Those you just named?

Yes, along with George and Coulson Tough, and there may have been one other MND person, maybe it was Ed Lee, who was then president of The Woodlands Corporation. I'm not sure about that, but I do remember that Coulson was a member. We would come down here to The Woodlands and George would convene a meeting of the executive committee of the Board, which he chaired, as he did the Board. He would look at us and say, "Has anybody got any ideas on what HARC can get started on?" I put two projects on the table that had hit my desk at the Engineering Experiment station. And remember that the Texas Engineering Experiment Station, so-called TEES, was created in 1914, really, in my judgment, as kind of the publicly funded agency attempting to do for Texas what other such technology-focused organizations would later do for regions like Silicon Valley, Route 128 and the Research Triangle Park. In the five years I directed that agency, what became evident to me was that if you go back to those three I just mentioned, Silicon Valley, Route 128 and the Research Triangle Park, and also look at Stanford and SRI, the common thread is they are all private. There is not a publicly funded entity in the lot. From Texas's point of view, I was heading our competitive agency.

Competitive with HARC?

No, competitive with those I just mentioned. Competitive with the other states in attempting to have Texas benefit from technology like California, Boston and North Carolina have. One of the reasons I was intrigued to seriously consider George's request to me that I come to HARC was my conclusion that you really can't play this game as a publicly funded institution. If Stanford or MIT were to cut a deal with Exxon, the governor would have a hard time calling them and asking them what they've done with taxpayers' money to help a company like Exxon. I'm not picking on Exxon, just using it as an example. I got a lot of those calls. Political forces are a great anchor to drag along in playing a fast-response game in a free-market system where you need to be responsive, you need to be able to make decisions and work with the private sector in a timely fashion and not be encumbered by political ramifications.

You're saying that political considerations had to be a factor at TEES?

That's right. TEES is responsible for doing for the State of Texas pretty much the same kind of thing that HARC is challenged to do. But the real advantage, in my judgment, that HARC has is that it's private. It benefits from the same situation as Stanford and MIT and the Research Triangle Institute and SRI in that their resources are not politically encumbered.

I have the idea that the program concerned with growth issues was brought in fairly early in the creation of HARC. Is that generally correct?

The Center for Growth Studies, or now Global Studies? Yes, because there had been a program before there was a HARC. The program, as I understand it, dates to the pre-HARC time frame, when The Woodlands was created; about that time, The Woodlands Conference was created. Its purpose was to bring attention to thoughtful development and utilization of the earth's resources, managing development in harmony with nature and the environment, etc. So, hence, a natural place had a natural name, "The Woodlands Conference."

I think, just to put it on tape, that with the creation of HARC, The Woodlands Conference—which had been a child of the University of Houston and Mitchell Energy & Development—just seemed to fit in naturally and constructively with the research organization that was then being created.

It seemed to me that HARC was in a position between The Woodlands Conference events to have the intellectual energy to maintain ongoing dialog that would connect one conference to the next so that it just wasn't every two or three years 300 or 500 people got together for some part of the week and a lot of energy was pumped in to talk about the issues and somebody walked off with the $100,000 Mitchell prize and then we waited until the next one. It seemed that HARC could stimulate scholarly dialogue and input about the last conference and work toward designing the next one. So I said, "What we really ought to do is bring that inside HARC, have UT, Texas A&M, Rice and U of H collectively engage in all of this and find a person to head it." And I did that, and that's when I brought in Jurgen Schmandt from the LBJ School.

Let's move back. Please talk about the circumstances involving Harvey McMains and then your departure from A&M.

When I was on HARC's Board, probably sometime in '83, Harvey McMains was hired, I think on some part-time basis. He had been at UT. His background was in physics, purportedly with a Ph.D., but that didn't turn out to be accurate. He was brought on after I came on HARC's Board; my recollection is that for the first several meetings of the Board, Harvey wasn't on the scene. How I came to know Harvey was through HARC. I did not know him otherwise and I don't know how George found him. Maybe Gerry Foncken can enlighten us on that, because I think Gerry was the UT liaison person with HARC. I had suggested to the Board, to this executive committee, two projects that I thought were well-suited for HARC. Both of these came about as a result of my position as director of the Engineering Experiment Station. One of them was the supercomputer, because all of the computing in the A&M System reported to me. So I was asked by the chancellor to chair the computer committee. I had already run an experiment on the A&M campus that looked into the question of, "What is this supercomputer stuff that's coming out, should we go get one?" I had offered free supercomputing to the entire system and set up an account using the one at Boeing and the one at Colorado State. I had already realized that to buy one of those for one campus was not an efficient use of funds, but to share a computer remotely is one of the most natural things that you can do.

We're talking early '80s?

We're talking 1983. The other project was the Super Collider. It had hit my desk as a rather large-scale, national scope, multi-billion dollar project that the State of Texas should take on. It had been brought to my attention by a guy named Peter McIntyre, professor of physics at Texas A&M. He was very aggressive, bright, excited about this project. He had talked about it to Frank Vandiver, who was then president of Texas A&M, and Frank had sent him to see me. Peter was a tenacious kind of guy and you finally had to look into what he was talking about. When I looked into it, I realized that it was a project that the A&M system could not compete for alone. Those in the competition were federal laboratories, Fermi, Berkeley, Brookhaven. The only way I could visualize the State of Texas marshalling any credible critical mass of talent to compete for such a project would be if we pooled all of the talent that we had collectively at all our institutions of higher education. The only mechanism in place for doing that was HARC. You know the character of an academic institution: You're not going to go to the other one to do the collaborative work, they just don't trust one another, and HARC presented a neutral ground opportunity.

At this time you're still with A&M?

Yes, I was still director of the Engineering Experiment Station. And I put that on the table as a project. I thought that if HARC could muster the resources to pull the team together, put a credible accelerator group together to go after some of this $20 million of magnet design money, that it would be a good kind of collaborative project to launch the institution. So bringing a supercomputer in and going after the Super Collider were two projects that I put on the table as projects that I thought HARC could get started on.

And when you say "put on the table," what do you mean?

I presented it to the executive committee of the HARC Board of Directors. Usually the operative question at any of those meetings was, "Does anybody have any ideas about what HARC can get started on?" And I said, "Yes, I have two. Go get a supercomputer and network it to all of our universities. That would help. And we can rally the leadership of the State of Texas politically to compete for the Super Collider, organize the talent that we have in the state, and put it together in a critical mass and give us the basis for competing." In 1983, very few people had heard of a Super Collider. And I wouldn't have had it not been for the fact that we had a couple of high-energy physicists on campus who were in on the upstream conceptualization of all this stuff. Peter McIntyre's an aggressive guy. He brought Sheldon Glashow, the Nobel Prize winner, down here to talk to George. He was working with Pat Zachry in San Antonio. There was a lot of energy poured in, trying to stimulate interest in this thing. When I got back to the executive committee of the A&M System, which I served on, I said, "Gentlemen, I have tried to give away A&M's position in taking the lead on the Super Collider project because we can't do it." I just made that call and there was this uniform sigh of relief around the table, "Well done." I guess it was about that time that George decided that there was a guy from U of H who could lead our computer activity. We launched the effort together after it was announced that there would be four supercomputer centers created around the country. A proposal from HARC to the NSF for one of those four national supercomputer centers was prepared and submitted.

At this time does HARC have a physical facility?
No. Well, it's got some rental space, on Grogan's Mill. That's where Harvey McMains set up the first HARC office.

And there was a handful of staff, is that right?
That's right. There were maybe three or four people. There was a secretary, and once the SSC magnet contract from DOE was awarded, Ann Sherman from the University of Houston came in as HARC's contracting officer. There was a secretary and there was a lady named Terry Witt, who was the first bean counter. So when I came to HARC, when the Harvey McMains fiasco occurred, it was, as the British would say, "We had a little drama." I got a call from George about 10 o'clock at night saying, "Oops, there's going to be a big announcement in the *Post* tomorrow." And as it turns out, this guy doesn't have the credentials he claimed and he's been terminated. David Gottlieb, then U of H representative on HARC's Board, was asked to sit in as president on an interim basis. I don't know if the search for HARC's really first full-time president had been launched at that point or if it was launched shortly after the Harvey McMains situation.

My recollection is that it was after.
That's what I think, too. So there was a period in there, after Harvey left,

that, as a Board member at A&M, I'd get calls from David Gottlieb about the stress he was having dealing with some of the high-energy physicists. Most of the time, it was related to some of the A&M high-energy physicists. David would call and just need somebody to talk to, and so I was it. It was in the fall of '84, maybe early fall like September, that George Mitchell came to me and said he would like to nominate me as one of the candidates for the presidency of HARC. I hadn't even considered that. I was fully engaged, fully employed and really busy, but I'd been in that job long enough to learn what I talked about earlier, about the disadvantage of being publicly funded. We had not been sitting on our behinds. We had turned a $3 million-a-year agency of the state into a $30 million-a-year agency of the state and that was in four-plus years. I was pushing the envelope, I was aggressive, trying to get this state to wake up, and in doing that, you're lucky if your friends remain the same, but your enemies accumulate. That's just life, change is tough on a lot of people. I was pushing hard at it. My first reaction was, I told George I was flattered, I appreciated the thought and I would think about it. George is no shrinking violet, he follows up. So finally I agreed to think about it seriously and then he told me he wanted my answer by November, for the November Board meeting, because he wanted to present the candidate. At George's initiative, the Board had hired Heidrick & Struggles to do the search. They were bringing candidates in from all over, and there was a search committee. Paul Howell was on that committee. I don't know who else was on it, but I know George and Paul and probably two or three other Board members. I took about 10 days in Scotland in October, where I really just sat down in the Highlands and thought about the future and the past and the present and integrated all of that and decided I could probably go to HARC and in 10 years have an institution that's far more valuable to the state than if I stayed as director of TEES for another 10 years and kept trying to "turn the tanker." I asked myself, "Why don't you go captain the speed boat and see what happens?" I agreed and in November, I came to the Board meeting, was introduced as the new president-elect, and I started spending a couple of days a week here as a consultant until January when I came on full time.

Take a moment to describe the organization you found when you came here, what was going on, what had been accomplished, what clearly needed to be accomplished.

It was in a pretty sad state. The good news was that we had won the magnet development contract. We were in competition for the SSC. HARC had been successful in getting maybe as much as $4 million a year out of a $20 million pot to design the magnets for the SSC. The Texas Accelerator Center had been established as the first operating division of HARC but that "first" is debatable–John Margrave claims that he and Mike Barry from Rice really created the first activity of HARC in the laser center. But those two things came on simultaneously. The laser center came as a result of the commitment that HARC be placed where both proprietary and classified research work could be done, which typically was not comfortably done, even if it was permitted, on a university campus. So HARC would provide that kind of service and that kind of strengthening for the community, a place to do that kind of research.

Where was TAC located at that time?

TAC was located over in the warehouse off of Timberloch. You cross Grogan's Mill going back to the west and then turn back south and it was on that street. It was warehouse space that we had turned into the accelerator center. It was a real beehive of activity, consuming about $4 million a year of contractual support. I'd say roughly half of that was being consumed at HARC in the accelerator center and the other was subcontracted out to A&M, UT, Rice and U of H. So the universities were benefiting already from HARC's presence in that they were participating in that project. But what I found structurally and organizationally was a real mess. We had a $4 million contract but there was no accounting for those dollars that was going to be auditable by anybody's stretch of the imagination, from a DOE auditor point of view. Harvey had just not had the experience and discipline, nor had David, of dealing with that level of contract. The communications with Russ Huson and his team were not good. Russ was the director of the Accelerator Center. Russ had come from Fermi Lab. Now that we talk on it, I can see where things got off on the wrong foot, and that happened before I got here. The tradition there is that DOE gives money to a group called URA–George Mitchell is currently on the Board of URA, the University Research Association. The University Research Association has oversight of Fermi Lab. What they really do is they pick the director and pass through DOE money so that the employees can be paid a wage beyond what employees of DOE can be paid. DOE skirts around that wage limit so they can hire bright people, who really work at a university, like employees of a contractor, like University of California, Los Alamos, or AT&T, now Martin Marietta.

I never realized that.
That's the reason they do it, Joe. So they can pay them about twice what they can pay a GS-17, or whatever a government service job pays. Otherwise they wouldn't get anybody who can think to go work at Los Alamos. Those people would rather be tied to the University of California than be a government employee working for DOE. The tradition there and the tradition Russ brought with him was that these places are autonomous and that the parent organization is just a convenient way to get money to them. That was a real interesting time. I began to see why David Gottlieb called me on a regular basis while I was still at A&M when he was trying to make some sense out of all that. I immediately brought in Dan Davis, who had been my chief financial officer at the Engineering Experiment Station. He had all of that knowledge and background and ability to deal with auditors to get HARC in a position so it would not be embarrassed by failing an audit on a multimillion dollar government contract. And Russ Huson had no discipline for that, he had no experience with it, he'd just never run something like HARC. But yet TAC, in his mind, was bigger than HARC and HARC was just a convenient way to get money to him and he structured the whole thing in the classical federal laboratory high-energy physics mode, and thought of it that way to the day he left here and fought being structurally a part of the institution. That's part of what I found and I didn't really realize what I had found. It took me about two years to sort through it till I began to understand, because I didn't know that tradition. I had to learn that after I came here, because I had never been in a federal lab. I really didn't understand all of what I've just shared with you—I learned that the hard way. I paid for all those good tidbits.

So TAC was far and away your largest operation at the time?
Yes, and remained that for three or four years.

At that time were there several specialized centers as there are now?
No. TAC was just becoming operational. I guess we'd have to go back in history to be sure when the contract actually came in, but I think we would find that it happened mid-'84. I can trace it to then because I started trying to get my arms around the institution in November of 1984 in preparation for coming in as HARC's first full-time president. In January of 1985, that's when I began to see, whoa, we've gone far enough down the road, we probably were six months or so into the contract and coming up on an audit in early 1985 and the institution wasn't ready for it. TAC was here, Margrave

was on HARC's Board from Rice and involved in the laser activity with Mike Barry. We started into the dialogue about, "Gee, we ought to move the growth studies into HARC as opposed to just leaving it as a Woodlands thing," so to speak, to try to get this continuity that we spoke about earlier.

On that one, I'm almost sure it was under Harvey's wing by the time you got here.
Really? OK. It may have been conceptually, but functionally it wasn't, unless Harvey was running it himself.

He had hired a young woman, actually a Ph.D. in English literature. The Center for Growth Studies did not exist. It was a late creation of HARC, but at least the conference part of it had started.
It makes sense. What we ended up doing, and I think that happened probably in '85 or '86, was to establish the Center for Growth Studies as a HARC operation. I recruited Jurgen to head that up and he still heads it today. A Geotechnology Research Institute was established, we recruited Manik Talwani to head that and he heads that today and holds a joint appointment as the Schlumberger Chair in Geophysics at Rice. Jurgen still holds a joint appointment at the LBJ School of Public Affairs at UT Austin. We created a Materials Center that John Margrave heads. We created a Space Technology and Research Center that David Norton, who came from Texas A&M, headed. He now holds the VP for research position and Andy Blanchard heads that center. Olin Johnson from U of H came in to head our Computer Center. Almost immediately after putting some organizational structure in place, I started getting some management support to deal with the reality that we had multi-million dollar government contracts and we had to be able to respond professionally to that. I went to Washington, to the National Science Foundation where Eric Bloch was then the director. Eric had been a long-time friend of mine. HARC had made the proposal, as I mentioned earlier, for the supercomputer. We had competed in the top eight, and four of those centers were awarded. When I got here, I reviewed the reviewers' comments about our proposal, so I just called Eric and told him I wanted to come chat with him. I went up to D.C. and I sat down with him and asked when they were going to fund another one and he said he wasn't sure, they didn't have it in the budget, etc. So I asked him if I went out and got a supercomputer on my own, without any help from the government, would he commit to $1 million a year support from NSF, to use a million dollars-a-year worth. He said, "Well, if you can get one on your own,

we would probably do that." I came back and pulled together a computer advisory committee. I took two of the top computer science people, faculty from our four founding institutions. Olin Johnson from U of H was one of the eight and I asked him to chair that committee. Then I sat down and wrote an RFP and sent it to every supercomputer manufacturer in the world and said "We are going to acquire a supercomputer, and if you would like for us to consider your system, please contact Dr. Johnson and schedule a time when you could come in and convince us that it's the one we ought to get." That was an interesting period also. They all came. I had no money. I knew it would be $20 million to play, I knew how to get it and I'd already run the experiment. I knew that A&M, UT, Rice and U of H collectively could use half of the time available. My strategy was that since we were tax free, I could sell tax-free bonds. Once I got everybody interested in selling us one, then I could say, "Look, here's the plan, here's how we're going to finance this. We can't possibly use more than half of the time, which leaves the other half available. Since you have the best supercomputer in the world and everybody wants it, we will join with you to sell the other half and that revenue stream will pay off the bonds to buy the computer, but you're going to have to guarantee the bonds." It worked. When we decided to go with NEC and get the SX2, when NEC got right down to that financial structure, they decided they didn't want the exposure on the public bonds, so they agreed to loan us the money at the same rate, and the agreement was we'd sell the other half to pay the debt off.

Did you run into some flak for getting a Japanese computer?
Yes.

Did that amount to a hill of beans?
Not really. At the time, there was a lot of media drama. But this is probably worth mentioning: In 1981, Mark White was governor of Texas. He pulled together 15 people from around the state one Sunday evening at the mansion to talk about how we were going to get MCC. I was there, I met Ross Perot for the first time, a few other folks were there. Once we got through talking it out, Ross ended up being asked by the governor to head the business and I was asked to head up the technology. We went after MCC and that's how I met Bobby Inman, and we were successful. I had hosted for Bobby Inman, at his request, in my home, a dinner to introduce him to the microelectronics people. He confided to me in my office at A&M one day. He said, "Skip, the reason I was successful in the CIA was that I knew where

all the talent was. To be successful at MCC, I've got to know where all the talent is, and I don't know where the talent is. Can you help me?" I said, "Sure, we'll start with dinner and I'll introduce you to the first group and that network will expand and they'll all help." So we did that. Three years later at HARC, guess who's bringing in a Japanese supercomputer? So I picked up the phone and I called Bobby at MCC and said, "I need to talk to you about something." I told him the story and he said, "Skip, if I were you, I'd just go ahead and do it and batten down the hatches and get ready for the flak." I said, "Bobby, that's what I'd planned to do. I didn't want to do that without talking to you about it first and without personally giving you a 'heads up' that it's going to happen so it doesn't come as a surprise." And he said, "I appreciate the call, but I understand completely, and carry on as far as I'm concerned." It was also something that the university presidents and the university would be attached to and would get some heat from. We discussed it at the Board as well, but we proceeded. Yes, I had some very interesting calls. But anyway, that came and went and the relationship with MCC is still wonderful and we will install the new SX4 in April of '96.

They've kept the supercomputer pretty busy?

Yes, the real value has been to our Geotechnology Research Institute and our 3D seismic research. The SSC used it extensively in the early phases. Once it was decided to build the SSC in Texas up in Waxahachie, before they got their computers, they used this one remotely for the first year or so. Computer technology changes so fast. We have one-fourth of this building currently now with the SX3. When the SX4 comes in, it'll go in an office and we'll relieve thousands of square feet.

Thank you, Skip. Let's resume in a couple of weeks.

* * * * * * *

Today is Wednesday, August 30, 1995. This is Joe Kutchin doing the second of the interviews with Skip Porter in his HARC offices in The Woodlands. Last time we covered your background, your days at A&M, talked about the Arthur D. Little study that was kind of fundamental in the development of HARC, and talked about some of the early days. One of the things we spent some time on was the supercomputer and its role, and we talked about the situation with McMains, leading up to the vacancy of that office. Certainly one of

the things to talk about now at some length is the SSC, Superconducting Super Collider, because I think you had brought it to where McIntyre and Russ Huson had been doing some work at Fermi. Tell the story of HARC's role in bringing the Superconducting Super Collider to Texas.

The whole Super Collider project was bubbling up in the early '80s in the high-energy physics community. It hit my desk, as director of the Texas Engineering Experiment Station, in about '82 or '83. The question was, what role can Texas A&M play? This was being driven by Peter McIntyre, who was at that time the only high-energy physicist on the faculty at A&M. When I decided I needed to look into the project, where was it coming from, who was behind it, what scope would it be, where would the competition come from, what could be Texas A&M's or the State of Texas's role, etc., it became clear that if Texas wanted to participate in such a project, and certainly if we wanted to have the project located in Texas, it became clear that we didn't have a single existing institution with a critical mass of talent capable of competing and being "meaningfully involved." We could have individual faculty members from Texas institutions working on the project, but somewhere else. Once I came to that conclusion, it hit me, being a member of HARC's Board, that HARC could be the vehicle.

Was that prior to McMains?

Yes, probably just prior. We'd really have to go back and check months, but it's in the '83 time frame. I brought the idea to an executive committee meeting here at HARC that George was chairing.

There were two representatives from each of the universities?

No, it was one representative on the executive committee. I think there were two on the Board, but just one on the executive committee. As I recall, it was John Margrave from Rice, Gottlieb from U of H, Jerry Foncken from UT Austin.

UT was in?

Yes, UT was in at that time. I represented Texas A&M. Then there was George, Coulson Tough and Ed Lee. Ed was then president of The Woodlands Corporation. I simply said, "Hey, there's a major project called the Superconducting Super Collider. There's no way that Texas can compete for this thing through one of its existing institutions. It's going to take the collaboration of all of them and the talent that exists in each of them to get a critical mass and then it could still be weak. The only mechanism I know that exists that's in place that might pull that off is HARC. So if HARC is looking for something to do to get started, that in my judgment is a viable option, but it's big and it's going to take some serious resources to play the game." Peter McIntyre had brought Sheldon Glashow from Harvard down a couple of times. Sheldon was one of the big ones in the community of high-energy physics. A Nobel Prize winner. As it came to pass, he met with George more than once, as I recall it, made some rather fantastic arguments about detecting oil and gas in the ground with some of the particles. Apparently, he's a good salesman, too. Finally, did a proposal to design the magnets in competition with the other interested groups, which were primarily Fermi, Brookhaven and Berkeley, three federal laboratories. HARC made a $1.6 million commitment to bring in matching money. Then Russ Huson was hired from Fermi in Batavia, Illinois, to come and join the team. I had said to Peter back at Texas A&M, in my office at TEES, "Hell, Peter, how many high-energy physicists do we have?" He said, "Well, it's just me but I know this one other guy in Illinois who said he would come." I said, "Good, I'm beginning to get the drift of where we are." So Russ was hired to come and join the team.

He was hired by HARC or A&M?
A&M, and the details of that I don't know because I was still at A&M. Then Harvey McMains was coming on the scene about that time. I think that project was getting some focus and some energy and some reality, because by the spring of '84, people were showing up and the contract was in hand and the work was beginning . . .

You're saying that A&M got the $1.6 million?
No, that was George's commitment to HARC against having this group of Texas talent come through HARC to position itself to put Texas in the mode of being able to successfully compete for some of the $20 million of Department of Energy money that was available to begin designing the magnets for the Super Collider. We got about $4 million a year for several years out of that pot. In order to write credible proposals, what George had said was, "Look, if you guys will get your proposal together and write it and submit it, and if it wins and you actually get the contract, HARC will put in $1.6 million to support getting this thing going over the next two or three years."

Until that contract was awarded, the Texas Accelerator Center did not

exist?

That's right. The Texas Accelerator Center was actually created at HARC for the purpose of bringing a place at HARC into the high-energy physics community, where the talent could come together. There was literally about one professor at each of Rice, U of H, Texas A&M and UT Austin. There were about four guys in the state who even knew how to spell SSC. Then they hired Russ out of the magnet operation at Fermi to come in, and he became the head of the Accelerator Center. I think that was a decision made by the group of these other four faculty members who really had full-time teaching positions back at the other institutions. Russ was hired as a faculty member at Texas A&M, but really in a full-time research position in the Accelerator Center. I think maybe he had some teaching responsibilities, but the nuances and the details of that, I really don't know. David Gottlieb, finally after the McMains debacle, picked up the operation on a day-to-day basis. That was some time in '84, but it was at that time, as we talked last time, that the Board said, we'd better get somebody here as president to run this thing. That's when George asked me, and the November 1984 meeting of the Board was when I was elected president of HARC, effective January 1, 1985. I started coming in November, one or two days a week, to work with David Gottlieb, who was sitting in the position of trying to keep all this stuff going. The contract with the Accelerator Center had kicked in, in the spring of 1984. So we probably had done six months or so of work on that. It was coming up on audit time of that first six months, dealing with the question of how are you spending this $4 million we're sending you, etc. Books were being kept on the back of envelopes, and it was an interesting situation. Probably in February or March of '85, I brought in Dan Davis, who had been the chief financial officer at the Engineering Experiment Station, and said, "Put some sanity in our structure here. We've got to get ready for the DOE audit." We began to put a project and contract accounting system in place. The image they got was that the Accelerator Center was an autonomous entity that used HARC like Fermi used URA, as opposed to being an operative division of HARC. It took me a while to understand where they were coming from, but all that is part of the history and stress and strain of getting something going in different cultures and different people from different places trying to come together to make something go. I remember the day the DOE people came to HARC to actually execute the contract. We had an A&M, a UT, a U of H and a Rice professor sitting around the table; they were the management group. I'll never forget the acronym PMG, the Program Management Group. This was in the culture of these high-energy physicists. They were all debating in real time, as I was told by either David or Harvey, I can't remember, which piece of the contract would go to A&M or Rice, U of H or UT. Finally, the DOE person said, "Hey, we're here to give one contract." Maybe it was the A&M guy who said, "Well, OK then, A&M will just take it. Isn't that what HARC is here to do?" Even at that moment, it was hard to make this leap of collaboration. It's just tricky, it's a mine field. As George and I often said, "It's like walking through a mine field trying to get people to work together, particularly in the academic community." That's just reality, that's the way it is.

So finally the contract did get signed and TAC developed magnets which were in competition with those developed by the other three?
Fermi, Berkeley and Brookhaven. That went on for a year or two, then the three federal labs were asked to form a collaboration on the so-called cosintheta magnet. Then there were two programs being funded by the DOE, the cosintheta and the superferric, which was the HARC design through TAC. Those two competing approaches then were funded by DOE for another couple of years or so. Truly dramatic, just gut-wrenching drama and competition. But both magnets worked. If you ask one team, they would say the other one didn't, but actually both magnets worked. But for whatever set of reasons, the magnet review committee chose to go with the cosintheta magnets. Cosintheta is a mathematical expression that relates to the theory behind the design of the magnet, as opposed to the superferric, which relates to the superconducting magnet which was based upon a superferric iron basis for design. Those became shorthand notations to refer to the competing approaches. At the end of that, once the magnet competition was finished and the design was over with, there was a radical change in funding. It went from $4 million a year to essentially zero, but there was a huge struggle and participants had a sense of, "God sent me here to do this work and now that we've worked for four years to get all this critical mass together, you aren't really going to let us just go away, are you?" Meeting after meeting with the high-energy physics community and the Department of Energy about what other work could we be doing . . . designing linear accelerators and high-speed rail, levitated magnetic trains, etc. We started looking for ways to make use of our special magnet technology, and that's when we discovered the self-shielding nature of the superferric design, which had a good medical application, and we hooked up with Baylor and got going in magnetic resonance imaging and improved diagnostics for medical applications. I came up as a physicist in low-temperature solid

state physics and migrated and mutated into semi-conductor device physics and engineering, and went from the pure world of science to the applications world of engineering. It was amusing how the scientists at Harvard always looked down their noses at MIT, because Harvard was clean and pure and MIT was that grubby, engineering, grease-under-the-fingernail place down the street. I grew up in the academic world, where I was recruited into a pre-engineering program by the physics department after they realized I could pass the exams in physics. And comments were made to me, "God, Skip, you don't want to go over and be a grubby old engineer. They only understand how, they don't understand the elegance of why. So why don't you stay over here and really be a true intellect in the scientific world and understand why?" They sold me, and I did that for the next 10 years of my life and then finally as the caterpillar mutates to the butterfly, I went back to engineering. So it was really interesting to watch guys who had come very much in the tradition of "holier than thou" scientists to design this wonderful Super Collider, to break apart a quark, to find the mysteries of the origins of the universe. Then they found themselves in this reality position, complaining that these guys aren't going to give us any more money to play this game. What can we do now? And here we go into mag-lev trains, we go into magnetic resonance imaging, and you watched guys like Russ Huson, who was internally difficult to deal with—manage is not in his vocabulary—get excited about it, be pretty creative about it. But you'd never point out to him that he was doing it, it would have just killed him, you would have lost him. He would've just stood on the sidelines and applauded. The worst thing that can happen to some wonderfully elegant, esoteric work that you've done is for society to find some really valuable, practical application for it. Just horrible, just ruins all the purism of it. But here it really began to work that way. It's just amazing how it gets in the blood, it gets to be fun, you begin to see, you begin to thrive on it, and these guys did it. What's also interesting is, they're smart, they're good at it in some dimensions, and in other dimensions, they're disastrous. Trying to manage, that is the real challenge of this institution. Managing to keep a balance, a reasonable perspective, and keeping the place open and intellectually stimulating and with a sense of exploration and willingness and ability to take a risk and fail forward, so to speak, as opposed to just having to find the practical application and chase the money—those are big challenges in managing an institution like this one. The reality of the world today, 10 years later, is that damn few places, particularly this one, have the freedom just to pursue what's interesting.

Let's go back to the SSC. You reached the point where it was a very good effort, but for whatever reason the other magnet was chosen. Then you say you and the institution went through these attempts to apply what you knew. But there were also huge amounts of HARC involvement in the political effort to bring the SSC to Waxahachie. Talk a bit about that, please.

Yes, we were the G2 headquarters for the State of Texas. Because we knew here what any community that wanted to compete needed to know, and that was, what is the competition doing? The way we knew that was through the professional network of scientists, because they talk to one another. That's the tradition, and we talked to the Russians and the Chinese, so to speak. This got started when Mark White was governor. Then Bill Clements came on board and then Ann Richards, and now it's George W. Bush, and he gets to finish it. I was with him last week and he said, "I'm closing that place, Skip." It was a week ago today. During the campaign he said, "I want to come back to you after I'm elected and find out what to do with that Super Collider site." And when I met with him last week on another issue, he said, "I know I didn't come back to you," he had remembered it, "but I just decided to cut it off. It's a dry hole and we're capping it." It's going to cost $230 million, a $230 million dry hole.

But it was a remarkable effort.

See that thing right there? That's probably the only copy of it left. All those 14 white-bound volumes. There was $300,000 appropriated by the state to put this document together. UT did the geology, Texas A&M did the engineering, Rice, I think, did the economics, and the University of Houston did the environmental impact. Those may be inverted, but I think that's the way they were, the last two. And HARC provided the oversight and coordination. There was a letter which I transmitted to Governor White. He sent this whole set of documents to DOE, under his cover letter as the governor of Texas, and this whole set of documents was used by DOE on the Hill to help sell the SSC as a project in the United States.

Is this the one that evaluated eight different locations in Texas?

Yes, and there was a subsequent one that had 11 or 16 possible sites, and we agreed we'd come down to one. How that happened is the following: There was a science and technology council that George Mitchell and I both served on that was actually at the state level. I was able to say, "Hey, there's this SSC project. HARC is probably going to be involved in this, but if the State of

Texas wants to compete, this is going to happen in an off year—the State of Texas legislature is on a biennial meeting system—so, we'd better get some legislation passed now that creates an operative body that can deal with this issue, one that has legislative authority to act on behalf of the State of Texas." And we got the TNRLC, or the Texas National Research Laboratory Commission, established. Bill Hobby, who was then lieutenant governor, helped with all of that. It was just fortuitous that HARC had those connections and contacts. That was a nine-member commission, appointments to be made by the governor. Governor White never made any appointments. There really wasn't any pressing need for it to move ahead. He lost in the next election and Clements came in. Then all of a sudden, site selection was on the scene, there was a need for action to be taken. I got a call from Clements: "Come over here and tell me what an SSC is." I flew over to Austin and sat down with him and went through what I've just told you. He said, "Well, golly, who ought to be on this commission?" With his appointments secretary sitting there, I put out a bunch of names. He said, "Now, Skip, do you want to be on this?" I said, "No, I don't want to be on this because I don't want to in any way put HARC in a position of looking like it's in a conflict of interest where it could also be funded and supported and engaged in the R&D side." And he said, "Well, I'm going to create a technical advisory board to the commission and I want you to serve on it." I replied, "I can serve on that and I will." And I did. I had recommended that he make Fred Busey, former president of Texas Instruments, chairman of the commission. He ended up making him chairman of the technical advisory group. And Peter Flawn became chairman of the commission; he's former president of UT Austin. This was '85. The commission was named, town hall meetings were held in Austin and any community that wanted to argue they had the right site. And I believe there were 16 different sites in Texas where proposals were put together for siting the Super Collider and the technical review committee had the responsibility of selecting and prioritizing the top three in order. And we did that. There were about 15 or 20 of us on that group. Fred Busey chaired that. We sent those recommendations to the commission and that was the process by which the final Waxahachie site was selected for submission.

Had Texas been selected yet?

Oh, no, Waxahachie was submitted by Texas, and a number of other states submitted other sites. But because it was such an economic down period in the '80s, everybody was looking for any kind of economic boost, and having a $10 billion project move into your community was a pretty good idea.

There was a lot of interest and there were a lot of places in Texas that thought they could handle it, but in reality there weren't. Subsidence was the problem here in the Houston area. In the Panhandle area and the Midland area, the problem was lack of an international airport and having a community that could absorb 2,000 people, etc. Steve Weinberg, also a Nobel laureate at UT Austin, was on this technical committee. He got very upset that Austin hadn't even put in a proposal. Austin didn't want it. There would be too many people trampling around in the Austin Chalk over there. So they stirred around and got a proposal in at the last minute, but the geology in the Waxahachie site and its proximity to Dallas International Airport, the fact that Dallas is nearby and could absorb newcomers, the availability of cultural amenities that were needed for this community, all came into play. Really, Dallas and Houston were the only two places that really made sense, and Houston had the problem–or the site had the problem–of subsidence. During all that process, I was getting calls from the governor about, "What's California going to do? What's Illinois doing? What's New York going to do?" I could just call up the guys at the accelerator center who had been talking to their colleagues in those different federal laboratories and we could answer the questions. So we were kind of the G2 headquarters for what was going on. Finally, we put a forced march together to help communicate to the citizens of the State of Texas that they needed to tax themselves a billion dollars and throw that billion dollars in the pot as an incentive. It worked. It happened in the old classical American way, a lot of Rotary Clubs and Kiwanis Clubs and luncheon and breakfast clubs that we talked to about what this project could mean.

Did "we" include you and others from HARC?

Oh, God, yes, we made talks all over the state. And then the competition was launched, the site was announced and Bill Clements did a masterful job. One of the things that helped him to be governor at that time was his experience at DOD and his political savvy of the Washington scene. He did damage control in what could otherwise have been, in my judgment, a naive approach. The Academy of Engineering and Sciences formed a site review and selection team that made the decision based upon its technical merits. There was always the suspicion that George Bush, being president, brought it to Texas. I think it was one of those cases where, practically, it didn't hurt that he was, but in public perception it did. But in point of fact, those of us close to the process knew that the people making the recommendations wouldn't be swayed, that they would look at the merits of the issue and make a recommendation. Then once it was selected to come to Texas, in the process of the construction and all the wrap-up and start-up, Ann Richards became governor. Then in trying to save it from the budget cutters, I got pressed into the SWAT team to fly around to every state. This was just a year-and-a-half ago. We picked out every senator from every state and put him in the "yes," "no" or "undecided" column. All of the undecideds were Number 1 on the hit parade. People knowledgeable about the project were asked to go to those states and talk to the leadership that influenced those senators to see if we couldn't get 36 votes.

That was coordinated out of Austin, right?
Yes, it was coordinated out of Ann Richards's office and that of Jane Hickey and the Washington Texas office and the Department of Commerce.

HARC and its people played a role in that?
No, we just supported it. I think I was the only one at HARC who was physically involved, getting to get on airplanes and flying with all these different people. There was a professional team hired to put together the presentation and we were handed an agenda. We knew when we came on what we were going to say. Finally, there were 37 positive votes and there was calling and celebratory anticipation of "we did it," and then within a week, the House killed it.

Tell me again what you think are the important things in HARC's history. We talked about the supercomputer and about the SSC. Those clearly have been high-profile steps along the way. What other things should be known about HARC? What about your campus?
One hundred acres of land are committed for the development of the campus. We've developed about 11, with three buildings and parking lots and infrastructure in place. About 160,000 square feet, 160 or so people in place doing about $16 million a year in research volume right now. Today in '95, there are five major focus areas of HARC and that's energy, medical and environmental applications. That's three of them and maybe I should comment on this, coming to the fourth. About three years ago, I was asking myself, "What's so great about HARC?" Part of the answer is that whether you're punching seismic energy through the ground, listening to the reflections, gathering all that data, or whether you're punching magnetic energy through the body and looking at the transmitted signal, or bouncing microwaves or lasers off of the earth to see deforestation and acid rain or coastal

erosion, what you're doing really, simply put, is you're taking data and turning it into a picture. And the mathematics doesn't know the difference, it doesn't care whether it's in energy, medicine or environmental application, but the need is the same and the problem is the same and the problem is you've got far more data than any supercomputer can digest, and besides it's expensive to digest it all. So you'd like to use a lot less of the data but still get a high-quality picture. That is kind of the core technological theme running through our energy, medicine and environmental applications. What we need to do is become known as the world's best institution that knows how to take data and turn it into a picture no matter what your application is. What we need to do is solve the problem of how can we take a lot less data and still get a really high-quality image that doesn't lose anything. And so we created our information technologies division and started working on digital image compression, and that's the more recent breakthrough. That's the history and the logic behind the new division.

You were on "three."

Three is energy, medicine and environmental applications, four is information technology, and fifth is what I call science policy. An example is the global commons kind of project that we have with the Academy of Science that manifests itself at HARC in our Center for Global Studies. It is the whole issue of how we have to pay attention to how science and society influence one another. I think there are three or four quick examples of that. In my judgment, we in the science and technology community in this country have done a great disservice in that we didn't communicate with society about the nuclear power reactor. So nobody is going to build one today, even though it's the cheapest and probably safest way to deliver electrical energy. But society is afraid of it and we've regulated it to where it costs so much that we can't turn them out like France can. That's an example. George Mitchell likes to point out, "There are 5 billion people on the earth today. In another decade or two, there's going to be 10 billion people. If we can't manage with 5 billion, what the hell are we going to do when there's 10 billion? What are we going to do about these global issues of health, natural resources, population, environment?" A dimension of the personality of HARC that I'm very much interested in, and this is very much shaped by George's historical thinking and my own personal experience, is that it's foolish and folly to think that the public understands the implications of fast-moving technological change. Most people are scared of their VCRs, most people have that blinking 12 on it. You just can't keep forcing this stuff in a society that doesn't understand technology. Things that you don't understand you are usually afraid of. If you're afraid of them, then you'll want them to go away. I don't want to see America develop an attitude of, "Let's burn all the books." So that's why I think it's really important that this institution be sensitive to the impact of science on society. It's easy to just get caught up in the elegance of the technology and not think about where it fits and how people will receive it.

In 10 years plus, what are the one or two things that you are the most pleased about in your experience with HARC? What do you feel really good about?

That we survived. It's amazing that you could have started the only new non-profit research institute in the United States at a really down time in the economy and lived to tell about it. I think it was Jim Pickering, president of the University of Houston, who said at a retreat last year, "Hey, this is not trivial: Just the fact that HARC is still around is amazing." Other than that, there's something going on here, it's a feeling we're special, that we're doing about the right thing at the right time. It's unusual that so many of the people who are here have been here as long as they have for such a young institution. A lot of the leadership has been here, essentially, through the life of HARC. Since I've been here as president, all four of the founding universities–Rice, U of H, Texas A&M and UT Austin–are now on their third or fourth president. About the time I get the president of one of those campuses informed about what HARC is really about, he's gone. At A&M it was Vandiver, then Mobley and now Bowen. At U of H it was VanHorn, then Barnett, then Pickering and now it's Goerke, that's four. At UT it was Flawn, then Cunningham and now Birdall. At Rice it was Hackerman, then Rupp and now Gillis. I've worked with at least 12 different presidents of the four founding universities of HARC.

Where is HARC going in the future?

I read something today that said, "Trying to deal with the changes in any company or industry today is like trying to fly an airplane while you build it." That's very appropriate. My belief is that staying small and high-quality, focused and responsive and fast is where we have to go. Our capital is intellect. The value is in the human mind. Getting bigger, physically, is no longer an objective of mine. Having access to top talent wherever it is and being able to utilize it effectively and have the network and contacts with the private sector so that the value can be gotten in an ever-decreasing window of opportunity because of competition and rapidly changing technology is the challenge—that's important. All the rules are out the window. Now you write four proposals for what you need to keep your team going in any area of research because you know you're going to hit one out of three and so you add a fourth one just for insurance, and you've got your next year's budget covered on the statistics. We just got a $700,000 contract from NSF for our Rio Grande water, Mexico-U.S. project. It's a miracle. There were 1,400 proposals and 20 got funded. As we speak I don't even know if the federal government is going to be open. We know it's broken, they may just go ahead and close it. We just talked about the SSC. They simply closed it. Whether NASA is going to continue to function or not, who knows? Every federal laboratory that I know of is looking for a mission. The Russians aren't coming today. We're trying to denuke the world, not nuke it, and so these billion-dollar laboratories are about to turn a lot of people on the street. We've just received Texaco's geochemistry laboratory, we're remodeling a huge piece of this building, they've donated it, equipment, intellectual property and people to HARC and fully funded it for two years. We've agreed with Texaco to get three or four other majors in the next two years to help support it. Maybe there will be five of them paying 20 percent of the cost, rather than five of them paying 100 percent of the cost. They'll all still get the same amount of value in technology. The competitive advantage of having your own has dried up and gone away. No one can afford to do it alone any more, so HARC becomes a comfortable, fair, level playing field. Once again, five of those companies can pick up 20 percent of the cost and still have the intellectual product coming. I think that's good, I think that it's a win, win, win collaboration. This is hard, there's nothing easy about this. You read that book up there, *SRI, The Founding Years*. They struggled for 40 years to get there and it's hard, it's tough. George and I met last night. Right now HARC is probably in the most exciting, strongest technological position in our energy, our environment, our information systems and our geotechnology areas that we've ever been in.

The global commons project must be tremendous.
Absolutely.

So that means strong activity in all your five basic areas, doesn't it?
That's right. The one that's probably the shakiest right now would be the medical. In terms of the magnets, we've just about run that string out. But there's excitement about the value of the technology, the opportunities on

the horizon. I told George this last night, "George, if you take a snapshot of HARC as a venture company, like all these up and down Research Forest Drive, we'd look pretty good." In that sense, there's been about $30 million pumped into HARC over 10 years and we've sold $75 million worth of product. Most of these venture companies don't have a product yet. I've got about $5 million of short-term debt that's associated with doing $100 million worth of business over a 10-year period. And that's about to eat my lunch. I've got $10 million worth of long-term bond debt that we have on a 30-year amortization to pay for the buildings. That's not a problem. But at a moment when, institutionally, the buds are on the bush and the blooms are going to be beautiful, we just need to put in some more fertilizer and grow this baby. And yet times are tough, resources are difficult to come by and I sure wish I had been lucky enough, smart enough, clever enough, better manager enough not to have accumulated $5 million worth of debt over the last 10 years doing $100 million worth of business. I don't know how many times I've gone out to ask for money from a foundation to support us and they've made the comment that they like the whole concept but they have a strategy of not putting their money into something that isn't going to be stable, and why would anybody start an organization like this without an endowment? And where is the security for HARC? To come within 5 percent of your budget in a year isn't too bad, but come within 5 percent of your budget over a 10-year period, running $100 million through it, I can argue to myself when I want to be lenient on myself, that that isn't too bad. But it still doesn't diminish the size of the $5 million debt. Paying that debt off, carrying that debt when you have no endowment and no sustaining support other than what you can generate by your wits in a non-profit environment where nothing goes to the bottom line is tricky as hell. So there are positives and negatives. I think that our commitment three years ago to solve the problems on image compression has produced fantastic work. In the collaborative sense, we've got Charles Chui, distinguished professor of mathematics at Texas A&M. With this image compression capability, you get a really high quality picture where you've used 1/300 of the data that you have. It is important to get it into the marketplace. We've created HARC Technologies, Inc., our for-profit subsidiary–Bill Anderson is president–as the front door to get this technology to the private sector. We created that in July of last year. So we've got a lot of things going, a lot of new dimensions.

With some breathing time those are going to start turning into cash

flow, positive cash flow.

That's right. But you've got to have the breathing time. HARC is ready to come out of diapers, but you don't want to give it the keys to the car yet, it would probably have a wreck. So you need to still nurture it along some and help it understand the ways of the world and put some more energy and time and guidance and resources into it to get to where finally you say, "OK, here are the keys to the car." I'd say that's another decade away, and if I'm right about that, if in another 10 years those resources come in and the kid is looking healthy and mature and you kind of feel like it can fend for itself on the street so it doesn't need a guardian to go with it, it will have done it in record time compared to any of the others that are around today. That will only have happened because of the sustaining, constant, unwavering support of George. The $30 million of venture capital money that's been invested in HARC so it can sell $75 million worth of its product has predominantly come from George. The other thing, though, that I think has happened, not nearly to the extent that George or I would like and need, is that in the last three years, other members of the community have begun to recognize HARC and take a little pride of ownership and actually be willing to take it out on Sunday afternoon and take care of it.

On your Board you've got a lot of great names in the community.

And a lot of real interest. A lot of these people are busy and successful. We're not without our stress and strain—the stuff jumps up and bites you—for instance, the compression technology lawsuit. This drama that's going on was a dimension that I hadn't even considered going into marketing the stuff. So there's always stress, there's nothing easy about this. It's fun, it's worthy and it's needed, but it needs a few more parents, some godparents, it needs some friends, some aunts and uncles to join in the nurturing here and get it on down the road. But its career path looks pretty good.

What haven't we covered that needs covering, anything?

I come to my impression by listening to others, that HARC is playing an important role, not only in the value that the $75 million worth of product we've been selling and all the people and impact and energy that goes with that, all of that is digestible and real. But I think beyond that, there's an aura of establishing Texas and Houston with a widely held perception of intellectual content, of substance, of depth, of ability to think and act. All that adds a level of credibility, of value that goes beyond just having good bricks and mortar and pretty streets and a low crime rate in The Woodlands. It brings a sense of accomplishment, and we have to recognize that this is something that George saw and has stayed consistent on.

Skip, thanks for a very enlightening view of HARC.

Michael H. Richmond

This is Joe Kutchin, the date is Wednesday, August 16, 1995, and I'm in The Woodlands office of Michael H. Richmond, who is executive vice president of The Woodlands Corporation. Why don't you just give a bit of your professional and personal background before you came to Mitchell.

I grew up in the City of Baltimore, Maryland, and went to high school there, then went to the University of Maryland where I graduated in 1969 with a degree in accounting. I signed up to work for Coopers & Lybrand–it had some previous names prior to that, but that's its current name.

You are a CPA?

I passed the CPA exam while at Coopers & Lybrand, and I am a CPA. I still pay my dues. I had various clients at Coopers & Lybrand; one of them was the joint venture development between Connecticut General and the Rouse Company for the development of Columbia, Maryland. I was at Coopers & Lybrand only three-and-a-half-years but I moved up to be one of the lead people on the account. That's how I got my background in real estate development.

What years were you at Coopers?

From 1969 through 1972. I left there in September of 1972. I did an internship with them during the summer of 1968. I was recruited here because at that time Mr. Mitchell was formulating a management team and one of the people I worked with from the client perspective was Sam Calleri, Salvatore Calleri. He asked me to come down and be the controller. He had been hired by Mitchell but hadn't been here a very long time when he asked me to come down. He was, maybe, the vice president of finance or senior vice president. I don't remember the precise title, but probably chief financial officer.

He was the person at Columbia that you worked with from Coopers?

Right, he was one of the client representatives. He represented Howard Research and Development. That was the name of the joint venture, one of the developers of Columbia. And I chose to do so and arrived somewhere around the middle of September, which was approximately five to ten days before groundbreaking for the whole Woodlands. Groundbreaking was on

September 20-something, 1972.

How old were you at that time?
I was 24. I didn't think I'd be here this long but 23 years later almost to the day, I'm sitting here talking to the esteemed Joe Kutchin. Do you want me to go through my career here?

What were you hired to do here?
I was hired originally to be the controller for the real estate division, which, I believe then was called Mitchell Development Corporation of the Southwest, or it was called George Mitchell & Associates, I don't remember. And also The Woodlands Development Corporation—because The Woodlands Corporation was a separate entity at that time because HUD required it. It was a free-standing corporation that sold the bonds, did the initial financing—there was a wall all around that company. Affiliated party transactions were scrutinized and had to go through HUD.

We're talking about September of '72? By this time the company had gone public, although not for long, but had not yet been given the HUD guarantee, is that correct?
No, the bonds were sold a week before I got here. I think they were sold September 6, or funded September 6. That was $50 million, and when I walked through the door we had $26-$27 million left in the bank, which when you look back to 1972 was one heck of a lot of money. But it didn't take us long to spend it.

Well, actually, one of the things I learned that I didn't know before is that HUD required that the land be fully paid for, which really meant there was a lot less than $50 million.
Right. What was left was approximately $26 million after paying off the debt, or the majority of the debt, on the land.

Other than with Sam Calleri, did you have much contact with the Columbia folks who came here?
No, I did not know Len Ivins very well, or Bob Grace or Vern Robbins. Then Larry Kash came and David Lawrence.

Were they both from Columbia?
Yes. They were in a different division, I believe. Larry and David were in the

mortgage business. Rouse had a mortgage company, and they came down to set up a mortgage company, which is now Mitchell Mortgage. It used to be Mitchell Mortgage & Development Company; its function was to help arrange financing. They were the ones that arranged the original $12.5 million loan on the conference center, the information center, the swim center, the old wharf. We wound up spending probably close to $30 million to complete that whole complex. That was one of the things that caused us some financial problems at the end of 1974, early 1975.

The idea of getting a new community opened kind of depended on that kind of establishment.

There was a philosophy. It was a very strong philosophy. Due to the distance from downtown Houston, The Woodlands at that time was perceived as south Dallas. Sometimes today if you talk to people who work inside the loop, they still believe The Woodlands is so far. It's really not, but that's their perception. Back then, our belief was that the whole amenity package, the whole concept of what this place was going to be, had to be in place early on. There was a great desire to have that imprint up-front, to be visible and be part of the market. It was a very bold step, very expensive step. And I don't know if we had to do it over again whether we would do it the same way. I probably would think hard about doing it that way. But as a result, we were able to attract the Houston Open in 1975. We had barely completed the golf course. The concept of the conference center was that it was there to attract visitors, to get name recognition, because it was a destination and it was a new concept in conference centers, because there hadn't been that many of them prior to that. That philosophy of bringing in traffic and exposing The Woodlands was part of the early-on strategy.

What did you find when you arrived here?

We were in One Shell Plaza at the time. The construction people in 1972 were the only ones out at the site working on the roads and utilities and getting it all built. We moved out of One Shell in April of 1974, which was six months prior to opening. It was just a tremendous amount of work starting in 1972 and two years later opening the project with all this infrastructure. Not only the land development part but the building development part. It was an unfortunate time, probably, because it was a time of a lot of other construction activity in Houston and labor was hard to find, the quality of labor was hard to find. It's not unlike the oil industry that went into its boom years and drilling became very costly and very inefficient because the people doing it weren't capable of doing it well, and they had to redo it. That was a little of the era back then, where you paid a lot and got very little in the way of quality just because of the nature of the marketplace.

What are some of the specific things you were involved in?
I spent a great deal of time in the general functions of accounting, but I spent a lot of time at HUD, with the processing of paper work. During this period, I moved from controller to treasurer and wound up having to borrow some of the money when we ran short. I remember very distinctly in 1974, working with Ted Nelson, who was our director of sales, the day we sold the first 20 lots in The Woodlands to Jerry Kirkpatrick. We sold him 20 lots for $8,000 a piece, which was $160,000. They were in the Millbend area. It was the first sale I ever made. I remember the evening that we made it, we got a check from Jerry down at One Shell Plaza for $2,000 and Ted and I probably spent $2,000 Xeroxing the check in celebration of the first sale. It's real interesting to know that Jerry, 21 years later, is still building homes out here, still has a great reputation and has made a tremendous commitment and contribution to this community. But that has been somewhat typical of what we've found, that a lot of people who have been involved have stuck with it and they make major personal, financial and emotional commitments. I'd like to add one thing: 1974 was an unusual time, because at that time there was a credit crunch. Most people don't remember that, but the banks didn't have any money to lend. There was an absolute credit crunch. Interest rates had gotten up to 12 percent, which seems high today, but wasn't high, compared to when it got to 20 or 21, but at that point, at 12 percent, there was no money. So in order to get the first model home park built in The Woodlands, we had to go to the old Allied Bank of Texas and borrow enough money—I think it was $1.5 million. We borrowed it and lent it to the builders so they could build their model homes so we could open The Woodlands in October. I remember working with Walter Johnson, who came out to tour the property. He was the president of the bank and is now president of some other bank in town. We stood in the rain and I explained what The Woodlands was going to be and how we made a commitment, that we were obligated and we had the builders obligated. But just to get that allocation of funds to open The Woodlands was a tricky test back then and obviously an important one, because that's what got us going.

Say something about the management that was here when you came. Some people regard it as having been a bunch of mad geniuses, and

others have been less generous.

It was a flamboyant group. Len Ivins was a visionary, a dreamer, an impact person. The group had a feeling of invincibility. They conceived more from the gut than from the hip, but weren't great on day-to-day operations. They liked to party, they worked hard and they played hard, and they were spiritual and emotional leaders. They had a dream of making this place special. They weren't good at cost control, they weren't good at day-to-day operations and management, but some people will look back on it and say it was that outlandish commitment early on that gave the project an identity. Others will say that it was a waste of money, excessive expenditures that weren't necessary. And having been here and looking back on it, I'd say it was probably in between. The success of The Woodlands has been its ability to differentiate. We worked hard over the years to differentiate ourselves in risky areas, but risky ones that were well thought out. Maybe they were risky and less thought out and less controlled than they should have been. But it's easy to look back on things and say there were big mistakes, they spent money, they partied, they had a good time. They had a desire, they had a perception of what this place would be and to some degree set a tone for it at that stage. There were problems, but the big problem came when we ran out of money in 1974, the month after we opened. We could not pay all the bills that we owed from getting the project opened.

And you were right in the middle of bill paying?

You always have images in your mind where you can sort of wind the camera back and clearly remember and visualize things. I remember how much we owed—I'm going to guess it was $9 million, and we didn't have a nickel in the bank. There were all sorts of rumors that we were going under, and that was just real estate. Energy owed some more, too. Every Friday I would go downtown and meet and see who we could pay, who we couldn't pay. Mr. Mitchell had the philosophy, "Well, just pay them something." So we made a lot of partial pays and kept it going. We got to the point at the very end where we didn't know how much we owed because the attitude was, "Get the project open." There were no controls at the very end because everyone wanted to get it open, and there were a lot of people who authorized expenditures, and the bills rolled in later. We talked to as many people as we could, we tried to be honest with them, told them we would try to get them paid, and if we couldn't get them paid, we called them back. A big vendor that we owed the most money to was Austin Bridge. I think we owed them $3 million. They had built all the roads coming into The Woodlands and I think at the very end, Mr. Mitchell had to pledge some of his personal stock to secure their loan. We eventually paid them out over time. We paid everybody out over time. But it was a real struggle. I used to come back from lunch or from being away and there would be 10-20 messages from vendors calling to complain and to try to find out when they would be paid. We had this vendor control lab of about three or four people who controlled all this and communicated and dealt with all the people. Everything eventually got corrected in May of 1975, when the energy market tightened up and Chase Manhattan Bank, our bank, gave us some additional financing that allowed us to pay off these vendors.

By that time the OPEC embargo was in place.

Right, it was right about then. I clearly remember May as being a critical time period when the turnaround occurred. One policy we adopted was that if anybody was out of town, they would get paid the last. The small people that would be put out of business, we paid first. And then we worked through the rest.

Who established policies like that?

I did. We learned how to deal with it. I will always remember the tenant sweeper—the one that goes around and cleans up the parking lots—was from Minneapolis and we kept them at the bottom of the barrel, but the painters and people who helped remove the trash, companies that were labor intensive and had to make the payrolls, we would try to pay quickly. We tried to pay as many of the small ones as possible because if we were going to be screamed at, we'd rather be screamed at by fewer of the bigger ones than hundreds of the smaller ones. Just a matter of practicality. I remember clearly getting a call very late one Friday evening after I'd had a bad week and it was the guy who put the marble tile in the health spa area of the country club. I think we owed him $20,000 or $40,000. We just couldn't pay him and he told me that he was going to visit Mr. Mitchell over the weekend, and he did. On Monday, Mr. Mitchell called me and we wound up giving the tile man a special dispensation to get him paid. I don't know if we paid him entirely or in part, but he showed up at Mitchell's house and did his personal collection routine. I'm sure he gave Mr. Mitchell a very good story as to why he needed to be paid.

By this time most of the Columbia team had turned over . . . by the spring of '75?

Leland Carter, who was president of the energy division, became president of The Woodlands Corporation, and we went through a massive layoff and staff reduction. I believe that Len Ivins and Calleri and probably Bob Grace were no longer here. I think Larry Kash was still here and David Lawrence and also Vern Robbins. So some stayed and some went. It was greatly thinned down.

The number to the order of 200 comes to my mind. Does that ring any bell with you?

I don't really remember, but it wouldn't surprise me. It was extensive. Of course, we had built up a big organization as a result of the amount of money we spent to get opened and after we were opened, we didn't need nearly as many development people, construction people, so 200 wouldn't surprise me. I was more worried about paying the vendors than keeping count of the employees.

From corporate headquarters you never were viewed as part of the Columbia people, is that right?

Yes, I would say I was. There was a great animosity back then between energy and real estate and corporate. I think the corporate people or energy people, and maybe they were one and the same at that time, felt that The Woodlands could break the company, that it was not a wise investment, that energy was a better investment. As a result of that and personalities, and maybe the lifestyles of some of the people, there was a great animosity. And as is typical when there is that type of relationship, the people who are fighting don't always fight between the levels that dislike each other. They start going for the underlings. I found myself having a corporate responsibility, being in the finance area, going to meetings and leaving the meeting and then finding out something happened in a negative manner and began to wonder, "Was I at the same meeting?" So I was kind of in the middle. It was a very uncomfortable position. There was no help or assistance. Where I got the most assistance was in the bill-paying side because the company had to pay its bills and Budd Clark and Bob Newton back then were involved in that and understood the picture. But it was a terrible political situation. The part I find most interesting is that when it all cooled down a couple of years later, both sides had completely turned over. There wasn't anybody from the original real estate management cast, and the same was true for corporate and energy. The warring factions both were obliterated or disappeared, and that's when things began to settle down. I'm sure that there were some legitimate concerns and legitimate feelings, but it almost became a child's game, and all the children got put in detention.

So there you are, a young man who moves to the Houston area to undertake a job as controller, then you find yourself soon as treasurer in a company that was extremely cash short.
When I came in here in 1972, there was $26 million in the bank. In 1974 we were $9 million or more in the hole. It was a big, radical change in a short period of time. We were in default under loan agreements, we were in default with HUD.

And your function was really directly involved in dealing with those defaults, is that correct?
That's right. I wasn't the only one. I'll give you another crystal clear image in my mind. We were broke. As part of the HUD agreement, Chase had supplied a letter of credit for $3.5 million to fund what was perceived to be a gap—a shortage of money. Of course $3.5 million was hardly the right amount, but it was a portion of it. Sam Calleri had gone to Texas Commerce Bank and applied for a loan that was secured by a Chase Manhattan Bank letter of credit. I'll never forget the day that he got the call that they turned that down, which really put us in even more of a bind. We went up to Provident National Bank in Philadelphia, which is now PNC Bank or the PNC Bank Group, and within three or five days, they funded us $2-something million dollars against this Chase letter of credit. That really helped us get over one of the severe bumpy portions. And we got some help. We got a lot of people who cooperated, a lot of people had patience, a lot of people were nervous.

All right, let's go to spring of 1975. The extreme difficulty of the past year, more or less, was behind you. There was a change in management of The Woodlands; it was a smaller operation. Was your responsibility changed?
I don't remember all the dates, but I went from treasurer to vice president-finance and senior vice president-financial operations. All this occurred during the period from 1975 on up through 1986 or 1987, a year or two before Ed Lee died. When I moved into operations, I found myself, after having borrowed all this money, now having the responsibility of paying it back—which was terribly unfair. Most people don't have that privilege.

Would it be oversimplifying too much to say that from this period of

'75, your responsibilities were pretty conventional financial and accounting management?
Principally financial and financing. I did oversee and coordinate the accounting functions, but I was, particularly, in the later stages, the senior financial officer of real estate, although I didn't have that title.

Does anything stand out as a major accomplishment or major problem during that period?
I think we established our credibility in the market place, we established it with the banks, we established it with the insurance companies who lent permanent money on the buildings, we established it in the public debt markets with our utility district bonds, when we sold our first ones in 1975. We tried once in '75 and couldn't make it, but the second time we made it. We established our relationship with insurance companies and rating agencies that have really served us well–up to now we've sold close to $200 million worth of bonds. It's been a very important part of financing our growth over the years. I remember big transactions, like refinancing the conference center that helped us out with the banks, getting Chase paid off.

And that would be largely your initiative personally?
Larry Kash and I worked on it, but I wound up closing it and getting it done. I remember some of the non-Woodlands projects, spending time in Colorado selling and working on our Stonebridge Inn and Silver King Apartments and working through all that, and building projects out there. We began to establish ourselves in the marketplace, not only from a financial point of view but from a market-acceptance point of view.

Earlier you said something about differentiating The Woodlands from the rest of the market. Expand on that idea.
One of the things we've done is to develop a fairly substantial amenities program. We also have the desire and vision to not only develop a residential community but to develop a community that has a job base. Our goal is to have one job per household.

Give examples of the amenities base.
The amenities have been the country club, the golf courses, the parks and pathways–which are absolutely the Number 1 most popular amenity in The Woodlands–the recreation parks, establishing the swim center, the "Y," teen centers. The Pavilion would come in later and then the mall, developing the

whole infrastructure. Amenities not only include streets, roads and utilities, but churches, synagogues, schools, day cares. All the things that make it a true community rather than just a housing development.

You put emphasis on developing those things early on rather than late in development.

Right. We brought on as many as early as we could and kept it going. I remember the time when we got our first post office, our first bank, our first supermarket, Jamail's, how we lent money to Jerry Jamail to open the first supermarket. I remember how we loaned him $350,000 so he could put his fixtures in at the Grogan's Mill Center.

Was there a grand, master design to do those things?

No, I think we reacted just like we reacted to the need to borrow money to get to the builders to build model homes and open the project. That's the nature of the company. If something needed to be done, we figure out a way to make it happen.

That was against a much grander backdrop of the HUD plan?

By that time we had worked our way out of the HUD agreement. We got a settlement agreement and substituted a letter of credit from Manufacturers Hanover Bank, got all of our land released and unencumbered and got out from under HUD when they decided to close down the New Towns program. It was a mutually beneficial solution.

It seems that from the beginning, there's been kind of a master vision of how the town would develop.

No question about it. As for some of the other projects we worked, we obtained financing for the Section 8 rent-assist projects of Wood Glen and Holly Creek. The purpose was to develop low-end, affordable housing for people who would pay just what they could afford. We went through the whole spectrum of housing. A lot of people think of The Woodlands as a resort or a country club community, but when you really look at it, we have a significant amount of low and moderate affordable housing.

It seems to me that you and your position, from the time you came with the company, played a creative role in financing the project, and that was as important as deciding what building to build and what road to build.

I worked on sustenance, the ability to survive. One thing that I did begin to work on even when I was in finance that probably helped me a lot as I moved forward into operations was the development of the Research Forest. And when I was in finance I created the venture capital company.

When was that?

1984, '85. Houston was in a real economic depression and the oil industry was losing jobs. The whole city lost, over a period of several years, 200,000 jobs and it was a tremendous burden. To some degree, in some areas in the economy, we still haven't recovered. Here we were with all this land, and the question was, how do we absorb it if we're losing jobs in the community? The economic development group of the City of Houston put increased effort into economic development. In our business not only are you a developer, but you have to be an economic developer because you have to create the right environment and incentives for people to want to relocate and build something new rather than taking space where it already exists. We felt that nobody else was dealing with technology transfer out of the medical schools. To some degree, the universities had to absorb some responsibility for helping the economy recover. In the past, all the investment focus of this city was in oil and gas and real estate, but come '83, '84, those industries were flat in the the pits. Most of them were going bankrupt. And then the banks began to go under, so there weren't any opportunities there. We made an assessment and Mr. Mitchell had this vision of creating a research park. Some portion of our 25,000 acres should be set aside for research, and he personally worked very hard on creating the Houston Advanced Research Center. I focused more on the corporate side rather than the research side because back then, all the findings and all the technologies that were being created and developed in one of our greatest assets, the Texas Medical Center, the largest medical complex in the world by a factor of maybe two, were being licensed out to other parts of the country. There was an opportunity to try to keep some of that at home, and we worked with Baylor and M.D. Anderson and UT Health Science Center and, in effect through the venture capital company, helped them raise funds to start their first technology transfer companies.

Were you out making those contacts yourself?

I was working with the universities, working to marry their research to commercial opportunity. We talked Baylor into creating Baylor Center for Biotechnology, and that had a company spun off of it called Houston

Biotechnology. And then we had a group from UT Health Science Center out here, and that spun off LifeCell. And M.D. Anderson has spun off Argus and Argene, and all of a sudden, we began to develop facilities and space and help fund high-risk rank start-ups.

That was through the venture capital company and Marty Sutter?

Marty Sutter worked for me, and between us he did the day-to-day work and I was involved, and I still am involved, because we still hold some of those investments. We took a tremendous amount of risk, and I'll never forget the time when I was talking with Mr. Mitchell in the lobby of the conference center and I said, "You know, we really want to help the economic development. This city needs capital. And the capital direction should be in the form of medical because that is the strength of the city that's not being financed." He agreed at that point to commit the company to making available $15 million over a several-year period to start up the venture capital company.

He committed $15 million—that was financing to be available for leveraging other financing?

Right. I've never done the calculation, but I'm going to guess we've put out $15-$20 million which probably represents $200 million of investment. That began to create companies. We have had some that didn't do too well, we've had some that have hung on, are marginal, and we have some that appear to be doing very well. The medical industry and the biotech industry went through a great flurry, kind of like the oil companies, and those people could raise $10-$20 million just on concepts. And of course, there was a nose dive three or four years afterwards, and it's been very tough to get new financing for some of the new ones. There has been a consolidation in that industry, and some of ours are merged and combined. I think that in the next couple of years, there will be a recovery as the number of these concept companies gets filtered out and the ones that really have the right products merge or consolidate or get funding from major pharmaceuticals as they develop relationships. But when you look at it, we've got a research park that we didn't have, and as a result, we've wound up with companies that are outside the medical industry but do some form of research or technical thing. Hughes Christensen chose to locate out here as a result of this technology concept. They'd been in one location for 80-something years and they wanted, in effect, to rediscover their company. They felt a change in environment to a technology kind of environment would be one that would create the atmosphere to go through that evolution. And they are really happy and really proud of their facility. It's been an internal show piece for their company and they've now expanded it by bringing in another division from Utah. They're actually under way with the expansion right now. We have Pennzoil's research center out here that deals with their lubricants division. So there began to be a degree of diversity. We've got some companies that we didn't need to fund, that had their own money, come out here and locate. Also, we've got a mixture of others. Most recently, although they don't fit 100 percent, but psychologically they fit, is Tenneco. We're planning on building a Tenneco building for a brand new division that is going to be sensitive to their cost structure and their re-engineering of their company, providing what they call "shared services."

How many research-oriented, technology-oriented jobs are there?
I think when you add it all up, we've probably got employment there in the neighborhood of about 2,000.

How many different entities does that represent?
Probably 35.

And that continues to be a direction of development?
A direction of focus, yes, and I think it puts us in a unique position. Last year we opened a new building for GeneMedicine. Although there has been a slowdown and a stumbling and a retraction, we're finding that when it's going to happen in Houston, we have a good shot at making it happen here.

Does the university technology transfer company still exist? Are they still active?
Very much so. In fact, we're getting ready to move in a new company. GeneMedicine is only a year old, year-and-a-half old. We moved them in last year, and we're moving in a new one in the next 90 days. I think that process will continue. It will be much more selective. It will be much more carefully done, but I don't think that market is closed at all. I think it will come back and I think it will still be a strong presence, and I believe it will help us land some other things.

And that means jobs and other impacts on the economy?
We were successful in getting a telecommunications company, WilTel, out here and they're into research. We're actually going to lose them soon

because of a merger. Maybe our focus will be telecommunications, maybe it will continue to be medical, but it could be energy, it could be manufacturing, it could be different components, maybe it's electronics. The research park should never be single purpose. It should provide an environment for innovation, whether it's innovation in manufacturing, whether it's innovation in technology or in new ways to do business.

What roles do HARC and the new community college play?
All great assets. Tenneco has taken advantage of the smart jobs program in the state through the university. So all that is a very important part, and the more you get mass, the easier it is to attract more.

And the job of attracting really has been yours from the onset?
In the research part, for sure.

Thank you, Michael. We'll do another session.

* * * * * * *

This is Thursday, August 24, 1995, Joe Kutchin in the office of Michael Richmond for our second conversation. When we left off, you had just about wrapped up on the Research Forest. Tell me now about the Town Center, what the strategy is there, what the plan is and where it'll go from here, and where you are right now.
The Town Center is clearly the most exciting commercial development that will occur in The Woodlands. The research park will be very significant also. One point I'm not sure I made on the research park is that with 5,000 acres devoted to a business park–and that is a huge business park–you have to have multiple uses within that business park in order to be able to absorb that land over 40 or 50 years. The research park provides a unique use that can help absorb that much land. The size of what we're trying to create commercially is significant: 5,000 acres devoted to business is a lot of acreage devoted to business. I just want to put it in perspective.

Just for clear definition, the research park is part of the Town Center?
We segment it differently, no. The whole crescent of the commercial park goes from the Trade Center to the college park, transitions into the research park and then on the southern end is the Town Center. Those four categories, or sub-locations or divisions, make up the whole business complex, which is 5,000 acres.

The concept of these four sub-units, how did that get created?
At least three of them got created early on: the research park, the Town Center and Trade Center. When the community college came into being, College Park got created. It sets up sort of an education center that creates more of a college campus style which could have housing or other academic-type activities. But that's a central focal point and I'm not sure it's a specific division; still, that area could have the makings of what a small college town would look like. I went to school in College Park, Maryland, and we're creating College Park in The Woodlands.

But the other three, were they part of the original land plan?
They were, generally, very early on. The research park was focused on trying to create a Silicon Valley or a Research Triangle, and both of those areas were studied, along with Boston Route 128. Those are probably the three premier research park areas, or high-tech parks, in the country.

Now talk about the development of the Town Center. How important is that?
That's what we're focused on right now. And we're focused on it because the most critical event that has probably ever happened in The Woodlands from the commercial point of view really is the mall. The Town Center would not be a town center without a regional mall. When you look into the significant events in The Woodlands, just in total, the Hardy Toll Road and the mall probably are the two most significant infrastructure events that have allowed The Woodlands to be what it is. Hardy Toll Road has brought access so that the distance factor was alleviated, which made residential grow and also made business grow, and then the mall created the heart of the retail complex. If you look at malls around the City of Houston or around the country, they say if you build a million-square-foot mall, you will build another million square feet of adjacent retail around it, and that's what we're in the process of doing. We hope that will become the springboard to expand into other areas, principally offices, hotels, recreation. The Pavilion came on earlier, but all that fits together in a mixed-use town center.

Let's jump for a moment to the Hardy Toll Road. Your operation was not primarily in those negotiations, or am I wrong? I thought that was largely George's involvement.
George played a big role with Judge Lindsay and the Toll Road Authority. And Plato Pappas did, too. Ed Lee was the one who came up with the idea

of the signage. Many people have called the Hardy Toll Road "The Woodlands Toll Road."

Describe the development of the mall.

A study was done early on by a market research firm that said that creation of the mall would be the most critical commercial event to occur in The Woodlands. This was done in the '70s, I believe. And the mall would be the anchor for the whole Town Center. Knowing that and knowing how important a mall was even without that study, we started negotiations with all of the mall developers in the country. We met with DeBartolo, we met with Rouse, we met with Hahn, we met with Homart and at that point we also met with Federated Development Company, which was the developer for Greenspoint and other malls around the country. We originally made a decision to go with Federated because they were synonymous with Foley's, and Foley's was the lead anchor department store in the City of Houston, has been and probably still is. We actually had an agreement–they owned 70 percent and we owned 30 percent.

This is about when, Michael?

Late '70s, early '80s. At the very end, they would not commit to bring Foley's along and the whole agreement kind of disintegrated, and we sort of went back to square zero. In retrospect, that may have been good fortune, because Federated eventually got purchased and went into bankruptcy. With all the problems that arose when the department store industry started consolidating, we may have never built the mall. We were also competing simultaneously with Friendswood Development Company, which had identified a mall site on their property in the Conroe area. They had worked out an option agreement with DeBartolo, who was probably the largest developer in the country at that time. So we then went back and chose to make our deal with Homart. Homart was a Sears Roebuck subsidiary, or will be until later this year when they're sold. But we struck a better deal–we got a 50-50 deal with Homart, and although they didn't quite commit Sears, they felt pretty comfortable they could. We signed that agreement in 1982 and we put up a sign, "The Woodlands Mall Coming Soon," and right after we signed that agreement, the Houston economy kind of went into a free fall. At that point in '82-'83 is when Houston began to lose its 200,000 jobs. The malls were suffering, everything was suffering and our mall deal just got put on the back burner. There was just nobody interested in it because while malls were being built over the rest of the country where there was a lot of growth, the economy in Houston was in negative growth.

Meanwhile, was the Conroe possibility hanging over your head?

Yes, it was still hanging over our heads. We kidded each other that the only guy who got wealthy on the mall was the guy who repainted our sign up there every year and kept it in good shape. But later on, towards the end of the '80s when the job growth had returned and The Woodlands was more mature, there was a stronger focus on getting it done. Homart really picked up its efforts and we worked hard with Sears to get them bedded down. The department stores were very worried about transference. They were worried that if they built a mall here, they would deplete their revenues at Greenspoint, in particular, and some of the other malls. So they were always in their economics, deducting out what their revenues loss would be in their other stores, and that would always make the numbers not work. But finally, we actually made a commitment to Sears to pay them a decent amount of bonus money, or whatever you want to call it, an allowance not unusual in department stores for the most part, at least the initial ones coming in. We gave them the land and the parking and the pad and paid them a bonus, too. That got to be very expensive and we were at the bad part of the market, so the payments were much higher.

Was this still the '80s or are we in the '90s?

We're probably right around 1990. After we had been working with Sears, and this is probably getting back into the late '80s, the Homart people brought Bill Dillard into the picture. He came and did his review, and he made the commitment at that point that The Woodlands would be where he would put a store. That commitment from Bill Dillard is probably the single most important event in why we have a mall. Bill Dillard did not have the sensitivity of transference because he did not have a store at all at Deerbrook. His store at Greenspoint was one he had bought from Joske's. When he purchased Joske's, it was a very small store and he did not have a store at Willowbrook, or if he did, it was a small Joske's. He did not have a dominant store in the marketplace, so he did not have the sensitivity that the other stores had about transference. A Dillard's at The Woodlands would not affect one somewhere else.

Is he chairman of Dillard's?

His father is. His father is chairman, he's president. He's Bill Dillard, Jr., and there's William Dillard, Sr. Bill handles their real estate negotiations, and he said to us, "I'm coming here." Then it was a matter of taking the Sears commitment and Dillard's commitment and getting two more anchors. But at least we had the commitment and the knowledge that he was going to come here rather than go to Conroe. And Conroe could never get built without Sears and Dillard's, or at least it would be very unlikely. He was the first one to commit. It really was never in writing, it was just a commitment that he would repeat and repeat. It was a verbal commitment. I don't think we ever had anything absolutely in writing, other than maybe a paragraph from him. I'm not even sure we had that, but we kept in contact with him every year and met with him, and he was just waiting on us to get the rest of it done. The next one we got was Mervyn's. We wanted Penney's, but Penney's was not ready and Penney's has the reputation of coming in later, coming in after the fact. Then there was the cat-and-mouse game with Foley's because, clearly, Foley's would be a great addition to the mall, and then you would have your two premier stores, or dominant stores, in Houston, Dillard's and Foley's. We got into some sticky negotiations with them about price and finally on a ski trip in Telluride, we had Foley's at the table and Dillard's at the table, the site plan and Homart, and myself and Steve McPhetridge. And at that meeting one evening it was all laid out and the Foley's guy said, "OK, we're coming."

Could you give me a year on that?

That must have been '91 or '92. With that we knew we had a mall, and then it was a matter of getting the final design. The design was an interesting issue. We had done some rather unique designs during the early '80s—U shapes, crazy shapes—and finally we wound up with a much more traditional design that the department stores felt comfortable with. But we worked on the finishes, the volumes, the feeling of the mall.

When you say "we," who is that?

Homart and The Woodlands Corporation. We were more concerned about the architecture of the mall than Homart ever would be. Homart is not bad, but that's not their greatest sensitivity. Their emphasis is on being functional—they're not the most stylish of the mall developers. We hired a firm, ELS out of Berkely, California, that had done some malls that we thought were nice and innovative, and they were retained. They had actually been retained early in the game, way back in the early '80s. Eventually we got into the final design and tried to fit everything on the site. As much land as we have out here we never seem to have enough in the right spot. Foley's, to some accidental degree, when they decided their store size and how they wanted it to be visible, forced the mall to have a subtle bend in it that we had not planned on. They wanted it tilted up so that when you came along Lake Woodlands Drive, you would see the Foley's. That requirement has had one of the most dramatic positive impacts on the design of the mall. Although the mall curves, it's so subtle you don't even know it. And the sight lines it provides as you walk through break up the mall–it's really attractive. But that was not in the plan. It was a reaction to Foley's requirement, and we're thankful that they had that requirement because it helped us improve the quality of the mall. We went out to bids, the bids came in over budget, we had many weeks of negotiating change orders. Steve McPhetridge lost his little glass top to the mall above the carousel. We had planned on some outdoor seating, that probably wouldn't have worked, that we cut back on. We got into some really ticklish discussions with Homart on the amount of brick, the marble floors. They wanted tile upstairs, we didn't. We wanted to keep the brass rails. Another big issue was having a metal roof versus a composition roof. If you look at the requirements that we were successful in keeping, I think those are the architectural issues that really make the mall what it is. And our people, Coulson and Steve, Roger and myself and others, worked hard to keep design first class.

You just named the main people involved from beginning to end?
From the design side. Steve was the project manager. He represented our partnership interest. He wasn't here during the negotiations of the joint venture. That involved Doug Leonhard and me. He and I did the agreements but when it got into the execution of it, Steve was really our chief point person. He coordinated with engineering to make sure all of the infrastructure was there–the roads, streets, utilities . . . he coordinated with finance to make sure we got our loan done, he worked with Homart's architects and kept our people informed, helped force decisions, maintained the construction schedules, was our representative in the partnership. The operations have moved over to Eric Wojner's area since it's up and running, but Steve was really our project manager to get it built and open.

It's a little less than a year now.
October 5 was the ribbon cutting and the day it opened, and I don't know how to describe the opening. We had 80,000 people, balloons were all over the place, the people were lined up starting in the wee hours of the morning, the ceremony was not long but was touching and I'd have to say that

that was one of George Mitchell's proudest moments. When he cut that ribbon and saw that crowd and those balloons came out of their nets and the mall was opened, it was one of the most festive things—even Bill Stevens made the comment that he'd been to a lot of openings but he had never seen anything like this one. William Dillard, Sr., said the same thing, that it was the grandest opening he's ever seen and he was really proud of his store. He thought it was the most beautiful store he'd ever built.

And a year later?

A year later the mall is operating. In general, most of the tenants are doing well, some have already turned over and we're replacing them. The mall has been well received. We created a Town Center improvement district through the legislature that was funded by a sales tax to help with security. It was interesting to watch the community say they wanted it and then all of a sudden when it's upon them, they're saying, "Crime, traffic, I don't want this." I would have to say from a security point of view, we've gotten nothing but compliments. We've got mounted horses that patrol the lots in a very friendly but helpful way, we have a lot of visibility with the sheriff's department. In the old days, malls were nervous that if you showed that type of security it was an indication that the area had a criminal problem. Now people demand it, they want that feeling and we've gotten many compliments on the visibility of security and how people feel safe and secure going to that mall. Our job is to maintain that.

Are the anchors happy?

I think the anchors are very happy except Mervyn's. But I think Sears, Foley's and Dillard's are doing very well and a lot of the in-line stores are doing well. The real issue and the real opportunity is going to be the growth that occurs. I think the mall is going to be helped greatly by the new Landry's and, more importantly, by the theater. The theater and Landry's are now beginning to expand the scope of Town Center. It's gone from office to retail community services like the hospital and the municipal governments, and now moving in and sharing that entertainment concept with the Pavilion and the theater, which is the largest theater being built in the City of Houston, and Landry's on the Waterfront, which will be the largest restaurant here and I think the largest that has been built in north Houston. And now we're getting more generators of traffic and population and people. They're talking about the theater having anywhere from a million to a million-and-a-half people come though it. It's 4,200 seats, 17 screens. That will all bring people. We are creating destinations—the mall is a destination, the Pavilion. People will drive an hour-and-a-half for a performance at the Pavilion.

Depending on the performer, sure.

I believe people from the north will drive an hour to the mall, and I believe the theater and Landry's, since they will be unique to north Houston, will get at least half-hour to 45-minute drive times, because people will want to go to those unique places. And that's what we're creating in Town Center, a series of unique opportunities, reasons for people to come. The next big step is to get the mile-and-a-half waterway developed. I was in San Antonio yesterday looking at the San Antonio Riverwalk, which is just an outstanding place. But ours will be different. Putting our own Woodlands touch on it, I think, will create a core destination marketplace that will allow all our residents to take advantage of great opportunities to shop, office, recreate, enjoy themselves, and will really be paid for by the regional marketplace. But our residents will have it accessible to them and convenient to them, but its cost will be shared by a far broader marketplace. And that will become the commerce center of north Houston. I think the next step after that will become hotels and convention space and more people coming for destination activities. We can tie that in with the golf courses.

And the 5,000 acres provides enough land to do this?

Yes, plenty of land. The Town Center itself will be closer to 1,000 to 1,200 of those acres, but it will be the most dense, the most intensely developed and will probably have the tallest structures, particularly as it relates to the freeway. Creation of the waterway can tie all this together and be unique. The City of Houston has been talking a lot about something like Buffalo Bayou for the 20 years that I've been here. But we have the opportunity to make it happen. Whether we have pedestrian traffic on it or we have little buses or little vehicles that will help you walk along it, or water taxis or unique water vehicles, I think in lighting and landscaping we can create just a wonderful place. I think ultimately it will even further justify the recognition that The Woodlands has received for being an outstanding development, whether it be international awards such as it's won from the Urban Land Institute or through some other recognition. It will really be different and that's what's going to make it work.

I think that historically The Woodlands Corporation has had an

equity interest or complete ownership in many ventures and then you operate them for a while and sell them at a gain. Is that correct?
Let's talk about Town Center because Town Center is probably a good example of a combination of things. In some cases in Town Center around the Pinecroft Center, we've sold land to users and not maintained a residual interest, although we do own the Barnes & Noble and Marshall's and Blockbuster Video land and building.

You own the store?
The buildings, and lease it to them. What we looked at for the Town Center, particularly around the waterway and the core, is either owning an equity interest like we do in the mall, where we own 50 percent—we maintain control there although we don't operate it because that's not our expertise—or, as an alternative, we've also chosen to lease land rather than sell it. We lease land now to Macaroni Grill, to NationsBank, to Grady's and to Landry's. The theater we will build and own and lease to the theater, so we will retain the equity interest there subject to the lease. The ground lease allows them to put their capital in, and we get a percentage of the profits of the restaurant. The bank, we don't get a percentage of the profits but we get leases that increase in their rental rates through participation agreements, which makes our land more valuable and generates more revenue and provides an ongoing revenue stream. The Wood Ridge Center, across I-45, we bought from the Resolution Trust Corporation and renovated. I'd like to make a comment or two about that. We may sell 50-75 percent interest to cash out but still maintain at least a 25 percent interest in order to keep our hands in the overall activities of the Town Center.

And also enjoy a continuous income stream.
We've done that this year with 10 office buildings and research buildings with a partnership with Crescent Real Estate Investment Trust out of Fort Worth that has given us more capital to reinvest. But we are the managing partner and we have control, although we don't have absolute control any more because we have to respect their 75 percent interest. But it keeps us involved and it gives us the ability to recirculate our capital to expand. We've also worked hard to try to get others to help pay for things. The Pavilion has raised some money on its own to help that operation; the various utility districts have helped us get bond money that's been used for some of the infrastructure. And we're going after some federal grants to help with the development of the waterway. So we've looked at different combinations of

things. We have to do it for a profit or we're not going to be in business. I don't want to minimize that component of it; everything we do, we do with that in mind but we look at the short term, mid term, and long term. The purchase of the Wood Ridge Center is an interesting one. That center had been basically abandoned, in poor shape and had been foreclosed upon. RTC owned it and we wound up buying it. The main reason that I was able to get Mr. Mitchell to really be interested in buying it (although we made a great financial deal, and I don't want to minimize that aspect of it, because we did it right) is that it was at our front door. If we didn't control that–and we had lost a lot of control of a lot of the land along the freeway–then the X-rated joints or strip joints would naturally gravitate to that low-end space. What we did when we bought it was, we kind of put it on hold for two years until the mall got built. When the mall got built, we could then get a better tenant for the center. Now we've got Office Depot moving in and we're real close to a deal with Pier One Imports. So that whole center has now come back to life. We spent several million dollars in physical renovation, and now the place doesn't even resemble what it used to. It gives us a whole different feel for our front door. It's made the City of Oak Ridge, where it's located, very proud. It's a good investment deal for us and it's a good community development arrangement, and we're real proud of that.

The Woodlands development has created all kinds of parasite businesses outside The Woodlands that are enjoying the traffic created by The Woodlands.

We try to control as much as possible. When you spend $500-$600 million, you can't let others take all the easy opportunities. That's why we have interests in title companies and mortgage companies. Those generate current income while you're struggling to keep that big land base.

I was thinking of businesses along Sawdust.

We created opportunity for a lot of people. Those opportunities, I think, are diminishing because we're beginning to take better control of the situation and people are realizing that the mall and our new village centers and things like that are the places where commerce is really going to be taking off in their businesses. But we lost a lot of momentum for a while. I feel real confident that we're in a much better, stronger position than we've ever been. We're able to get much better pricing. Also this week, the president of J. C. Penney's department stores was in talking about coming into the mall. We're considering that right now, so we could have a Penney's within a year-

and-a-half, two years. And that gets me to this point: I think the best is yet to come because if we're successful in attracting office users, then we can begin to look at attracting the right type of executives. With more of those, we can look for a higher-end department store 10 years down the road. Let's be realistic. We're a family community, we're a middle to upper-middle income location. We're not quite the Galleria, although we have aspirations. The Galleria took 20 years to develop and has about 26 million square feet that's been developed. Our goal is 18 million square feet equivalent over a 20-30 year period. It's been done before. People would say when the Galleria started that it was in the middle of nowhere. We've become realistic and patient. We want to be in a position over the next 10 years to be able to attract a Nieman's or a Nordstrom's or a higher-end store, and then tie the mall into the waterway. We will have the flexibility for that to occur, to really tie it into hotels and create that type of synergy that's hard to relate to now. But when you looked at what we've done and what's coming, it's not as hard to picture as it used to be.

Talk a bit about corporate relocations.

Corporate relocations are the toughest. First of all, most corporations, like 99 percent of them, relocate within five minutes of where they were previously located. You hear a lot of press about GTE moving from Connecticut to Dallas or other companies moving. Those are the high-profile ones, but they represent a minuscule amount of opportunities. Companies are real sensitive to relocation costs these days. We have a heck of a time competing with existing space–downtown Houston, the Galleria have millions of square feet of available space. To reproduce that space and get a reasonable return would require the rents to be significantly higher than those offered by people that already have the space built who are trying to lease it out. We are in a real struggle. We've been successful in a couple of cases, like Allstate in Parkwood II, successful with Tenneco that we've just announced, with GeneMedicine, Hughes Christensen, but it is very hard. A corporate move disrupts their operations and in some cases the company bears the expense of having to move people. Plus we're not going to be the least expensive. We just can't be because of our capital constraints. So that is a real challenge that lies ahead. We are making a couple of proposals next week to major tenants or users, but corporate moves are our biggest challenge. We're going to have to pull out all the stops, to create financing, tax abatements, things which will permit us to lower our cost structure and still allow us to provide the best quality and make us competitive. Yes, I believe there will be a small premium people will pay, but there will not be a huge premium because their shareholders and stockholders and boards might be very uncomfortable. So we have a lot of work to get our structure down so that we can be competitive, but we know what our goal is, we know what our objective is, and we're going to have to see how easy it's going to be for us to accomplish it.

Is it a fair statement that commercial has been going pretty much at a very acceptable rate?

Retail is going great. We've been reasonably satisfied with our research technology park. Offices are the biggest challenge that remains. With the cinema and Landry's and the expansion of the Pavilion, entertainment and hospitality are doing well. We've now got a Drury Inn under way, we're talking to Marriott for a Residence Inn, and maybe there are some opportunities to do some more group-type space in our conference center or up here in the Town Center because we've been identified with the meetings market. We are also talking to a major corporation to create their learning center here which would be a result of our ability to service corporate learning needs. We already have the Exxon Conference Center which is a private conference center for Exxon Chemical. So we're expanding in a lot of different ways. The hospital just announced a $20 million expansion program, they bought nine more acres from us to put them in a position to have a major medical broad-based facility, both offices and clinical and beds. There are a lot of markets that are expanding; office clearly is the toughest.

Is there anything significant that you would want to say about company-owned rental properties in The Woodlands?

Apartments are extremely important. Apartments provide starter housing for homes, provide interim locations for people moving, and they also provide a home for those who cannot afford to live in houses but also will become a big part of the work force in service industries. We've learned that you don't want to build masses of apartments all in one location, and more importantly we've learned that we don't want absentee ownership because when a family lives in a house, they take care of it; when a person lives in an apartment, he or she takes care of the inside but the owners don't always take care of the outside. There would be nothing worse than to create apartments that go into a decline and then start affecting neighborhoods. So we have an active apartment program going, including some that are government-subsidized. So we have people who can literally pay nothing or people who can pay modest amounts, and we have people who can pay standard, reasonable rates. We have kept, except for the seniors apartments, the management and partnership interest in all of those units. Right now, we're 97 percent occupied and 99 percent leased, which is as much as you can get. We're evaluating two more apartment projects. They tend to be controversial in the community because people associate crime and traffic and problems with malls and they also associate overcrowding in schools and crime with apartments.

Would these be market-price apartments?
We're looking at one that is affordable, using tax incentives, and one that is market rate, upper end.

Your biggest apartment project is Grogan's Landing?
I believe it's Grogan's Landing, yes.

Has that been successful? That's one of the very early ones.
Yes, it was unsuccessful early on, but now it's very successful. We have spent a lot. We didn't build those–we had a partnership interest. Our partner did a poor job of construction, and we literally rebuilt them when we bought him out. The property is doing very well. It will never get to a level that we probably like. You just can't rebuild everything in it, but it works very well and we're pleased with it. It provides a great housing product for some of our residents who fit into that income profile. I would have to say that the management folks who work for us have done an outstanding job of maintaining the property and screening residents and taking care of the apartments so that they are an asset to the community.

Will there be more?
No question about it. We will have to do that to provide housing to accommodate our labor force. And also take into consideration that there are a lot of divorces and there will be a lot of new graduates. The demographics show there is a need for that type of housing. A lot of retired people move into apartments, whether they are widowed or not. That's just the way they live. Another big area of our business is the conference center, and it has become an outstanding facility. It was one that probably struggled as The Woodlands opened but really provided us a base of identity. I don't know how many times I go across the country and bump into the people who will tell you, yes, they've been to The Woodlands to play golf or to have a meeting. It's amazing what it has done from an identity point of view.

Do you remember the initial investment there?

The initial investment in the whole complex, which includes the athletic center and some of the retail and all that, I think was close to $30 million.

And all those were opened on day one, weren't they?

Yes. They were all open on day one for impact, and they did have impact—positive and negative. But now we have really focused on the conference market. We hold conferences for Fortune 500 companies, board meetings, weddings, bar mitzvahs, parties, reunions, everything. It is a tremendous focal point, it is a very profitable operation. We have physically redesigned it to accommodate these events. If you drive over there, you see a unique-looking facility that's kind of casual looking, but you walk into it and it is a first-rate facility. More than that is the service that our employees provide, because that's what our conferees pay for, and we've gotten just tremendous positive response on the meetings held over there and how we service our customers. One of the things I truly believe is that if you took a survey of people who don't live in The Woodlands they would probably identify The Woodlands with trees and golf. Golf is still one of the things we are most predominantly known for. Having a TPC course and having the Shell Houston Open here is a great asset.

Was bringing that out here under your wing?

No, but I've been involved in the TPC course negotiations and our partnership relationship with the PGA tour. We are now working on another TPC that will really be looked upon as a potential top 100 course in the whole nation. Our TPC course now is well known and well liked but we are looking at building the ultimate golf course for the Houston area and are actually beginning to talk about that right now. It won't occur until around the year 2000, but planning takes many years for those things to happen. Golf is clearly an element of our community that's important. It attracts people here, it attracts residents here and is a big business for The Woodlands Corporation. We now have about 1,600 members in our country club, with our four courses and an additional half course to open within the next month or month-and-a-half—the third nine in the Palmer that will be opened. It's a big thing for us. We just hosted the Texas Women Amateur Golf Tournament. A lot of people have played our courses.

Thank you Michael. We'll need another session

* * * * * * *

Today is Thursday, September 14, 1995. I'm in my office in the MND building in The Woodlands with Michael Richmond. This should conclude the series of interviews that Michael and I have had.

Tomorrow is my 23rd anniversary. I was brought down to The Woodlands as controller of The Woodlands Corporation, which had been newly formed and was the developer of The Woodlands and was an isolated subsidiary at that time because of the HUD agreements. I moved into finance later and then became a senior financial officer and spent the majority of my career doing that. As a senior financial officer, I got involved in a lot of deals and I got a lot of exposure to some of the operations. I worked with Doug Leonhard in the regional mall joint venture agreements, got involved in starting the venture capital company, got involved in deal making. So I had a lot of experience in that particular area. I probably did as many deals for the company as anybody, not routine transactions, but unique transactions. When Doug Leonhard resigned, Ed Lee asked me if I would move up into the operating side and head up commercial/industrial development. I believe that was somewhere around 1986. A year or so later was when Ed passed away and Roger Galatas moved up to president. I became executive vice president and assumed more responsibility at that time.

Is The Woodlands working?

Yes, I think it is. I'm more optimistic about it now than I've ever been. Home sales are increasing, we've had a fabulous year so far, got off to a slow start but it's been tremendous the last couple of months. The three months of the second quarter of this year, I think we averaged over 100 homes per month—haven't done that ever. In commercial activity, the theater is coming out of the ground, it's impressive. The overpasses are being built, the new bridge over Lake Robbins is out for bids, Landry's just opened and we are introducing the waterfront concept to people. There's a lot going on. The medical area is growing, Tenneco is under construction. There are a lot of challenges ahead.

What are they?

The biggest challenge on the commercial side is being competitive in pricing. It is more price-sensitive on the office side, particularly. Same thing with the industrial side. The retail and service sector follows the population. It follows the businesses, so it's not as price sensitive, it's more a numbers game. For residential, I think we have to worry about the fundamentals—

mobility, traffic, schools and security, amenities. We must maintain those strengths and add more product niches. We've done real well with townhouses, and expanding our markets will increase our volumes.

You talk about "we." The time is much closer now when "we" is no longer The Woodlands Corporation and, instead, is the citizens of the community. Is that going to affect the continued development?

Oh, sure, there is a natural tendency among residents that once they got what they want, or what they feel they need, they tend to go into a "Not in my back yard" NIMBY routine, or they try to exercise more input and influence. I'd say though, Joe, we've really had some good fortunes. Well, I don't think it's fortune because we've worked at it–that is, relationships with our various community boards, associations. I think they feel that we're doing something good, or we're at least sensitive. As Roger has pointed out several times, we live in this community, too. We know the people, we're friends with the people, we're not just a developer, we're interspersed with the churches, synagogues, the "Y's," the activities, we live it, we're here, we're not absentee owners. So we're more sensitive to the needs and wants of the community just because we're part of the community.

When do you see the community's being built out?

I think it will be 30 years. We will create huge milestones when we get more of the Town Center built out and when we get a true business center developed. That will mature it a lot more, and then it will be just a matter of filling in.

That is a longer period than originally conceived?

It is, but we also started out with 15,000 acres or so, and now we're at 26,000. We've actually acquired as much land as we've sold, or darn near that, so we've extended it by our own choice by the additions.

Is 150,000 population still the target?

That's the goal. A lot will depend on the markets and the densities we're able to achieve. We have to be more dense if we get to that number, but we'll just have to wait and see.

Have there been mistakes? What about Grogan's Mill Village Center?

Grogan's Mill Village Center was not the best-planned center. There was the issue of planning, and then there was the second issue of functional obsolescence. It was designed for a supermarket of 25,000-30,000 square feet, which was predominantly the size for supermarkets when it was built. Now they are 50,000, 60,000 square feet. We hope that later this year, in three or four months, we'll announce something that could totally revitalize the center and bring it back to a more vibrant life than it ever had.

Don't tell me you're discovering oil.
No, no, we have two prospects out there that would absolutely change that center and make it as good a center as we have anywhere.

In fairness though, that's a recovery of a situation that hasn't been good for 20 years.
No question about it, but you have to look at what happened. The Wharf itself was a disaster when they had the ice skating rink and indoor retail. Then we converted that to conference space, so it got corrected. It got corrected sooner than the retail. When Randall's bought out Jamail's supermarket, they had competitive sites and didn't want to do anything to harm them. Then we struggled with Gerland's. Kroger's was already in, so they didn't want to help us. Then when HEB came into the marketplace, Gerland's had the space tied up, so HEB chose to go elsewhere. Right now, we may be right at the best timing. A new chain is coming into town and they are looking for sites, and that's an available site. And they are not afraid of the competition. They don't have anything to protect. So, sometimes it's just a matter of being at the right place at the right time. We haven't been there in the past, but right now we may be. We may be really fortunate that old supermarket space exists because we're talking to another user that would change that into one of the finest facilities we would have out here. So we may get lucky. It's amazing how if you still have good locations and good sites, something will change.

Are there other things that come to mind that have not gone as well as expected?
Grogan's Mill was probably our biggest white elephant. The shopping center, not the neighborhoods. I think we've done better in our housing stock, gotten better quality, we've been more sensitive in our building designs. The crazy part of a boom period is not only do things go very well, but they get very expensive because there are shortages of labor, shortages of supply, and you have your worse quality. It's kind of what happened to the energy industry. The cost of drilling skyrocketed when prices went up, and now that

prices are down, the cost of drilling and the quality of drilling is probably a lot better and more efficient. When anything works, you probably make more mistakes, because you can't do anything wrong. Looking at other things that we could have done better, I'm sure there are, but not too many of them. If you were to ask the residents what they would like to see changed, I'm not sure there would be too many major things. The school districts we had to correct. That was already here, not something we caused. Our job as developers is to correct and to change. But other problems come when you make the mistakes yourself.

What about the concern of being annexed to Houston?
I don't think that would be very popular here. I think our residents and our businesses feel very comfortable with the way things are operating right now. I feel that we have a good situation where, in general, tax rates have actually come down, the quality of services has at least stayed the same, maybe even improved, while in other areas they're deteriorating. I think the systems that have been in place with the community associations and with all the various utility districts that have been created are working smoothly, and I think everybody is getting the benefit of it. I don't think people would really want it to change.

But George Mitchell has been very clear through the years about his expectations concerning The Woodlands vis-a-vis Houston.
There may be a better time later when the city expands closer to us and its services are more readily available. But just to add an appendage on to a city, I don't see that. Ultimately, annexation will probably happen unless something forces it not to happen. Right now, things are quiet, but you never know.

I know it's not a popular idea with the residents. Would you say that it's not popular with the business community, as well?
The business community is probably less sensitive to it, but put it to a vote and they'd favor staying status quo.

One of the commitments made to HUD was to try to mirror the socioeconomic makeup of Houston. Does that concern fall under your wings?
No. That's more of a residential issue. I don't know exactly what our mix of population is. I would suspect we are gradually increasing our minority pop-

ulation, but clearly we're not there yet. We have come a long way and worked hard to provide affordable housing. We've done some tax credit programs on apartments, we've been one of the early ones in the state to take advantage of them and have really done a good job. We're trying for a third one right now; we got turned down the last time, but maybe we'll get the next one. So we do have a concern about that. We have to do them properly. You can't be in it just for the tax credits and then walk away because a lot of the profits for the development company are up front. You have to have a desire to stay in. I think that one of the smartest things we've ever done is to maintain control of managing the apartments, so that absentee ownership doesn't allow them to deteriorate because of lack of supervision and thereby put a housing investment at risk.

It seems that a teacher in a Woodlands school probably couldn't afford to live in The Woodlands except in a rent-assisted apartment.
Maybe not. It would be marginal. That's not just in The Woodlands. That's a broader, universal problem. Housing costs are still difficult.

Are there going to be any more new towns built?
I don't think there'll be many large ones, if any, because the cost of carrying the land over such a long period of time is such a burden that it puts a lot of pressure and stress on profitability. You'll see people buy 1,000 acres, maybe 2,000 acres, and try to plow through it in five or seven years and turn it quick. Ours is, literally, a 50-year project. That's a long time to keep the far extremities of the land which are non-revenue producing. You've really got to create enormous values as you go along to offset those non-revenue producing assets. There will be a tremendous amount of difficulty, if not impossibility, to get financing for something that long term. Just because of the uncertainties, people feel more comfortable with a three-, five- or seven-year projection. When you start getting beyond five, you get into so many variables you have no idea what's going to happen.

When you say people, you mean lenders?
Lenders or investors or developers, whoever is going to be at risk.

And besides that, of course, is the huge front-end investment.
Unbelievable, again without any revenue.

Now, toward the end of 1995, what is The Woodlands situation

regarding large investments in the future?

I think you'll see our capital investment in pure infrastructure beginning to decrease. Now, you'll always have opportunities to build new projects, new office buildings, new apartments, new retail centers, but those generate income immediately. They are free standing, independent operations. They generate revenue very quickly from the time that you make your investment. The harder thing is getting over the hump of bringing in enough cash so that it's greater than your outflow in order to start recovering the initial investment. I believe we are there now, but we can't relax. It's got to get even better to make it worthwhile.

I remember your view was that in the final years of development, The Woodlands would generate large amounts of cash flow.

It has, too. That's why I'm encouraged that commercial development is growing as is the number of residential units. By broadening our product mix, we will absorb land faster and, thereby, recover or reduce interest costs and recover infrastructure costs that are already on the books. That's critical. We've got to lower that investment so we can then go to profits.

What have we forgotten to talk about?

I think there still remains a challenge to maintain the integrity of The Woodlands, and not let it get away from us.

What would cause it to get away from us?

I think if you lowered standards of development if you wanted to try to maximize short-term cash flow to the point where you just didn't care about quality, the development could go out of control. I think maintaining our standards–in fact, improving them in different ways–is what's going to make The Woodlands even more recognized. The residents who live out here have great pride. They talk about it, they get others to live out here. That quality needs to be maintained, and even enhanced. The fact that we do new things and open new places and give them more choices is what makes them like being here. I believe we need to keep that up, along with keeping an eye on security and education and mobility, the fundamentals of community life, to really make The Woodlands, when you look back on it in 20 or 30 years, a place that was well thought out, well conceived and more importantly, well executed. I hope that's what happens, and I know that's the way George envisioned it. For his risk taking and his confidence and his guts to allow it to happen, I certainly would hope the reward is there. I know he has told me how many times people have come up to him and thanked him for allowing them to live in this type of community, providing this type of community for their families and their children and their relatives. I know that should and does make him feel good, that not only are we running a good business, we're accomplishing something more than that.

Michael, thank you very much.

(In October 1997, Michael Richmond was named president and chief executive officer of The Woodlands Operating Company, L.P. The company had been known as The Woodlands Corporation prior to its acquisition by Crescent Real Estate Equities, Ltd., and Morgan Stanley Real Estate Fund II, L.P., earlier in the year.)

Philip S. Smith

This is Joe Kutchin and the date is May 7, 1996. I'm in The Woodlands office of Philip S. Smith, who is senior vice president, administration and finance, and chief financial officer of Mitchell Energy & Development Corp. I'd like to start out by asking for some of your personal background—where you are from, where you went to school, what you did before.

I graduated from Mississippi State University with a B.S. in accounting in May of 1957, worked for Arthur Andersen for one year, went in the army and served as a lieutenant in the U.S. Army Artillery from May of 1958 through May of 1960, at which time I returned to Arthur Andersen and was employed there until January of 1980. Then I joined Mitchell Energy & Development Corp. I was a partner in Arthur Andersen at the time I left them to join Mitchell and had responsibility for the firm's oil and gas industry competence program.

So you were, in effect, the chief oil and gas person at Arthur Andersen?

Right.

And that would be at the top of the Arthur Andersen energy function nationally?

Yes, it was a firm-wide responsibility.

You said you came to Mitchell in 1980. Do you happen to remember what year it was that Arthur Andersen became Mitchell's auditors?

It was in early fiscal 1973 or '74, I think. I was the initial engagement partner on the audit and served as such for about five to seven years. I had been off the engagement for one or two years before I joined Mitchell.

Who were the auditors before?

Peat Marwick Mitchell.

Do you know the circumstances which caused the company to change accountants?

I'm not totally sure.

Tell me what the circumstances were about the time you came. I remember Tonery was here.
The company was undergoing a reorganization at the time. Joe Williams had been brought in to head up the administrative function, which included finance, accounting, administration, personnel, data processing, legal and other related matters. I was brought on board as chief financial officer.

Tonery had just left or was he still here?
No, Bill stayed for about a year or two after I came, I think, and he handled the financing while he was still here. The structure was such that I reported to Joe Williams and Tonery continued to report to George Mitchell.

One of the very important periods of time in the company's history was in '74 and '75 when cash flow was so tight. Did you have any particular role then? Do you remember what brought that about?
As I recall, the company was in a liquidity crunch. One of the contributing factors was the opening of The Woodlands which had required significant amounts of capital and, therefore, debt. That caused the company to be in something of a liquidity crunch at the time. We at AA were involved from the standpoint of being the company's auditors but weren't directly involved in discussions with banks and financial institutions.

So during that whole period, the problems were essentially handled internally and you were an outside consultant?
Right.

Neither you, personally, nor Arthur Andersen had any particular role in dealing with the lack of cash?
Not directly. The agent banks at the time for Mitchell were Chase Manhattan and First Chicago. They were the primary financial institutions involved, but I did not directly participate in any discussions with them.

The change to Manufacturers Hanover had already occurred by 1980, is that correct?
Yes, that occurred sometime in the late '70s, I believe, the latter half of the 1970s. I'm not sure exactly what date it happened.

Do you still consider them the lead bankers?
Yes. Manufacturers Hanover was subsequently merged with Chemical Bank

and Chemical Bank was subsequently merged with Chase Manhattan Bank. It's now Chase Manhattan Bank–that's the official name, although many of the people there came from the original Manufacturers Hanover and/or Chemical Bank.

Over the course of the years, I've been able to identify a couple or three of the major things you were involved in. I expect you have more milestones in your mind, but some of the very significant changes that you presided over, probably initiated, were the outside financings—going to the public for debt. I'll come back to one or two others later. Tell me about the company's use of debt.

George has always had a philosophy of primarily growing the company through internal cash flow and to the extent that cash flow was not sufficient, debt was used. For years that was primarily accomplished through bank debt, although the company had done some financings with insurance companies–private placements through insurance companies–in the late 1970s. That vehicle continued to be used periodically for a number of years, up through the mid-1980s, as I recall, at which time the company–in fact, the whole oil and gas industry, as well as Mitchell Energy & Development Corp.–began to undergo financial stress as a result of lower oil and gas prices and things of that nature. That caused the private placement market to be more difficult to access and made it desirable to lower the dollar magnitude of the bank debt because it was out of kilter in terms of being too large a percentage of the total debt of the company. As a result, the company first accessed the public debt market in the late 1980s, I think around 1987, in the form of a public debt offering. I believe it was for a 10-year period, with First Boston Corporation being the lead underwriter and I forget who the co-agents were.

Goldman Sachs was in there?

It could have been Goldman Sachs, and Merrill Lynch could have participated. The cover sheet of the prospectus would say exactly who participated. The company was rated a BB credit for its first offering, which was a couple notches below investment grade. That was the company's first access to the public debt markets. After that, the public market became the company's primary source of longer-term debt, and private placements were phased out. A number of public debt issues were subsequently done, up to and including the current date, and bank debt became a smaller part of the company's total financing picture, which is customary in most large, publicly held companies. Normally a company borrows under its bank revolving lines of credit and then once those borrowings accumulate, they are paid down with proceeds from either public debt financings and/or equity offerings. The company has had only one equity offering since I've been here, which was in May of 1993, I believe, for about $120 million.

Is there any particular advantage in public debt? Does it cost less, usually?

Generally the covenants and terms are more flexible than they are in private placements. The rates, I would say, generally tend to be somewhat competitive with private placements, depending on market conditions at the time of issue. The biggest advantage is the broader market, particularly if you have a good debt rating. And as I said, the terms and covenants generally give a company more flexibility in its ongoing operations.

Was the move to the public market your initiative or was that kind of an agreement among you and others in top management?

Financing conditions were somewhat tight in the late 1980s. The company would have found it very difficult to access the private placement market, and our bank debt had reached a level where it would have been difficult to acquire additional bank debt or get additional bank commitments because most of our existing lenders had already extended the maximum amount of credit that they considered prudent to one individual company. So I determined that public markets were the best alternative source of capital, and that was concurred with by other top management at Mitchell and, of course, the Board of Directors.

Overall, historically, outside investors have looked at the heavy proportion of debt in our capital structure as a negative. In recent months or years, it has come closer to a better balance. Do you have a target?

What's considered prudent or desirable, insofar as the whole matter of debt-to-equity ratios and the absolute dollar amount of debt is concerned, is obviously in the eye of the beholder. It gets down to a matter of how you want to manage the company. Clearly, the company over time has tended to bring its debt-equity ratio more in line with that of other companies. There has not necessarily been a specific target amount. The company in recent years—in the mid '90s, '94, '95—got bumped up to a low investment grade debt rating by both Moody's and Standard & Poor's, which is a desirable goal for a publicly held company to attain, although it was at the minimum investment grade level. That clearly represented a major improvement in how the company's debt was viewed by the financial community. Prior to the recent North Texas water well litigation, the company had the potential to have been upgraded further. But that's presently on hold because of the uncertainty created by the water well litigation.

I remember that when I was employed by the company, debt was pretty consistently around the billion dollar level. It's been reduced a pretty good chunk from that level, is that correct?

Current debt level, I think, is a little over $800 million, with the prospect of being reduced further this year and the following year. I don't think it ever quite hit a billion dollars, but it was in the $900-plus million range. The absolute dollar amount of debt is less significant than the amount of earnings that you have to cover interest and service the debt, and the amount of equity that the company has, and the assets that it has to support debt. So, the dollar amount of the debt is not what's so significant, but it's the relationship of that debt to the company's earnings, cash flows and asset values that really governs what a prudent debt level amount is.

Maybe you've already answered this, but is the company tending toward an improved debt-to-equity ratio?

Yes, there has been steady progress during my 10 years at Mitchell to having an improved credit rating, and our credit ratios have generally shown an upward trend over time, realizing that nothing moves up in a straight line. There have been zigs and zags along the way, but the general overall trend has been improving.

At your retirement party, when somebody gets up and makes a speech about what a great guy you are, is that one of the accomplishments you'd like to have them talk about?

I think that would be a worthwhile thing to comment on.

You mentioned the debt offerings. And there was an equity offering. Tell the background of that, please.

The company, as you know, had only made one offering of common stock in the history of the company, and that was back in the early 1970s when it was necessary to issue equity in order to get a $50 million HUD loan guarantee to open The Woodlands. Since that initial offering, George had been very protective of his ownership percentage and did not want to have his ownership diluted and, therefore, was not inclined to issue additional equity in the company. That's why the philosophy has always been to grow the company through internal cash flow and debt. As part of a program to improve the company's financial position and increase its financial stability, credit standing and overall financial position, the company converted to two different classes of stock—a Class A stock, which had all the voting rights in the company, and a Class B stock, which was identical in all respects to Class A stock except that it did not have voting rights and it paid a slightly higher dividend than the Class A stock. The purpose of going to these two classes of stock was to permit the company to issue additional equity to the public in the form of Class B stock which would not dilute George Mitchell's voting percentage. Also, it gave the company the additional flexibility of having Class B shares that could be used to acquire other companies, that is, exchange that stock for other companies without affecting George Mitchell's control of the voting stock. The company to date has not done that, but that capability now exists. That is the background to the company's Class B common stock offering in May of 1993. And I believe that the two classes of shares were put in about a year before that. Those shares were created to give the company additional financial flexibility in the future.

But the offering of the B shares also raised some equity money that was very handy, I assume.
I think the proceeds from selling the class B shares were about $120 million. You can get the exact amount from the prospectus.

Although up till now use of the Class B shares in an acquisition has not happened, there's a reasonable prospect?
That option is now available to the company. George Mitchell would have been very reluctant to have issued voting shares before because it would have diluted his voting control. George still doesn't like the idea of issuing equity because he doesn't want to dilute his ownership, so there are really two concerns—voting control and ownership. Going to the Class A and B share structure effectively solved the voting control issue but still didn't do away with George's concern about diluting his ownership percentage.

My impression is that not too much earlier, the New York Stock Exchange would not have gone along with two classes of shares. Is that correct?

The question of voting and non-voting shares has been a point of controversy with both the New York and American Stock Exchanges. The New York Exchange's willingness to list nonvoting shares is a fairly recent development. They have certain criteria that had to be met in order to have those shares listed on the New York Stock Exchange. I forget exactly when those new rules were put into effect, but it is a relatively new development.

One of the questions that's been asked over 20 years concerns the separation of the real estate and the energy segments of the company into separate entities. This kind of thing has been talked about both inside the company and outside the company. Do you care to say what would be required before anything like that happened?

Analysts and others in the investment community have long held the belief that the parts would be worth more than the sum of the total. That is, two separate stocks–an energy stock and a real estate stock–would have more value than the current consolidated MEDC shares. Investors today tend to look for pure plays. They like to invest in an energy company or a real estate company but not a hybrid type of company, and so, as a result, many of our investors have expressed the thought that it would be in the shareholders' best interests to separate the two operations. The biggest issue in doing that, aside from George's willingness to do it or not do it, has been the manner in which the company has been capitalized. That is, the company has been primarily financed with debt as opposed to equity, and in order to separate the energy and real estate companies it would be necessary to recapitalize the real estate company so that it would have a much lower level of debt than what it currently has. Real estate has been able to carry that higher debt level because of a guarantee by the parent company, Mitchell Energy & Development Corp., which is indirectly backed up by the energy operations of the company. While the company's real estate operations have been able to support this level of debt as a part of MEDC, the financial markets would not support that level of debt if real estate were a stand-alone company because of the illiquid nature of real estate assets as opposed to oil and gas properties which are perceived to have a more assured income stream. There's a concern that the cash flows from the real estate land development operations are not as predictable as oil and gas revenues. So what would be required to accomplish a separation of the businesses would be a lower debt level on the real estate side. That could be accomplished by a portion of real estate's debt being assumed by the energy operations, but that would present problems because it would put an undue burden on the energy side. Another way to get the debt down would be to sell non-earning assets of the company, which the company did during fiscal '95 and '96. Other alternatives would be to sell additional equity in the company, either through an offering of MEDC common stock or by other means. Probably some form of additional equity would be needed if a separation were to occur in the near term.

You've had a big role, over the years, in moving the debt from the private market to the public market, preparing for an equity offering and creating two classes of shares. Is there anything else that you're especially proud of?

I would say that, hopefully, during my period of time here, the company's made a significant upgrading in the quality of its financial accounting and reporting procedures. Better data are produced, not only from a financial standpoint, which helps access the financial markets, but also from an operational standpoint, which facilitates the making of better operating decisions within the company.

The biggest cloud at present, of course, is the North Texas lawsuit. Presumably that's going to be handled. Do you see any great problems or opportunities on the horizon? Where is the company going to go in the next five or ten years?

Obviously, the whole industry—when I say the industry I'm thinking primarily of the oil and gas industry—has been and continues to some extent to undergo a period of adjusting to lower energy prices, although there are peaks and valleys along the way. Over the past two years, the company has undertaken a number of restructuring efforts. These have involved a number of sales of primarily non-earning assets that were held for long-term investment or development; a buyout of the company's long-standing contract with Natural Gas Pipeline Company of America which resulted in significant cash proceeds being received and an acceleration of the company's transition to a market-sensitive pricing environment for most of its natural gas production; and a lowering of personnel levels within the company as part of the transition to a market-sensitive pricing environment and lowering of cost. Undoubtedly this will, to some extent, continue to be a way of life within the industry. It's essential that companies in the oil and gas business get their costs down to an appropriate level. We've done this over the past two years, and it should put the company in a position where it will now be able to take better advantage of future opportunities which, I think, George has pretty well defined as being niche opportunities in the onshore

United States, as opposed to exploring opportunities overseas or more expensive exploration opportunities in the offshore Gulf of Mexico waters. Operations have been refocused, primarily in Texas and, to some extent, New Mexico and perhaps the surrounding states of Louisiana and Oklahoma. That will probably continue to be the case for the next several years, depending on the ultimate direction in energy prices.

It sounds like the next several years will be years of consolidation and cleaning up so that we, after these things are in place, can move ahead.

Hopefully, most of that has been done in fiscal '95 and fiscal '96, or at least the decisions have been made as to what will be done. Some of the execution in terms of some asset sales and other things has yet to occur, but the transition has been made, so I would say that will be a way of life as long as energy prices remain at their current level. But, hopefully, the majority of that is behind us and we can look more to future opportunities in the current fiscal year and future years, particularly once the major water well litigation is behind us.

What have I forgotten to ask you about?

I guess just an observation, which I'm sure you have already made, that Mitchell is one of the few remaining independent oil and gas companies that is still being run by its founder and one of the few remaining wildcatters. While there are still a number of independent oil and gas companies, for the most part they are fairly widely held in the public markets and have grown through access to the public equity markets, mergers, etc. So, the manner in which Mitchell operates with the primary shareholder being George Mitchell is somewhat unique in today's oil and gas energy environment. George represents one of the last of the breed of wildcatters who did well and prospered during the 1970s and '80s.

Thank you, Phil.

Coulson Tough

Today is August 1, 1995, and this is Joe Kutchin. I'm talking to Coulson Tough, senior vice president of design and construction for The Woodlands Corporation, in his office in The Woodlands. First, Coulson, please tell me something about your background.

I went to the University of Michigan and got a degree in architecture in 1951. After that, I worked in Detroit for about a year, got married and then we moved to California in 1952.

What did you do in Detroit?

I worked for a company called The Austin Company, a design/build company. I designed industrial-type buildings. In California, I worked as a draftsman and job captain for two architectural firms in Los Angeles.

You would have been how old then?

I was probably about 25. Then in '55, I took a job with UCLA's campus planning office as a project architect. There at UCLA I handled a large number of projects for the university for six years and then got a promotion to establish a new campus at Irvine, in Orange County, California. I was the second person there and was responsible for planning for the new university campus. I coordinated with all the consultants we employed to try to get the university opened, and we did open in 1965. I was there six years, and then I was recruited by the University of Houston to become vice president for facilities planning and operations and left the University of California at Irvine. In 1968, I came to the University of Houston. I was at the University of Houston for five years and had responsibilities for pretty much all the physical operations at the university, including real estate, building design, construction, planning, policy, etc. The U of H was charged with establishing two new campuses for the university. One was south of Houston, which became the Clear Lake campus. I negotiated for the land with the Friendswood Company, acquired the land, did the master planning for the campus and selected architects for the design of the initial buildings. After we got that going, I started looking to the north of Houston for land for a campus, and somebody told me about George Mitchell, who was trying to develop a new community and suggested we talk with him. So Dr. Phillip Hoffman, president of the University of Houston, who was my boss, and I began discussions with George Mitchell to determine if he would be willing to donate about

400 acres of land for the university.

So the university took the initiative, right?

Right, right, because the State–it's hard to believe these days–but the State Coordinating Board for Higher Education stated that we should establish these campuses. In subsequent years, the State took a position contrary to that. We began discussions with George Mitchell and his staff, which was very small at that time. They were officed at One Shell Plaza. I had a number of meetings with Bob Hartsfield and Jim McAlister. I didn't get into too many discussions with Len Ivins or Sam Calleri at that time, Plato Pappas a little and then George Mitchell.

Was the group from Columbia here?

Yes. Len Ivins, Sam Calleri, Bob Grace, Bob McGee and others were there. They started in '72 with Mitchell, and I started in May of '73. After further discussions we had the U of H Board of Regents visit The Woodlands to look at the land and potential sites.

What were the issues in the discussion?

The issues were the location of the land, was it well situated, would it drain well? We agreed on a site that was on Lake Woodlands. Of course, Lake Woodlands wasn't built yet. It was a beautiful site, 400 acres. But it was contingent on approval by the State Coordinating Board for Higher Education. And in later years, contrary to their position, they decided that we should not be establishing new campuses in the State of Texas.

So they went 180 degrees?

They went 180 degrees. Many of the campuses that were established in the state were done for political reasons more than real educational needs and population growth. I think the Coordinating Board had some justification in taking that position, but it was unfortunate for Montgomery County because I think we could justify a campus here.

Let's go back—when you were at the university, you were working with Hoffman, and I take it that the people at the company were more than receptive to the university's initiative.

Yes. They were very receptive. Mr. Mitchell came to the U of H and made a presentation to the Board of Regents at the university. Jim McAlister was at the meeting. They made a fine presentation. Mr. Mitchell was very anxious to get the university in The Woodlands due to the effect the campus would have on the quality of the community's development.

Had the university done any particular studies?
Yes, there were some studies on population growth and enrollment trends, but a lot of that had really been done by the State before and that's why they took the position that there should be these campuses.

OK, talk about the circumstances under which you left the university and came here.
I had been negotiating for the university, dealing with people with The Woodlands Development Corporation. Discussions had gone well. I had an opportunity to talk to Mr. Mitchell a few times. I was very impressed with him, and he indicated that if I was ever to leave the University of Houston, to give him a call. So about three years later, I did that, and as a consequence, I left the university and joined the Woodlands Development Corporation in May 1973.

In what capacity did you come to the company?
I came as vice president for community development.

You reported to whom?
I'm not quite sure if I reported to Len Ivins or Bob Grace. Or maybe both. It was a very awkward time. The development of The Woodlands was going at a frantic pace. When I arrived, a lot of construction was already under way, but I hadn't been a part of the initial planning and it was kind of awkward to try to catch up.

What were the major projects?
I began by reviewing the master plan for The Woodlands and then some of the other plans for development, particularly the Executive Conference Center. I remember when we had meetings on the interior design of the Executive Conference Center and all of the furniture. Steve Harrison, who was the person in charge of Terradevco, which had the contract with Mitchell to develop the conference center, was proposing that his wife do the interiors, and we had a presentation by them to Mr. Mitchell and myself. Mr. Mitchell asked me to get somebody else involved after that.

So we're talking now 1973, 1974?

Yes, that was early 1974, because The Woodlands conference center opened up in October 1974. But things were very awkward for me because I realized after I joined the company that Len Ivins and some of the others there evidently weren't too happy with my coming on board. At least, that's how it appeared to me because I believe I was just ignored in a lot of things and cut out, which in the end wasn't all that bad anyway because I wasn't a part of that clique. I mean they had all these big parties going and spending a lot of money wining and dining and I was never a part of that scene.

That was essentially the entire group imported from the Columbia new town in Maryland?

Yes, pretty much. Then they had brought in a lot of other people, probably from the same area. I can think of some of them. And there were a few local people, and when I say "local people," I mean those who probably came from the Houston area like myself. Some could fit in and some others could not. So anyway, that went on until we opened in 1974, and shortly after opening there was kind of a hiatus, and the next thing we knew, Len Ivins, who was president of the company, was leaving. That was about December 1974.

Were you a party to any of that?

No, I wasn't a party to any of it. I hardly knew what was going on. Corporate was calling people in to meet with them and terminating their employment. There were a number of people who were terminated shortly after we had opened in October of 1974. There was a lot of shuffling going on. I hardly knew what my position was. I had people transferred to my department who had been in the building development part of it before, and then I was told later to lay them off. It was a very difficult time.

Do you know the circumstances that went into all this shuffling?

The Company had experienced very difficult financial conditions due to cost overruns in the building construction with the Wharf, the conference center.

Can you put some dollar numbers on any of this?

I can't, but I know that the overrun in the Wharf and the conference center was many millions, because plans were being made and changes were being made in construction. I know and I've heard all kinds of stories where something would be built and the next week it would be torn down and rebuilt for something else, and a lot of emotional decision-making was going on at the time. As a result of that, the company had spent a lot of money, had a lot of overruns and got into pretty serious financial condition. I would imagine that George Mitchell was told that the people that he had were not the ones he should continue with. Decisions then were made by him or others to cut back on the staff. So they started cutting back, and they let Len Ivins go and Bob Grace left, Sam Calleri stayed for a while. That was in 1974. In 1975, Leland Carter, the president of Mitchell Energy, came in to run The Woodlands. He had me doing many different things. I felt like I was going to be a utility outfielder at the time.

During this time the U of H campus was on the back burner?
It was on the back burner, but Mr. Mitchell continually tried to push it forward by arranging meetings with members of the Coordinating Board, particularly Commissioner Ashworth. I remember meeting him myself and touring him around The Woodlands. Dr. Hoffman made a presentation to the State Coordinating Board requesting approval of the campus. It was not approved due to objections by State Representative Jimmie Edwards and others. There was a lot of politics involved. I think Hoffman really got caught unaware of the politics and the seriousness of the situation.

The U of H Board supported . . .?
Yes, the U of H Board supported the new campus request.

You were not involved in the political considerations and the backing and forwarding on that?
I was very much involved, doing things under Mr. Mitchell's direction in terms of meeting with people and helping to coordinate things, but no, I wasn't dealing with the people in Austin. I know that Mr. Mitchell made a big effort and we then entered into a series of agreements to give the university more time to try to get the campus established and we had agreements that called for performance or completion by a certain date because planning of The Woodlands was moving along. We had to actually move the site maybe once or twice.

OK, to simplify, the university had proposed this over time, the Mitchell organization said yes, we endorse it, we'll provide the land, so there was never any doubt about the 400 acres being available?
No, not at all. The University of Houston and the Mitchell organization

were all for it. They really wanted to get it done. That was about 1975. Development resumed towards the middle of the year, and Mr. Mitchell talked to me about building two more buildings like this office building, Office Building I. They became Office Buildings III and IV, which are right across the street on Timberloch Place. So that was really the beginning of the first part of our building program after opening, and I think those opened in 1976. I've got this list of completed buildings attached. We did those buildings in 1976, and then we started on a few other projects like the tennis clubhouse at the country club. We built that clubhouse building, expanded the tennis courts and then continued Grogan's Landing Apartments, did those in 1976; the Exxon Conference Center, did that in 1977. It just kept going from about that time period.

Would you be more specific when you say you had projects? What was your responsibility?
I was the vice president in charge of the building development department. At one time we had about 14 people on the staff. And we were responsible for doing the coordination of planning, design and construction for all the buildings, seeing them through from concept through completion of construction and turning them over to the property management department. Later, we resolved problems that occurred after construction for the next year or two. As I look at this list, it's clear we had a good, steady program of building development that went from 1976.

There were some that must have been standouts. Talk more about those.
One of the more interesting ones was the Exxon Conference Center. We worked with the Exxon Chemical people and constructed the conference center on Woodlands Executive Conference Center land. It was a turnkey job for them. It turned out to be very successful.

What was interesting about it?
It was very unique. It was the first time we had done a small conference center for a company that was going to use that conference center exclusively for their own use and training.

When was that done?
That was done in '77-'78. There were two phases to it.

And you didn't design it?

No, I had hired two different architects to do the design, but we managed the building development process from beginning to end to get it done, awarding the contracts, seeing it through construction. We had everything done so all they had to do was turn the key and walk in and everything was there, completely furnished. All the interior design was done and the kitchen was ready and operational, so it was a pretty interesting project, not a real large project but interesting. It was a high-quality project, too. Another interesting building was the MND Building, which was opened in 1980. I remember that when we started on that, we had earlier started on designing a building out by what is now Lake Robbins. But it was so far out at that time in terms of infrastructure—roads and utilities—it was decided to drop that site. And the thinking was that we should then build an office building that could be leased out to others later, and the Mitchell company would then build its corporate office building over by Lake Woodlands.

The original design on the Lake Robbins site was considerably larger.

Considerably larger. It was a corporate office building. We dropped that idea and went with the idea of doing a building that would be almost like a speculative office building, although you wouldn't believe it when you see it. The Mitchell Energy people wanted a building that was about 150,000 square feet and they wanted to have large floor areas, so we came up with that concept of creating the two wings of the building that split it up with that connecting atrium, and that gave them the floor area they wanted but broke the building down into areas that we could lease in the future. The thinking was that in five or eight years the Mitchell company would build a corporate office building and the MND Building would then become a lease building.

Who designed that?

James Sink was the architect.

He's done a lot of work for the company?

Not a lot. He designed the MND building and the Millside Office Building. Another interesting project was the Bank One Building because we had to get the bank established in phases. We built the parking lot first and moved a portable building onto the site so the bank could get established. We worked very closely with them and got that going so they could get into operation. We started construction of the building, got it built, moved them into the building. It worked pretty well.

When was that?

That was 1980 also. We had a lot of work in 1980, nine buildings that I can count. We were really busy. That was, of course, before the recession which started about 1982, came back up again a little bit, but then about '83, '84 it really went down, and if you look at this list, you can see that there wasn't much done in '83 and '84 or '85 and then construction starts picking up again in '86, '87, '88. But about other interesting buildings . . . there are so many of them. The TPC Clubhouse building, not a large building but an interesting one because of the character of the building, being on the TPC Golf Course. Also, a very successful design for the building use.

What about the character of the building?

I think the character of the building worked well. We constructed it in two stages. Because they wanted to have a golf cart storage facility first, we built a golf cart storage building with a flat roof. They used a temporary building for the starter shack. I ramped up the parking lot to make it level with the roof of the golf cart storage building. A year or two later we built the clubhouse on top of the golf cart storage building. It worked out successfully. Moving on in years, we got involved in what is now called The Woodlands Memorial Hospital. We worked closely with the planning department preparing a master plan for the hospital. This became a big issue when another group wanted to establish a hospital adjacent to the Woodlands. Ed Lee (president of The Woodlands Corporation) and others wanted the hospital within The Woodlands, which created a thrust for getting that hospital created sooner than planned. Recently, we sold some additional land to them in The Woodlands to create a campus. In addition to projects for The Woodlands Corporation, I was involved in project management for others. One significant project was the Houston Advanced Research Center. Mr. Mitchell had decided to donate 100 acres to establish that center, and my department became very involved in that right from the beginning. We worked with Skip Porter and Dan Davis on the master plan. We hired 3D International to do the master planning, but I supervised that process to get the master plan done and then to locate the site for the first building.

Master plan does not necessarily mean architecture does it?

No, it is also called physical planning.

But as it worked out here the company did both?
Yes, that's right. The master plan located the roads, the buildings in the future, parking lots, etc. So we did that and tried to create a concept for a small campus.

Is there anything extraordinary about that, that you feel is a highlight?
Just the fact that we were working on a campus, which would be like a small university campus, was fascinating to me. I liked that. Of course, I was working with Mr. Mitchell at the time he was thinking of establishing a high technology center. This was before the concept of HARC began.

Talk in some detail about that.
The high technology center idea came to Mr. Mitchell, I believe, after he read an article in Fortune Magazine, and I'm going to guess that must have been in the early '80s. There was an interesting article on high technology in the United States. A Dr. Ternam of Stanford University was instrumental in getting the high technology center started at Stanford because some of his students in the engineering school there after World War II became successful and wanted to get land to begin their business. Two of them were Packard and Hewlett. There were other students, too, who came to the university trying to acquire land. As a consequence, they developed a high technology center. This was happening also in Boston and other areas in the country. Mr. Mitchell was interested in creating such a center in The Woodlands.

Was that a precursor of the Research Triangle?
The Research Triangle began earlier in North Carolina, I can't remember when. It was slow in getting started. It was another example of the kind of centers that we were going to try to emulate. Route 128 in the Boston area became a corridor for that kind of development. The Research Triangle near the University of North Carolina and North Carolina State and Duke University became another. We had many meetings. Mr. Mitchell approved of hiring the Arthur D. Little Company to work with us in developing recommendations for establishing an organization. That's what led to establishing HARC.

You were one of those who visited Ternam?
Yes, with David Gottlieb–he was with the University of Houston and we

were working with the U of H to try to get them involved here. After we met there, we invited Dr. Ternam to come here and meet with Mr. Mitchell and others and talk further about how the technology campus started at Stanford and what we might do here to get one established. But as I said, that led to employing Arthur D. Little. And then we also hired Arthur D. Little to work on the idea of establishing a Medical Center at The Woodlands. That didn't work as well. We had many meetings with representatives of the Texas Medical Center. The Baylor College of Medicine people were the most active, but the concept didn't pull together as well as the HARC program.

Arthur D. Little in about '82, '83, you think?
Yes, I'd say yes, that's about when.

Let's come back. I want to talk a little more about the growth program which to some extent was a precursor of HARC.
Yes, it was.

And I know you were involved in that as was I.
Right.

At the time of the dealings with Stanford and the planning, was Harvey McMains here as head of HARC, do you remember?
McMains came here in the early days of setting up HARC. We had office facilities in a building on Grogan's Mill Road, which was originally the Family Health Center for The Woodlands. When the hospital was established, that building no longer served that purpose. The University of Houston extension program was conducted in the same building. McMains and his staff worked there for several years.

I was just trying to get straight in my mind whether Harvey came after your dealings with Ternam of Stanford.
Yes, he came after that. And around the same time—this must have been around '75 or '76—this book came out by Dennis Meadows, *Limits to Growth*.

Yes, that was actually in the '60s, I think.
Mr. Mitchell was concerned about the limits to growth proposition. You and I worked with Mr. Mitchell to establish a series of conferences dealing with the issues of the limits to growth.

Before getting into that, let's get back to major projects that you were involved in.
The Pavilion should be one, along with HARC and the John Cooper School. I took over the design and the planning of the original school building.

The Cooper School predates the Pavilion, if I remember correctly.
Phase I of the Cooper School was in 1987.

Talk about the circumstances back then.
The planning for the Cooper School had been in discussion stages for some time, and their problem, of course, was trying to raise money.

When you say "their problem," who is they?
This was a small group that consisted, I think, of John Cooper. I think Marina Ballantyne was involved at that time and others who wanted to establish a private school.

Did the initiative actually come out of the company?
I don't know if it came out of the company. It may have come from Mr. Mitchell. Planning started, then it was stopped, some drawings were done and shelved for some time period. I was asked to take it over and try to improve the design and manage the design-construction process.

When was this?
This must have been around '86, '87.

So the matter of the rationale for creating the Cooper School was really something that happened before you got involved?
Right.

I'd had the impression, and I may be wrong, that there was a feeling among top management of The Woodlands Corporation that a school comparable in quality to St. John's or Kinkaid would be necessary as The Woodlands developed.
Right. The thrust was to try to provide a private school here that would have the reputation of a Kinkaid and others, that would appeal to residents who were moving into The Woodlands who were looking for that type of school, one that would give more personal attention to the students.

Please talk some more about the Cynthia Woods Mitchell Pavilion and the planning for that.
That was a project that (there are different versions) got started in some discussions between Richard Browne and George Mitchell or Roger Galatas, probably in the late '80s. The story I heard was that Browne had tried to get an amphitheater started similar to what you see now at the 18th hole of the TPC golf course where the spectators stand on the slope and watch the players come in. Browne supposedly said, "We can create an outdoor amphitheater here for a million dollars." Well, that outdoor amphitheater became $7 million due to many changes in the design program what with putting a roof over it and fixed seats, etc.

I heard from Phil Hoffman that the idea came in a conversation which involved him, Ben Woodson and a third person, but I forget who that was.
Yes, I think you're right. I don't know which came first. Phil Hoffman and others suggested that The Woodlands be the summer home of the Houston Symphony. I remember being in one of those discussions.

The third person, I think, was a banker, but I forget which one.
Yes, it was John Cater. I remember their coming out and meeting with Mitchell. That may have been the start of the Pavilion idea, and Mitchell may have talked to Browne to try to come up with a site. The story I get is that Browne talked Galatas into doing something like that, but I'm not quite sure where the direction came from. Browne got under way in trying to plan it and had a large number of consultants involved, and it wasn't moving very well. Roger Galatas asked me to take hold of the thing and get things straight. He said, "We'll never get it built the way we're going." So I kind of took charge and terminated the architects and engineers and got some local people going.

When?
We opened in '90, and this is the fifth year, so that must have been just about '89, because we were on a crash schedule.

You say you terminated the architects and engineers.
Yes, they had engineers from New York City and Kansas. The engineer in New York City was Horst Berger, who we retained to do the fabric roof. They had Ellerbee-Beckett, an architect from St. Louis, and they had some

civil engineers from Houston and I don't know who else. I started working with the Pace Concert people, particularly Rodney Eckerman. So I worked closely with him to try to get their recommendations on what all we should do. It was difficult because we had never done a pavilion before. We hadn't had the opportunity to do much advance studies as to what we should be doing. Later on we visited a couple, but by that time it was too late to make big changes. We were being advised by Rodney, who had the viewpoint of the commercial productions that Pace would be booking, and he directed us towards that kind of a design. At the same time, Mr. Mitchell was getting input from others and saying we should construct a ballet stage, ballet floor, and have an orchestra pit. We had a tight construction schedule and a lot of rain. I remember buying a pair of rubber boots just to walk into the meetings with the contractor. We were building an orchestra pit and we had to tear down a concrete wall and make a bigger pit, and we did the stage floor two or three times. But we got it done and got it opened in, I think, April of 1990.

That's just about right. I remember that one of the bad bits of advice, and I'm not really quite sure of the source, was that we did not need an orchestra shell. As it turned out, there's no way that the Houston Symphony would have come out here without it.
Yes, there was a lot of bad advice. Rodney also said we didn't need a loading dock. I thought we ought to have a loading dock. I couldn't imagine how we would operate otherwise. So we did do a loading dock. We had to expand it later because it wasn't big enough, but he didn't think we needed it. But there again, their direction was solely from commercial ventures and programming. Looking back at all that experience, with all the advice we were getting, we could have done it much better if we'd had more opportunity to study what was being done elsewhere, and if I knew how much Mr. Mitchell was really going to want that facility and continue to put money into it to improve it, we would have done things differently. We got stuck with a low budget and, unfortunately, like a lot of these things, you start with a poor budget concept, but the program keeps expanding, and then you start doing things which are cheap to try to get it all done within the program. In the end, we did a fairly good job, but looking back at it, we could have done a better job if we'd had a more realistic budget and a better understanding of what the program was going to be. Also, Mr. Mitchell's real interest didn't become real clear to us until much later.

I think in fairness he didn't realize his interest until he saw how successful it was and how much of a magnet it turned out to be for the whole community.

From the design standpoint, constructing the hill was a major undertaking. I can remember all the trouble we had trying to do that because we had to excavate the earth elsewhere and deposit it.

That was Plato Pappas's job, wasn't it?

Yes, we assigned that to Plato because they were building streets elsewhere in The Woodlands and digging ditches so they would bring the earth from those sources. But, of course, we were controlling it from the design standpoint to get the height and the site lines and the shape and everything the way we wanted it. It was a big job because we had to control the compactions so we wouldn't have failures in the earth fill. A similar hill had been built in Dallas, and after a big rain a large portion of the hill slid away. We've never had that problem. I think we did a good job.

About Horst Berger?

He was the design engineer for the fabric roof. He had done some in Saudi Arabia and elsewhere. He's very knowledgeable, but a little difficult to work with. We got the job done.

What about some other projects, some other problems?

The restraints, the schedules, the budgets and the things we went through in the early years, like doing the building for Superior Oil, were difficult. I remember that because the biggest thing that happened to us in those days was the Superior Oil Building. And I think we built the shell of that building for about $25 per foot. Our leasing people were dealing with Superior and, I mean, it was down to negotiating nickels and dimes. They were continually giving things away to make the deal but my budget wasn't increased. They added another elevator and marble with the same budget.

Is $25 per foot a lot or a little?

It's a little. The Parkwood II office building, which was built a year ago, was $47 per foot.

But there's a difference in years.

I know, but still it was a cheap building. We did learn a lesson that it doesn't pay to do poor quality buildings, or cheap buildings. However, I remember how desperate we were to try to make that deal because it was a big thing for us to try to get a building that size–80,000 square feet office building, with a single user like Superior Oil. That was a real coup for us.

Let's finish off the Pavilion with some information about the expansion. It went through four years of increasing success and a fair number of sellouts each season. The decision was made to increase the capacity by 3,000 with contributions coming from various places. What is it that you dealt with?

When we were advised they wanted to expand the Pavilion, we had to decide how to do it. We did several planning studies. I decided to employ Morris Architects who have done similar work. We worked with them and with the Pavilion staff on determining how we could expand the fixed seating. The additional fixed seating would be outside the fabric roof cover. We added about 1,900 fixed seats and we added about 1,500 plus-or-minus seating spaces on the grass hill. The grass hill was expanded, and we built structures to be kind of like bookends to the hill to provide space for more concession stands, points of sale, expanded toilet rooms. We expanded the north plaza. I wanted to create a big entry drop-off at Lake Robbins Drive, which I think worked out well for buses and individual automobiles to drop people off there. We created a ticket window there and improved circulation and space, particularly in the north plaza for the Pavilion. It's been a successful improvement.

OK, and the "we" is essentially your department?

Right.

And I also know that through the years, largely at George Mitchell's initiative, there were constant improvements.

Yes, even a year after we opened. We opened probably in April of '90 and by October of '90, Mr. Mitchell, primarily, and probably Kirk Metzger, who was the first executive director, and you were looking at needed improvements. One of the things that had to be done was to provide air conditioning to the stage area because the Houston Symphony required that as a part of their agreement to perform at the Pavilion.

It was air conditioned when we opened.

There may have been air conditioning, but it wasn't adequate, so we had to improve that and the distribution system because they had specific temperatures that needed to be maintained.

And that was especially difficult because of the presence of the shell that had to be there.
Right. We had problems getting the shell in. Later we added fans over the audience seating area. It worked out pretty well. Anyway, we made a number of improvements with each year. We would start in about November and try to finish by April 1 of the next year, which is always a bad time to build because everything was outdoors and it's always a wet time of the year, but we made significant improvements each year. This last one was a major improvement. We spent about $3,200,000 on the expansions I discussed earlier.

You said $7 million on the original, $3.2 million on the addition and then there must be another million or two . . .
Probably, because I think Mr. Mitchell had approved budgets of maybe $180,000, $200,000 a year.

Actually some of them came in higher than that. I remember the shell alone, which came from a grant, not from Mr. Mitchell, was to the order of $125,000 or $175,000. So when you've got improvements like that over the course of time, several more million must have been added. And I would say $10 million for the original is pretty conservative.
Yes, I agree. I think it's a much better place than it was in the beginning. I wanted to get away from the wood construction and get into better, more permanent materials that look better and are easier to maintain.

* * * * * * *

Today is August 3, 1995, and this is Joe Kutchin in Coulson Tough's office for a second interview. Coulson, talk about the beginnings of the growth conferences as you experienced them.
As I recall, Mr. Mitchell began some discussions regarding the book written by Dennis Meadows and his wife called *Limits to Growth*. Mitchell was very impressed with the book.

This would have been '73, '74? Basically, I think you and I represented the company. George had already made an arrangement with the University of Houston by then, hadn't he? Do you remember the

development of that?
I remember that you and I met at the University of Houston to begin discussions with Allen Commander. I think that was in connection with the first conference.

Roger Singleton was the university man on the second conference.
I can remember meetings in Allen Commander's office at the University of Houston. I think we then decided we wanted to have someone to manage the conference in addition to the U of H and Allen didn't like that, but we brought in John Naisbitt, who later wrote some extremely successful books about economic and social trends. I think Dennis Meadows recommended Naisbitt to us.

Do you remember how Dennis first got involved?
I think Dennis got involved because Mr. Mitchell had him down here and we had some discussions, and the idea of the conference grew out of the discussions. Mr. Mitchell decided that he would put on a conference and have a prize. And I believe the prize idea came from Dennis Meadows. A prize for papers that would be delivered at the conference. There was a lot of work, we put in a lot of hours, lots of meetings, both with George Mitchell and Dennis and the University of Houston and all the people we had to contact. Dennis felt that he couldn't run the conference and we couldn't run it. I don't think he had faith in having the University of Houston run it. Anyway, Naisbitt was recommended. I remember after we got into discussions, Naisbitt made predictions regarding the attendance which we couldn't believe. He had a curve factor that was really impressive because I can remember checking with him myself after getting nervous about the numbers because of the response we had at that time. It turned out that he was pretty close to the projection as to how many people were going to attend. It was amazingly close for the number of people based on the curve that he had developed for conferences. Naisbitt was a little difficult to deal with but not terribly so, could have been worse.

Certainly Allen Commander was smoother and equally difficult by the time everything was done.
Yes. I don't remember Allen Commander contributing a lot to the conference. He was the university representative for a while, but as the conference unfolded he seemed to fade out of it. I don't remember him being involved. I remember you and I working real hard.

Yes, and we had our PR agency considerably involved in publicizing it.
Naisbitt's people did a good job. The first conference took place in '75, held at The Woodlands Inn, very successful, good attendance.

Three hundred-plus. And at the time there was very little in the way of accommodations for guests in the area, so we had some people staying 20 miles away and busing them. The people on the program were really all-stars. There are copies of the program listing the participants and the subjects. We can get that. Do you have the feeling that the conference shaped any aspect of the future of the company? I feel, for instance, that it was one of the clear predecessors of HARC.
That became a large program element in what is now HARC. I don't know if that contributed to the establishment of HARC. I recall clearly that Mr. Mitchell was very impressed with this high technology center concept and was particularly focused on the Research Triangle in North Carolina, the corridor in Boston and Stanford University. That was coming from a research and development type of approach, rather than from the limits to growth approach. But that might have become blended in his mind later as the conference became successful. He may have felt that a program devoted to global problems that we were dealing with in these conferences could be handled by HARC, if HARC was established. But I don't think that became clear right from the beginning.

My view is a little different. As HARC was being established, the growth conference first seemed to me to be a natural element of what HARC was to be dealing with. Also when HARC got started, there was nothing programmatic really in existence, so growth issues gave them something ready-made.
I think it was a matter of timing. I always talk about the high technology center as the original concept for getting that going. But we had been through conference Numbers 1, 2 and 3 before HARC really got going and then they started to move in. About that time I dropped out. They had a staff and started taking over the coordination part.

I remember David Gottlieb was representing the university.
We worked with Allen Commander, then with Roger Singleton. He was involved in that meeting we had in Paris, which was a follow-up. Then Roger dropped out of it and David Gottlieb became involved in the third confer-

ence. We're going back about 20 years.

I just had the impression that it's a unique element of the way MEDC and George Mitchell look at the way business should exist.
Mr. Mitchell always felt that more businesses in the country should get involved in the concerns expressed by these conferences. But I don't know if that was ever successful. He got a few companies involved, mostly those who were friends of his before anyway.

We've talked about the Pavilion, we've talked about HARC. I know you did a lot of work for Mitchell the company, and also Mitchell the man in Galveston.
The work in Galveston has been very interesting historic building restoration work. As I recall, Mr. Mitchell called me on the telephone in '75 or '76 and mentioned historic restoration in Galveston and said I should meet with Peter Brink and that he, Mitchell, wanted to buy a building. I knew where Galveston was, I didn't know who Peter Brink was and I didn't know what building George wanted to buy. He said, "Go look." So I went through that process, met Brink, toured the buildings, selected a building, called Mr. Mitchell up and recommended a building which became known as the T. Jeff League Building on the corner of 23rd and Strand. He said, "OK, we'll buy it," and he bought it. We then employed Ford Powell & Carson. I did not know of them at that time. They were San Antonio architects. But Ford Powell & Carson had done the Mitchells' home in the Riverside area of Houston, which is south of the University of Houston area, prior to their home in the Memorial area. And they were very impressed with O'Neill Ford, who was alive at that time. I remember the first meeting we had with them was in the living room of their home, in Memorial, to meet with Ford Powell & Carson to start talking about the restoration of this building. It was really a significant event.

This was in the Mitchells' own interests, not the company's?
Right, what is now called GPM, Inc. That was the first building of what became a series of buildings. I have a list of buildings, which is not current, but there are 16 of them that the Mitchells own and have restored in Galveston. The cost that I have is current as of December 1993, but at that time the investment for purchase and the cost for remodeling and historic restoration is about $49 million. Incredible program. It may be the largest program in Texas and maybe in the United States for one owner purchasing a group of

buildings in one area and restoring them.

Which of those do you consider the most interesting?
I think the T. Jeff League Building. That was interesting because it was the first one I got involved in and that's now 20 years ago. We opened that up in 1979. It was a difficult project because we were learning things. Even though Ford Powell & Carson had some experience, they didn't have the experience they have today. We were learning as we went along.

Talk about some of the problems that you had.
Structural problems, holding the building up while we were trying to do the restoration. How to deal with the cast iron columns, how to reinforce them and still maintain them, having to go and do color sampling and testing to try to meet the Texas Historical Commission's standards for restoring the buildings back to their original colors and character, learning to deal with the Texas Historical Commission and then also the National Park Service people in the regional office and also in Washington, D.C.

What state was the building in when you started, and what did it finally develop into?
It was deplorable, like most of those buildings. There was an antique store operating on a portion of the first floor. The rest of the floors were abandoned, with debris all over. Some of the buildings we purchased later were even worse than that building. It was at least standing, although it needed a lot of structural reinforcement. We came in and created the Wentletrap Restaurant on the first floor, several small retail shops on the first floor adjacent to the restaurant. On the second floor, we created offices, a large atrium through the center of the building. On the third floor, we created the Top Gallant Room, which is a ballroom, and also a small office area and a small kitchen. It was a very successful restoration. But I think probably the most significant one was the Tremont House, which was an old warehouse-type building.

When does that building date from? Do you remember?
I think the T. Jeff League Building is from about 1873. The Tremont came a little later, and we opened that building about the same time as the San Luis Hotel, which was 1984. But the Tremont House was a much larger building than the T. Jeff League. We took an old warehouse building, where there were automobiles parked on a portion of the bottom floor. When I walked

THE WOODLANDS CORPORATION
COMMERCIAL/INDUSTRIAL DESIGN & CONSTRUCTION
LIST OF COMPLETED BUILDINGS

	Building Name	Address	Year Completed	Architects
	THE WOODLANDS PROJECTS			
	The Woodlands Athletic Center	11111 Winterberry Pl.	1973	Terradevco, Inc.
(1)	**Industrial Building 1**	2319 Timberloch Pl.	1974	Richard Fitzgerald & Associates
(1)	**Information Center**	2120 Buckthorne Pl.	1974	Bennie Gonzales
(1)	**Lakeside Building**	2220 Buckthorne Pl.	1974	Edward D. Stone, Inc.
(1)	**Office Building 1**	2201 Timberloch Pl.	1974	3D International
(1)	**Office Building 2**	2203 Timberloch Pl.	1974	3D International
	The Wharf	2230 Buckthorne Pl.	1974	Edward D. Stone, Inc.
(1)	**The Woodlands Executive Conference Center and Resort (WECCR) Initial Building**	2301 N. Millbend Dr.	1974	Edward D. Stone, Inc.
(1)	**Industrial Building 2**	2407 Timberloch Pl.	1975	Richard Fitzgerald & Associates
	The Woodlands Country Club (WCC) Tennis Clubhouse	2301 N. Millbend Dr.	1976	McKie & Kamrath
	Exxon Conference Center	2450 N. Millbend Dr.	1977	Clovis Heimsath & Associates

Exxon Conference Center Ph. 2	2450 N. Millbend Dr.	1978	Cate & Castillion Associates
Office-Warehouse Buildings	2408 Timberloch Pl.	1978	Burleson & Watson
Office Building 3	2204 Timberloch Pl.	1978	3D International
Office Building 4	2202 Timberloch Pl.	1978	3D International
Grogan's Mill Village Center	7 Switchbud Pl.	1979	Albert C. Martin & Associates
Village Square Apartments	2301 S. Millbend Dr.	1979	Wahlberg Wright Waite & Associates
2002 Timberloch Building	2002 Timberloch Pl.	1980	The Klein Partnership
Bank One Building	1400 Woodloch Forest Dr.	1980	Golemon & Rolfe
MND Building	2001 Timberloch Pl.	1980	James M. Sink Associates
Railmart 1	310 S. Trade Center Pkwy.	1980	W. W. Duson Associates
Railmart 2	410 S. Trade Center Pkwy.	1980	W. W. Duson Associates
Tech Center 1	9391 Grogan's Mill Rd.	1980	Burleson & Watson
Tech Center 2	9393 Grogan's Mill Rd.	1980	Burleson & Watson
Superior Oil Building	10200 Grogan's Mill Rd.	1980	ARENCO, Inc.
The Woodlands Public Safety Building	9951 Grogan's Mill Rd.	1980	W. Irving Phillips Associates
Woodstead Building	1610 Woodstead Ct.	1981	Paul Paul + Madrid
Parkwood Building 1	10077 Grogan's Mill Rd.	1982	Golemon & Rolfe
Research Forest Plaza North	9450 Grogan's Mill Rd.	1982	James Falick/The Klein Partnership
WECCR Halfway House (West Course)	40 Southgate Dr.	1982	Langwith Wilson King Associates
Millside Building	2170 Buckthorne Pl.	1983	James M. Sink Associates
Panther Creek Village Center, Ph. 1	4775 W. Panther Creek Dr.	1983	James A. Bishop & Associates, Inc.

	Pine Circle Office Building	1600 Lake Front Circle	1983	Frank Burleson
	Tournament Players Course (TPC) Clubhouse	1730 S. Millbend Dr.	1984	Langwith Wilson King Associates
	Woodlands Professional Center	1120 Medical Plaza Dr.	1984	Brooks/Collier
	Woodlands Prof. Ctr. Parking Garage	1120 Medical Plaza Dr.	1984	Brooks/Collier
(2)	**Memorial Hospital—The Woodlands Phases 1 and 2**	9250 Pinecroft Dr.	1985	Brooks/Collier
	Houston Advanced Research Center Office Building	4802 Research Forest Dr.	1986	3D International
	Venture Technology Center 1	3606 Research Forest Dr.	1986	Richard Fitzgerald & Partners
(3)	**John Cooper School Phase 1**	3333 Cochran's Crossing Dr.	1987	McKittrick Richardson Wallace
(3)	**Surgimedics/Texas Medical Products**	2828 N. Crescent Ridge Dr.	1988	Jim Carter & Associates, Inc.
	Venture Technology Center 2	3200 Research Forest Dr.	1988	Richard Fitzgerald & Associates
	Texas Golf Hall of Fame/TPC Patio	1800 S. Millbend Dr.	1989	Wahlberg Wright Waite & Associates
	Venture Technology Center 3	8701 New Trails Dr.	1989	Kirksey Meyers Architects
	WECCR South Wing	2301 N. Millbend Dr.	1989	Hoover and Furr
(3)	**Cynthia Woods Mitchell (CWM) Pavilion**	2005 Lake Robbins Dr.	1990	Sustaita Associates AIA, Inc.
(3)	**John Cooper School Gym/Classroom**	3333 Cochran's Crossing Dr.	1990	Kirksey Meyers Architects
	Palmer Course-Temp. Clubhouse	100 Grand Fairway	1990	McCleary-German Associates, Inc.

	Project	Address	Year	Architect
(3)	**South Montgomery County Community Center**	2235 Lake Robbins Dr.	1990	Langwith Wilson King Associates
	The Woodlands Athletic Center (WAC) Tennis Pro Shop Addition	11111 Winterberry Pl.	1990	Ken Anderson & Associates, Inc.
	WAC Renovations	11111 Winterberry Pl.	1991	Ken Anderson & Associates, Inc.
	Bank One Building Entry Plaza and Lobby	1400 Woodloch Forest Dr.	1992	Hoover and Furr
	Panther Creek Village Center Phase 2	4775 W. Panther Creek Dr.	1992	Heights Venture Architects
	TPC Clubhouse Expansion/Remodel	1730 S. Millbend Dr.	1992	Langwith Wilson King Associates
	Venture Technology Center 4	8665 New Trails Dr.	1992	Kirksey Meyers Architects
	Venture Technology Center 5	4200 Research Forest Dr.	1992	Kirksey Meyers Architects
(3)	**John Cooper Middle/High School**	3333 Cochran's Crossing Dr.	1993	Kirksey Meyers Architects
	WCC Expansion and Renovation	2301 N. Millbend Dr.	1993	Harry Golemon Architects Inc.
	Blockbuster Music	1335 Lake Woodlands Dr.	1994	Hodges & Associates
	Cochran's Crossing Shopping Center	4747 Research Forest Dr.	1994	Heights Venture Architects
	Custom Homes Sales Office Renovation	3535 E. Panther Creek Dr.	1994	JKL International
	GeneMedicine, Inc.	8301 New Trails Dr.	1994	Watkins Carter Hamilton Architects
(3)	**John Cooper School Phases 2 and 3**	3333 Cochran's Crossing Dr.	1994	Kirksey Meyers Architects
	Parkwood 2 Building	10055 Grogan's Mill Rd.	1994	Kirksey Meyers Architects
	Pinecroft Center Phase 1 Common Areas (Target)	1100 Lake Woodlands Dr.	1994	Lichliter-Jameson & Assoc., Inc.

	Project	Address	Year	Architect
	Pinecroft Center Phase 2 (Marshall's and Barnes & Noble)	1120 and 1310 Lake Woodlands Dr.	1994	Heights Venture Architects
(3)	**South Montgomery County Annex Building**	9909 Grogan's Mill Rd.	1994	Hall/Merriman Architects
(4)	**The Woodlands Mall**	1201 Lake Woodlands Dr.	1994	ELS Architects
	WAC Interior Renovations	11111 Winterberry	1994	Kirksey Meyers Architects
	Wood Ridge Plaza Renovation	27100-27900 I-45 North	1994	Morris Architects
	WECCR Gift Shop	2301 N. Millbend Dr.	1994	Gorman & Associates, Inc.
	WilTel 2 (Venture Tech Center 4)	8665 New Trails Dr.	1994	Kirksey Meyers Architects
	Pinecroft Center-Phase 3, Common Areas (Service Merchandise, Toys 'R' Us)	1410 and 1420 Lake Woodlands Dr.	1994	Lichliter-Jameson & Associates, Inc.
	CWM Pavilion Renovations	2005 Lake Robbins Dr.	1995	Morris Architects
(3)	**Fire Station No. 3**	1522 Sawdust Rd.	1995	Tackett Lodholz Architects
	Public Safety Building Renovation	9951 Grogan's Mill Rd.	1995	Merriman Holt Architects
	WECCR Expansion and Renovations —1994, 1995	2301 N. Millbend Dr.	1995	Morris Architects, JKL International
	Pinecroft Center — Linens 'n Things	1360 Lake Woodlands Dr.	1995	Quorum Architects
	WECCR Lobby Display	2301 N. Millbend Dr.	1995	Habitat

	Woodstead Office Building Lobby Renovations	1610 Woodstead Ct.	1995	C/I Design and Construction
	Office Depot—Wood Ridge Plaza	27500 I-45 North	1995	Hodges & Associates
	Venture Technology Center 6 (Tenneco)	8401 New Trails Dr.	1995	Watkins Carter Hamilton
	Palmer Golf Course–Club House Addition	100 Grand Fairway	1996	McCleary-German Associates, Inc.
	Cinema Parking Garage	1500 Lake Robbins Dr.	1996	Walker Parking Consultants
	Shell Learning Center	2230 Buckthorne Pl.	1996	Morris Architects
	Pier 1 Imports—Wood Ridge Plaza	27900 I-45 North	1996	Morris Architects
	New Information Center	2000 Woodlands Pkwy.	1996	Morris Architects
	WECCR Laundry & Service Building	2301 North Millbend Dr.	1996	Morris Architects
(3)	**Fire Station No. 4**	7900 Bay Branch Dr.	1997	Tackett Lodholz Architects
	Grogan's Mill Village Center Renovation	7 Switchbud Pl.	1997	Morris Architects
	WECCR — Fairway Pines	2301 N. Millbend Dr.	1997	Klages Carter Vail Associates
	Parkside Apartments Phase 3	10601 Six Pines Dr.	1997	Wahlberg Wright Waite & Associates
	Pinecroft West – Jason's Deli	3400 Lake Woodlands Dr.	1997	Heights Venture
	Aronex Pharmaceutical, Phase 1	8707 Technology Forest Pl.	1997	Kirksey and Partners
	Venture Tech Center 7 (Telxon)	8302 New Trails Dr.	1998	Kirksey and Partners
	Venture Tech Center 8	8708 Technology Forest Pl.	1998	Kirksey and Partners
	Alden Bridge Village Center	8000 Research Forest Dr.	1998	Hermes Reed Architects
	Town Center Office Building 1	1450 Lake Robbins Dr.	1998	Gensler
	Pinecroft West Phase 2 (Ross, Office Max, Cost Plus, GLA)	1700 Lake Robbins Dr.	1998	Hermes Reed Architects

Mobil Building 1	1200 Timberloch Pl.	1998	Gensler
Lexicon Expansion (Mouse Facility)	4000 Research Forest Dr.	1998	C2HK Architects
Town Center Office Building 2	1330 Lake Robbins Dr.	1999	Gensler Houston
Venture Tech Center 10 (Splitrock)	2777 N. Crescent Ridge	1999	Kirksey and Partners
Venture Tech Center 11-B (Genometrix)	2700 Research Forest Dr.	1999	Kirksey and Partners
Waterway Plaza One & Garage	10003 Woodloch Forest Dr.	2000	Gensler
Maersk	8686 New Trails Dr.	2000	Watkins Hamilton Ross
Sterling Ridge Shopping Center	6700 Woodlands Pkwy.	2000	Hermes Reed

(1) Constructed prior to October 1974. Not done by C/I Design and Construction.
(2) C/I Design and Construction involved in initial planning only. Separate owner.
(3) C/I Design and Construction were Project Managers for owner.
(4) C/I Design and Construction involved in design as partner with Homart Development.

GALVESTON PROJECTS			
Pirates' Beach Commercial Center	FM 3005	1975	Paul Paul + Madrid
Galveston Country Club Renovation	14228 Stewart Rd.	1977	Paul Paul + Madrid
San Luis Hotel	5220 Seawall Blvd.	1984	Morris-Aubry Architects
San Luis Condominium	5220 Seawall Blvd.	1985	Morris-Aubry Architects
Stars on the Seawall Renovation	53rd and Seawall	1991	JKL International
Pier 21	2200 Harborside Dr.	1993	Ford Powell & Carson

Pier 21 - Harbor House Cafe	No. 28 Pier 21	1993	Ford Powell & Carson
Pirates' Beach Sales Office Renovation	FM 3005	1993	Kirksey Meyers Architects
Randall's West End Restaurant and Bar	13706 FM 3005	1993	Kirksey Meyers Architects
San Luis Hotel Conference Center Addition	5220 Seawall Blvd.	1993	Harry Golemon Architects Inc.
San Luis Hotel Spoonbill Renovation	5220 Seawall Blvd.	1993	JKL International
Galveston Country Club Kitchen Renovation	14228 Stewart Rd.	1994	SDT Architects
Inn at the San Luis Renovations	5222 Seawall Blvd.	1994	JKL International
Landry's Seafood Restaurant at the San Luis	5310 Seawall Blvd.	1994	Morris Architects
Pirates' Beach Gate House	Pirates' Beach	1995	C/I Design and Construction

CAPE ROYALE PROJECTS			
Cape Royale Sales Office Renovation	306 N. Cape Royale Dr.	1991	Ken Anderson & Associates, Inc.
Cape Royale Golf Clubhouse Renovation	203 N. Cape Royale Dr.	1992	Kirksey Meyers Architects
Cape Royale Guard House	Cape Royale	1994	Merriman Holt Architects

As of September 22, 2000

through the building originally to inspect it, some people had built plywood shacks on the second and third floors and were living in there. It was fascinating. We employed Ford Powell & Carson as the architects again and began a series of intensive meetings that went on for a year just in the design process, then began construction probably about early summer of '83. We were under way and the contractor had been removing a lot of old stucco off the brick when Hurricane Alicia came into Galveston in August of 1983. The brick was so soft and absorbent that it was absorbing water from the rainfall, and it appeared the whole front wall of the building was going to collapse. The contractor called everyone's attention to it, and the structural engineer came out and directed that a number of huge pipe braces be placed on an angle from the street up, and that's how we held up the building until they could come in and put new foundations under the columns at the front to hold up the wall. It was pretty exciting. So we survived the hurricane and just moved on with the work. It was probably finished in 1985—we got a portion of the building opened earlier, but it was really finished in '85. It opened with about 120 rooms, a large central atrium, nice restaurant and a few small shops.

Mrs. Mitchell was involved?

Yes, Mrs. Mitchell was involved in the interior design of the building with Ann Gray, who is an interior designer from Chicago. She had done historic work on a hotel in Chicago and Washington, D.C. She was married to a man who owned some hotels in Chicago. Ann Gray worked very closely with Mrs. Mitchell on the interior colors and materials. I know Mrs. Mitchell particularly picked out the tile that went in the bathrooms of all the guest rooms, an Italian tile, which was a big changeover in cost because we were already under way on construction when that decision was made. The hotel was done beautifully, the interiors were gorgeous, very faithful to the time period, probably the 1870s-1880s, and is still admired as one of the best restoration jobs in the state. There were 20 buildings purchased and restored. The square footage in all those buildings, as I have recorded, is 680,700 square feet. They range in area from 7,000-8,000 square feet to 182,000 square feet, which is the Hotel Galvez. I've got a list of all the buildings in The Woodlands completed by the Commercial/Industrial Design and Construction Department of The Woodlands Corporation (attached). These are projects which I've been involved in since joining the company in 1973. For the most part, I was directly responsible for all the projects listed. Some of them, however, are peripheral to our ownership, such as the

Cooper School, although I was very much involved in the initial buildings for the Cooper School. Same way with the Woodlands hospital, which is now called the Memorial Woodlands Hospital. Also the Houston Advanced Research Center. I think I've got that listed as a project because it's one that we were brought in as project managers to assist them in getting it off the ground in design and construction.

Any other observations you may have about projects that we haven't talked about? What about the San Luis?

Yes, I was very much involved in the design and construction of the San Luis Hotel. It was a Woodlands Corporation project in Galveston. Construction started in 1983. The concrete shell of the building was up when Hurricane Alicia came in. There were no exterior walls erected at that time. There was some air conditioning duct work that had been installed and that was completely blown away by the hurricane, so they had to start all over on that. It didn't have any structural damage, however. And we were able to maintain the schedule and recover and open that building in 1984. That was one of the largest buildings we did. It has 15 floors, about 238 rooms. We were almost done building the San Luis Hotel when it was decided that we were going to build a condominium project immediately adjacent to it. We employed the same architects, Morris Aubry of Houston, to design the condominium, and we designed it to be architecturally compatible with the hotel so it appears as one building, although there's a slight difference in dimensions if you look at it closely. The San Luis Condominium was completed in 1985 and the hotel in 1984. It was a huge project, too—15-story buildings built right along the seawall. In fact, they were built on top of old seashore coastal artillery bunkers that were built probably in the 1930s for coastal protection for the United States. In fact, Mr. Mitchell told me he had been involved in pouring some of the concrete on top of those bunkers. He knew a lot about them, and one of the things I had to do was get in there and do some testing of the bunkers because they were covered with sand, and also clean them out. During the design process, we even had a design completed for space within the bunkers, but that was never developed. The third floor of the hotel is up high, because the bunkers were so massive and so big it would have been too expensive to try and remove them. So we just built the main body of the hotel behind the bunkers, then we created a large entrance walkway over the top of the bunker. The circular driveway arrives on top of the bunker. There was a large depression between the main bodies of the bunker, and that's where we built the swimming pool. The swimming pool is laying almost right on top of that concrete bunker.

You've been involved with the company for about 23 years now, which is a significant part of your career. The Woodlands certainly has to be the most important of all the things you've worked on. What are you proudest of insofar as your own involvement in developing The Woodlands?

I think The Woodlands is outstanding, and I think the original concept for maintenance of the existing environment, the trees and the vegetation has been the strongest element in planning for The Woodlands. The layout of the streets is important. The creation of neighborhoods by cul-de-sac design creates a sense of neighborhood and association of the people living in those streets. Also important is the design control that we've exercised since the beginning through the Development Standards Committee–I served as a chairman in the early years and have been on the committee ever since. We've reviewed every building and site development that has been done within The Woodlands. It's always been difficult because there are always conflicting demands; I think we've done a good job in trying to preserve the natural environment, particularly vegetation corridors along the streets, even though we still have to build large buildings and those require driveways and surface parking areas, etc. We've exercised controls on that and also on signage and just overall beautification. I think those are tremendous accomplishments. I feel when I drive off I-45 and drive into The Woodlands, on Woodlands Parkway, it's like coming into Camelot. I think Grogan's Mill Village, which is the first village, shows the best results of the kind of work that we've done, because the vegetation was preserved and has been restored. There was a better job done in the earlier years than in later years because things were done much slower at that time, The Woodlands was just getting started and a number of the people who were involved in the early years were very committed to what we were trying to do in the preservation of the natural environment. But I think The Woodlands is environmentally the most successful large project of its type in the United States, if not the world.

What have we failed to talk about that should be covered?

The other thing I've enjoyed about being involved in The Woodlands is living in The Woodlands and being a part of the community. I remember in the early years, some of the people who were here, particularly those who came from Columbia, really didn't feel like they should live in The Woodlands. I think being a part of the community is really one of the benefits of this kind of development, and as a result I've been very involved in community activities. I got involved in the local Chamber of Commerce, and later became its president, was elected to the school board twice and served as a trustee for six years. Many employees have been involved in various activities which represents, I think, a type of social interest that people who get involved in the planning/design of communities like this have.

Why has The Woodlands been successful while other new towns haven't?

Several reasons. George Mitchell's tremendous interest in making it a success and his personal dedication and financial commitment to making it a success. The ability of The Woodlands Corporation to use the parent company, Mitchell Energy, as a source of financing. If it hadn't been for the parent company and its financial ability, I don't think The Woodlands would have made it, even with Mr. Mitchell. Mr. Mitchell personally could have done a lot but if he didn't have the financial base of Mitchell Energy I don't think it would have been able to happen because the investment cost in an undertaking like this is incredible.

The idea is that over the latter years of the development, it all turns around and that money is supposed to come back. Isn't that right?

Yes. It's a long-term game and that's a problem. So many things can happen adversely during the development process that it makes it very difficult. There can be changes in legislation that affect the development, there can be changes in market forces, the economy, the weather, so many things can happen that will affect the success of a development which requires a continual pace of development to ensure the cash flow in order to retire debt and finance the operation.

Would you agree that we probably have reached the point where that turnaround is occurring?

I would say yes. In fact, The Woodlands has probably been more successful in the last few years than it has been since its beginning. In the last five years, The Woodlands has been the Number 1 developer for new home sales, selling over 900 new homes in a year, and this year we probably will even exceed that. I think our commercial development has really taken hold. While we've had commercial development before, it hasn't been of the magnitude that we experienced last year with the opening of the regional mall and the

Pinecroft Shopping Center. We also opened Cochran's Crossing Shopping Center, we purchased the Wood Ridge Center, renovated it and opened it in October 1995. With that and the other developments like the restaurants and retail shops, it's been a tremendous push into commercial development, and I think that's going to be the impetus for further commercial development in the Town Center.

Thank you, Coulson.

Reverend G. Richard Wheatcroft

Today is Tuesday, August 13, 1996. This is Joe Kutchin and I'm in my office in The Woodlands, in the MND Building. I'm with the Reverend G. Richard Wheatcroft, who is presently a resident of The Woodlands and, in a longer perspective, is one of the founders, if not the founder, of Interfaith of The Woodlands. First, tell me something about yourself. Where you're from, where you went to school.

I was born in Quincy, Illinois, moved to St. Louis, Missouri, at the age of two. My father was in the newspaper business and worked for several newspapers in St. Louis. I grew up there and went to public schools there, went to Washington University in St. Louis from which I graduated in 1939 with a bachelor of arts degree, with a major in economics and psychology. Then I went to Union Theological Seminary in New York for two years in preparation for ordination into the Episcopal ordained ministry. That was in 1940 and 1941. I spent my last year at the Virginia Theological Seminary at Alexandria, Virginia, which is an Episcopal school, and graduated from there in 1942.

Is graduation equal to ordination?

After graduation I was ordained in St. Louis and was assistant at a large suburban church in Clayton, Missouri, which is a suburb of St. Louis. I spent two years there and then the Bishop asked me to go to Kirksville, Missouri, which is 90 miles north of Columbia, Missouri, up in the northeast part of the state. It's primarily a college community. The Kirksville College of Osteopathy and Surgery is there. That's the birthplace of osteopathy. Kirksville is a town of about 10,000. I stayed there for five years, 1945 through 1950, and came to Houston in 1950.

In Kirksville you had your own congregation?

Yes, a small one. It ministered principally to the college community. Then Bishop John E. Hines—who was later the presiding bishop of the Episcopal Church, but at that time was in charge of the missionary work of the diocese of Texas which comprises 57 counties in east Texas—invited me to come to Houston to start a new congregation in the Memorial Drive area, at 345 Piney Point Road. We had about 25 people. We arrived in May of 1950 and stayed for 40 years. St. Francis Episcopal Church. During that time we also established a parochial day school which now has almost 500 children in it, from kindergarten through the eighth grade, and a child day-care center, which serves about 300 families in the community for child care on an hourly or all-day basis.

You say 40 years. Those were which years?
1950 through 1990, and in May of 1990 I retired because I reached the mandatory retirement age.

And during this period, your congregation had gone from zero to about 1,800?
Yes, to about 1,800 communicants, which is what we call our members.

Is there a bigger one in Houston?
Oh, yes, there are about four that are much bigger. Then in May of 1990, I retired and moved out to The Woodlands, principally because I had some history with development of the town and knew what a wonderful place it was and we're not sorry we did.

The Mitchell family were members of your congregation?
Yes. I don't remember the exact date they came but it must have been about 1960, whenever they moved to the Memorial area. They lived not far from the church and I think I baptized most, if not all, of their children. They would come to church and fill up two full pews with the 10 kids. As I said before, I've known them for many, many years and still have close relationships. We have dinner with them a couple of times a month and keep pretty much in touch, particularly with Cynthia.

Is the Interfaith idea unique to The Woodlands?
The way it developed I think it probably is, although I can't be sure about that because I'm not familiar with what's going on elsewhere in the country. The way the Interfaith concept started was really out of a conversation with George Mitchell the first time I was aware that he was in the process of planning The Woodlands. I think it was at a coffee hour one time following the service on a Sunday morning. We were talking in the parish hall and he said, "I'd like you to give some thought to what we might do in The Woodlands which might be a creative response to the issue of land use and money that the churches spend for buildings and parking lots and so on, to see if we couldn't come up with something that would be helpful both to the town and to the churches." George and I began to kick that around and finally

one day he asked if I would be willing to organize and head up a committee of representatives from the major denominations in the Houston area, all of which would ultimately be involved in The Woodlands in one form or other, to see what we might come up with in terms of some cooperative planning. With that, I began to contact heads of judicatories: Catholic, Methodist, Presbyterian, etc., all the majors including the Jewish community, and got a committee organized, about 15 people. W. D. Broadway might be able to help me remember names of those people. And we began to meet. This was in 1969-1970. George sent the whole group on a trip to Columbia, Maryland, and to Reston, Virginia, to visit those two planned communities to see what they had done in terms of religious institutions. Incidentally, we called ourselves the Religious Institution Planning Group, or "RIP" for short. I worked about a day a week, I guess, with the consent of my vestry, or Board, of the church. They allowed me the time to spend about a day a week down at the Shell Building on this project.

Do you want to tell about some of the things you found at Columbia and Reston?

At Columbia–I don't remember how old that community was at that time, pretty new, five or six years–they had come up with the concept of multiple-use buildings for religious groups. As a matter of fact, they had built one large building that could be used as a sanctuary and various groups used it on Sunday mornings at various times. That had some appeal in terms of land use and the problem that churches spend too much on buildings and parking lots and land and so forth. That was one of George's concerns, too. He said, "If we could come up with some idea for getting three or four churches to jointly use a facility, it would save them a lot of money and they could use their money for programming to meet the needs of people." And Columbia had done that.

Are you saying that all the churches used the one building?

All the churches that were there at that time were using that building. There weren't too many. I can't remember what they did in Reston. I think they were trying to do somewhat the same thing. We tried to sell that idea to the judicatories in the Houston area and it just didn't work. They all wanted their own places.

That's a word I'm not familiar with—judicatories.

That's like the diocese of Texas. The diocese of Texas is a judicatory.

It would be something a few steps up from the local congregation?
Or the presbytery or the synod or whatever the groups call it. The one building idea wouldn't fly here, so we ended up in the traditional mode of each church buying its own land and building its own buildings, etc., although several churches like the Baptist church and the Lutheran church on Grogan's Mill Road built next to each other so they could share some parking space and so forth from time to time. In the long run that didn't work either. The Lutherans felt that it was a bad location so they sold it to the Baptists and bought land and moved to another location. So George's original idea just was not practical in this part of the country where every church wanted its own facilities. I don't know what has subsequently happened in Columbia or Reston in terms of this concept. It probably was too idealistic and it had complications—who would get the 8 o'clock time and 10 o'clock time and so on. It was just not practical. That's what we worked with for a long time and finally it turned out to be just impractical, so we went the traditional route.

Was this RIP group dealing directly with George?
No, we were dealing directly with one of the groups responsible for institutional planning for the new town.

The Interfaith idea has not dealt solely with land and land use.
No, it developed. Land was our initial concern.

It sounds like early on you knew you weren't going anywhere with the idea that was in effect in Columbia. So basically you went back to a more traditional means of building the religious institutions. Was your committee charged with fostering that development?
Yes, we were designated as sort of the middle man between the denominations that wanted to come into The Woodlands and The Woodlands Corporation, and that still is in effect. Any church that wants to buy land in The Woodlands has to go through Interfaith.

What does that accomplish?
The first thing is that if a local congregation wants to come into The Woodlands, they contact Interfaith and Interfaith helps them as sort of a mediating force between them and The Woodlands Corporation in terms of location, size of property and pricing of land for churches, trying to help keep the prices within reason. Incidentally, that reminds me of when I came

to Houston and to this church out on Memorial. We bought land for $3,000 an acre and now it's worth a half-million an acre. Can you imagine that? So Interfaith still has a mediating position between the local congregation and The Woodlands Corporation. If the local congregation joins Interfaith–and most of them do–Interfaith does this without charge. If they don't, Interfaith charges them for the services.

I assume The Woodlands Corporation would be very hospitable to the religious community.

Yes, and I think they have been. There's a terrific variety of religious institutions in The Woodlands. Everything from mainline to people on one fringe or the other and so on. I don't think we have any Buddhist or other such groups here but we probably will in time because we're becoming a multicultural, multireligious society.

I know that a number of congregations would meet in other facilities before they had their own building.

They meet sometimes in schools, sometimes at Interfaith, at the Interfaith building, sometimes at other churches.

Does Interfaith play a role in helping that?

Yes. The Interfaith building was built partly for the purpose of providing what they refer to as incubating space for new congregations. They can only accommodate so many at one time in the space they have set aside for worship services and classrooms and space for education purposes, but they have served that purpose well over the years. There are usually two or three congregations using the Interfaith space at one time. Sometimes it's just one congregation, sometimes two or I think as many as three. They have to juggle their time.

What are the other things that Interfaith does?

The child care facility. That, of course, all developed after I was no longer involved.

Did they do that in Columbia, do you recall?

No. I don't think so. I think that was probably one of Don Gebert's contributions to Interfaith. But as soon as a number of local congregations, maybe two or three to start with, became established in The Woodlands, the people on the judicatory level withdrew, like I withdrew as soon as the Episcopal church was established out here and Bob Gipson was the minister. I withdrew from the whole thing and he was the Episcopal representative to the Board. So there was this gradual turnover until the Board was all local people, no longer anybody but the local level.

In terms of the Episcopal group, when did you withdraw?

It must have been probably in 1977. So I withdrew and by that time they had employed Don Gebert. I don't know when Don came but he was the first executive director and he is the one who really developed what Interfaith is doing today. He shared in the process. A lot of other people were involved, of course, including the Board. They established the child development center which has a wonderful reputation and they built a new building a couple of years ago.

What is the child development center?

That's the school. I don't know how far it goes up now, but it's in great demand. Interfaith has a senior citizen advocate on their staff, they participate in Meals on Wheels for senior citizens, they have a program on employee assistance, EAC, and I think they have contracts with various businesses in The Woodlands that help fund EAC and also they receive some federal and state money. They have a budget of about $6 million a year. A lot of it is grant money that they get from the federal government and the Houston-Galveston Area Council. They provide a lot of services for people—helping them get jobs, evaluating their potentials. They have a very extensive program.

Was this kind of thing conceived while your committee was still active?

No, that all developed later with Don Gebert and the local Board. In other words, I think they just responded very brilliantly to the emerging needs of the community, and they're still trying to do that. Then they put out a community phone directory every year. I think they get about half a million dollars revenue from that, and that helps to fund their programs.

I know The Woodlands Fund gives them money from time to time.

I'm very pleased with what has developed out here. Interfaith brought the churches together in a cooperative venture and I think it's been very successful in terms of helping to build a community within The Woodlands.

Your goal initially, requested by George Mitchell, was to develop a means where facilities and land would be shared. It didn't take you long to learn that that wasn't going to fly.
Shared facilities we called it. Not here, and I'm not sure it would any place else, either.

But at least in one or two cases here, the attempt was made and then later abandoned. The organization that you helped give birth to has developed into one that does all these good things you just talked about.
And is the center of the community's interfaith relationships and service to the community. Interfaith does things that individual churches could not do by themselves.

Such as?
Well, this employee assistance program, for example. An individual church can do a school, but it takes a lot of doing for one individual church.

And it would be next to impossible for a smaller congregation.
But they all can share. The senior citizen program is a wonderful program. There are quite a few seniors living at Tamarac Pines and some of those low-cost public housing facilities. It's a wonderful program, having a full-time person who relates to those institutions and the people who live in them.

Do you know whether there are comparable programs elsewhere?
Houston has what used to be called the Houston Metropolitan Ministries.

I know there's a Northwest Assistance Ministries program.
Yes, there are various programs that do some of the things that Interfaith does, but nobody I know of does all the things that Interfaith does. There are a number of institutions that do a lot of things in common with Interfaith.

I have an impression that Interfaith probably does more and better. It may be in better circumstances with a more defined community.
I think probably the best one in this general greater metropolitan area is Houston Metropolitan Ministries. I think they call themselves Interfaith now. That pre-dates Interfaith here, although they did nothing in terms of land use planning or anything like that. It was purely social services activi-

ties. It's a fine institution.

But since you slowly withdrew from Interfaith, your involvement has dropped considerably.

Yes, although when I moved out here in 1990, W. D. Broadway got me on the Board right away, as sort of an anchor to the past, as he put it. And I was on the Board for a three-year term and then I did another one-year term. I just went off last year. I was on the executive committee, so I've had a lot of relationships with Interfaith since I moved out here. I'm no longer on the Board, but I keep in touch with W. D. He was the pastor of the Baptist church down on Grogan's Mill, which is just a couple of blocks from where I live. Before he was elected director of Interfaith, he left that local church and went to the development office of one of their seminaries, I think, in Fort Worth.

Before he came to Interfaith his congregation was where?

His was one of the first churches established in The Woodlands and he was the pastor of that church. At first, there was a man from Dallas who represented the Southern Baptists on our committee. Then when the first local Baptist church was established here and W. D. Broadway came as pastor, then the Dallas man withdrew and W. D. went on the Board. So he's been on the Board a long, long time, and then subsequently was elected the executive director.

For a period of time this was essentially a volunteer group. I assume at some point you saw that this was more than a volunteer group could handle.

Yes, and that resulted in the hiring Don Gebert.

Did Mr. Gebert come early?

Yes, pretty early. I think he came shortly after I withdrew from the Board. I think the Board found him in Pennsylvania. But I think it was shortly after I withdrew from the Board and Bob Gipson, a local man, was on board the day they hired Don.

Say again when you withdrew from the board. That sounds like it would have been mid-70s.

That's right, mid-70s.

Is Interfaith successful?
I think so. I think it's really fairly unique in the country as far as I know.

If you had it to do all over again, would you have gone down some different paths, done anything different?
I don't think so. It seems to me you have to learn by doing some things wrong. That was inevitable. And when George said, "Gee, I'd like to go this way," I know from experience it's hard to say, George, that's not going to fly. I had some direct experience with the bishop of the Episcopal church here, who just never did understand the concept of The Woodlands. He felt it was just a suburb. In their planning, The Woodlands Corporation had certain areas set aside for church sites. The bishop didn't like that, so he paid commercial prices for the property on Panther Creek and Woodlands Parkway for the Episcopal Church and bought a good bit of land. Well, the congregation didn't develop as it should have. They had to sell some of the property back to The Woodlands Corporation because the bishop paid commercial prices for this land and that was totally unnecessary. I've had some experience with the judicatory head not understanding what we were doing and not being content with pieces of property that The Woodlands Corporation had set aside for churches, all of which was excellent property, well located. Nobody had any problems with it. Sometimes the church wanted more land than The Woodlands Corporation was willing to give up, but you know, that's negotiable.

Anything you want to say that you haven't said?
Don't think so, except I'm grateful to George for asking me to get involved in this thing because it was one of the most interesting and exciting things I think I've done. Just to be involved, even in a minor, minor way, as I was in the process of helping to develop a community like this, was just a wonderful experience. I've met some wonderful people.

Thank you very much.

Bruce M. Withers, Jr.

Today is Tuesday, November 14, 1995. This is Joe Kutchin, and I'm in The Woodlands office of Withers Enterprises with Bruce M. Withers, Jr. Bruce for many years was the man in charge of natural gas processing and natural gas transmission at Mitchell Energy & Development Corp. Bruce, to start with, just give a fairly brief résumé of your background. I know you're from Beeville, Texas.

Actually I'm from a little town called Tuleta, which is 12 miles north of Beeville, Texas. I graduated from high school in Longview, Texas, in 1943 and joined the U. S. Air Force shortly after that. After finishing my tour of active duty with the Air Force in 1946, I enrolled at Texas A&I University at Kingsville, Texas. I received a B. S. degree in petroleum and natural gas engineering from Texas A&I University in 1950. Upon graduation, I went to work for a small gas utility company in Waxahachie, Texas. I stayed there for three years and then joined a brand new company called Wilcox Trend Gathering System, which had just announced that they were going to build a brand new natural gas gathering and transmission pipeline system, beginning at a point near the Texas-Mexico border and spanning all the way up to Provident City, Texas, where the gas was going to be processed and then delivered into Texas Eastern Transmission Corporation's pipeline system that went all the way to the East Coast. By the way, Texas Eastern was the majority owner of Wilcox Trend at that time. Wilcox Trend later became a wholly owned subsidiary. After completion of construction and placing the new pipeline system in operation, I moved to the office headquarters in Cuero, Texas. I stayed with Wilcox Trend for the next three years, until Tennessee Gas Transmission Company called me in mid-1956 and offered me a position with their newly formed gas processing group in Houston. I stayed with Tenneco for the next 18 years. In late 1973, I received a call from George Mitchell, whom I had known for several years (since Mitchell Energy became a part owner in a gas processing plant I built for Tenneco near Palacios, Texas). The purpose of George's call was to see what it would take to get me to join Mitchell Energy Corp. to build a gas processing division for him.

Tell a little bit more about what your responsibilities were at Tenneco. Why did George want to go to you instead of someone else?

I joined Tenneco as a senior plant engineer from Wilcox Trend and progressed upward over the next few years to manager of their gas liquids project development and process engineering. At Tenneco, we had built eight gas processing plants, including three of the first gas processing plants built on the new Trans Canada Pipeline System in Alberta Province in Canada. I had specialized in all types of gas gathering and gas processing operations with Tenneco prior to joining Mitchell. George was a partial owner of Tenneco's Leabo plant down near Palacios, Texas. He attended several of our plant owners' meetings where we would discuss the progress of the plant. He told me that he had talked to a friend of mine, Bill Hudson of Butler, Miller & Lents, who was doing all of the gas processing work on a consulting basis for Mitchell at that time. George asked him who would be a good person to come over to build his gas processing operation, and Bill Hudson and his boss, Max Lents (both of whom were very active, as I was, in the Gas Processors Association at that time), mentioned my name. George said, "Well, I know Bruce Withers," so he called me and that's how we first got together. It was not an easy decision to leave a large company like Tenneco to join a small independent oil and gas company like Mitchell Energy. I did not join Mitchell until February 1, 1974, so you can see it took me a while to decide, but George was patient with me. As it turned out, I'm sure glad he was.

Then I beat you . . . by a month.

I remember that our employment dates were close together. Guess it just took me longer to make this very difficult decision. I told George, "I'm not looking for a job . . . Tenneco is treating me very well." You know what he said? "I know that, I just want to know what it will take to get you over here with me." The thing that intrigued me most and really caused me to make the final decision to join him was George's foresight. His question after he explained what he wanted to talk to me about was, "Bruce, is there any way we can build a small portable gas processing plant that can be easily moved from place to place?"

The ones you had worked on at Tenneco were not portable?

No, they were not, and that is why George's question just turned on a lightbulb in my head. I had had this idea for over a year and had tried to get Tenneco interested in it. Tenneco was such a big company and worked on such a big scale, they were not interested in my idea. If you had 50 million cubic feet of gas a day in one spot, with a 20-year reserve life, they might be interested in building a new plant. No one had really thought about small, portable plants that could be easily and economically relocated to a new location when gas reserves played out at their initial location. This idea was dear to my heart, and all of a sudden, here I was talking to George Mitchell, who was chairman of the Board of an independent oil and gas exploration and production company, who had this same idea and asked me if it could be done. Needless to say, we "clicked" on this idea and worked well together from that time on.

Each of you arrived at it independently?

Yes, completely independent of each other. We had not talked about such an idea at all prior to that time. That was the key to our getting together. I was very impressed by his foresight, and my comment was, "Yes, it can be done, George, and I've been wanting to do this for a long time." He also understood the implications of Tenneco's size and the magnitude of the plant they had to look for. At any rate, that's how we got together. It took a while to get our financing arrangements lined up because, as I mentioned earlier, Mitchell was a small, independent company at that time. I'm not certain about the employee head count then, but I am reasonably sure that it was less than 750 people.

What did you find when you came to Mitchell? What were the circumstances that you ran into?

Actually, I found that the opportunity was there to do just about anything that I felt big enough to do that made any economic sense. Tenneco was operated almost by committee at that time and to get anything done, I'd have to present my story to several people. I remember my first comment to George was, "If I come over, I want to know who I will report to. I'm not a politician or committee man. I like to get things done and I'd just like to know who I have to convince or report to in order to do what I want to do." George very quickly answered, "Me." I said, "OK, that's fair enough. I'll give this some thought and let you know." As a matter of fact, I gave it a lot of thought, and so did my wife, before we jointly reached our decision. Over and over, I kept coming back to the long-range vision that George had, how our thoughts meshed so well and how he promised me that I could do about anything that I wanted to do if it made sense. The end result was, I couldn't turn down such a challenge and I joined George Mitchell. That's the way it worked from the time I joined him until the time I left. He didn't second-guess any of my decisions, he just let me do what he hired me to do. That's what I enjoyed the most, and that's why we became a successful leader and a pioneer in the installation and operation of the small, portable turbo-expander, cryogenic gas processing plants.

As I recall, at that time there was the big plant at Bridgeport, and Decatur was there.

Well, that was a long time ago, Joe, but if my memory serves me correctly, I think there were at least three other small mechanical refrigeration plants (not highly efficient) that were running at that time. Bridgeport, which was operated by Warren Petroleum, was in operation at that time. Mitchell had no employees in that plant. I think there was a plant at Lone Camp and one at Ranger and perhaps one other, but Decatur doesn't ring a bell. I do know there were about three of those little refrigeration plants that Byron Green and Leland Carter had installed on the Southwestern Gas Pipeline system. They were very small mechanical refrigeration units, not any of the turbo-expander plants that we were interested in building. We were going after highly efficient plants for the recovery of ethane and heavier hydrocarbons from the gas streams being processed.

Allen Tarbutton joined at the same time you did?

No, Allen came over about 30 days after I joined Mitchell. Allen had worked for me for 13 years at Tenneco. I hired him upon his graduation from Texas A&I University in 1961. He also has a bachelor's degree in natural gas engineering. Allen actually worked for Tenneco a couple of summers prior to his graduation. Allen was transferred around to work at several plants in the field before I moved him into the Houston office as senior process engineer. When I told Allen I was leaving Tenneco to join Mitchell, he said he wanted to go with me. As bad as I needed and wanted him, I told him I just couldn't take him with me at that time because I really didn't know what the future might be at Mitchell. Besides, I explained that Tenneco was a fine company and had been good to both of us, and I just didn't want to do that to them–that is, take one of their best engineers away. His comment was, "If you don't take me with you, I'm going to quit anyway." My response back to him was, "Well, that's a different situation. If you do quit Tenneco, then come over and I'll offer you a job." So, that's what he did, and he joined me about 30 days later. We both operated out of one office until we found a spot to move him. You might remember that, Joe, up on the 39th floor, right next to George's office.

Talk a little more about the supply that you expected to find, the diversity of small streams really, and the market for the liquids.

Let me go back. Before I made the decision to join George, I asked him about his gas processing potential and just what it was that he wanted to do.

He said he wanted to process every bit of gas that he produced. He had several opportunities and he gave me a list of some of them. I reviewed them with Morris Thompson and asked him about the gas volume, the liquid content of the gas, reserve life, and other necessary criteria. I saw some opportunities right off. As a matter of fact, after giving Tenneco three weeks notice of my plan to leave, I was working with Randall Engineering Corp., which at that time was probably the lead designer and packager of these small turbo-expander plants. So, we were working on about five specific prospects during these three weeks, going over engineering drawings, mechanical specifications, plant recovery efficiency, throughput volumes and other general requirements. After Allen came aboard, that made two of us so we doubled our efforts, and in a matter of days we actually went out for bids for 10 plants at one time. Five were designed for a range of 10 to 15 million cubic feet per day, and five more were designed for a range of 20 to 30 million cubic feet per day. I wanted a "Plan A" and a "Plan B" so I could just call up and say, "I want X number of Plan A or Plan B, as soon as possible." Actually, we bought 10 the first rattle out of the box because we found that many opportunities. I might also add that we had the first four or five in operation in approximately six months, and all 10 on stream in about 12 months from the day we arrived.

This was at Mitchell Energy?

Yes. Tenneco was not interested in these small plants at all. We had the first plant running at Lone Camp in August 1974, as I recall. You talked about the supply and the demand for natural gas liquids. Our "crystal ball" indicated there would be an increase in ethane demand. It was just beginning to be a very valuable petrochemical feedstock and that's what these expander plants were designed to do, increase the efficiency of ethane recovery. As I said earlier, for some time I had been intrigued by the idea of the small portable plant, and Randall Corporation seemed to be the leader in cryogenic design technology, process skid design and best delivery time, so we went to them for our first bids to get our plants built. I could see that the petrochemical industry was sort of doing a "flip-flop" from their earlier thinking on plant feedstocks. In the late '60s and early '70s, the chemical plants were designed to use crude oil and condensate as the main feedstock for their cracking process. In '73-'74, if you will recall, the first oil embargo took place. Some of the major petrochemical companies had already been looking to get away from the heavier end crackers (such as crude oil, condensate, naphtha and gas oil) and going toward the lighter ends (such as ethane and propane) because the prices for these natural gas liquids were considerably cheaper. So that's what prompted our decision to go after a higher percentage recovery of ethane and propane from our gas processing plants. We figured that if the chemical companies were going to build new petrochemical plants to crack the lighter ends, there would be an increase in demand for ethane and propane. All of these ideas just seemed to fit, and since Mitchell had a lot of gas in small quantities that was not being processed at that time, we implemented our very aggressive plan to process all the gas that Mitchell had, as well as future production of Mitchell gas. Fortunately, we made the right decision at the right time and got a head start on our competition. After we were well on our way to processing all of Mitchell's gas, we began to go after opportunities to process what we called "third party," or non-Mitchell, gas.

Was the bulk of that in North Texas?

The biggest opportunities were along our Southwestern Gas Pipeline system in North Texas, yes, but we also had one at Seven Oaks (about 70 miles northeast of Houston) that was one of the first plants we built. We built one at Barton Chapel, two at Galveston and one in Sutton County, which is out in West Texas. As I stated earlier in our discussion, my recollection is that we had all 10 of these plants located and running within 12 months after I joined Mitchell on February 1, 1974. So you can easily see that we were on a very fast track. As you know, over the next 18 years we built a total of about 57 of these plants and that's what we had at the time that I left Mitchell. We were very aggressive and very successful, and the prime reason was that George Mitchell had this vision and we just went right after it full speed. Once again, we were able to do things quickly, because we could get a quick decision out of George. I could walk into his office and within five or ten minutes, I could walk out with a yes or no answer, but I don't ever remember a no. One other major factor that contributed highly to our success was the people we were able to get to come to work with us. We were pioneering a new technology on these cryogenic plants, unfamiliar to a lot of operations personnel, but we were fortunate to get some real good people who were dedicated to learning new things, and we began a training program that was very successful and we became very proficient pretty quickly. A lot of people were looking at Mitchell during that period of time, wondering how we were building so many new plants so fast and the key was that George was backing us 100 percent, and we had a team of loyal, dedicated people who loved the challenge we put before them.

We became the 13th largest in the country, as I remember.
Yes, that's right. We were the 13th largest producer of natural gas liquids in the United States at a peak production rate close to 55,000 barrels per day by the time I left in August 1991.

As I recall, too, the North Texas gas was particularly rich.
Yes, the North Texas area was the spot to be in for gas processing. However, the plant we built up at Seven Oaks in East Texas was good, rich gas, too, but you are correct—the better gas was in the North Texas area and that's where Mitchell had a very large acreage position. Our Galveston production was much leaner gas, but we had a large enough volume to justify a new cryogenic plant there. By the way, this project resulted in another Mitchell first. Our decision to recover a high percentage of ethane from our inlet gas stream caused our recovered "raw mix stream" to have a higher vapor pressure than could be transported via any existing tank trucks at that time. Tank trucks had not been designed or approved by the Department of Transportation at that time. You could not haul what we call a demethanized stream in any of the existing over-the-road transportation vehicles. Only way to move this demethanized (high vapor pressure) product stream was by pipeline. But because Galveston is an island, there was no way to economically justify building a liquids pipeline to the market. We had prospects for two good plant locations on Galveston Island, but because of the lack of transportation there seemed to be no economic justification to build the plants. In my search for alternatives I contacted Mississippi Tank Company, whom I had heard was looking at a tank wagon design to handle high vapor pressure product. They said they had a design, and as it turned out, the design was being done for Warren Petroleum Company in Tulsa. I also found that John Lesch at Warren was in charge of this project. John was a good friend of mine, so I called him to discuss my problem and my need for a couple of high pressure tank trucks. John verified they were trying to get a couple of these same tank trucks on the road, too. So we kind of joined forces and were able to put the first insulated, high vapor pressure tank trucks on the road. You probably recall that we owned two of these trucks which allowed us to build the two plants in Galveston and one in Sutton County, Texas. Otherwise, none of these plants would have been economically feasible. We utilized those trucks for a couple of years, but both George and I were uncomfortable about hauling that volatile liquid over the road and especially over the Galveston bridge. George would call me every so often and say, "Bruce, are you sure we've got enough insurance to cover

these trucks?" Eventually we sold our trucks to a third party and got out of the business and just paid the fee for transporting our liquids. But the point I was trying to make is that when I presented this story to George, he liked that idea, too. He always liked to do something new and creative and often without questioning, he would answer, "Yes, if you think that's what you need, go on and do it." So that's three plants we would have never built without pursuing a new type of technology and having the aggressiveness and the guts to go out and do something like that. Between Warren and ourselves, I think we put the first five trucks in this type of service.

Other technology . . . did you or the company play any special role in furthering turbo-expander technology?

I think what we did was not necessarily a technological advance but more of an operational improvement. For instance, we played a role in the way our skids were designed and laid out, with easier accessibility to maintain and operate equipment on the skid, such as the pumps, motors, changing the motors, changing the expander wheel and that sort of thing. The better design made operations more efficient. After we got the first couple of plants on stream, we began to make suggestions to Randall and he worked with us to move some things around, locate them in different places to improve operational and maintenance efficiency. We did not make any contribution as far as the chemical process side of the equation is concerned. The expander design and so forth were done by other companies, too. Randall did not design the expanders themselves, they only put the package together. At the time, Roto-Flow was the lead manufacturer of the turbo-expander, and later on, as the technology progressed, other competitors began to surface. We wanted certain ideas built into our skids, and I think ultimately Randall began to use some of those ideas to upgrade their skids, as well. The suggestions came from some of our field people's suggestions to Allen Tarbutton while he was visiting with them and asking for their input on improved designs.

Does this oversimplify things? You and George had an agreement on the concept and that concept was not rocket science, it was doing something that nobody else seemed to do. The technology was there, the streams were there, so your job was to make everything come together. Is that oversimplifying your role?

That's right on the nailhead, Joe. That's exactly what happened. We utilized existing technology to accommodate our opportunity.

Is there anything during your course of years at Mitchell that you're especially proud of?

Yes, I'm very proud of the people that we were able to bring together at Mitchell. My philosophy at that time was, and still is, that people are the best assets that a company can have. I've always enjoyed people, respected people and tried to motivate them in such a way to let them know that they were key players of a team organization. I'm very proud of what we accomplished with our Mitchell team. Beginning with Allen Tarbutton, Carl Springer, Bruce Toellner, Fred Crum and some of the others who came early on, we built one of the finest teams in the industry, which became well respected within just a few years. A lot of people would call us from time to time and ask us many questions related to gas processing problems, and we developed a good reputation very quickly. It was not anything special that I did, but it was the people that I was able to surround myself with. As a matter of fact, I remember one of the first things I told George during our early negotiations was that I did not expect to hire a lot of people initially, but the ones I planned to hire would be the best I could find. You know what he said? "I couldn't agree with you more." Another thing I've been most proud of all through the years was the close personal relationship that I was able to develop with George himself. He believed in and trusted me and I respected and admired him. I also trusted his judgment and together I just think we made a pretty good team. As I said before, we built 57 plants during the time I worked for George and we built most of them in the first 10 or 12 years I was there. And if you stop to think about it, that's putting four or five plants on stream each year. We kept our people busy and I think that was the secret to our success. They all felt like they were an important member of our team, that they were making a real contribution, and they were. They worked hard and we didn't have a bunch of people who didn't have anything to do.

Do you remember the gas liquids production volume when you came and when you left?

I've got an old annual report here which shows that the company's gas liquids production for the fiscal year ended January 31, 1974, was 13,400 barrels per day. I joined the company the first day after that fiscal year-end. During the year I left, which would be the fiscal year ended January 31, 1992, gas liquids production was just shy of 50,000 barrels a day. That's four times as great as it was when I started. As you know, the volume dropped off a little in recent years, and I believe the latest annual report I recently received showed NGL production at about 48,000 BPD.

Was there any one deal that was especially gratifying, hard to pull off, anything come to mind on that?

Well, without having a lot of time to think about this, Joe, offhand I'd say probably one of the first and best acquisitions that we made early on was the Winnie plant and its gathering system over near Beaumont. We bought it at a very reasonable price from a division of Allied Chemical through our relationship with Sparkman Energy at that time. Also buying into Tejas Gas Corp. was a good deal, even though it was a hard one to pull off. The third major pipeline system we acquired was the Ferguson Crossing Pipeline Co., which was a sizable system located in the heart of the Austin Chalk area where the gas was very rich, in fact richer than in North Texas. This was a significant acquisition and is still growing today. In addition to these three, we were also very aggressive and successful in expanding our own Southwestern Gas Pipeline System in North Texas.

Tell why that pipeline system is important to the gas liquids business.

The Southwestern Gas Pipeline system is really important to Mitchell Energy because it is an aggregator of a significant, rich gas supply that keeps on being developed. That was most important because we wanted to process every bit of the gas that we could gather and sell off of this pipeline system. It was quite extensive and really covered the North Texas area very well. It gave us significant opportunities to buy the gas at the wellhead, take it into our pipeline system and allow us to process it before selling to customers. For example, down at our Lone Camp plant we had a real opportunity to install a turbo-expander plant because we had high pressure gas to be delivered down to a low-pressure burner system for the big power plant owned by Brazos Electric Power Co-op. There they burned gas at a very low pressure, so we didn't have to recompress the residue gas back up to pipeline pressure. This was a low-cost installation and ideal for us to install a couple of our first turbo-expander plants. We paid the producers on the basis of their wellhead gas volumes, so we owned and controlled the gas after it got into our pipeline system, thus providing us the opportunity to process gas and recover the valuable natural gas liquids.

Tell what some of the typical deals might be with the producer.

Actually, Joe, there really isn't a "typical deal" in this business, even though this may sound strange. For example, each processing deal is made based upon the liquid content of the gas, the volume of the gas to be processed and the location of the gas to be processed. The producer is paid on a very equitable formula, where the richer his gas is and the more volume he has, the more he is paid for the liquids that are recovered from his gas. This liquid payment from our processing plant allows the producer to enhance the total value of his gas at the wellhead. This was good for the producer and good for Mitchell, and it was a good selling point that we had on processing third party gas. Most of the gas we bought or processed was owned by smaller producers who did not want to spend their money to build a gas processing plant or install gas compressors to compress their gas or build a pipeline or make other such capital expenditures required to make their gas marketable and deliver it to market. Smaller producers would rather spend their resources on lease acquisition, exploration and drilling prospective wells. What we brought to a producer was the opportunity to get all of these services performed by Mitchell and at the same time enhance the overall wellhead value of his gas by processing it and paying him a percentage of the natural gas liquids value recovered by our processing plants. As I recall, the percentage payment to the producers was somewhere in the 60 percent to 70 percent range. In fact, back in the early days of the Bridgeport plant (which Warren originally built back in 1949 or 1950, I believe), this percentage payment was closer to 45 percent to 50 percent. Both of these percentage ranges depended on the liquid content of the producer's gas, as well as the volume and location of the producer's gas.

Be more specific. Who gets what?

The producer would get the 60 to 70 percent. That is, he would get that percentage of the value of the liquids recovered from his gas, less the transportation and fractionation charges required to market the liquids.

That's really typical?

That's fairly typical right now, but again it depends upon the things I mentioned earlier, plus the physical location of the plant is a key factor. For example, if the plant is far removed from the end user of those liquids, you get less netback due to increased transportation cost. The percentage has to be variable so that gas processors can afford to build the plants. Sometimes, after payout we would enhance the payment to producers. Each individual plant prospect has its own individual characteristics and requires different things to put a deal together. What we tried to do was build a reputation of being a fair trader with our producers. Sometimes tough, but always fair.

What about bad experiences? I remember College Station was a bear.
Yes, that was a difficult situation due to plant location primarily and not economics, but we did resolve it satisfactorily. Joe, I only remember one bad deal, and probably the worst deal I ever got into, and that was the Wilson Pipeline system acquisition. This was a situation where several of our producer friends thought they had discovered a good, new field and were beginning to develop it in the Austin Chalk trend near Giddings, Texas. The Chalk play was just getting started at that time.

When would this have been?
This had to be early on. I would say probably between 1977 and 1979, somewhere in that range. The Wilsons (small producer in Midland) had a gathering system, compressor station and one little plant ... that is, they called it a plant. Based on our evaluation, it wasn't all that good. They approached us to buy their system, but they didn't want cash money. Instead they wanted Mitchell stock. As you know, George was not enthusiastic about issuing new shares of Mitchell stock, so we went to the public market and bought our stock (which was selling at a high price at that time) and traded Mitchell stock for their system. Conoco, Exxon and some other large producers had a big acreage position in the area and it looked real good to all of us. Unfortunately, that area of Austin Chalk is one where a lot of producers and a lot of processors went down hard, and we were one of them. I believe that's the worst deal I ever made and since I'm the guy who sold George on it, I have to take all the credit. Those guys sold me a bill of goods. It was in the Austin Chalk area, however, and we wanted to establish a presence there. The gas was also very rich but there just wasn't enough of it. We operated this system for a good while and eventually sold it, but it sure wasn't one of my better trades.

What about the future of the industry?
I think the future is bright and that gas processing will always play a major role in the future development of our nation's energy resources. On occasion, our industry has been squeezed by the combination of low prices of liquids and high prices for natural gas. That's just one of the hazards of doing business. Any time you are in a commodity business, you have to worry about the volatility and fluctuation in commodity prices. Nevertheless, I still believe, as I always have, that natural gas liquids are such a valuable commodity, with a wide range of utilization, that they will always be worth more when sold as a liquid than they would be worth on a BTU

equivalent sold in a gas stream. I have to warn you, though, Joe, I have been accused of being the most optimistic guy in the gas processing industry, probably next to George himself.

Still true?

Still true. I have seen many ups and downs over the years, but we always survived all of them and right now our industry is doing well. Starting in the last quarter of '93 and early '94, margins got squeezed terribly and plants were shut down all over, especially straddle plants (built across transmission lines) where the processor has to make up fuel and shrinkage value on a BTU equivalent. On lean gas streams and low margins, economics won't justify processing, so our industry went through some lean times. Now things are much better. The gas prices are lower, liquid prices are sufficient to make recovery worthwhile and all plants are running again. In addition, as stated earlier, over recent years the petrochemical industry has designed and built all their newer plants to utilize natural gas liquids as their prime feedstocks. So for a lot of reasons, I still think the future of the natural gas liquid business is good. I know I warned you about my eternal optimism but I haven't been let down too many times over these many years. As a word of caution, however, I believe you need to be in it big or you should not be in it at all. You must have plant flexibility and sufficient liquid volume to carry you through the lean times. But overall, I think the natural gas liquids business is here to stay and it's always going to be.

And the small plants? Is the small turbo-expander still valid?

You bet. The small turbo-expander is one of the most efficient means of recovering gas liquids from natural gas. There's not been another better mousetrap built to date.

On a BTU basis, liquids have been historically selling at what? Two-thirds the price of crude, something in that neighborhood?

Yes, that's pretty close. Normally, two-thirds is pretty good and even up to the low 70s as a percentage of crude value. However, this comparison is not as relevant as it used to be because the relationship between crude oil and natural gas has become disassociated in the last few years. It used to be that when crude prices were strong, you could always pretty well depend on the advantage of extracting liquids. But right now, the key factor to gas processors is the price of natural gas from which you extract these liquids, more so than the price of crude oil. I think perhaps this disassociation between crude and natural gas may last for awhile and will not be nearly as important to the gas processor in the future as it was in earlier years. Good point, Joe, I'm glad you brought that up.

You spent 18 years or so at Mitchell. What caused you to leave?
Well, leaving Mitchell Energy was one of the toughest decisions I've ever had to face in my entire business career because I thought so much of George. He is such a great guy with a lot of vision, and I truly admired and respected him very much. I really didn't want this job from the beginning, but they worked hard on me to come and take over the chairmanship of Trident NGL, Inc. I turned them down two or three times, but they kept coming back. I had talked to George about this from the first time they called. He said, "Maybe you should talk to them but don't do anything until you talk back to me." Finally, they just embarrassed me by calling and saying, "The least you can do is come up here and talk to us and see what we have to offer." I told them straight off I wasn't interested in the job, but I would come up and talk to them, which I did. I guess the thing that made me respect and admire George even more was when I sat down and counseled with him about the offer they finally made and what they wanted to do. He said that he would do almost anything to keep me from going, even as far as matching their offer. We had a long and emotional discussion that day in his office, and he finally said, "Bruce, I want to keep you. I don't want you to go, but I have to tell you, I don't see how you can afford to turn down this opportunity and challenge. It's a good one and they need you a whole lot worse than you need them." I was sitting in his office downtown and I never will forget that day. I sincerely appreciated his support and counsel on this important decision in my life, and we're still very good friends. I would do anything in the world for George Mitchell, even today. I remember another thing he told me that day was, "Any time these guys decide they want to bail out or something, let me know and we'll try to put these two companies together if there is any way we can do it." I told him I would keep that thought in mind. As a matter of fact, soon after my investors told me that Trident might be an acquisition candidate, I had lunch with George and reminded him of our conversation that day in his office. He seemed very interested, but it turned out that the value we placed on Trident was apparently higher than he felt he could go at that time. We still both agreed that it would have been an awfully good plan if we could have put our two companies together. We'd have been Number 1 in the liquids business, even above Exxon and Phillips. Pardon this digression, Joe, but I just remem-

bered our discussion on this that day. As I said, leaving Mitchell was a tough decision and I didn't make it quickly. I stalled Trident from along about mid-May until about the middle of July, but finally did agree to take the job and actually joined them on August 1, 1991. The thought of starting a new company was intriguing and exciting for me. Few people have an opportunity in a lifetime to start a brand new company from scratch like that but even so, I never would have left if I hadn't thought I could leave George Mitchell on a friendly basis, and he allowed me to do just that. I never gave up the long range thought of putting our two companies together some way, and I did what I promised George I would do when the time came. Unfortunately, when the time came, Mitchell Energy wasn't in the right financial position or mindset to help make it happen, and I understood that. It would have been a beautiful combination. We would have been a very strong company.

Just to get some housekeeping straight: At Mitchell you were president of the Transmission and Processing Division and senior vice president of the parent corporation. Then, you left to become chairman and chief executive officer of Trident NGL, Inc., which later was merged with Natural Gas Clearinghouse in March of 1995.
Yes, that is correct. When Trident and Natural Gas Clearinghouse merged, the two companies became subsidiaries of a parent corporation, which is now NGC Corp. Trident is a major division and it still maintains its name and identity. Natural Gas Clearinghouse, which is the gas marketing and electric marketing side, became the other main division and maintained its name and identity. I remained as chairman and CEO of Trident NGL, Inc. and also became vice chairman of the new parent company, NGC Corp.

Bruce, thank you very much.

Paul Wommack

This is Joe Kutchin on Wednesday, November 15, 1995. I'm in The Woodlands offices of Paul Wommack, who retired on February 1, 1987. You had been general counsel and senior vice president of the parent company, Mitchell Energy & Development Corp. Give me a few minutes of your pre-Mitchell days, where you're from, where you went to school, a little bit of personal information.

I was born and raised in Texarkana, up in Bowie County in Northeast Texas. I went to high school at Texas Senior High on the Texas side of Texarkana, graduated from high school in 1938. Then I went to Texarkana College, which was then just a junior college, for two years, from 1938 to 1940. Then I worked as a secretary for two years at the new prison that opened in Texarkana in 1940. That was for the Bureau of Prisons Department of Justice. My last job there just before going into the military was secretary to the Classification Committee that interviewed and classified incoming inmates. In August of 1942, I enlisted in the Army. I was 20 years old just five days after the Japanese bombed Pearl Harbor and was classified 1-A in the draft. So, I decided to enlist in the Army in August of 1942 and went from the reception center at Little Rock, Arkansas, to the overseas staging area at Indian Town Gap, Pennsylvania. By October of '42, I was in Casablanca, French Morocco, working in what was called back then the Army of the United States to distinguish it from regular U.S. Army personnel. By this time, I had been promoted to corporal. In fact, this happened at the reception center in Little Rock when the officers there discovered that I knew how to take shorthand and type. There was a shortage of people who had those capabilities and that's when I was promoted to corporal. I went from there to Casablanca and served three years overseas in the North African and European theaters of operations. I spent six months in Casablanca, six months in Oran, Algeria, approximately one year in Naples, Italy, and approximately a year in Marseilles, France, and from there came home to be discharged in late 1945. Then I applied for admission to Rice Institute, as it was called in those days, to complete my undergraduate degree. I graduated from Rice in 1948, and then took the Doctor of Jurisprudence degree at South Texas College of Law in Houston immediately after Rice, and was licensed in December of 1950. I have been practicing law in the Houston area ever since, either privately or in a corporate legal department environment.

You came to Mitchell when?

I actually became an officer and employee of Mitchell on January 1, 1968. Prior to that time I had been associated with Smith and Fulton, an outside law firm that did a lot of Mitchell legal work, from 1964 until 1968.

What were your responsibilities when you came here?

It was to organize and begin the function of an in-house legal department at Mitchell Energy because up to that time, the corporation had no in-house legal department as such but relied on small law firms such as Smith and Fulton to do most of its legal work. Of course, some outside legal work was done even in the early days by the Houston firm of Vinson & Elkins. It was my job to start up the legal department and I was made a vice president. There were no lawyers on staff. The legal department initially was myself and one secretary. Within a couple of years or so, we added David Bumgardner, who had been in the lease administration division of the company prior to that and had gotten his law degree by going to school at night, just like I did, at South Texas.

Do you remember when David came?

I would say probably about 1970. The next lawyer to be added was Homer Penn, who did a lot of work for the company when it was going public, which happened in 1972. From that we just grew to the present situation in the legal department at Mitchell.

At the time you left, how large a staff was there?

At one time, we had about 24 lawyers and legal assistants and secretaries. I'm not sure, but we probably had to reduce the staff some in view of the dramatic drop in oil prices that happened in the early 1980s. We had probably 15 employees in the legal department when I retired.

Your background had been primarily real estate-oriented?

Yes. As soon as I passed the bar and was admitted to practice in Texas, my work for the most part consisted of the practice of real estate law.

How did you deal with the energy requirements of the company when you came?

Homer Penn was an energy lawyer, and we had two or three people who were added there to help with the energy side. In the meantime, though, until that was put in place, I guess Smith and Fulton did all the energy legal work for the company.

What kind of things were you involved with in the early years here?
The first thing was the purchase of the Grogan-Cochran Lumber Company. Actually I had gotten involved in that before I came on staff. I was associated with the firm of Smith and Fulton at the time Mr. Mitchell purchased Grogan-Cochran.

You said a moment ago that he purchased the company and not directly the land.
Yes, May 11, 1964. Later there was a trade for land as security rather than just the claim the prior owners had on the stock. That was the thing I got involved in right away, plus efforts to start a legal department and get it organized properly along company operational lines to serve the company's legal needs.

Let's come back to that land purchase. That is the basis of The Woodlands as it exists today. How many acres? Tell more about that.
Actually, there were about 49,000 acres involved in the Grogan-Cochran Lumber Company. But as I recall, only about 6,000 or 7,000 of those acres of land actually became a part of The Woodlands and the rest was timberlands, to the northwest and north of The Woodlands development.

Tell me the story about being an hour on the phone in Austin.
The company felt like they needed somebody on hand in the office of the Secretary of State in Austin to keep track of any filings that could have been made or might have been made against the Grogan-Cochran Lumber Company up until the very minute the exchange of stock took place. I sat in the Secretary of State's office for over an hour with the phone to my ear, watching to see if anything was filed. Of course it wasn't, but the company needed to be sure of that, that nothing was filed against Grogan-Cochran at the last minute.

Were you involved in any of the negotiations?
That time in Austin was all I had to do with that purchase. Other lawyers were involved, like Vinson & Elkins attorneys. I don't remember specific names, but Tom Fulton was involved from a title standpoint and my part of that was strictly this standby need at the Secretary of State's office just in case anything might be filed against the company that would affect the title.

Were you involved in the planning of The Woodlands? Did George have The Woodlands in mind?

I was not involved in the early days in the planning of The Woodlands. The only thing I got involved in was the supervision of lawyers doing the real estate legal work for The Woodlands in the early days. I did some of the work myself. At one time, there was a separation within the legal department. The real estate lawyers moved out to The Woodlands as soon as the development was begun and an office arrangement made in the first office building, called OB 1. The real estate lawyers were out here in The Woodlands, and the energy work and the corporate work was done downtown from the One Shell Plaza. It was later reunited when the company moved to The Woodlands, back in the fall of 1980.

What role did you have, or did you have a role in the HUD filing?

Some of the very early HUD filings we did in the legal department and I was involved in that. That was at a time when it was a fairly simple thing to file for HUD approval for subdivision of land. This was in the early '70s. I was involved in that, as well as Dave Bumgardner. Dave was experienced in both real estate and energy legal matters. He and I, together, were involved in some of the meetings with HUD personnel in Washington in the early days. I attended as a representative of the company, with company personnel, to discuss the project agreement between the United States Government and The Woodlands and anything that involved, from a legal standpoint, the development of The Woodlands. After the bonds were sold following the agreement by the government to guarantee the payment of those bonds in the total amount of $50 million, I attended the closing of the sale of those bonds and also worked with HUD on the actual grant of the HUD guarantee of the bonds.

Most people have described the experience of working with HUD as very bad. Is that your opinion?

It was not pleasant all the time, to say the least.

In what way?

They were energetic young people for the most part who had a social agenda in connection with the development of The Woodlands, and sometimes it was difficult to get them to accept some practical considerations instead of just looking at the social aspects of The Woodlands development. There never was any doubt about what Mr. Mitchell wanted, of course, which was for The Woodlands to set a sort of standard for new community development with the government's help. That was not really an issue, but just working out details made it very difficult. Most of these people were extremely young. Of course, I wasn't all that old, either.

Does any particular issue stand out in your mind?

Not really, Joe. It's been so long ago that I have difficulty pinpointing a particular issue. Later on, most of those negotiations between the government and HUD, which were ongoing until the bonds were paid off, were handled by Robert Hinton, who had been made general counsel of The Woodlands, assisted by Dave Bumgardner and several others.

Hinton was on our payroll?

Yes, in the early days, as general counsel for the real estate phase of the company's operations.

The preparation of the application was essentially the responsibility of The Woodlands people. McAlister, I know, had a role, and Bob Hartsfield. Did you have an oversight responsibility?

Not really. Back in those days, because they had their own legal department, they operated pretty much on their own.

Would that have been a typical kind of arrangement?

The real estate activities of the company were a foreign thing to the Mitchell company, which had always been an energy company. It sort of made sense, at least after onsite office facilities were available, to put the real estate people out here where they would be closer to the operations people. They didn't require a lot of supervisory involvement of the downtown corporate legal department and energy lawyers.

Just using my own recollection, I remember there were a lot of itsy-bitsy picky kinds of concerns in the dealings with HUD. One of them was related party transactions.

Yes, we had to be concerned about that, of course, and we had a minor problem with the connection between the company and Stewart Title Company, which was opened in The Woodlands, originally called Stewart Woodlands Title Company. There had to be a policy issued in favor of the government arising from the government's guarantee of the $50 million in bonds. In other words, they wanted a $50 million mortgagee's policy because they were taking a deed of trust lien on all of the land involved in The Woodlands. Mr. Mitchell had joined with Stewart Title in establishing a title company in The Woodlands. In view of his interest in the title company and his interest, of course, in the company, they required that we get that policy written as a sort of home office issue at Stewart Title Guaranty Company, rather than by the new Stewart Woodlands Title Company, which was owned 50 percent by Mitchell and 50 percent by Stewart Title Guaranty Company in Houston. So, we worked all of that. Dave Bumgardner and I spent quite a bit of time in New York and Washington working out the quirks in the title situation to the total land involved in the original project to be sure that title was good and we knew all the exceptions as to each individual tract that comprised The Woodlands acreage.

You say each individual tract.
Each individual tract in The Woodlands. I believe Mr. Mitchell's figure is about 300 separate transactions by which we acquired land that comprised The Woodlands initially.

Almost all of those, except the one big one, would have been relatively small transactions.
Yes, they varied in size from half-acre, two-and-a-half acres, to the Sutton interest which was a 2,000-acre interest in 4,000 acres. Robert Mann had the other half-interest in that 4,000 acre tract. It was acquired early on, actually before the development of The Woodlands began.

I presume that Charles Lively was one of the guys most active in lining these tracts up.
Yes, Charles was the representative of the company in contacting owners of the various tracts in the area of The Woodlands, and Mr. Mitchell, of course, was the one who would set the terms and purchase prices that the company was willing to pay. Charles was the contact between the company and the property owner in practically all of those transactions, and Mr. Mitchell directed the process, and I did the legal work in the very early days of the acquisition of these various tracts of land. Charles Lively told me one day he wanted an option agreement. Mr. Mitchell wanted to tie up as many tracts as possible under an option with right to buy. Charles wanted an option agreement that would not be longer than one page, front and back. We designed an option agreement to comply with that. All of these tracts were bought in the name of Mitchell & Mitchell Land Development Com-

pany, which was the early predecessor of The Woodlands Corporation. We discussed at one time buying some of this land in the name of trustees so the word wouldn't get out as to who was buying and why. Mr. Mitchell insisted on buying in the name of the company because he wanted the people who owned the land to know that some responsible company was the one buying the tracts. So that's the way we did it.

That would be unusual, wouldn't it?
Yes. Ordinarily I would think that, both then and now, any big potential land developer going into an area to block up land would rather not have too many people know who it is and what they wanted to do with it. But that wasn't true out here.

What else about The Woodlands?
In '72 when we sold the bonds, we had all this block of land put together by then, but we had to give a deed of trust lien to guarantee the government title to the first lien on all of this property.

You had to give the government that lien?
Right. Secretary of Housing and Urban Development.

So your role was providing support for the various transactions in all the things that were happening in creating The Woodlands.Were you active in the energy side of the business?
No, I was not.

What do you remember about the opening of The Woodlands?
'74 was the opening of The Woodlands, I believe in October of '74, and there had been some very bad weather conditions that were not conducive to real estate development and that slowed down the beginning of construction in The Woodlands. But by '74, I believe, we had completed The Woodlands Inn and Conference Center and some housing in the area.

You were talking about the rush to get it completed.
Yes, George wanted The Woodlands Inn and Conference Center to be in place before they started talking to people about locating businesses out here and buying homes out here and not some promise that some structure like this would be erected in the future. So that's what was done.

What about lighting work areas by helicopter?
Len Ivins, who was the president of The Woodlands at that time, used lighting by floodlights from helicopters towards the end of construction of The Woodlands Inn and Conference Center to get it completed by the date on which Mr. Mitchell wanted it completed, which it was.

Others have talked, very carefully, about working with the Columbia group. Do you have any comment about that?
For the most part, my work with them was very pleasant. I worked directly with Leonard Ivins when we made a contract with Gulf States Utilities Company for them to come into The Woodlands and supply electric service. The people who came here from Columbia who had worked, I believe, for the James Rouse Company, shopping center developers and major real-estate developers, were all responsible people. I remember Vern Robbins and Len and a number of them that I felt were very, very capable and very pleasant and easy to work with.

We've come to '74 and very rough sailing because of cash. I know that Bill Tonery told me at one time, it must have been in early '75, that he had papers ready to go to the courthouse to file for Chapter 11. Do you have any recollection about that?
I didn't have anything to do with that and if that's true, he probably used Bob Hinton to do that.

He also may have exaggerated. My guess is he would not have gone to real estate to do that.
Probably not. If that decision was being considered, the legal work would probably have been done by Vinson & Elkins. That would be my guess. But I was never involved in any plan to take the company to bankruptcy. There was a cash crunch after most of the money from the sale of the bonds was disbursed to meet various company requirements. That was when we sold 100 million board feet of timber to the Georgia-Pacific Company. I worked on that contract, which became the base for all of the company's timber operations even to this day. Georgia-Pacific was later replaced as the buyer by Louisiana-Pacific, which is the current major company in this area due to, as I understood, a spinoff required by the government because Georgia-Pacific had gotten too large and Louisiana-Pacific was spun off.

Was this 1975?

Joe, I don't remember. Records ought to have a more definite date on that, but I don't remember. It was back in the early days of the development of The Woodlands, and I remember that I would attend meetings of company people and representatives of Georgia-Pacific and then sit up at night and draft a contract to have it ready the next day for review again. This took some time to negotiate. But finally we reached an agreement with Georgia-Pacific. Morris Thompson was involved in that, Max Newland was involved in that, and I had to depend on other people to give me the technical advice I needed to prepare the contracts. I had never drawn a contract of that magnitude before involving timber purchase or sale. Mr. Mitchell, of course, was involved. There may have been others.

How much money?
I think the original purchase was either 50 million or 100 million board feet at $50 a thousand.

Was that the immediate source of cash?
Yes, in addition to whatever infusion of capital that might have been supplied by the energy company back in those days.

That's about the time, too, of the embargo and the spike in prices that I've always thought more than any other thing kept the company whole.
As you may have been informed by other people you've talked to, a great deal of the proceeds of the $50 million bond issue was used to pay off liens on property that we had bought on the company's credit to block up the land. Of course, we had to clear the title of all those liens as a part of the closing with the government on the bond issue.

I have heard that from Budd. I've never understood that $50 million really amounted to a relatively small amount for operations.
Right, exactly. It was a lot of money. We had substantially leveraged the purchase of land by paying as little down as we had to and signing a first-lien note for the balance. Those had to all be paid off in connection with the closing of the government guarantee of the bond issue.

The Woodlands got started and got over the hump of this tight-money situation and we're talking now about mid-late '70s. Talk more about some of the major events in succeeding years.

There was one major transaction by which the company acquired several thousand acres of land in an exchange with Champion Realty. We worked out a deal, exchanging some of the land we had acquired from Grogan-Cochran Lumber Company for some of the land that Champion owned on the west side of The Woodlands. Dave Bumgardner did most of the legal work. I started that and then Dave took it over and did most of the work on that phase. Then we purchased—I'm not sure whether any of this land is in The Woodlands, but it did provide us an additional timber source—about 7,000 acres from the University of Texas. This was timberland in areas surrounding The Woodlands, and I worked on that and closed that deal. This was the place from which most of the early timber was sold in the transaction with Georgia-Pacific, from land acquired from the University of Texas.

The things that were happening in energy were essentially under Homer?

Yes, and Smith and Fulton was available to represent the energy side of the company. After Homer left in '81, we hired Tom Battle from Exxon to be the general counsel on the energy side of the company. Later, of course, Battle succeeded me as general counsel of the parent company when I retired back in '87.

You mentioned the Champion deal. Anything else that comes to mind?

As far as I can recall, Joe, those were the major ones, the acquisition of the University of Texas block of land, the exchange with Champion, and contracts under which we joint-ventured some land with Champion in real estate developments outside The Woodlands: West Magnolia Forest, Clear Creek Forest. I was involved in the drafting of those agreements.

I know just from sitting in energy management meetings with you that you were certainly more than well-informed about energy events. Does anything come to mind in that area?

Yes, after Homer Penn left in '81, I was named corporate secretary, which was a position that he had held prior to that, and as such, I attended all Board meetings of the company, and management meetings. Through that means I was kept apprised of energy matters and somewhat educated in energy matters. I was the only company lawyer who attended management meetings and also the only lawyer who represented the company at annual stockholders meetings, directors meetings, and this type thing.

Any specific energy issues or deals that seem important to you?
Not that I would have been directly involved in, Joe. During that time, particularly up to '81, Homer Penn was the one mostly responsible for in-house energy legal matters, aided by the Smith and Fulton firm, in day-to-day energy legal work.

And for SEC work and other such specialized things you'd have gone to Vinson & Elkins?
Yes, Vinson & Elkins. Taking the company public was handled by Vinson & Elkins. Homer Penn did a lot of work in-house on that and I might have been involved peripherally, but I didn't have the major responsibility.

Do you have any observation about the company that you've known extremely well for 30 years? How would you characterize it?
The thing that would come to the front and the surface was all the activity the company was involved in, the excitement, the increase in company areas of involvement from just purely an energy company to the very viable and very active real estate development company—that's what interested me. There never was a dull moment in company operations, especially after it became a serious real estate developer. That was my chosen field and I loved every minute of it because it was exciting and I had a lot to do with the purchase of land at The Woodlands. For instance, one Labor Day weekend, which I believe was in '72, Dave and I and one secretary met with Robert Mann, who owned the remaining undivided one-half interest in the land along the freeway where the mall is now located, and Bank One and other business interests. George had been negotiating with Mr. Mann. We spent the entire Labor Day weekend working on the transaction and had to bring Mr. Mitchell in from the beach, where he was vacationing at Galveston, to make some last-minute decisions. To make a long story short, Mr. Mann had met us at the Exxon hangar at Intercontinental where they had a conference room. We met there and hammered out the contract. Mr. Mann was trying to get away to make a trip to Washington and that was the deal that we hammered out that weekend for the purchase of Mr. Mann's undivided one-half interest in 4,000 acres. That was a very exciting, nerve-racking type of work until we got his agreement. Everything about The Woodlands has been very, very exciting to me, even though later on we had a pretty sizable staff that was primarily responsible for The Woodlands. I was still involved and just enjoyed every minute of it. The company was growing and up until the crash in the oil industry back in the early '80s, there was plenty of money to do

what everybody wanted to do. Those were very exciting times, and even during the hardship times we went through, we were still moving forward in the overall objective of developing The Woodlands. We originally planned, as you probably know, to have it fully developed by 1992. That was the date on which the final installments on the bonds were due, but we had two recessions during that period from '72 to '92 which slowed things down. If we had moved according to the projected plan, we were supposed to have 150,000 people living in The Woodlands by the fall of 1992, but in looking back and considering the overall economy of the Houston area, I think the company did a remarkable job in carrying out development. Even though we didn't achieve the population goals, the quality of the development has been beyond reproach.

Speaking of that, were you involved in George's desire to be in the extraterritorial jurisdiction of Houston?
Yes. We hired a small firm in Houston that was a specialist in this field to handle that. Dave and I were involved very much in asking Houston to extend the ETJ to cover most of The Woodlands area.

Why don't you talk about your understanding of George's motives in doing that?
As I recall, we didn't want The Woodlands to just become a bedroom community for Houston. We wanted it to be a vibrant, complete development with housing of all levels, retail, business and light industry. Foremost in his mind was what had happened to the west of Houston where there were a number of small communities with zoning boards telling the developers how to develop their property. As I recall, the main reason that he wanted this to be in the ETJ of Houston was so that type thing wouldn't happen in The Woodlands and so development could continue with the developer calling the shots on how to develop the individual tracts in The Woodlands, and how to divide it among residential, commercial, light industry and whatever you expect to find in a complete city.

That apparently has turned out to be an unpopular notion among residents. I still think it's the right thing.
I'm a resident here. I've lived here since 1979 and I think we made the right decision. No question in my mind. But always you have some people who diverge in opinions, to which they're entitled, but fortunately for us they are in the minority.

I'm not sure that's correct.
Well, it may not be, but at least so far it has accomplished the goals we set.

I don't know that you would remember the Saturday morning after the rig went down in the Gulf. Four or five men died in that accident and you and Joe Williams and I and two or three others were at One Shell Plaza. The NBC bureau chief wanted to get out on the production platform that, I think, was nearby. And I said, "Oh my God, I've got to deal with the lawyers, lawyers always say no." I talked to you and you simply sat down at a typewriter and in about 47 seconds you pounded out a release which the NBC guy signed without even looking. I thought, OK, this is not going to be such a bad job after all. I've admired you for that ever since. You've always been a guy who said, "Yes, let's find a way to do it," rather than simply saying no.
I appreciate that. I've always had a goal in mind in legal work to try to be as direct and factual as possible and not use strictly legalese in preparation of contracts, so that both sides would come out of deals feeling reasonably happy. That's the policy I followed in doing Mr. Mitchell's legal work.

What have you been doing since your retirement?
For one year following retirement, I served as a consultant to the company and after that then I just decided to open an office for private practice of law. In the beginning, I was accepting work from any source I could get, but I became more and more involved in representing GPM, Inc., which, as you know, is the company that does Mr. Mitchell's private investment work and that sort of thing. I'm exclusively involved in the practice of the law and do not take any outside work anymore. That situation's been going on ever since the one year as a consultant.

Thank you, Paul.

Jack J. Yovanovich

Today is Thursday, July 13, 1995. I am Joe Kutchin, retired vice president of communications of Mitchell Energy & Development, and this morning I will interview Jack J. Yovanovich, who is senior vice president of land for Mitchell Energy Corporation. Tell a little about yourself before you joined the company—where you're from, your education, and so on.

I was born in Fort Worth on November 17, 1935. Went to high school in Fort Worth, later on went to Arlington State College, which is now a part of the University of Texas system, and majored in geology. After that, in 1957, I transferred to the University of Texas, got my degree, which was a bachelor's in geology, in 1959. It was a pretty tough time for the oil and gas business. There was very little demand for geologists or technical people in that area. So I worked a year on a seismograph crew and doing other things and then decided I'd go back to UT and get a master's. Well, actually, I decided I was going to get an accounting degree and I started in the business school. There I found an old friend of my father who everyone called "Mister Z." His name was Zlatkovich and he talked me into going to graduate school, saying that another undergraduate degree wasn't any better than having just one, so I decided to go into graduate school and got a master's in petroleum land management from UT. That was 1962. And then in '62, I went to work for Continental Oil Company, originally on their training program, so I moved around a few times. During the first year, I actually started in Durango, Colorado, was transferred to Oklahoma City, then later on I went to Abilene, and from Abilene I was transferred over to Midland. It was in 1965, I guess, or thereabouts, before actually I got transferred to Midland. I saw an ad in the paper, in the AAPL magazine, from Mitchell Energy. It appeared as if they were just looking for a field landman. I had been out away from what I considered the mainstream because everybody thought about Houston as being the big oil center, so I was really curious to see what kind of salary they were paying compared to what was going on with me at Conoco, and so I sent in a letter inquiring. That was when I was in Abilene, then I was transferred to Midland and then I got a call from Mitchell Energy. I came down to Houston on an interview in '65 and I met with Leland Don Brooks (then head of the land department) and Morris Thompson. It was in the old Houston Club Building and they interviewed me about 10 o'clock, and by the time they finished it was about 11. They didn't invite me to lunch or anything. They just turned me loose on the street by myself and I think my flight back to Midland on Saturday was 3 or 4, so I just wandered around Houston until it was time to get to the airport and go back to Midland. Then, after awhile, I got this offer for $10,000 for this job. And when I accepted it, I told my boss in Midland, and he apparently checked with the Houston office of Conoco and said, "I don't know, you maybe ought to reconsider. I'm not sure you want to go to work for George Mitchell & Associates, Inc." I'd already made a commitment so I decided to come down, and then when they were first bringing me around to introduce me to people, they took me to Budd Clark. He was one of the first ones to meet and he looked at my employment papers and my salary was to be $10,000, but it was broken down into monthly increments of $833.34. My first contact with Budd was "Hell, we ain't doing that, we'll round it to $830 a month," which meant it was less than $10,000 a year. Even as a kid, I'd thought of reaching a plateau of making $10,000 a year. Everybody thought at that time that if you made $10,000, it would be really great, and here it was Budd cut me back in one little swoop, not even thinking one way or the other about it and I was kind of ticked. But I'd already given up my job at Conoco, so I couldn't do anything about it. So I just stayed with Mitchell, and then it's funny because later on I'd gotten a nice increase from George, and Budd has always thought that I used that to get even with him 'cause he screwed me on my initial salary.

What job did they hire you for?

They actually hired me for assistant manager, and that's what's kind of surprising because initially the job sounded like a field job, but then they were looking for an assistant manager and at that same time they had Don Brooks, who was manager of the land department, and they had John Curtis, who was assistant manager. So for the first five or six months, we jockeyed around about my title. Curtis was sometimes assistant to the manager and then one of us was assistant manager and then later on it was switched around to where I was assistant manager and Johnny was assistant to the manager and it created all kinds of havoc except that Johnny was really a nice guy and at some point down the line he came to me and said, "Jack, it's very obvious that they're planning bigger things for you." He said that he recognized that and he was going to move on and go somewhere else. And he ultimately left and then later on I was made manager of the department and then Brooks went into real estate and then I made vice president and later on senior vice president.

Let's move back. Do you remember the month if not the date that you started?
August of 1965. And this year in August it will be 30 years.

Please describe the company at the time, what its business was.
First it was GM&A and that was back in the days when they were primarily operating in North Texas. Johnny Curtis really more or less handled the North Texas operations through the Bridgeport office at that time. He had a guy named Bill Hutto up there and then later on Jack Brackett was in charge of that office; but they all reported to Curtis, and Brooks kind of handled basically the Gulf Coast. The company was also active in the Gulf Coast. They had properties down in the Valley around Hidalgo and in through there. The business as I saw it in those early days was primarily confined to North Texas and the Gulf Coast and South Texas area, so there wasn't much in Midland for Mitchell, or anywhere else at that time.

Was there any real estate business then?
Not really, although it was somewhere about that time that I think George started putting the block of acreage together around where The Woodlands is right now. Brooks went over and worked in that area. Charles Lively, at one time a landman for the oil and gas group, was transferred over there and helped with the acquisition of the properties, but in terms of my job, it was primarily focused on the oil and gas business, and Brooks, and later on Lively and that group, worked on picking up Woodlands area acreage with George and Frank Karnaky and some of the old brokers in those days. I'm sure it wasn't too long after I got here that Don Brooks went over into real estate, and I'm sure he was probably handling some real estate in those early days, but I think in those early periods is when they started.

From the beginning your main responsibility has been land acquisitions for oil and gas?
It's been strictly oil and gas.

But you've gone beyond straight land acquisitions, you've been involved in general acquisitions, as well, haven't you?
Primarily producing property acquisitions. And that's part of what we do now, too, we handle the producing acquisitions. At one time, we even had within the land department a specific group that handled nothing but producing property acquisitions. 'Even had an engineer in this group and vari-

ous acquisition coordinators. That has been disbanded basically because we've been shrunk down, and we've let the land guys handle their own acquisitions as they come up in their areas. The funny thing was that for the first three months that I came to work for Mitchell, I thought I'd made the biggest mistake in the world and I was pretty miserable. After that, it seemed that I just kind of got in the swing of things, and this is the strange part and it's not a self-serving statement, but it's just the way I felt in those early days, after we got in the swing of things, I could hardly wait to get back to work every morning. And I hated to see 5 come along and people going home because I wanted to continue to work. It was just that exciting. Not only that, we were leasing in the Galveston Townsite, but I was working all day in Houston and I tried to get Brooks to let me work in the evenings leasing at Galveston after I got off the office job. He didn't want me to do that. He really didn't want me going to Galveston and working two jobs. Of course, later on I became manager of the department and I did a lot of other things in Galveston and Crystal Beach, like getting permits to drill. We even had an election at Crystal Beach that George told me we couldn't win, but we did. The election was about supporting offshore drilling at Crystal Beach.

Start at the beginning. LaFitte's Gold Field at Galveston is partly onshore and partly offshore. And that was the late '60s, early '70s?
Early '70s. The Galveston play had the full force of the company. By that I mean it had me working on it, it had George working on it, had the geologists working on it, the legal department, the engineers and everybody involved, all were involved trying to get permits to drill and not only offshore but onshore, too.

That's because George felt good about the geology?
Well, yes, there was a big structure but also because it was Galveston, and you'd be operating within the city limits. At the same time that the Galveston play was going on, up on Bolivar we had also bought some leases, state oil and gas leases that were offshore, and what had happened was that Crystal Beach extended their city limits way into the offshore area and incorporated our oil and gas leases. They put restrictions on drilling and required approval of permits to drill. The time that we were trying to do this drilling was during a big environmental movement and everybody was opposed to offshore drilling. We had the full use of all the company's resources to work Galveston, but at that same time I was working Crystal Beach on my own. In other words, Crystal Beach was my sole responsibility while Galveston

seemed to be the whole company's responsibility. Crystal Beach City Council met on Friday nights at 7 p.m. and that's when I got our drilling permits approved.

How much acreage was involved? And wasn't there something special about getting the drilling permit?
It was our 176 Field, and if I had to guess, I'd say we had 6,000 acres offshore. But an interesting part was that the City Council met on Friday nights at 7 p.m., so any time you had to get a drilling permit you had to meet with them at that time. I had been meeting with various groups, such as the church groups, men's groups, women's groups, and making talks on the energy crisis and the need to support offshore drilling. Depending on which group you were talking to, you might be meeting at church or at a saloon or beer joint–no matter. What we needed was support for offshore drilling. They ultimately had a referendum on whether they wanted offshore drilling. And George told me that I could never win such an election at the height of the environmental movement, and it just made me that much more determined. We won the election.

You say you became a PR man?
Yes, in a sense. They had an election at Crystal Beach and I was partly responsible for the outcome. It was just a referendum, I mean it was to give the Council members guidance. One Council member or alderman that constantly opposed the drilling was a guy named Land. He had a grocery store. And so one night, a Friday night, I went to a Council meeting to get a drilling permit and I took my wife and my son with me. Jack, Jr., and Betty sat in the audience as I made my spiel for my drilling permit and, of course, the Council members had their opportunity to speak up, and this guy, Land, just really attacked me from one side to the other, and the whole time that he was attacking me, I never voiced any kind of counterattack at him because I knew that even though there were other aldermen on my side, if you were to attack one of their own it probably wouldn't go too well for you. So Betty (Mrs. Yovanovich) told me later on that my young son Jack said, "How can he talk that way to my daddy?" And the guy had all of his say, and I answered all the questions very calmly, and then the City Council voted–I don't remember for sure how many aldermen were present–but he was the only dissenting vote. I always won approval for every one of the drilling permits, like six to one, because Land was the only one who constantly opposed us, and so at Crystal Beach I always got all of my permits. We even donated to the Crystal Beach

Volunteer Fire Department. In addition, we had a beauty contest and a parade. We had two young girls who we sponsored as "GM&A's Beauties" in the contest. So I kind of felt like Crystal Beach was my own little project. We often also worked the Galveston play at the same time as the Crystal Beach problem, but with a lot of people. George and I bought the piece of property where the San Luis Hotel is located. It was 23 acres. We were trying to get approval for offshore permits at Galveston and we were getting so much protest that the City Council said, "Go find you a piece of property on land and directionally drill." So we found this 23-acre site. It was owned by Bert Wheeler, but it was handled by a lawyer for him under some corporation name, like Palmetta. I think Ralph Balasco was the lawyer's name. George and I met with Ralph Balasco at his South Main office to talk to him about buying the 23 acres. In those days, George had a little bit of a twitch in his eye and he didn't drink too much caffeine. We were talking to Balasco about the piece of property and George was sitting on the edge of his chair while Balasco kept filling George's coffee cup. George was getting excited during the conversation as we were trying to buy this property and the coffee was not helping the twitch. Finally, it didn't look like we were getting anywhere. I was sitting there with George and just listening to George and Balasco. George was doing most of the talking, and I finally saw an opening so I said, "Well, Ralph, would you be willing to do this? George, would you be willing to do this?" They both said yes and I said, "Well, why don't we do that deal?" So we made that deal and we walked out to the parking lot, and George said, "You know, it helps sometimes when two people go to make a deal, doesn't it?" or something like that. We ended up buying that piece of property, seems like to me, for about $1 million. It wasn't very much.

This is where the hotel is?

Yes. Then we went back to the City of Galveston and proposed wells on a portion of the 23 acres. This was about 1972. We gave a rendition of the property, including our proposed drilling site and how it would look with a shielded drillsite and how we would put a hotel on it. This was long before the San Luis ever came about. I've even got pictures in a scrapbook of the plan presented at the Galveston City Council. Then all the people who lived around our new site came before the City Council and protested about the new site. So the City Council said "Go back offshore." The City of Galveston was responsible for our purchasing the San Luis site and then they denied drilling from the site.

Never did drill there?
Never did. They allowed us to go offshore and finally agreed to give us a permit to drill three wells offshore provided we didn't produce any oil from there and provided that we took our gas ashore at about 113th Street. So we had to lay extra pipeline at an increased cost to produce the gas. We could never have more than three producing wells offshore. If one of them went dead you could have a replacement.

These were the low Christmas trees?
Yes. Before we ever drilled a well offshore at Galveston we felt like vacuum cleaner salesmen because Howard Kiatta, Joe Ramsey, George Roberts and I had to talk to every business along the seawall at Galveston about our offshore drilling program. I took Howard Kiatta and we had a spiel that we were going to give and I was really concerned. You walk into a business a lot of times, they may not have time to listen to you, or maybe they don't want to listen to you. The first place I went into on the seawall was a grocery store that was owned by a guy named Milosovich. And the reason I went there is I knew he would listen to us and we would get our whole spiel out, wouldn't get rejected and we would start on a positive note.

That's because your name is Yovanovich?
Right. Anyway I started there, and we made that whole beachfront, and surprisingly every place that we went into listened to our complete story. We had a little brochure we gave them, and when later on I met up with Ramsey and Roberts and we talked about all the places they had been, everybody had listened to their spiel except at a service station where some guy drove off before they'd actually finished. But everywhere we got good acceptance. But like I said, we, Howard and I, even reminisce about that today, about how we felt like Hoover salesmen just walking in on people with our little magazine that we took for information purposes about our drilling offshore along the seawall.

Yes, I remember that magazine. Were you in charge of preparing it?
This was done as a group effort. I think Joyce Gay was involved in that, too.

So the company finally got going with its three wells.
Yes, and also we had problems drilling, even in the townsite, primarily because you had to drill north of Broadway, which was in the heavy industrial area, and because part of the structure was south of Broadway, so you

had to directionally drill to get to the area of interest. There were all kinds of problems with subsurface easements, and we came up with a lot of ideas on how to handle that problem in the land department. We also got involved in helping the city draft an ordinance to allow us to operate in the city. So we were instrumental in getting that done, worked on it quite a while. It was an all-consuming project. I remember we bought about 9,000 leases in the townsite. Just to tell you how consuming it was, George would come around almost three and four times a week, sometimes twice in a day, and always tried to control every bit of the leasing. Now you think about this: You're leasing lots that are less than, well, some of them, around 1/10 of an acre, some of them 2/10, and George is coming around wanting to know what you're doing here and how you're doing it, why you're doing this and why you're doing that and he was driving me and my guys crazy. We would start to try something new, but before you'd even give it a chance to develop to see whether it was working, George would come in with another suggestion. So, before we could try out one idea, he'd have another one and kept on, and my guys were just going crazy. So one day I finally said, "Look, George, you've got to leave us alone. You know, I can look at you right now and I can't tell whether your hair is growing. But if you wait a week, OK, then I can tell what progress has been made." And I added, "I'll tell you what, every Monday we'll give you a written report, and if you want to change anything we'll change it and we'll work it that week and if you want to change it, we'll change it week to week. But we can't do this blow-by-blow and daily making changes to where we're not even seeing if the new ideas are working," so he agreed with that. We finally got all of Galveston put together and got our wells drilled and the field has been there ever since.

This would be about 1972 that it was brought under control?
Yes.

Are we doing much production there now?
We still have some production in the townsite and still have some of the offshore production, but it's down on its last leg at this point.

OK, so LaFitte's Gold was a good field but not a really super one, is that fair?
Yes. It was an interesting project in that there were a lot of problems that we solved. The structure was there, no doubt, and other people were aware of it, but nobody would tackle it. When George saw the structure--I think he

got it from interpreting some seismic that Gulf or somebody had shot—and being from the Island himself, he figured, well, you know, we could do something. As a matter of fact, we even ran a seismograph crew at midnight down the esplanade on Broadway. Then we had to settle all these lawsuits where people claimed that their property had been affected. We even had to settle with one guy, 'cause he claimed that we cracked the foundation of his house but he had grass growing out of the cracks, so we knew that we didn't have anything to do with his foundation.

Actually the company was kind of a pioneer in the environmental wars, showing that you could produce oil and gas without harming the environment.

Right. As a matter of fact, Galveston was a big tourist area and there was always the idea, well, that offshore rigs were going to hurt their tourism. But George was always convinced that you could protect the environment and do the drilling. We talked about Crystal Beach. They incorporated in July of 1970, approximately 60 square miles of the Gulf of Mexico. Now figure that out. They couldn't give any services, but the reason they incorporated was to include our leases in their city limits and ultimately they were charging us $7,500 for a permit to drill. Unheard of! I told you earlier that I thought we had 6,000 acres. I see here from this article where we had at that time, Mitchell and our partner, Diamond Shamrock, jointly owned approximately 10,000 acres within the incorporated area. All state-owned leases. Even before tackling Galveston, George didn't go about it haphazardly. He sent me and one of the engineers to California to see how they were doing their drilling, how they hide their rigs.

Yes, I remember those pictures. But you never did do it like that?

No, we didn't have to, primarily because it was not really California. We did buy a warehouse in Galveston north of Broadway, a cotton warehouse. We knocked the roof out of it and slant-drilled some wells from there. We were never really drilling in the areas where people were living—our rigs were located either at railroad tracks, cotton warehouses, that kind of location.

So over time you produced a lot of hydrocarbons and provided a lot of revenues for both Crystal Beach and Galveston?

Yes. Crystal Beach produced a while but it wasn't any big success and probably it didn't make a lot of money. It was part of an offshore program we had with Diamond Shamrock where we were the operator and they were our partner, and also Natural Gas Pipeline Company of America funded a lot of the drilling to get to call on the gas.

That's reasonably rich gas, isn't it?
I don't think the Crystal Beach gas was.

No, I mean the Galveston area.
I think so.

I just had the idea that I've seen liquids plants there.
They had a liquids plant somewhere off Stewart Road, I think.

Why don't we wrap up now, and I'll come back another time.
I might tell you about one other thing. Somewhere in my early history here, I made the comment to George about you couldn't do something. And he showed me you could.

Isn't that scary about him?
Then when I was working in Crystal Beach and we were going to have that election he told me I couldn't win it and I decided to show him I could. And it was all, I think, motivated by that first time when the situation was reversed, so I was determined that I was going to do it and I did. A lot depends on where you are at a given time and where the company is at that time. In the early days, the reward was really for a guy who was fairly aggressive and would work hard. The company rewarded that kind of behavior. I've often said, "You take the early days of Mitchell and put in a landman who would have been competent with a major company, he would not have been competent with Mitchell."

That's not only true of land, I think. The whole feeling of the company—ever since I've known it—is entrepreneurial.
I've often said that people who were competent in big oil companies, if we'd put them in here they would not have flourished. Our company rewarded a certain kind of guy. In my case, I didn't want to fail no matter what. I'm almost like Jimmy Conners. I don't know whether it's winning or just that failing is so bad to me that I don't want that side of the deal.

The other side of the coin is that you're not unreasonably penalized for failure, you're actually rewarded for making the effort.

I don't like to fail. It is a side of me that I don't want to do. You always felt like, in this organization, it's not like in a big company where maybe your boss is not aware of what's going on. The chairman of Conoco, the president, they didn't know whether I screwed up in Midland or Abilene, where I worked for Conoco. But it's like here at Mitchell, George's eyes were on me every minute, like he knew everything that was going on. It was like there was that much more pressure in my mind to perform and to never lose. That is my perception of it.

Well, anybody who knows you knows you're competitive as hell.
So I didn't want to lose because of always wanting George to be satisfied with what I had done. That puts a lot of pressure on you a lot of times.

But there's a reward that goes with that.
The thing that happens to you then is that you've got to realize that as the company gets bigger and bigger, it becomes more and more like a major company. And that same aggressive, entrepreneurial style that you had in those early days that got you where you are, you can't operate that same way in later years. You've got to start adjusting and changing. If you stay like you were in the early days throughout the whole time, I think it hampers you from progressing later on because the company is changing. And if you don't change, you get left behind.

You go back for a number of years before me, but even in my time we've become a completely different company, for better or worse. Clearly it is just not as much fun.
No, it's not and in those early days it was exciting. No question about it. And you felt a degree of accomplishment. Although I said I still enjoy coming to work, it's not like when I told you I could hardly wait to get here the next morning and it was like, "Hey, man, we need to get back on this baby and get it going." It was really a happy time. Even my wife told me at times during that period, "You have got to remember you've got a young son who is growing up and you hardly see him," which was probably true because of all of the activity that was going on in the office and at Galveston, but it was fun.

I want to talk about your views and experience in North Texas.
North Texas is an area that requires constant diligence and effort. In other words, it's not like some Gulf Coast prospect generally where you drill and

get a dry hole and that's finished and then you go off and think about something else. Or where you get a producer, because generally if you get a producer you may only have five or six wells at the most to drill and it's finished and you then go on to something else. It's like it's neatly tied in a package. North Texas isn't that way. It goes on and on and on. It's like you're constantly adding to the information and you've got all these wells to remember, you've got all these leases, you've got so many leases that are expiring, so many problems with pooling and everything. It's like an area that requires a lot of attention and diligence, yet doesn't give you really the day-to-day satisfaction that you might get from working a Gulf Coast property. You kind of get burned out in North Texas, much like a guy gets burned out if all he has to do is work titles. If you know anything about oil and gas and land work, if all a guy ever does is check titles to find out who owns what in the courthouse, it gets old after a number of years. What is the exciting part about land is cutting deals and negotiating trades, getting wells drilled, seeing the results. You don't get the same satisfaction with North Texas land work.

It may be boring, but it's paid your salary and mine for a while.
Yes, I acknowledge that, but there's something to be said for a person's feeling some satisfaction and that's something that you don't always get in North Texas. And Property Administration is a lot like that. They do a lot of work that goes unnoticed. They never get any compliments, any pat on their backs. They do a very responsible job and have a lot of work to do, but no credit for anything.

It's just not the nature of the job, whether it's here or anywhere else.
That's true.

* * * * * * *

First, let me say this is Joe Kutchin interviewing Jack Yovanovich, second installment, on Thursday, July 20, 1995, in Jack's office in The Woodlands. When we talked the other day, you compared Mitchell's main businesses to a phenocryst. The dictionary defines that as a conspicuous, usually large, crystal embedded in porphyritic rock. Your idea was that, over time, a number of the company's accomplishments are kind of equivalent to such crystals.
The point that I was making by that analogy was that when you look at the company as a whole, the events that have taken place over the 30 years that I've been here are kind of, more or less, vanilla except for various fields or

projects that were a real success for us, and those kind of stand out. I kind of picked those things out, comparing them like big crystals within a fine ground mass, and the ones that I picked out were, maybe, Galveston—I don't know how successful Galveston was, but it was a big project for this company—and then others were the Limestone County properties, the Polk County properties. From the Polk County properties, one great success was the Ike Smith well, which probably produced in excess of 30 Bcf of gas.

Since when?
Since its inception, which is probably somewhere in the late '60s, I would guess. It's been a while, maybe late '60s or early '70s, somewhere in there, but it's produced a lot of gas for us. Big discoveries or big projects haven't been that many over the years.

Is that somewhat by design?
Well, if you could, every day of the week you'd find you a Limestone County, but there are a lot of areas that we went into where we kind of played things close to our vest—North Texas, Gulf Coast—and then we began to expand. We broadened out as far as Ohio and then in the Clinton sand and we probably spent a lot of money there. We really didn't make money. The Ohio play was to be like a Boonsville because you had a lot of productive formations of the Medina or Clinton across a broad area. It was close to a market, so the idea at one time was that it would be like another Boonsville. Its marketplace is the heavily populated area in the East. But we never got to the point where the cost of drilling those wells would be low enough so that you could make a good profit based on what you could get for your gas. So we nursed and piddled along in that area over a period of time and nothing really ever developed of any great consequence, and that was kind of one of the areas that we shut down. And we moved into the Rockies, and worked the Rockies for a long time.

That's been successful, hasn't it?
Well, not really. We've kind of pulled back out of the Rockies and basically what we've got in the Rockies now is just the Hell's Hole area, about 70,000-75,000 acres there. We're developing some production. Another area that we put some money into that wasn't too successful for us was the Austin Chalk. We played that area, probably put in close to $40 million.

We were late in that game, weren't we?

We were late in getting in that game, but it's still one that we got into regardless. The big items have turned out to be North Texas, Limestone County, Polk County. We used to have some properties early on when I came here that we've had for a long time in South Texas, like in Willacy County the Oberg Geis units that we were doing a lot on. And then, we even went out to Whittier, California, and got involved there. We've kind of pulled back from all of those areas, and we're back now concentrating on the core areas, which are going to be East Texas, North Texas, Gulf Coast. We even got involved in New Mexico and got some pretty good production there. But we've pulled back in that area, too. We're now trying to market our undeveloped acreage in New Mexico, but we're going to keep our existing production and produce it ourselves, even though we've closed our Midland office.

Are you saying that our "bread and butter" continues to be North Texas, plus Limestone?
Yes, plus East Texas, the Gulf Coast area.

These are specific fields, aren't they?
No, they're broader than the fields. In North Texas we'll continue to develop our existing fields, but in East Texas we're also looking for new prospects. We're also trying to buy production in those areas, much like Lake Creek, which is an old area that we ended up buying and where we've done a lot of drilling and done real well.

Was that quite recently?
Yes. Well, we went back and bought some of that stuff that hadn't been worked recently. We've had that for some time and we started drilling and developing it a number of years ago. We've enhanced the value of those properties through our drilling and completion techniques to the extent that we've got a nice prospect there. This is the same area where we're buying Mobil's production at Lake Creek. We're supposed to close on that on August 15 for $22 million. We're buying property there for further development. That's another one that's been pretty good. I guess that over the 30 years that I've been here, we've probably been through five or six exploration managers.

Has there been anything out of the ordinary?
Not really other than the land work is basically pretty much the same, buy-

ing oil and gas leases, curing titles. There are some areas where titles are more difficult and a lot of times that's in East Texas where you've got a lot of problems with titles, so titles become a little more difficult in certain areas. Even in North Texas where you've got all metes and bounds and small tracts, you've got a lot of different titles to examine. But basically, it all depends on how much competition you've got, how difficult it is to lease and how much you have to pay. Basically the work is all the same in checking titles and finding out who owns what and buying it.

You've been here 30 years. Can you tell me about the one or two or three things that you're proudest of that you've been involved in?
I think that over that period of time we've improved the land department. I think we've improved the image of the company over and above what it was in the early days. I've always tried to stress honesty and integrity in the land department, going on the premise that if you can't trust somebody, it's going to be difficult for you to ever work with them or get anything accomplished. I've always said that if you've got an honest man, you can always predict what he's going to do because he'll always act like an honest man and you'll know how to deal with him, but if you have somebody who might lie to you or you can't trust, then sometimes he'll act like an honest man and sometimes he won't. And so you can never predict what he's going to do. So I always wanted us to be up front, just to be open, and even if we do something wrong, screw up, be honest about it, because if you're honest you can get a lot of people to help you solve the problem, but if you bury it or hide it, then it just gets worse. I think the land department over the years has improved in that area, and I think, also, that the company has got a better reputation now than it had in the earlier days.

Are you talking about the land department now or the company in general?
That was a companywide statement. For example, let's just say that you commit to buy a lease from a landowner and your geologists change their minds and now they don't want it. Well, do you buy that lease or do you tell the land owner, "Hey, we don't want it now," when you've already committed to buy it? Now he doesn't have anything in writing, but you committed to buy his lease. Do you go through with it or not? It may have been in the early days we wouldn't go through with it, and you might end up with a bad reputation. I know when we were working on the Austin Chalk, one of the companies that I made a deal with told me that his president didn't want to do

any business with Mitchell because we had contracted in the early days for a dry hole contribution and apparently there must have been some differences on that, we wouldn't pay the dry hole contribution and this guy, the president of the company, had harbored ill feelings for 20 years about it. Well, this particular deal that I made with their land guy–I'd known him for a number of years and I cut a good deal. I could have paid him even more and still been within what we had decided we would spend for the deal. I told him, "You know that deal that your chairman up there is upset about? I'm going to give you an extra $5,000 on this deal and I want you to tell him we're making good on the dry hole contribution," and that kind of squared things.

Is that a company you still do business with?
Yes. In the early days there were problems with paying bills and being timely. I think I've told you that the first time when I accepted the job at Mitchell, my boss at Conoco checked out who I was going to work with and then came back and wanted to make sure I wanted to go to work for the company because the reputation wasn't that good. I didn't know how good it was, but things have improved over the years and now Mitchell has gone to where I think it's got a good reputation, and I think that's important for a company. Nobody likes to say, "Hey I work for so-and-so company," and they reply, "Oh, that company."

How did the improvement come about, what caused it to get better?
I think people. I think just improvement of the people and also I think that you have to be able to stand behind your guns on certain things. I think it was important for the land department to have a bigger input into things and opinions about how to do things. If you've got somebody who's willing to make a stand for what he believes is right, then everything follows from that. Everybody else feels confident. But if their boss won't back up things, how are the people underneath him going to back up things? I can remember one time that I farmed out some acreage and made a pretty good deal and George was just chewing me out on where I had the authority to farm it out. All the guys that had approved it in this meeting didn't say a damn thing–and they were all behind it. And I remember when I left that meeting, afterwards, that the exploration manager came up and said, "Jack, you know I'm behind you 100 percent." He didn't say anything when George was present. I was just getting chewed up, but one thing I can say that's true about George: He's chewed a lot of times but when I've said, "Well, George, this is

what I've committed to do and I'm not going to back out of it. If you want us to back out of it then you go tell the people," he'd say, "No, I'm going to stand behind you," and he's never undercut me. That's the part that I think is important: Will you stand behind your word, and if you tell somebody you'll do something, do it.

In the company's organization, you're right up there among the senior management. Does the person in charge of land have that relative position in other companies?

It's unique here in the sense that for this company, we have separate groups, like land and exploration and production that all are on equal footing. To me that is a better way to operate a company, more so than where they have put, let's say, land and geology under one exploration manager. The reason I think it's better, I liken it to our government where you've got executive and judicial and legislative branches. It's like they're checks on the other ones and so if something doesn't seem right, we can step up and say so and put our opinions in and not feel that somebody else is going to control it.

Over the years, with your relationship with George and with your experience, you have a bigger say in how things are run in this company than, say, a guy in charge of land at another independent.

I don't know exactly how all the other independents work. There are some strong land managers at other independents. Take Enserch. Their man who was in charge, their landman, is now president of the whole Enserch group there. So I would say that he had a pretty responsible role. One of the guys who was a district landman for Conoco ultimately became the exploration manager, having land and geologists under him, too. I would have to say in most cases where they've had exploration managers, there have been technical people in that job as opposed to, say, people from the business end or land end. But here, I'm proud to say that land plays a big role. I think it's because of our ability to negotiate good deals. It's a pretty big function, particularly in the independent. It's important to be concerned about not only the big picture, about watching costs, but even some of the smaller costs. At one time, we had about 4,000 employees, and if everyone just made one small mistake a year of $1,000, the bottom line has been affected by $4 million. And that's a pretty tidy sum of money. But, even today, somebody was saying about me, "He not only watches his own money, but he watches the company's money just like he watches his own." And I really believe that, and I've often said that I didn't care whether you were sweeping streets, whether you were the president of the company, or what job you had, you ought to treat that job with dignity and do a good job of it because that job provided you with a way to take care of your family and to do all the other things that you wanted to do in life. So whether you were sweeping streets or are up here in this job, all jobs are to be treated with that kind of dignity.

What kind of acquisitions have you been involved in?

We haven't done any great big acquisitions. In the early days, we'd put together some of those partnership programs. For three years, we had a partnership program going with outside investors who were really not investors. The general partner, who was Joseph Boneparth, looked after the investors. We were involved in those joint ventures in the early days. We did some of those things in addition to having our own internal investors like Bob Smith and the Oshmans, Riddells and those kind of people, so doing land work at Mitchell was quite different than doing land work at Conoco. In those early days we didn't have really good maps here, like when I was at Conoco. There, when you saw a lease on the map and it said "Conoco," you knew Conoco owned it. When you came here, you had a tract of land in the name of Mitchell & Mitchell Properties, Inc., you didn't know who owned it, you had to go to accounting and Art Burnett would tell you, "Well, this is owned by group so-and-so," which included various people with interests. Then we had the plant packages in North Texas that Clark created in those early days and very few people understood how they operated or what impact they had on the company, and to this day probably there are not a lot of people who have a good feel for those plant packages. We're still operating with those plant packages today and they complicate our operations in terms of what we want to do. The company is the one that, in effect, owns all the working interests in most of the cases in these plant packages. Every time it proposes a well, it has a bunch of participants in that plant package who are getting the bulk of the income and then when a well is proposed and we drill a well in a package, their income flips over and goes back to the company until the company can recoup all of the costs of that well. During that recoupment period, these people are getting less of the production from those wells and when that pays out, it flops back over again, and they start getting more and the company gets less. Well, you can imagine that if you are one of the people who farmed out to the plant and ended up with what we call a LORRI & DORRI, which referred to the overrides in the packages, you almost have to set up your own accounting system if you're going to try to keep up with how much the company is spending out here . . .

did it recover this, how much did it get, because if you've got 20 wells in a package and you drill one well, you don't have just one well paying out the cost of the well you just drilled, you have all of those 20 paying out the cost. And sometimes those payouts may last two or three months, you don't know, and they just flop back and forth all the time as new wells are drilled.

Computers make that more manageable?

Yes, I'm sure they do from the standpoint of the accountants' keeping up and maybe with the production and payouts, but in terms of land, if we get ready to drill a well and we want to take part of a lease that's in a plant package and combine it with another lease that would typically be one that wouldn't be in the plant package, we've now created another ownership for this well that's half plant package and half some other basis. And so, I'm saying those are a lot of problems.

Jack, thank you very much.

CPSIA information can be obtained
at www.ICGtesting.com
Printed in the USA
BVOW08s0949180717
489580BV00001B/29/P